NEW EYES TO SEE INSIDE THE SUN AND STARS

INTERNATIONAL ASTRONOMICAL UNION
UNION ASTRONOMIQUE INTERNATIONALE

NEW EYES TO SEE INSIDE THE SUN AND STARS

Pushing the Limits of Helio- and Asteroseismology
with new Observations from the Ground and from Space

PROCEEDINGS OF THE 185TH SYMPOSIUM OF THE
INTERNATIONAL ASTRONOMICAL UNION,
HELD IN KYOTO, JAPAN, AUGUST 18–22, 1997

EDITED BY

FRANZ-LUDWIG DEUBNER
Astronomisches Institut der Universität Würzburg, Germany

JOERGEN CHRISTENSEN-DALSGAARD
Institute of Physics and Astronomy, Aarhus University, Denmark

and

DON KURTZ
Department of Astronomy, University of Cape Town, South Africa

KLUWER ACADEMIC PUBLISHERS
DORDRECHT / BOSTON / LONDON

A C.I.P. Catalogue record for this book is available from the Library of Congress.

ISBN 0-7923-5075-8

Published on behalf of
the International Astronomical Union
by
Kluwer Academic Publishers, P.O. Box 17, 3300 AA Dordrecht, The Netherlands.

Sold and distributed in the North, Central and South America
by Kluwer Academic Publishers,
101 Philip Drive, Norwell, MA 02061, U.S.A.

In all other countries, sold and distributed
by Kluwer Academic Publishers,
P.O. Box 322, 3300 AH Dordrecht, The Netherlands.

Printed on acid-free paper

All Rights Reserved
©1998 International Astronomical Union

No part of the material protected by this copyright notice may be reproduced or utilized
in any form or by any means, electronic or mechanical, including photocopying,
recording or by any information storage and retrieval system, without written permission
from the publisher

Printed in the Netherlands.

TABLE OF CONTENTS

FOREWORD   xiii

I.   Global Structure and Evolution of the Solar and Stellar Interior

     DATA FOR PROBING THE SUN   1
        Y. Elsworth (invited review)

     HELIOSEISMIC DATA REDUCTION   13
        F. Hill (invited review)

     SEISMIC SOLAR MODELS AND THE NEUTRINO PROBLEM   21
        M. Takata and H. Shibahashi

     INTERNAL ROTATION, MIXING, AND LITHIUM ABUNDANCES   25
        B. Chaboyer (invited review)

     LOI/SOHO CONSTRAINTS ON OBLIQUE ROTATION OF THE SOLAR CORE   37
        L. Gizon et al.

     Posters

     HELIOSEISMOLOGY AND THE SOLAR NEUTRINO PROBLEM   41
        H.M. Antia and S.M. Chitre

     THE ART OF FITTING P-MODE SPECTRA   43
        T. Appourchaux and L. Gizon

     A COMPARISON OF LOW-DEGREE SOLAR P-MODE FREQUENCIES FROM BISON AND LOI   45
        T. Appourchaux et al.

     TOUCHING ON THE EFFECTS OF AN IMPERFECT WINDOW FUNCTION   47
        W.J. Chaplin et al.

     GONG SPECTRA IN THREE OBSERVABLES: WHAT IS A P-MODE FREQUENCY?   49
        J. Harvey et al.

     APPLICATION OF A NEW OBSERVATIONAL STRATEGY TO THE STUDY OF GRAVITATIONAL SOLAR MODES   51
        I. Martin Mateos and P.L. Pallé

     VAMOS: VELOCITY AND MAGNETIC OBSERVATIONS OF THE SUN   53
        M. Oliviero et al.

     SEARCHING FOR G-MODES AT THE SOLAR LIMB   55
        C.G. Toner and S.M. Jefferies

     AUTOREGRESSIVE MODELLING OF GOLF VELOCITY TIME SERIES   57
        R. Ulrich et al.

II. Solar Convection and Variation of Irradiance

LARGE SCALE CONVECTION AND THE SOLAR RADIUS 59
R.K. Ulrich (invited review)

STELLAR CONVECTIVE CORES 73
I.W. Roxburgh (invited review)

A SEISMIC MODEL OF THE SOLAR CONVECTIVE ENVELOPE 81
H. Shibahashi et al.

THE CONVECTION ZONE AND OSCILLATIONS 85
V.A. Baturin et al.

TOTAL SOLAR IRRADIANCE VARIATIONS 89
C. Fröhlich and J. Lean (invited review)

SOLAR IRRADIANCE VARIATIONS: THEORY 103
H.C. Spruit (invited review)

Posters

OBSERVATIONS OF THE LATITUDINAL VARIATION OF 111
THE SOLAR RADIANCE OF NON-ACTIVE REGIONS OF THE SUN
V. Domingo et al.

AMPLITUDE MODULATION OF RADIAL P-MODES FROM VIRGO 113
T. Leifsen, B.N. Andersen, and the VIRGO team

AN IMPROVED CALIBRATION OF THE MIXING-LENGTH BASED ON 115
SIMULATIONS OF SOLAR-TYPE CONVECTION
H.-G. Ludwig et al.

NONLINEAR INVERSION FOR THE HYDROSTATIC STRUCTURE 117
OF THE SOLAR INTERIOR
K.I. Marchenkov et al.

QUASI 10-DAY AND 4-DAY PERIODICITIES IN SOLAR IRRADIANCE 119
M.V. Niconova et al.

A $k - \epsilon$ MODEL OF THE LOWER OVERSHOOT LAYER 121
K. Petrovay

NUMERICAL SIMULATION OF PENETRATIVE CONVECTION: A PARAMETRIC STUDY 123
H. Singh et al.

III. Large-Scale Structure of the Sun

INVERSION METHODS 125
M.J. Thompson (invited review)

HELIOSEISMIC CONSTRAINTS ON THE SOLAR STRUCTURE 135
S. Vauclair (invited review)

SOLAR INTERNAL ROTATION 141
J. Schou and the SOI Internal Rotation team (invited review)

SOLAR ROTATION AND LARGE-SCALE FLOWS MEASURED BY   149
TIME-DISTANCE HELIOSEISMOLOGY FROM MDI
    P.M. Giles and T.L. Duvall, Jr.

LINE PROFILES AND ROTATIONAL SPLITTING OF INDIVIDUAL MODES   153
    V.G. Gavryusev and E.A. Gavryuseva

SPHERICAL AND ASPHERICAL STRUCTURE OF THE SUN: 157
FIRST YEAR OF SOHO/MDI OBSERVATIONS
    A.G. Kosovichev et al.

Posters

WHAT DO SOLAR F-MODE FREQUENCIES TELL US?   165
    H.M. Antia and S.M. Chitre

LOI AND GONG LOW-DEGREE ROTATIONAL SPLITTINGS   167
    T. Appourchaux et al.

A SEARCH FOR $\ell = 2$ ASYMMETRIES IN BISON DATA   169
    W.J. Chaplin et al.

LOW DEGREE P-MODE SOLAR CYCLE TRENDS FROM BISON DATA   171
    W.J. Chaplin et al.

COMMENTS ON THE INFLUENCE OF SOLAR ACTIVITY ON   173
P-MODE OSCILLATION SPECTRA
    L. Gizon

TIME VARIATION OF VELOCITY FLOWS FROM RING DIAGRAMS: 175
A FIRST APPROACH
    J. Patrón et al.

PUMPING OF ROSSBY WAVES AND VORTICES AT THE BASE   177
OF THE SOLAR CONVECTION ZONE
    E. Tikhomolov

OBSERVATION OF LOW-DEGREE MODES FROM SOHO/MDI USING OPTIMAL MASKS   179
    T. Toutain et al.

THE INTERNAL ROTATION RATE INFERRED FROM LOWL AND GONG DATA   181
    Li, Y. and P.R. Wilson

IV. Solar Small-Scale Structure

LOCAL PROPERTIES OF THE SUN'S SEISMIC EVENTS   183
    Ph.R. Goode et al. (invited review)

OBSERVATION OF SEISMIC EFFECTS OF SOLAR FLARES FROM   191
THE SOHO MICHELSON DOPPLER IMAGER
    A.G. Kosovichev and V.V. Zharkova

LINE ASYMMETRY AND EXCITATION MECHANISM   195
OF SOLAR OSCILLATIONS
    R. Nigam et al.

	EXCITATION AND DAMPING OF P-MODES Å. Nordlund and R.F. Stein (invited review)	199
	RAMAN SPECTROSCOPY OF SOLAR P-MODES M.P. Ryutova and T.B. Tarbell	213

Posters

EXPLORATORY SIMULATION OF SOLAR GRANULES: HOW SHARP IS THE CONVECTION/RADIATION TRANSITION? K.L. Chan and Y.C. Kim	217
ACOUSTIC IMAGING AND SUBSURFACE ABSORPTION STRUCTURE OF SUNSPOTS H.-K. Chang et al.	219
EXCITATION OF OSCILLATIONS - AN UPDATE OF BISON DATA W.J. Chaplin et al.	221
ARE SOLAR P-MODES CORRELATED? T. Foglizzo, R.A. García, and the GOLF team	223
THE INTEGRATED MAGNETIC FIELD OF THE SUN SEEN BY THE GOLF EXPERIMENT R.A. García, T. Roca Cortés, and the GOLF team	225
TEMPORAL BEHAVIOR OF SOLAR P-MODES FROM GONG AND GOLF EXPERIMENTS V.G. Gavryusev and E.A. Gavryuseva	227
ON THE FORMATION OF LINE PROFILES OF SOLAR P-MODES I.W. Roxburgh and S.V. Vorontsov	229

V. Asteroseismology I

NEW DEVELOPMENTS IN ASTEROSEISMOLOGY D.W. Kurtz (keynote talk)	231
THEORETICAL ASPECTS OF ASTEROSEISMOLOGY J. Christensen-Dalsgaard (invited review)	245
SEISMOLOGY OF THE ZZ CETI STARS J.C. Clemens (invited review)	253
ASTEROSEISMOLOGY IN ACTION: PULSATING HOT WHITE DWARFS S.D. Kawaler (invited review)	261
REPLACING COLOUR BLINDNESS WITH DEPTH PERCEPTION J. Matthews (invited review)	269
roAp STARS THROUGH THEORISTS' EYES - EXCITATION MECHANISM A. Gautschy and H. Saio (invited review)	277
A SEARCH FOR SOLAR-LIKE OSCILLATIONS IN α CEN A T.R. Bedding et al. (invited review)	285
RESULTS FROM THE HIPPARCOS MISSION ON STELLAR SEISMOLOGY L. Eyer and M. Grenon	291

SLOWLY PULSATING B STARS: NEW INSIGHTS FROM HIPPARCOS 295
C. Aerts et al.

Posters

THE ACOUSTIC CUT-OFF FREQUENCY OF roAp STARS 299
N. Audard et al.

ASTEROSEISMOLOGY WITH THE SPACE MISSION COROT 301
A. Baglin and the COROT team

BISECTOR VELOCITIES OF Hα IN THE roAp STAR α CIR 309
I.K. Baldry et al.

THE SINGLE LIFE OF RAPIDLY OSCILLATING Ap STARS 311
S. Hubrig et al.

MAGNETIC PROPERTIES OF RAPIDLY OSCILLATING Ap STARS 313
G. Mathys and S. Hubrig

SEISMIC DETECTION OF BOUNDARIES OF STELLAR CONVECTIVE REGIONS 315
M. Monteiro et al.

ON THE SEISMIC SIGNATURE OF THE HeII IONIZATION ZONE 317
IN STELLAR ENVELOPES
M. Monteiro and M. Thompson

THE PROCYON CAMPAIGN: OBSERVATIONS FROM KITT PEAK 319
C.A. Pilachowski et al.

DISCOVERY OF NON-RADIAL PULSATIONS IN THE WHITE DWARF 321
PRIMARY OF A CATACLYSMIC VARIABLE STAR
B. Warner and L. van Zyl

VI. Asteroseismology II

ASTEROSEISMOLOGY OF δ SCUTI STARS: OBSERVATIONS 323
M. Breger (invited review)

DELTA SCUTI STARS: THEORY 331
J.A. Guzik (invited review)

THE DISCOVERY OF NON-RADIAL GRAVITY-MODE PULSATIONS 339
IN γ DORADUS-TYPE STARS
K. Krisciunas (invited review)

PULSATIONS OF OB-STARS: NEW OBSERVATIONS 347
D. Baade (invited review)

B STAR PULSATION - THEORY AND SEISMOLOGICAL PROSPECTS 355
W.A. Dziembowski (invited review)

THE EC14026 STARS - PULSATING HOT SUBDWARFS 361
C. KOEN et al. (invited review)

EC14026 STARS: THEORETICAL CONSIDERATIONS 367
G. FONTAINE et al. (invited review)

ASTEROSEISMOLOGY FROM EQUIVALENT WIDTH: A TEST OF THE SUN 375
    C.U. Keller et al.

Posters

ASTEROSEISMOLOGY OF THE β CEPHEI STAR 12(DD)LACERTAE 379
    W.A. Dziembowski and M. Jerzykiewicz

RECOVERING THE PULSATION VELOCITY DISTRIBUTION ON STELLAR SURFACES 383
    J. Hao

A HIGH SPEED PHOTOMETRIC SURVEY OF NORMAL AND PECULIAR A-TYPE STARS 385
    O.M. Kurtanidze and M.G. Nikolashvili

LINE PROFILE VARIATIONS IN THE SPECTRA OF THE γ DOR STAR HR 2740 387
    E. Poretti et al.

THE LIGHT CURVES OF DOUBLE-MODE CEPHEIDS: THE CO AUR CASE 389
    E. Poretti and E. Pardo

ASYMPTOTIC DESCRIPTION AND THE DIAGNOSTIC PROPERTIES
OF LOW DEGREE STELLAR P-MODES 391
    I.W. Roxburgh and S.V. Vorontsov

SPECTRAL LINE PROFILE VARIABILITY AS A PROBE FOR ℓ AND m 393
    C. Schrijvers and J.H. Telting

ASTEROSEISMOLOGY OF β CEPHEI 395
    H. Shibahashi and C. Aerts

PULSATION OF PMS $1.8 M_\odot$ STARS: A THEORETICAL INVESTIGATION 397
    D.M. Suran

OSCILLATIONS OF LONG-PERIOD VARIABLES 399
    D.R. Xiong et al.

OSCILLATIONS OF HB RED VARIABLE STARS 401
    D.R. Xiong et al.

VII. The Solar Atmosphere

CARBON MONOXIDE AND THE TEMPERATURE STRUCTURE
OF THE SOLAR ATMOSPHERE 403
    T.R. AYRES (invited review)

HIGH-FREQUENCY SOLAR OSCILLATIONS 415
    S.M. Jefferies (invited review)

THE INFLUENCE OF A MAGNETIC FIELD ON RADIATIVE DAMPING
OF MAGNETOATMOSPHERIC OSCILLATIONS 423
    D. Banerjee et al.

PHASE RELATIONS AND OTHER DIAGNOSTICS OF
SOLAR ATMOSPHERIC STRUCTURE AND DYNAMICS 427
    F.-L. Deubner (invited review)

THE NEW CHROMOSPHERE 435
    M. Carlsson and R.F. Stein (invited review)

Posters

HIGH FREQUENCY SIGNAL IN GOLF DATA R.A. García, P.L. Pallé, and the GOLF team	447
OSCILLATION OF THE MAGNETIC FIELD IN AN ACTIVE REGION T. Horn and J. Staude	449
DYNAMICS OF THE DEEP SOLAR PHOTOSPHERE AT SUPERGRANULAR SCALES A. Nesis et al.	451
MHD OSCILLATIONS OBSERVED IN THE SOLAR PHOTOSPHERE WITH THE MICHELSON DOPPLER IMAGER A.A. Norton et al.	453
PHASE SPECTRA SEEN FROM SPACE Th. Straus et al.	455
OBSERVATIONAL RELATIONSHIP BETWEEN MESO-SIZED CONVECTION AND 5-MIN OSCILLATION IN THE SOLAR ATMOSPHERE S. UeNo and R. Kitai	457
ON THE CHROMOSPHERIC BEHAVIOUR OF PHOTOSPHERIC Mn 539.47 nm SPECTRAL LINE I. Vince and S. Erkapić	459
A SPATIAL ANALYSIS OF LOCAL SOURCES OF OSCILLATIONS Y. Yan et al.	461

Miscellaneous

DISTRIBUTIONS OF THE SOURCES OF THE MAGNETIC FIELD DURING THE DOUBLE MAGNETIC CYCLE E.E. Benevolenskaya	463
TIME VARIATION OF THE GLOBAL SOLAR MAGNETIC FIELD INFERRED FROM THE SUN'S SHADOW AS SEEN IN 10 TeV COSMIC RAYS L.K. Ding et al.	465
NUMERICAL SIMULATION OF A TWISTED SOLAR CORONA S. Parhi et al.	467
CORONAL HEATING MECHANISM IN THE PRESENCE OF A FLOW: A NUMERICAL APPROACH S. Parhi and T. Tanaka	469

VIII. Synthesis of Solar-Stellar Seismology – Meeting Summary 471
S.M. Chitre (invited review)

Index 483

Foreword

The cover picture of this volume displays one of the most advanced products of helioseismic research: a view into the deep interior of the sun, revealing its distinctly non-uniform rotation throughout the entire depth of the convection zone. Just over 20 years ago, the first successful helioseismic experiment disclosed an increase of rotation velocity in the uppermost one dozen megameters below the photosphere. The stunning progress in depth and detail highlighted by the cover diagram (and by others shown in this volume as well) was made possible by considerable advances in instrumentation, by the development of powerful analytical tools and, foremost, by the involvement of new brains of enthusiastic proselytes and newcomers to the field, increasing nearly exponentially in number every year. New branches of research widened the scope of "uranoseismology", as e.g. time- distance seismology (the promising avenue towards small-scale and short-time variability), atmospheric seismology (a new look at strange phenomena we have always seen, but hardly understood), and finally the growing observational assault on hundreds of individual stars which are either manifestly or supposedly oscillating - i.e. asteroseismology. The formation of numerous solar and stellar observing networks and, ultimately, space missions like SOHO have greatly promoted the potential of this science.

This steady progress was accompanied by workshops, colloquia, and symposia in quick succession. Why should we have planned for yet another one? The previous reunions were chiefly for specialists to talk to and discuss with other specialists in the field. It was strongly felt by the organizers of the present symposium that with the brand new results from busy space missions and humming networks at hand, we should seize the opportunity to present all this to the community of astronomers at large, as represented in Kyoto at the 23. General Assembly of the IAU. Thanks to the excellent cooperation of the invited speakers, the result of their efforts to amalgamate the introductory habit of several of the invited talks with fresh topical research can now be put before the interested reader. Unfortunately, a few invited talks on "Progress and Puzzles in Helioseismology" (keynote talk given by T. Brown), "Recent Results in Time-Distance Helioseismology (T. Duvall), and "Interaction of Convection with Pulsation" (P. Goldreich) are missing in these Proceedings for reasons beyond our control.

It is in the name of the previous Organizing Committee of Commission 12 of the IAU on Solar Radiation and Structure, that I very gratefully acknowledge the assistance of the Scientific Organizing Committee of this Symposium (K. Chitre, J. Christensen-Dalsgaard, V. Domingo, Y. Elsworth, C. Fröhlich, D. Kurtz, J. Leibacher, J. Provost, and H. Shibahashi), and foremost of the two co-editors of the Proceedings, Joergen, and Don, who were immensely helpful whenever practical advice and mental support were needed. As a co-chair of the SOC, Don clearly excelled in promoting the idea of this Symposium most energetically. Many peoples' names would have to be mentioned to thank all the colleagues who supported the organization of this meeting locally and admistratively. It is with deep gratitude that I put here in lieu the names of J.

Andersen (Assistent General Secretary of the IAU) and T. Fukushima (Chairman of the Japanese Local Organizing Committee of the IAU GA).

John Leibacher, in concluding the last session of the meeting, prognosticated that this event would be remembered by all attendents. I am sure that he was referring also to the friendly hospitality and the harmonious spirit of the Japanese culture we, the participants, had experienced.

Würzburg, February 1998　　　　　　Franz - Ludwig Deubner

DATA FOR PROBING THE SUN

YVONNE ELSWORTH
School of Physics & Astronomy, University of Birmingham
Edgbaston, Birmingham B15 2TT UK
E-mail: ype@star.sr.bham.ac.uk

1. Introduction

The observations of solar oscillations provide an unrivalled, precise way of probing the solar interior. In this paper, I consider the observations and their interpretation in terms of the physics of the Sun. The oscillations that we are concerned with here are the so-called p modes, i.e. oscillations for which pressure is the restoring force. The modes for which gravity is the restoring force have yet to be unambiguously detected on the Sun. The observations are made either as Doppler velocity or as intensity and are, in general, very small effects. To get an impression of the precision required, consider that in integrated velocity the total signal is $\sim 1\,\mathrm{m\,s^{-1}}$ with the strongest individual modes being about 15–20 cm s^{-1}. The weakest, detected modes are of order a few mm s^{-1}. When this signal is measured as a Doppler shift, v/c is a few parts in 10^{11}. The observations are made by a variety of instruments on Earth or in Space which can be simply divided into those which observe the Sun as a star and those which image the solar surface into many pixels Although there are many different observers using many different techniques, in all cases one is analysing light emitted from a region relatively high in the atmosphere of the Sun. When one considers how these measurements can be interpreted in terms of the solar oscillations, two issues arise:

1. Roughly where in the solar atmosphere are the lines formed?
2. How different are the heights of formation for different lines?

There are three lines which are widely used. They are lines of sodium (IRIS and GOLF), potassium (BiSON and LOWL) and nickel (GONG and MDI) and we will now consider their formation height in the solar atmosphere. The reference point against which we will specify the height at which the lines are formed is given as the height where the local temperature equals the effective temperature of the Sun. This closely corresponds to optical depth equals unity. The temperature minimum is at about 515 km above this level.

As indicated in Table 1, the nickel and the potassium signals come from very much the same height whereas the sodium signal is formed significantly higher in the atmosphere. To see an implication of the height differences, we need to consider the sound waves themselves. The Sun acts as a resonant cavity for sound waves. The cavity

TABLE 1. Formation height for several common observing lines

Element	Height above zero (km)	Wavelength (Å)
Sodium D1&D2	500	5,893
Potassium D1	300	7,699
Nickel I	300	6,768

is limited by upper and lower turning points which are a function of the particular mode being observed. The upper turning point occurs when the vertical scale of the waves becomes comparable with the scale on which the vertical stratification is changing. This occurs where the frequency of the wave is approximately equal to the acoustic cut-off frequency. The acoustic cut-off frequency increases steadily from the deep interior, where it is small, towards the surface. Hence the upper turning point is located deeper for a low-frequency mode than for a high-frequency mode. For example, a mode whose frequency is 1 mHz would have an upper turning point at $0.987 R_\odot$ and one at 4 mHz would turn closer to the surface at $0.999 R_\odot$ ($R_\odot$ is the photospheric radius where the temperature equals the effective temperature.)

At very high degree, the upper turning point is a function of both degree and frequency, and the the simple numbers quoted above cannot be used.

As the excitation of the modes is believed to occur in the very upper layers of the Sun (within a few hundred km of the surface), the observed widths and strengths of a particular mode of oscillation will depend strongly on the location of the upper turning point. Comparison of the numbers in Table 1 with the turning points shows that, in most cases, the region from which the signal comes is actually outside the sound-wave cavity and we are detecting an evanescent wave. This has consequences for the observed strength of the signal. The density of the solar atmosphere drops with increasing height and the oscillation signal is hence stronger if the line is formed higher up.

So, we expect that oscillations measured with the sodium line are stronger than for those measured with potassium. The relative strengths of the two signals have been measured by Fossat (private communication) to be a function of frequency changing from no difference at all at about 1.5 mHz to a factor of about 1.5 in power at 4.5 mHz. We should thus expect that the raw spectra will look different if measured in the different lines.

The lower turning points vary strongly for the different modes considered. The lowest ℓ go into the core and the highest ℓ are confined to the surface. According to asymptotic theory the inner turning point varies as ν/L (where $L = \sqrt{\ell(\ell+1)}$). One mode which has its inner turning point at the base of the convection zone is 3 mHz $\ell = 50$ (Gough, 1990). At lower frequency and the same ℓ-value the modes are totally confined within the convection zone. On the other hand, at lower degree, a mode with $\ell = 1$ and $\nu = 3$ mHz will get within $0.06 R_\odot$ of the centre. Only $\ell = 0$ modes probe the very centre of the core. Thus it is an extremely powerful attribute of solar oscillations that the volume of the sun sensed by different modes varies enormously.

However, consideration of the inner turning point of a mode does not tell the whole story for the importance of modes in determining the structure of the Sun.

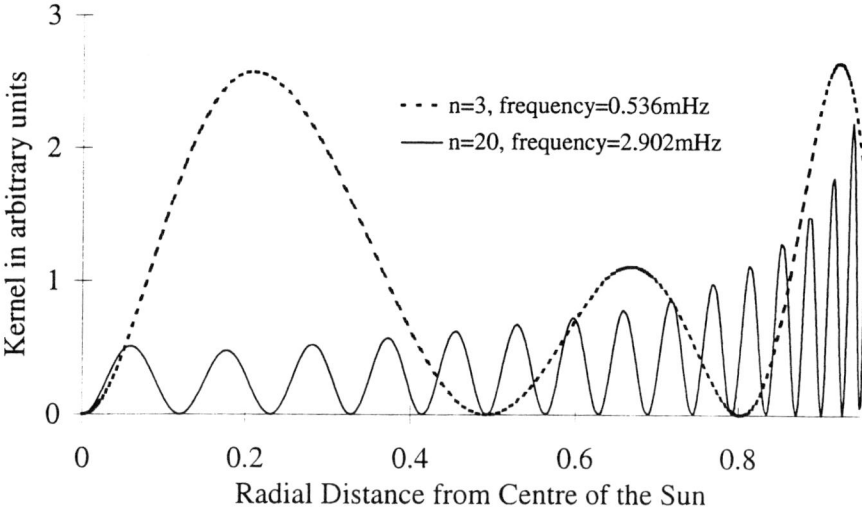

Figure 1. Typical kernels for c^2 for low-order, zero-degree modes. Note their differing sensitivity in the core

Asymptotic theory predicts that the higher the frequency of a mode of given ℓ, the closer to the centre of the Sun will it turn, but, one must also consider where the energy of a mode is principally concentrated. Figure 1 illustrates that although the mode of higher frequency does have an inner turning point which is closer to the solar core, the lower frequency mode has much more of its energy in the inner regions of the Sun. These considerations indicate that low-order modes of low frequency are very effective at sensing the conditions in the solar core. One must also take into account the fact that low-frequency-mode peaks are much narrower than high frequency ones and so can be measured much more accurately. Detecting modes of low degree and low frequency remains one of the very important goals of helioseismology.

2. Seismic Sun

The oscillation frequencies are sensitive to the conditions inside the Sun. What the oscillations are really measuring the speed of sound as a function of the mode which can be interpreted as a function of radius and latitude in the Sun. The sound transit time is $\int dr/c$ from the lower to the upper turning points, where c is the speed of sound. The time that a mode spends in a particular volume, and hence its sensitivity to the physical conditions there, depends inversely on the speed of sound. Originally this led to the suggestion that p-mode oscillations could not be used to sense the core. However the very high precision of the measured frequencies (up to 1 in 10^5 near the centre of the spectrum) means that we can, in fact, say quite a lot about the core. G-modes would be still better but they have yet to be detected. A picture of the Sun built up from helioseismic data may be termed the seismic Sun. Helioseismology gives direct constraints only on the mechanical properties of the Sun (i.e. determined by pressure, density and gamma). If the equation of state is assumed, or where the solar plasma is fully ionised, the sound speed constraint gives a constraint on the

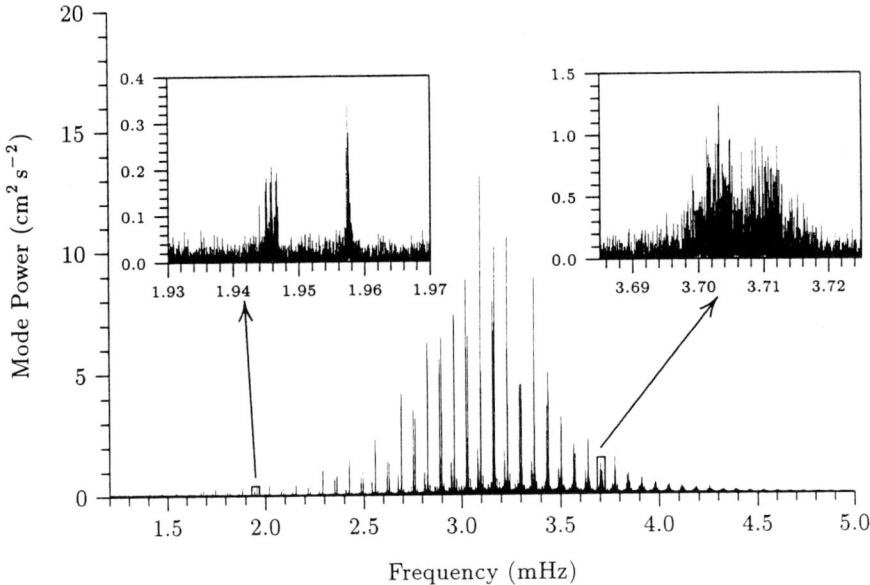

Figure 2. Sixty-four-month BiSON power spectrum.

ratio of temperature to mean molecular weight. To be able to extract information on the thermal structure or composition additional input such as the equation of state, nuclear reaction rates and opacities are required. An important additional point is that a seismic model has the advantage that it looks at how the Sun is now, and does not require one to be able to accurately follow the solar evolution.

3. Errors in theory and in practice

So given that the possibility that the frequencies can be used to extract direct information concerning the interior of the Sun, we should explore what limits the accuracy of the frequencies.

A spectrum taken from 64 months of data from the 6-station world-wide BiSON (Chaplin *et al.*, 1996) network is shown in Figure 2. It illustrates the typical range of power and frequency of the low-degree solar oscillations. The 64-month BiSON spectrum shows lines that can be measured from about 1.1 mHz to 4.7 mHz.

The spectrum can be divided into three regions: the central region where the signal levels are high; the low frequency end; and the high frequency end. Each region presents its own problems. In the middle of the spectrum, where the signal-to-noise is high, the potential precision is high but great care must be taken with the model used to fit the data (Toutain & Appourchaux, 1994). At very high frequencies and high ℓ there are problems in isolating the modes because the widths of the individual components increase substantially. There is also the issue of overlapping spatial aliases in resolved-Sun data. The key issue at very low frequencies is the signal to noise. If the mode is essentially unresolved by the observation, all the oscillatory power is put into the same frequency bin. Under this condition the time required to detect the mode above the background (say a signal to noise level of three) depends on the relative strengths of the background and signal. Once the mode is stronger than the background, the signal to noise ratio will increase linearly with observing time. If the data set is longer than a lifetime and the mode is spread out over more than one bin in the frequency spectrum, then the signal-to-noise improves only very slowly as more data are collected. To detect the mode at all, a good general rule is that one must detect a mode in one life time. It is fortunate that the low-frequency modes also have long lifetimes and so long integration times (many months) can be used.

Ground-based data nearly always have gaps. Breaks in the data in the first instance give rise to sidebands at a frequency that is characteristic of the typical interval - daily breaks lead to structure at $11.57\,\mu$Hz. But breaks in the data also give rise to general noise. For nearly 100% data coverage one has to turn to the data measured on the SOHO satellite. The GOLF 8-month spectrum (Lazrek et al., 1997) has noise levels of order $50\,\text{m}^2\,\text{s}^{-2}\,\text{Hz}^{-1}$ at 1 mHz. In their early spectra, taken before the failure of the rotator, the noise level is about a factor of two lower. Direct comparison with BiSON is a little difficult because the data have been taken at different wavelengths but at these frequencies direct comparison should be possible. BiSON solar noise (Elsworth et al., 1994) estimates from cross correlations of short data sets, and also observed levels in a one-day spectrum on a good day at their best site, indicate that the noise level is at about $20\,\text{m}^2\,\text{s}^{-2}\,\text{Hz}^{-1}$ at 1 mHz. Their 32-month spectrum has a power level of $70\,\text{m}^2\,\text{s}^{-2}\,\text{Hz}^{-1}$ at 1 mHz. Some of the difference between the two estimates of the noise is due to the effects of the atmosphere but the lack of a 100% fill is also relevant. Simulations suggest that the noise level at 1 mHz will rise by a factor of 10 when the fill changes from 100% to 74% (Chaplin et al., 1997). In reality, tests on real data suggest that the presence of solar noise may limit the degradation. In so far as these simulations are applicable, they suggest that the gaps are a significant source of noise in a computed spectrum. Typical fills achieved by the BiSON observations are of order 80% and the GONG observations are more like 90% over long periods. Given the indications from simulations, gap filling should be a very profitable procedure. Many of the gaps in network data are quite short and so should be amenable to data-filling techniques (Brown & Christensen-Dalsgaard, 1990).

The normal fitting procedure is to fit to a Lorentzian line shape with either χ^2 2 d.o.f. or gaussian statistics as appropriate. Using the correct statistics has been shown to be very important (Toutain & Appourchaux, 1994). Early indications that the lines were not well described by a symmetric function came from the high-ℓ data and this is now extended to a wide range of modes. It is now clear that at least some of the systematic differences between different data sets are dependent on the shape of the lines. The line-shape effects are different between intensity and velocity, giving rise to frequency discrepancies of low-degree modes as measured in intensity and velocity of about $0.1\,\mu$Hz with intensity giving higher frequencies than velocity. This is discussed elsewhere (Toutain et al., 1997; Appourchaux et al., 1997). Solar noise (see Figure 3) is the ultimate background against which the measurements must be made (Elsworth et al., 1994; Underhill & Isaak, 1997; Fröhlich et al., 1997). However, in reality one must consider also other sources of noise such as instrumental, photon shot noise, and atmospheric noise. The relative importance of these various noise sources depends on the region of the spectrum and the type of observation.

Theoretical estimates of the errors can be made. A simple derivation of the expected frequency error is given elsewhere (Elsworth et al., 1994) as is a more precise derivation which takes account of the signal to noise in the data (Libbrecht, 1992). Simply, assume that a mode is detected in one life-time. A crude estimate of the error on this measurement is one half of the line width. One then continues to observe and

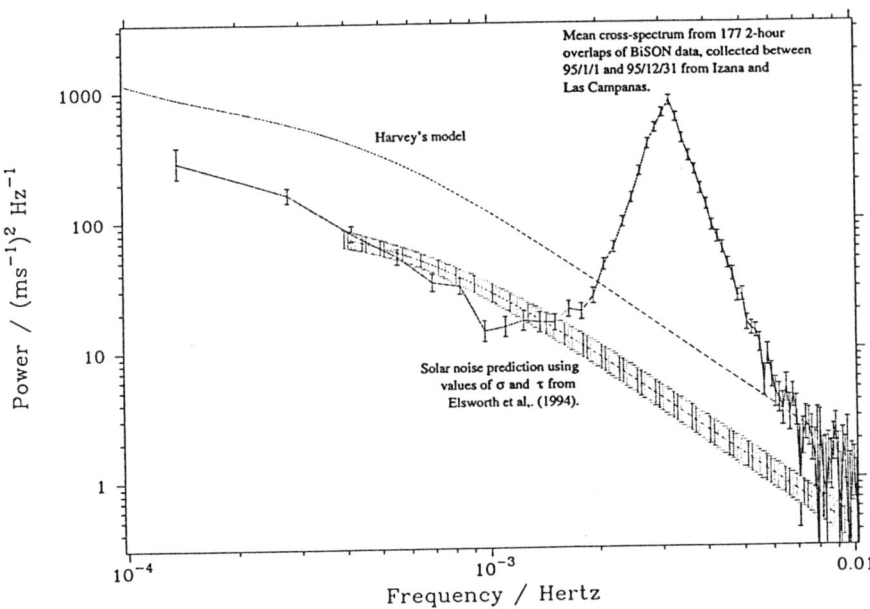

Figure 3. Solar noise spectrum as measured by BiSON using cross-correlation techniques

amasses data for n lifetimes. The data are adding incoherently and so the uncertainty in the frequency can be expected to have improved by $\sqrt{n}$. Putting this together we find that the error = $\sqrt{(\Gamma/4\pi t)}$ where Γ is the full width at half maximum of the line and t is the total observing time. This is the best that the error can be expected to be: poor signal to noise increases the error.

There are several points to be made here. First, if the signal to noise is good then the error on the frequency estimation is a slow function of improving signal-to-noise. Second, the error will scale with the square root of the line width which is itself a strong function of frequency. We can check how close the real data get to this prediction by looking at some existing datasets.

Figure 4 shows the line width and frequency errors as a function of frequency for BiSON, GOLF and MDI for $\ell = 0$.

On the same graph are shown the predictions of the formula above. The BiSON spectrum is taken over 32 months and there is a prediction for a dataset of this duration. The GOLF spectrum is taken over 8 months and the MDI over 12 months. In all cases, the observations are close to the predictions in the centre part of the spectrum where the signal to noise is good. At the extremes of the frequency range there are still improvements to be made. In the case that the error in the frequency determination is greater than the linewidth, we can say that there is probably not a secure detection of the mode. Where the error is close to the theoretical prediction, improving the signal-to-noise in the data will only have a marginal effect on the frequency accurary. The main improvement will come from having much longer data series. This is illustrated by the comparison between the BiSON errors and those from the other two, shorter spectra. The scatter in the measurements may indicate

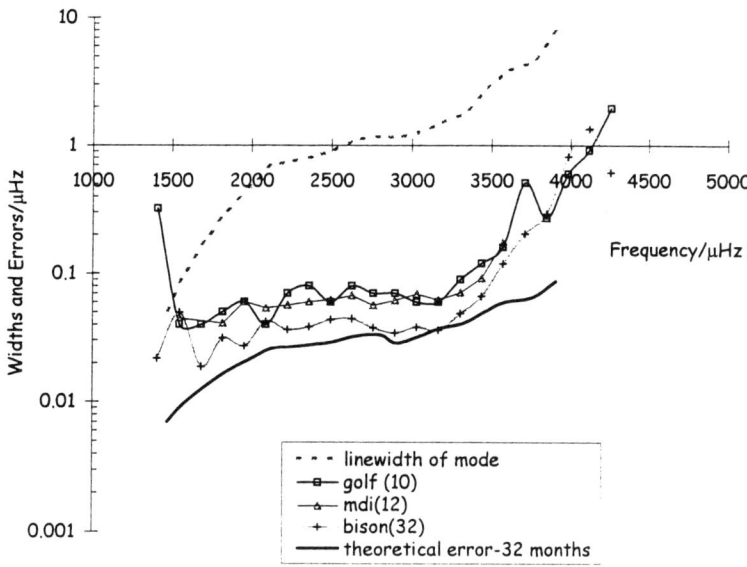

Figure 4. Frequency errors for $\ell = 0$, for different data sets

that the constraints used in the fitting are not quite right. In specifying the error on a parameter in reality, one has to consider internal and external (i.e. taking into account the data scatter) errors.

In this discussion we have not yet considered frequency changes due to solar magnetic effects in the short and long term. The solar cycle has been shown to cause the mode frequencies to vary with time (Libbrecht & Woodard, 1990; Elsworth *et al.*, 1990). Currently, we are near the bottom of a solar cycle and hence recent observations have not been much influenced by the effects of solar activity. An important point for the consideration of errors is how long a data set can be considered before the variation in the activity levels on the Sun must be considered. As far as the frequencies are considered, we specify that the bin width of the fourier transform should be greater than any frequency shift. At the bottom of the solar cycle, this constraint still allows us to use a 32-month spectrum.

4. Standard Model and its Physics

When setting up a standard solar model, the primary requirement is to make the model look like the real Sun. By this we mean that we must match the mass, radius and luminosity of the current real Sun. We also put in the physics that we believe to properly describe the behaviour of the solar interior. We do not perfectly understand the interior of the Sun and so there are areas for further development of the theory. In the energy-generating core there is the problem that the observed neutrino rates do not match the predicted rates. In the convection zone there is the general lack of a good analytical description. In most models, mixing length theory is used. There have been improvements recently in the treatment of the superadiabatic layer and turbulent convection. Another parameter that has to be supplied is the opacity as a function of radius. There have been significant changes recently in our knowledge of the opacity (Seaton et al., 1994; Rogers & Iglesias, 1992). The high precision of the solar oscillations frequencies required an improvement to the opacities so that the data would more closely match the theoretical predictions. A spin off has been that the changes made have allowed an improvement to the modelling of other stars and have removed certain long-standing problems. Low temperature opacities required in the outer layers of the Sun remain a problem.

One of the early successes of helioseismology was that it allowed an accurate determination of the depth of convection zone. Changing other physics will often require the depth of the zone to be changed. However, it has been shown (Basu & Antia, 1995) that a change in the convection zone by any means is not enough to improve the results. The base of the convection zone is believed to be at $0.713 \pm 0.001 R_\odot$ if one does not consider the effects of diffusion of helium and other heavy elements. This changes to about $0.711 R_\odot$ for model S (Christensen-Dalsgaard et al., 1996) when the effects of diffusion are included.

The equation of state determines the relationship between P, ρ and γ. The solar plasma is an almost ideal gas with gamma, the adiabatic exponent, equal to 5/3 everywhere except in the ionisation zones. Away from this condition. we need to know gamma. Recently, the opal project (Rogers & Iglesias, 1992a) has produced a new way of calculating the adiabatic exponent. This is somewhat of an improvement on the previous method (Mihalas et al., 1988). However there is still considerable discussion of the two methods and particularly how one generates information which is valid and accurate enough to cope with a range of stellar conditions.

Spectroscopic measurements of the abundance of helium in the Sun are very uncertain because the absorption lines are not in the visible region of the spectrum. Observing from above the atmosphere at short wavelengths gives the coronal levels which then have to be linked to photospheric levels. Again this is not an easy problem. The helium abundance can be obtained helioseismically from the variation in the adiabatic index of the solar material in the second helium ionisation zone (at depth of 15,000 km). The results from a variety of methods do not agree (Vorontov et al., 1991; Kosovichev et al., 1992; Antia & Basu, 1994) but are in the range 0.24 to 0.25 helium by weight. This is compatible with solar evolution theories only if helium settles out of the envelope into the radiative zone (Noerdlinger, 1977; Cox et al., 1989; Wambsganss, 1988; Christensen-Dalsgaard et al., 1993). In the absence of settling the present day helium abundance in the solar envelope would have to be about 0.27 to 0.28 to satisfy the solar luminosity constraints.

5. Comparison with observation

The physics in the standard model can be used, with ideas of stellar evolution, to get a picture of the Sun for which oscillation frequencies can be computed. These can be compared with the real observed frequencies.

The agreement looks quite good at first sight if the predicted frequencies are marked on a spectrum. However, the data can plotted in a more sensitive way on a so-called echelle plot. In an echelle plot, the spectrum is divided up into slices of a fixed length which are then stacked above each other. If the interval is chosen to be about 135 μHz then the different radial orders are approximately above each other on the plot. The 32-month BiSON spectrum shown is compared with recent frequencies calculated by Guzik and Swenson (1997)in Figure 5. Note that the agreement is

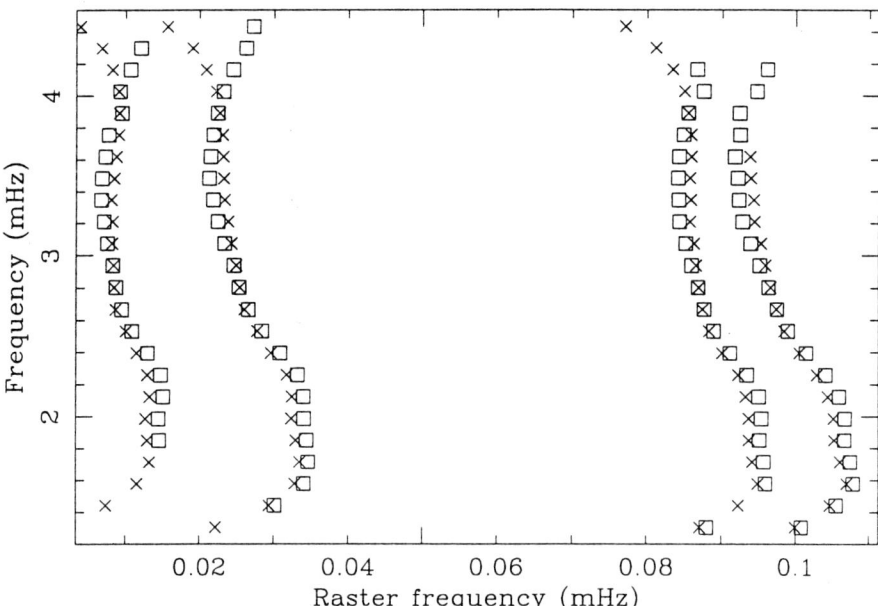

Figure 5. Echelle plot of a 32-month BiSON spectrum and theoretical frequencies from Guzik and Swenson. The squares denote frequencies measured by BiSON from 1994 May 16 to 1997 January 10. The crosses denote model frequencies (Guzik & Swenson, 1997).

good in the region about 3 mHz and deteriorates somewhat at lower frequencies and more seriously at at higher frequencies. At high frequencies, the lack of agreement is probably due to the difficulties in modelling the solar surface. Other models give even better low-frequency agreement. However, one of the great advantages of the Guzik and Swenson code is that it is analytic and can be easily extended to other stars.

Similarly with the high-degree data, the agreement between theory and experiment is a function of frequency. When comparing modes of very different ℓ it is useful to scale them by the mode inertia (Christensen-Dalsgaard & Berthomieu, 1991) which takes account of the volume of the Sun sensed by the mode and the ease with which the mode can be perturbed. Figure 6 (Christensen-Dalsgaard *et al.*, 1996) is taken from the issue of Science in which many early GONG results were reported. It shows frequency differences between the Sun as measured by the GONG project and a model (model C of JCD). The differences are scaled by Q_{nl} (the mode inertia). As is the case for the low-order data, the discrepancies between the real Sun and the model are most marked at higher frequencies. There is some residual ℓ-dependence which can be used to localise the region in which the physical description of the Sun is not accurate. In this particular case, there appears to be a problem at the base of the convection zone.

Small differences between data sets need to be understood and we need to know how to extract 'true' frequencies from asymmetric lines. There is also much more work to be done on solar models. However, one can say in conclusion that the disagreement between theory and observation, although at the few per cent level, is larger than the errors on the data and is hence very significant.

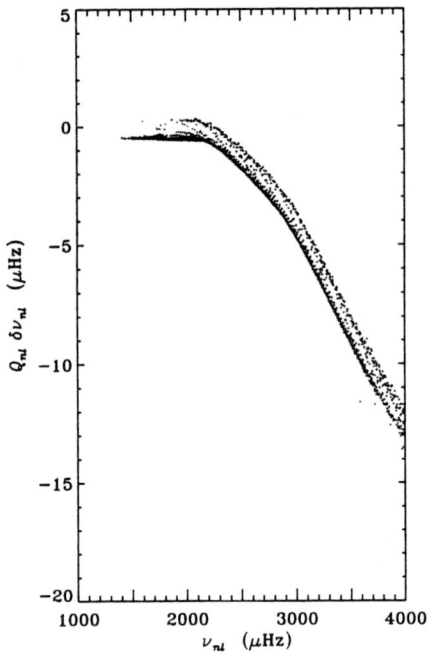

Figure 6. GONG frequencies scaled by mode inertia compared with model C

6. The Core

As has been apparent for over a decade, there is a problem with the rate at which the neutrinos from the Sun are observed at the Earth. At first it was felt that the problem was with our understanding of the solar physics of the core. However, the observed spacing between adjacent modes ($\ell = 0$ and $\ell = 2$ for example) are very close to the predicted spacing. All attempts to modify the theoretical separation in a way which will lower the expected neutrino flux have produced frequency separations which are excluded at high significance by the data. The importance of using the low-ℓ frequency differences is two fold. First the modes probe deep into the Sun with the $\ell = 0$ mode being deeper than the $\ell = 2$. Secondly, the only significant differences in the volumes sensed are in the interior of the Sun. The surface effects, which are very hard to model, have very little impact on the frequency differences (Elsworth et al., 1990). These same frequency differences can be used to explore the effects of gravitational settling of Helium and heavy elements. The data clearly support their incorporation.

The small and large frequency separations can be parameterized in a way which allows comparison of large and small separations to determine stellar ages since nuclear burning makes the core conditions very sensitive to the evolutionary state of the star (Christensen-Dalsgaard, 1984).

7. Solar structure inversions

Another approach to the problem is that of inversions. The data can be inverted to give the sound speed profile through the Sun. Other papers in this volume discuss the technique in much more detail. The inversions use not only the frequencies but also the errors on the frequencies to determine the solar parameters. An observational error which is unrealistically small will bias an inversion as will errors which are too large. It is crucially important to have realistic error estimates. Differential inversions address the problem of the difference in a particular parameter under consideration between the Sun and a model of the Sun whose physics is thought to be an approximation to that in the Sun.

The inversions tell a similar story to the one given above; the solar models match the data well but there are still differences and these are located at the base of the convection zone and in the core of the Sun (Turck-Chièze et al., 1997; Bauu et al., 1997). The sound speed is higher in the Sun than in the reference model just below the base of the convection zone. This could be due to the accumulation of excess helium below the convection zone (Gough et al., 1996). Model S does not include mixing below the base of the convection zone. If there were mixing then the helium abundance locally would be reduced, decreasing the mean molecular weight, increasing the sound speed and thereby bringing the Sun and the model into better agreement. There are other ways of achieving this, for instance, selective changes in the opacity (Tripathy et al., 1997). Changing the opacity alters the depth of the convection zone which has to be adjusted to meet the constraint that the modelled Sun must look like the real Sun. In the core, the inversion errors get quite large because the number of modes that actually penetrate into the region is rather small. The result here is quite sensitive to the particular mode set chosen. Only the lowest-degree modes sample this region and even those sample the core for a very short time because the sound speed is relatively high in the core. The results in the core are also very sensitive to the low frequency data.

8. Summary

Although it is attractive to be able to predict frequencies which accurately match those observed, the real goal is to understand the physics of the solar interior. For the Sun, we have access to a huge amount of data. For other stars, we have a much more limited set. Not only does the faintness of other stars make the problem more difficult, but the range of modes observable is much more limited. In trying to understand how we can interpret future asteroseismic data it is constructive to see what information can be extracted from the Sun using similar data sets to those we can expect to have for stars.

References

Antia H. M. & Basu S., 1994, *ApJ*, **426**, 801.
Appourchaux T., Chaplin, W. J., Elsworth, Y., Isaak, G. R., McLeod, C. P., Miller, B. A. & New, R., 1997, these proceedings.
Basu S. & Antia H. M., 1995, *MNRAS*, **276**, 1402.
Basu S., Chaplin W. J., Christensen-Dalsgaard J., Elsworth Y., Isaak G. R., New R., Schou J., Thompson M. J. & Tomczyk S., 1997, *MNRAS*, in press.
Brown T. M. & Christensen-Dalsgaard J., 1990, *ApJ*, **349**, 667.

Chaplin W. J., Elsworth Y., Isaak G. R., McLeod C. P., Miller B. A. & New R., 1996, *Sol. Phy.*, **168**, 1.
Chaplin W. J., Elsworth Y., Isaak G. R., McLeod C. P., Miller B. A. & New R., 1997, these proceedings.
Christensen-Dalsgaard J., Proffitt C. R. & Thompson M. J., 1993, *ApJ*, **403**, L75.
Christensen-Dalsgaard J. & Berthomieu G., 1991, in: *Solar Interior and Atmosphere*, eds. Cox A. N., Livingston W. C., Matthews M., Univ. of Arizona Press, Tuscon, p. 401.
Christensen-Dalsgaard J., 1984, in: *Space Research Prospects in Stellar Activity and Variability*, ed. Praderie F., Paris Observatory Press, p. 11.
Christensen-Dalsgaard J. et al, 1996, *Science*, **272**, 1286.
Cox A. N., Guzik J. A. & Kidman R. B., 1989, *ApJ*, **342**, 1187.
Elsworth Y., Howe R., Isaak G. R., McLeod C. P. & New R., 1990, *Nature*, **347**, 536.
Elsworth Y., Howe R., Isaak G. R., McLeod C. P., & New R., 1990, *Nature*, **345**, 322.
Elsworth Y., Howe R., Isaak G.R., McLeod C. P., Miller B. A., New R., Speake C. C. & Wheeler S.J., 1994, *MNRAS*, **529**, 537.
Elsworth Y., Howe R., Isaak G. R., McLeod C. P., Miller B. A., New R., Speake C. C. & Wheeler S. J., 1994, *ApJ*, **434**, 801.
Fröhlich C. et al., 1997, *Sol. Phys.*, **170**, 1.
Gough D. O., 1990, in: *Progress of Seismology of the Sun and Stars*, p. 302, eds. Osaki Y. & Shibahashi H., Springer-Verlag, Berlin.
Gough D. O. et al., 1996, *Science*, **272**, 1281.
Guzik J. A. & Swenson F. J., 1997, *ApJ*, **491**, in press.
Kosovichev A. G. et al., 1992, *MNRAS*, **259**, 536.
Lazrek M. et al., 1997, *Sol. Phys.*, in press.
Libbrecht K. G. & Woodard M. F., 1990, *Nature*, **345**, 779.
Libbrecht K. G., 1992, *ApJ*, **336**, 1092.
Mihalas D. M., Dappen W., Hummer D. G., 1988, *ApJ*, **331**, 815.
Noerdlinger P. D., 1977, *A&A*, **57**, 407.
Rogers F. J. & Iglesias C. A., 1992, *ApJ*, **401**, 360.
Rogers F. J. & Iglesias C. A., 1992, *ApJ Supplement.*, **79**, 507.
Seaton M. J., Yan Y., Mihilas D., Pradhan A. K., 1994, *MNRAS*, **266**, 805.
Toutain T. & Appourchaux T., 1994, *A&A*, **289**, 649.
Toutain T., et al., 1997, *Sol. Phys.*, in press.
Tripathy S. C., Basu S. & Christensen-Dalsgaard J., in: Proc. IAU Symp. 181, *Sounding solar and stellar interiors*, eds. Schmider F.-X. & Provost J., Nice Observatory, France, in press.
Turck-Chièze S. et al., 1997, *Sol. Phys.*, **175**, in press.
Underhill C. J. & Isaak G.R., in: Proc. IAU Symp. 181, *Sounding solar and stellar interiors*, eds. Schmider F.-X. & Provost J., Nice Observatory, France, in press.
Vorontsov S. V., Baturin V. A., Pamyatnykh A. A., 1991, *Nature*, **349**, 49.
Wambsganss J., 1988, *A&A*, **205**, 125.

HELIOSEISMIC DATA REDUCTION

F. HILL
National Optical Astronomy Observatories
National Solar Observatory
PO Box 26732
Tucson, AZ 85726-6732, USA

1. Introduction

Helioseismology seeks to infer the properties of the solar interior using measurements of the global normal mode oscillations as a function of spherical harmonic degree ℓ, azimuthal order m, and radial order n as observed on the solar surface. The frequencies, $\nu_0(\ell, m, n)$, of the modes are influenced by the physical conditions of the solar plasma through which the p-mode (pressure) waves propagate, while the power, $P(\ell, m, n)$, and line widths, $\Gamma(\ell, m, n)$, provide clues about the excitation and damping of the oscillations. These mode parameters are extracted from observations of the solar surface using a long and complex data reduction procedure. It is thus important to precisely describe the steps in the reduction and to assess the influence of the choices on the resulting inferred physical conditions.

A comprehensive and detailed discussion of the reduction process is outside the scope (and page limit) of this review, but can be found in Hill *et al.* 1998. Thus, this paper will provide only a top-level overview of the processing chain, and will then focus on two topics which significantly affect the inferred internal solar structure and dynamics. The first topic is the determination of the geometry of the images, in particular the estimation of the radii and the position of solar north of the images that provide the basic data. The second topic is the estimation of the mode parameters from the multi-dimensional spectra that emerge from the basic reduction processing.

2. Processing Overview

A conceptual outline of helioseismic data processing is shown in Figure 1. The core step of the reduction is the decomposition of a time series of images of the solar surface into a set of spherical harmonics. The observations are typically obtained in either Doppler velocity or intensity; here Doppler velocity images from the GONG project are used to illustrate the process. The decomposition can be expressed mathematically as a matrix multiplication:

$$[V_t] \times [Y_{\ell,m}] = [A_{t,\ell,m}]$$

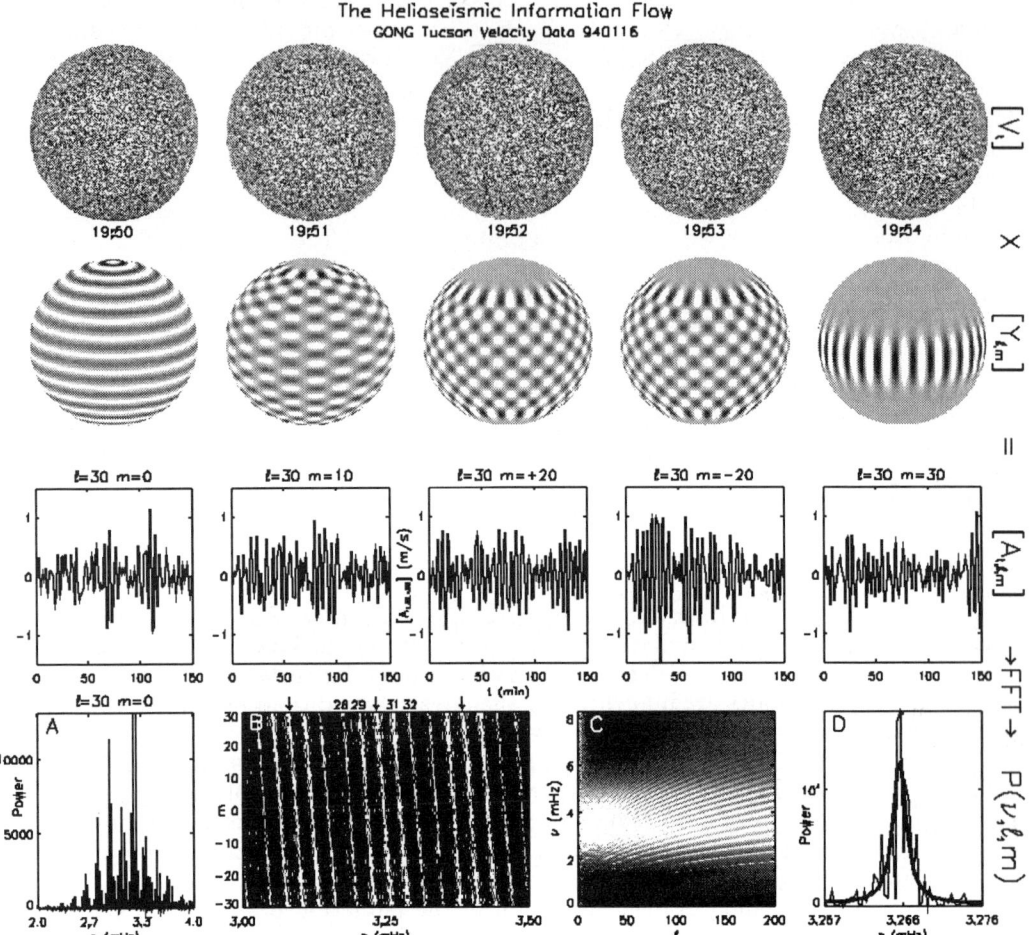

Figure 1. A conceptual outline of helioseismic data processing

where $[V_t]$ is a vector representing a series of velocity images V obtained at a set of times t; $[Y_{\ell,m}]$ is a matrix of spherical harmonic functions for a set of ℓ and m values with

$$Y_{\ell,m} = P_{\ell,m}(\cos\theta)e^{im\phi},$$

$P_{\ell,m}(\cos\theta)$ are the Associated Legendre functions, θ is the heliographic colatitude, ϕ is the heliographic longitude; and $[A_{t,\ell,m}]$ is a matrix of spherical harmonic coefficient time series.

In Figure 1, the top row shows a small five-minute portion of GONG Doppler velocity images to illustrate $[V_t]$. These images have been high-pass temporally filtered to remove non-p-mode velocities, and show the typical center-to-limb variation of the primarily radial oscillatory velocity field. The second row provides images of sample spherical harmonic patterns $[Y_{\ell,m}]$ for $\ell = 30$, with a zonal mode ($m = 0$), a sectoral mode ($m = 30 = \ell$), and three tesseral modes ($0 <| m |< \ell$). Two of the tesseral modes differ in the sign of m, and thus in the spatial phase of the harmonic. The third row show small portions of the five time series $[A_{t,\ell,m}]$ corresponding to the spherical harmonics in row 2.

In practice, the decomposition is not performed as a matrix multiplication. Since the individual image geometry varies temporally, this would require the generation of a complete set of spherical harmonic images for each image. It is much more computationally efficient to remap the input images onto a heliographic θ, ϕ grid, perform a fast Fourier transform (FFT) in the ϕ direction to obtain the m decomposition, and then apply a one-dimensional $P_{\ell,m}$ transform in the θ direction.

After the images have been projected onto the set of $Y_{\ell,m}$ functions, a temporal FFT is applied to each time series to produce a power spectrum $P(\nu, \ell, m)$. The bottom row of Figure 1 shows four representations of these spectra. Panel A is the one-dimensional slice $P(\nu, 30, 0)$. Panel B shows the two-dimensional image $P(\nu, 30, m)$ constructed from a set of one-dimensional slices. This panel shows three sets of slanted ridges with a central arrow indicating the ridge corresponding to the target spherical harmonic $\ell = 30$. Each ridge set corresponds to a specific value of n. The shape of the ridges is a depth-averaged measurement of the rotation of the solar interior. Most of the individual ridges within each set are spatial sidelobes that arise from the spherical harmonic decomposition of the data over a restricted area on the sphere. This results in the leakage of power from ℓ values adjacent to the target ℓ. In the figure, each n set contains five ridges, corresponding to $\ell = 28, 29$, the target 30, 31, and 32; these are labeled in the central set.

Panel C in the bottom row is a two-dimensional spectrum constructed by averaging for each ℓ an image as seen in Panel B along the ridges (so-called m averaging). This results in a set of one-dimensional spectra $P(\nu, \ell)$ which are then combined to make the image known as an $\ell - \nu$ diagram in Panel C. Each of the curved ridges in this diagram corresponds to a single value of n.

Panel D isolates a single peak in the p-mode spectrum at $\ell = 30, m = -30$. The spectral line profile shows considerable jagged structure due to the stochastic driving of the solar oscillations. Superimposed on the data is an estimate of the smooth limit spectrum obtained from a maximum-likelihood fit to the data. The parameters of this estimate, in particular the central frequency ν_0, are then used to infer the internal solar structure and dynamics.

It is evident that there is a long and complex sequence of data reduction steps between the observations and the inferred internal conditions. It is thus not surprising that the data reduction can substantially influence the final results. A few of these effects are discussed in the next sections.

3. Image Geometry

The geometry of the full-disk images used for helioseismology varies continuously. Terrestrial atmospheric refraction imposes an essentially elliptical shape with a diurnal variation in major axis, minor axis, and angular orientation. The image generally moves and rotates on the detector. Finally, the annual orbital motion of the earth creates temporal variations in the solar B_0 angle (the inclination of the solar rotation

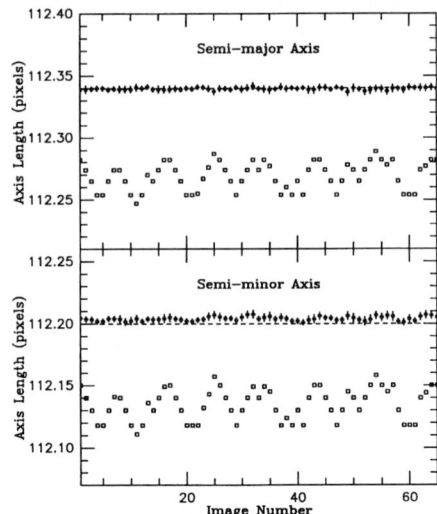

Figure 2. The image dimensions from the Hankel zero routine (filled circles with $1-\sigma$ error bars) and the Laplacian limb fitter (open squares) compared to the true dimensions (dashed lines) for a time series of artificial images with varying MTF.

axis with respect to the ecliptic), P_0 angle (the angle between the solar rotation axis and terrestrial north), and apparent radius. These changes are accounted for in the remapping stage of the spherical harmonic decomposition, but accurate measurements of the geometry of each individual image are essential to reduce subsequent errors in the analysis. Two of the most critical measurements are the apparent radii of the elliptical image, and P_0, the position angle of the solar rotation axis.

3.1. RADII ESTIMATION

It is essential that the scale of ℓ of the oscillation spectrum be accurately calibrated so that the inversion techniques that infer the internal structure correctly associate the kernels with the data. In turn, the ℓ scale is calibrated by measuring the radii of the images. Then, since ℓ is proportional to the solar radius, an error in the radius is linearly related to an error in ℓ.

The measurement of the apparent semi-major and semi-minor axes of the observed elliptical image requires both a definition of the solar limb, and a determination of its position. Usually, the limb is defined as the zero crossing of the second derivative of the solar intensity profile and is frequently determined by computing the Laplacian of the image. Tests with simulated images show that this method systematically underestimates the radii by a part in 10^3. A more sophisticated method based on the Hankel transform of the image has been developed by Toner and Jefferies (1993).

As seen in Figure 2, this method reduces the systematic error by an order of magnitude. In addition, the method also provides an estimated modulation transfer function (MTF), and a measurement of the limb-darkening function of the images. The MTF provides valuable information on the scattering, seeing, and focus quality of each image, and is an essential ingredient of the GONG merging scheme.

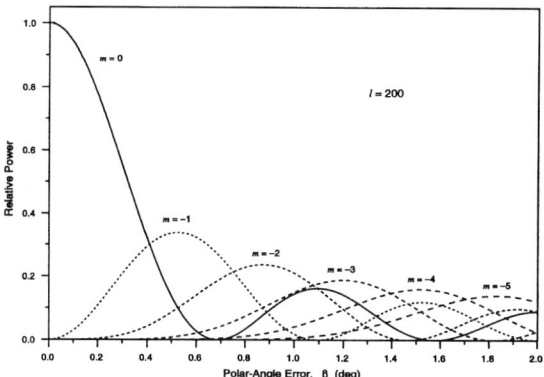

Figure 3. The relative redistribution of power from the $\ell = 200$, $m = 0$ mode into other values of m as a function of the error β in the position of the solar rotation axis.

3.2. P_0 ESTIMATION

The amplitude of the spherical harmonic transform $A_{t,\ell,m}$ for a given image at time t depends on the the coordinate system θ, ϕ. A rotation of this coordinate system changes the relative amplitudes as a function of ℓ and m. If the sun did not rotate, then the oscillation frequencies would be degenerate in m, the well-known transformation of spherical harmonic amplitudes between rotated coordinate systems could be applied, and there would be no need to determine P_0.

The sun does indeed rotate, and the dependence of the rotation as a function of depth and latitude is of fundamental importance for astrophysical fluid dynamics. It is thus crucial to determine P_0. Figure 3 illustrates the redistribution of power in the $\ell = 200$, $m = 0$ mode as a function of the error β in P_0 (Kennedy 1997). This figure shows a β of only 0.2° can reduce the power in the mode by 30%. The effect increases with ℓ.

The effect can be detected by reducing a set of images assuming different values of P_0 (Kennedy and Williams 1997), and it may be possible to eventually refine Carrington's solar rotation elements using helioseismic data. For a helioseismic network, the relative angular orientation of the images can be determined from cross-correlations of simultaneous images, and an optimal set of rotations found around the network. The absolute orientation can then be set by a drift scan at any station in the network (Toner and Harvey 1997).

4. Peak Fitting and Mode Parameter Estimation

The extraction of the mode parameters from the multi-dimensional spectra is the last step in the reduction of helioseismic data. It is also the least well-defined, and currently an area of vigorous research. There are several factors that complicate this step:

1. The stochastic excitation creates a spectrum with noise that does not decrease with increasing time series length
2. The spatial leakage produces a large number of extra features in the spectrum
3. The mode parameters have a temporal variation over the solar cycle

4. The line shape is not symmetric and the underlying physical model is not yet understood
5. The spectrum is qualitatively different when observed in velocity or intensity

The stochastic excitation of the modes adds noise to the observed spectrum as seen in Panel D of Figure 1. This random process dominates the spectrum, and is proportional to the power in a particular frequency interval. The underlying statistical distribution of the process is not Gaussian, but is instead approximates a χ^2 distribution with two degrees of freedom that must be explicitly incorporated into a fitting procedure that seeks the limit spectrum of the oscillation also seen in Panel D of Figure 1. A maximum likelihood minimization method must be used rather than the more common least-squares approach (Anderson, Duvall, and Jefferies 1990).

Spatial leakage was mentioned in Section 2 and can be seen in Panel B of Figure 1. It arises because the spherical harmonics are not orthogonal over a portion of a sphere, that is

$$\int_{\phi_1}^{\phi_2} \int_{\theta_1}^{\theta_2} Y_{\ell,m}(\theta,\phi) Y^*_{\ell',m'}(\theta,\phi) d\theta d\phi \neq 0$$

when $\mid \phi_2 - \phi_1 \mid < 2\pi$, or $\mid \theta_2 - \theta_1 \mid < 2\pi$. In addition to leakage in ℓ, there is leakage in m which introduces a spurious correlation between modes with nearly equal m. Finally, there can be n leaks, where the peak from a mode with a different n (and hence a very different ℓ) actually overlaps the target peak. This complex leak structure must be incorporated into the model of the spectrum that is fitted to the data, and the calculation of the relative amplitudes of the spatial leaks for a specific observational set is a large computational task in itself (Schou 1992, Howe and Thompson 1997). The leakage problem was even more complicated prior to the space and network observing strategies that eliminated additional temporal sidelobes. There is some evidence that the leakage matrix is most effectively used in fitting the complex Fourier amplitudes of the oscillations rather than the power spectrum (e.g. Schou 1992; Appourchaux, Gizon and Rabello-Soares 1997). However, these methods currently do not produce frequencies for individual values of m but instead project the m dependence onto a set of orthogonal polynomials.

Precise measurements of $\nu_0(\ell, m)$ are essential for further advances in helioseismology, and these in turn require very long time series of observations spanning several years. During these long periods the values of ν_0 (as well as P and Γ) change as the sun progresses through its activity cycle, degrading the precision of the measurements. The cycle dependence of the mode parameters is itself an active area of research.

The underlying functional form of the oscillation limit spectrum provides another complication. Virtually all fitting methods have so far assumed that the spectrum has a symmetrical Lorentzian line profile arising from the stochastically excited damped harmonic oscillator model of the oscillations. However, observations clearly show that the peaks in the spectrum are actually asymmetric in shape as a function of ν. The cause of this asymmetry is thought to lie in the physical details of the source mechanism (Jefferies 1997). At the time of this review, the correct physical mechanism is a research subject. Since it has been demonstrated that fitting the wrong line shape to the data results in systematically biasing the measured frequencies, it is feasible that prior results of helioseismology may need revision in the light of more accurate models of the oscillation spectrum.

Another indication that earlier results may need revision lies in the recent discovery of the qualitative difference in oscillation spectra obtained simultaneously in Doppler velocity and total intensity (Harvey et al. 1997). GONG spectra in these two observables show very large (as much as 50 μHz) differences in the apparent positions of the oscillation ridges above 5.5 mHz. In the 3-mHz band, the differences are smaller (a few tenths of a μHz) but still significant. Since the modes must have the same central frequency regardless of the physical variable they are observed in, the definition of a mode frequency is brought into question. The explanation of the differences probably lies again in the theory of the mode excitation, which also must explain the observed flip in the sense of the asymmetry between oscillation spectra observed in velocity and intensity.

It is evident that there are several difficult issues that complicate mode parameter estimation, arguably the most critical link in the helioseismic data reduction chain. The development of methods to improve this step is proceeding along a number of paths. More accurate calculations of the leakage matrix, including the contribution from the small but non-negligible horizontal component of the oscillatory velocity field, are underway. Fitting methods that consider more than one spectral dimension and instead simultaneously fit in both ℓ and ν have been successfully used. These methods are being extended to the much more challenging full 3-dimensional fit. The physical model describing the line shapes and background is currently undergoing rapid development. It is likely that simultaneous fitting of both the velocity and intensity spectra, as well as the $(V - I)$ phase difference spectra, will be necessary to get the correct frequencies.

Substantial progress has been made in applying more advanced spectral analysis methods to helioseismic data. These methods seek to reduce the realization noise component of the spectra that arise from the stochastic excitation. Homomorphic deconvolution promises to actually separate the realization spectrum from the limit spectrum (Baudin and Hill 1997). Multi-taper spectral analysis, coupled with wavelet denoising, has proved to be very effective in reducing the noise in the spectrum and improving the results of fitting the peaks (Komm *et al.* 1997). Figure 4 illustrates the smoothing of the spectrum that results from applying these techniques. Tests with GONG data indicate that the application of a multi-taper analysis in tandem with a wavelet denoising increases the number of good fits by 50%, and decreases the formal variances of the estimated frequencies by a factor of 10 without introducing any systematic bias into the results.

5. Conclusion

The processing of helioseismic data is a lengthy and complex chain of steps that can affect the inferred internal solar structure. In particular, the determination of the image geometry and the fitting of the features in the spectra can introduce systematic errors into the estimated frequencies of the modes. A vigorous development activity is currently underway to improve the mode fitting methods, particularly in the physical model of the asymmetric line shape observed in both intensity and velocity. The outcome of this activity will substantially improve our knowledge of the solar interior.

6. Acknowledgements

This work utilizes data obtained by the Global Oscillation Network Group (GONG) project, managed by the National Solar Observatory, a Division of the National Optical Astronomy Observatories, which is operated by AURA, Inc. under a cooperative agreement with the National Science Foundation. The data were acquired by in-

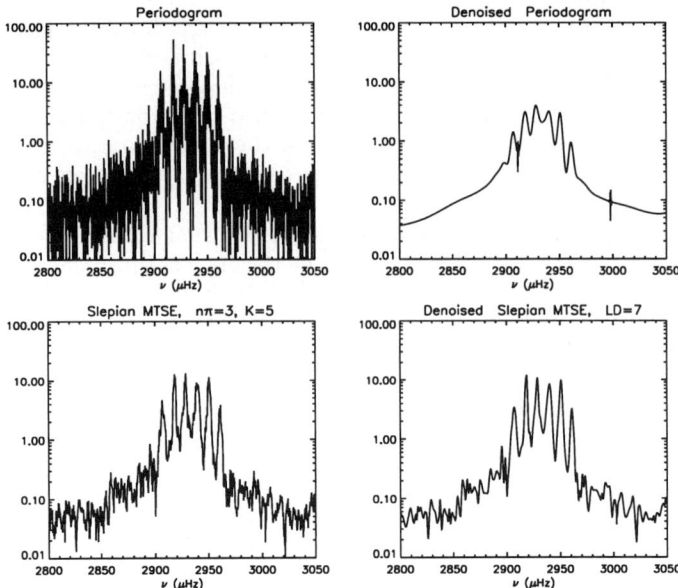

Figure 4. A comparison of a helioseismic spectrum created without any smoothing, with a Slepian multi-taper, a wavelet denoising, and both smoothings.

struments operated by the Big Bear Solar Observatory, High Altitude Observatory, Learmonth Solar Observatory, Udaipur Solar Observatory, Instituto de Astrofísico de Canarias, and Cerro Tololo Interamerican Observatory.

References

Anderson, E.A., Duvall, T.L. Jr., and Jefferies, S.M.: 1990, *Astrophys. J.*, **364**, 699.
Appourchaux, T., Gizon, L., and Rabello-Soares, M.-C.: 1997, *Astron. & Astrophys.*, submitted.
Baudin, F., and Hill, F.: 1997, *Bull. Amer. Astron. Soc.*, **29**, 893.
Harvey, J.W., Hill, F., Komm, R., Leibacher J.W., Pohl, B., and the GONG Team: 1997, these Proceedings.
Hill, F., Anderson, E.A., Armet, D., Chavez, E., Erdwurm, W., Goodrich, J., Harvey, J.W., Hubbard, R.P., Jefferies, S.M., Kennedy, J.R., Ladd, G., Leibacher, J.W., Pintar, J.A., Pohl, B., Toner, C.G., Toussaint, R., Williams, W.E., and Wing, T.: 1998, *Solar Phys.*, in preparation.
Howe, R., and Thompson, M.J.: 1997, *Astron. & Atrophys.*, submitted.
Jefferies, S.M.: 1997, these Proceedings.
Kennedy, J.R.: 1997, *Solar Phys.*, in press.
Kennedy, J.R., and Williams, W.E.: 1997, *Solar Phys.*, submitted.
Komm, R., Gu, Y., Hill, F., Stark, P., and Fodor, I.: 1997, in *Proc. Cool Stars*, in press.
Schou, J.: 1992, *On the Analysis of Helioseismic Data*, Ph.D Thesis, Aarhus University, Appendix 4.
Toner, C.G., and Harvey, J.W.: 1997, in *Sounding Solar And Stellar Interiors*, Proc. IAU Symp. 181, Ed. G. Berthomieu and F.-X. Schmider, in press.
Toner, C.G., and Jefferies, S.M.: 1993, *Astrophys. J.*, **415**, 852.

SEISMIC SOLAR MODELS AND THE NEUTRINO PROBLEM

M. TAKATA AND H. SHIBAHASHI
Department of Astronomy, School of Science, University of Tokyo
Bunkyo-ku, Tokyo 113-0032, Japan

Abstract. We determine the structure of the solar radiative zone with the imposition of the sound speed profile and the depth of the convection zone obtained from helioseismic analysis. We discuss the neutrino fluxes and capture rates using the resultant seismic solar model. We find that the seismic solar model cannot resolve the solar neutrino problem. The hydrogen and helium profiles of the Sun are obtained as a part of the solutions. We find that hydrogen is reduced in the core as expected in the theory of stellar evolution.

1. Introduction

The solar neutrino problem is a long-standing problem in astrophysics and physics. The detected solar neutrino fluxes are substantially less than the theoretical predictions. This discrepancy implies either (i) the physics of neutrinos is not fully understood, or (ii) something is wrong with the solar modelling. The theoretical predictions have been made by using the "standard" solar models. However, these models have the following problems. (1) They are not completely consistent with the results of helioseismology. (2) They are constructed with the assumptions concerning the history of the Sun, which may not be fully justified. (3) They are constructed by adjusting efficiency of the convective energy transport as a free parameter, which is not determined uniquely in the current theory of convection. In this paper, we depart from the standard construction of a solar model and construct a solar model by using as many of the experimentally well measured quantities, including the seismically determined sound speed profile (Takata and Shibahashi, 1997), as possible. We compute the neutrino fluxes based on the resultant solar model and compare them with the observations. First, we determine the sound speed profile of the Sun from the observations of solar oscillations. We then construct a solar model by solving the basic equations governing the stellar structure with the imposition of the determined sound-speed profile and with a constraint of the depth of the convection zone, which is also determined from the sound-speed profile. This method has the following advantages. (1) The model is consistent with helioseismology. (2) The model is a snapshot model of the present-day Sun so that we need few assumptions about the past history of the Sun. (3) We do not care about the treatment of convection since we treat only the radiative core by setting the outer boundary at the base of the convection zone.

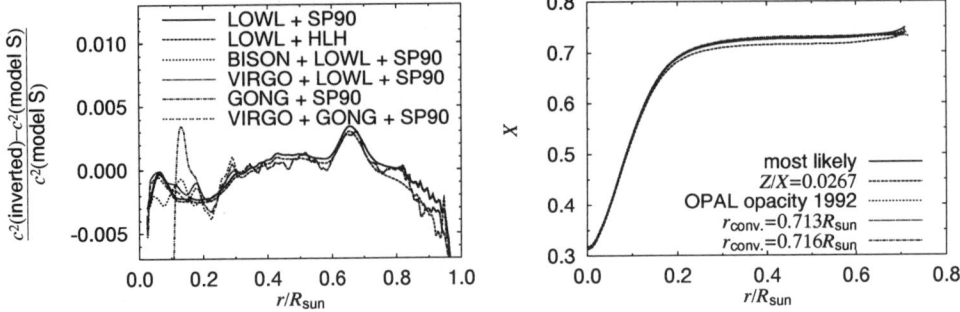

Figure 1. Inverted squared sound speed (left) and hydrogen (right) profiles.

2. Making a Seismic Solar Model

We perform an asymptotic inversion to determine the sound speed profile using the data from LOWL (Tomczyk et al., 1995), GONG (Hill et al., 1996), VIRGO on SOHO (Fröhlich et al., 1997), BISON (Elsworth et al., 1994), HLH (Bachmann et al., 1995), and the observation carried out in 1990 at the South Pole (Jefferies et al., 1995). To eliminate the possibility of spurious results, we calibrate the results of the inversion of the observed frequencies using the inverted results of the theoretical frequencies of the same modes of a solar model. Both the statistical error and the systematic error are as low as 0.1% for $r/R_\odot > 0.3$ and 0.3% for $r/R_\odot < 0.3$. The sound speed profiles determined from various combinations of these data are shown in the left panel of figure 1. The ordinate is the relative difference between the determined squared sound speed and that of the model S of Christensen-Dalsgaard et al. (1996). Almost all the curves are consistent each other except that the results deduced from the GONG + SP90 (South Pole 1990) data show peculiar behaviour near the center. A conspicuous feature is a hump near $r \simeq 0.65 R_\odot$. The general feature of the present results are almost consistent with the recent inversions made by other groups (e.g. Basu et al. 1996, Kosovichev et al. 1997). The location of the base of the convection zone is determined from the fact that the function $W \equiv r^2/(GM_\odot)\, dc^2/dr$ becomes almost constant in the convection zone. The present results are consistent with Gough et al. (1996)'s estimate, $r_{\rm conv} = 0.709 R_\odot$. Once the sound speed profile is determined, we solve the basic equations governing the stellar structure with the imposition of the determined sound speed profile. The hydrogen and helium profiles are obtained as a part of solutions as well as other thermodynamical quantities. We assume, as the first step, that Z is constant through the whole Sun. Note that recent evolutionary solar models including metal diffusion (e.g. Bahcall and Pinsonneault 1995, Christensen-Dalsgaard et al. 1996) are found to be more consistent with the sound speed profile obtained by helioseismology than those without metal diffusion. We will try inhomogeneous Z profiles later. The inner boundary conditions are $L_r = M_r = 0$ at $r = 0$. The outer boundary conditions are set at the base of the convection zone, and they are (1) $L_r = L_\odot$ and (2) $\nabla_{\rm rad} = \nabla_{\rm ad}$. The latter means that the neutral stability against convection holds at the base of the convection zone. We also impose that the relative abundance of heavy elements to hydrogen is equal to the photospheric value,

TABLE 1. Neutrino capture rates and fluxes.

	Seismic Model	Experiments
Cl [SNU]	$9.4^{+1.6}_{-1.9}$	2.28 ± 0.23*1
Ga [SNU]	137^{+7}_{-8}	$69.7 \pm 6.7(\text{stat.})^{+3.9}_{-4.5}(\text{syst.})$*2 $69 \pm 10(\text{stat.}) \pm 6(\text{syst.})$*3
^{8}B [10^6 cm^{-2} s^{-1}]	$6.7^{+1.3}_{-1.5}$	$2.80 \pm 0.19(\text{stat.}) \pm 0.33(\text{syst.})$*4 $2.51^{+0.14}_{-0.13}(\text{stat.}) \pm 0.18(\text{syst.})$*5

References. — *1 Homestake (Davis, 1993), *2 GALLEX (Hampel et al., 1996), *3 SAGE (Abdurashitov et al., 1994), *4 Kamiokande (Fukuda et al., 1996), *5 Super-Kamiokande (Totsuka, 1997)

$Z/X = 0.0245$ (Grevesse and Noels, 1993), since the matter in the convection zone is homogeneous due to mixing. We adopt the OPAL equation of state (Rogers et al., 1996), the OPAL opacity (Iglesias and Rogers, 1996), and the nuclear reaction rates compiled by Bahcall and Pinsonneault (1995).

3. Results and Discussion

The hydrogen profile of the most likely seismic solar model is shown by the thick solid curve in the right panel of figure 1. To see the influence of the input physics upon the results, we vary the value of Z/X, the depth of the convection zone, and the opacity. The results are also shown in the right panel of figure 1. The central hydrogen is really reduced in all cases as expected in the standard theory of stellar evolution. We estimate the age of the present-day Sun ($t_\odot$) from the amount of hydrogen which was converted to helium. Assuming that the average solar luminosity during evolution is $0.85 L_\odot$, we obtain $t_\odot \sim 5.5 \times 10^9$ yr. This is the helioseismic determination of the solar age and is independent of the conventional meteoritic determination. We also obtain the helium profile. The estimated surface helium abundance (Y_S) is between 0.23 and 0.25. This value is consistent with the other helioseismic results determined from the variation in the adiabatic exponent in the HeII ionization zone (Degl'Innocenti et al., 1997). Though the value itself is not very accurate, it should be emphasized that the present method is a new one of determining Y_S. Table 1 lists the neutrino capture rates and fluxes of the seismic solar model and those of experiments. We estimated the errors in the theoretical prediction by taking account of the uncertainties of various input parameters including the sound speed, nuclear reaction rates, opacity, chemical composition, screening effects, position of the base of the convection zone and the systematic errors of inversion process. As seen in table 1, the theoretical values based on the seismic solar model are still significantly larger than the observed ones. It should be stressed that we are free from the standard evolutionary processes and have constructed a snapshot model of the present Sun with fewer assumptions. Nevertheless, the large discrepancy between theory and experiments still exists. Hence, we conclude that the astrophysical solution to the solar neutrino problem is unlikely.

There is another kind of inversion technique called the non-asymptotic inversion, which is based on the variational principle. Some of the data used in the present paper have also been inverted by the non-asymptotic method (e.g., Basu et al. 1996, Gough et al. 1996). An advantage of non-asymptotic inversion methods is that we get directly one more physical quantity such as ρ in addition to the sound speed profile. The overall features of the non-asymptotically inverted sound speed profile are similar to the present results. However, the central density obtained in the present study is higher by several percent than the result of the non-asymptotic inversion. The cause of this difference is not yet clear. One possibility is the difference in the assumption on the Z-profile. As a numerical experiment, we depart from the assumption of constant Z and increase Z in the radiative core while we keep Z/X constant at the base of the convection zone. We find out how much we need to modify Z so that the inverted density near the center in our method matches with the non-asymptotically inverted density. The result is $Z_c \simeq 0.04$. This value is much larger than the standard values, and we think that the difference only in the Z-profile is unlikely to be the cause of inconsistency in the inverted density profiles. The non-asymptotically inverted density profile should be independent of the input physics such as the equation of state, opacity and nuclear reaction rates. On the other hand, the density profile obtained in our method is dependent on them. The apparent discrepancy of the density profiles may imply that something is wrong in the input physics.

References

Abdurashitov, J. N. et al. : 1994, *Phys. Lett. B* **328**, 234.
Bachmann, K. T., Duvall, T. L., Jr., Harvey, J. W., and Hill, F.: 1995, *Astrophys. J.* **443**, 837.
Bahcall, J. N. and Pinsonneault, M. H.: 1995, *Rev. Mod. Phys.* **67**, 781.
Basu, S., Christensen-Dalsgaard, J., Schou, J., Thompson, M. J., and Tomczyk, S.: 1996, *Bull. Astr. Soc. India* **24**, 147.
Christensen-Dalsgaard, J. et al.: 1996, *Science* **272**, 1286.
Davis, R., Jr.: 1993, in Y. Suzuki and K. Nakamura (eds.), *Frontiers of Neutrino Astrophysics*, Universal Academy Press, Tokyo, p. 47.
Degl'Innocenti, S., Dziembowski, W., Fiorentini, G., and Ricci, B.: 1997, *Astroparticle Physics* **7**, 77.
Elsworth, Y., Howe, R., Isaak, G. R., McLeod, C. P., Miller, B. A., New, R., Speake, C. C., and Wheeler, S. J.:1994, *Astrophys. J.* **434**, 801.
Fröhlich, C. et al. : 1997, *Solar Phys.* **170**, 1.
Fukuda, Y. et al.: 1996, *Phys. Rev. Lett.* **77**, 1683.
Gough, D. O. et al. : 1996, *Science* **272**, 1296.
Grevesse, N. and Noels, A.: 1993, in N. Prantzos, E. Vangioni-Flam and M. Cassé (eds.), *Origin and Evolution of the Elements*, Cambridge University Press, Cambridge, p. 15.
Hampel, W. et al. : 1996, *Phys. Lett. B* **388**, 384.
Hill, F. et al. : 1996, *Science* **272**, 1292.
Iglesias, C. A. and Rogers, F. J.: 1996, *Astrophys. J.* **464**, 943.
Jefferies, S. M., Duvall, T. L., Jr., Harvey, J. W., and Pomerantz, M. A.: 1995, private communication.
Kosovichev, A. G. et al. : 1997, *Solar Phys.* **170**, 43.
Rogers, F. J., Swenson, F. J., and Iglesias, C. A.: 1996, *Astrophys. J.* **456**, 902.
Takata, M. and Shibahashi, H.: 1997, *Astrophys. J.*, submitted.
Tomczyk, S., Streander, K., Card, G., Elmore, D., Hull, H., and Cacciani, A.: 1995, *Solar Phys.* **159**, 1.
Totsuka, Y.: 1997, in A. Olinto, J. Frieman and D. Schramm (eds.), *Proc. 18th Texas Symp. on Relativistic Astrophys.*, World Scientific Press, Singapore, in press.

INTERNAL ROTATION, MIXING AND LITHIUM ABUNDANCES

BRIAN CHABOYER

Steward Observatory
University of Arizona
Tucson, AZ, USA 85710

Abstract. Lithium is an excellent tracer of mixing in stars as it is destroyed (by nuclear reactions) at a temperature around $\sim 2.5 \times 10^6$ K. The lithium destruction zone is typically located in the radiative region of a star. If the radiative regions are stable, the observed surface value of lithium should remain constant with time. However, comparison of the meteoritic and photospheric Li abundances in the Sun indicate that the surface abundance of Li in the Sun has been depleted by more than two orders of magnitude. This is not predicted by solar models and is a long standing problem. Observations of Li in open clusters indicate that Li depletion is occurring on the main sequence. Furthermore, there is now compelling observational evidence that a spread of lithium abundances is present in nearly identical stars. This suggests that some transport process is occurring in stellar radiative regions. Helioseismic inversions support this conclusion, for they suggest that standard solar models need to be modified below the base of the convection zone. There are a number of possible theoretical explanations for this transport process. The relation between Li abundances, rotation rates and the presence of a tidally locked companion along with the observed internal rotation in the Sun indicate that the mixing is most likely induced by rotation. The current status of non-standard (particularly rotational) stellar models which attempt to account for the lithium observations are reviewed.

1. Introduction

Li[1] is a sensitive tracer of mixing in stellar radiative regions as it is easily destroyed at temperatures above $\sim 2.5 \times 10^6$ K. For solar type stars, the Li destruction region is located below the surface convection zone in standard models. As a consequence, standard stellar models predict that Li should not be depleted at the surface of solar type stars. This is a rather robust prediction of stellar evolution theory, which has been known for 40 years (Schwarzschild et al., 1957). However, comparisons between the solar photospheric Li abundance and the Li abundance in meteorites show that the Sun has depleted a substantial amount of Li at its surface (Greenstein & Richardson, 1951). The solar Li depletion problem has posed a challenge to stellar evolution theory for 40 years, and the solution to this puzzle is still open to debate.

The Sun is unique in that helioseismic observations allow us to probe the interior structure and rotation of the Sun. These observations can put constraints on possible solutions to the solar Li depletion problem, but by themselves solar observations cannot uniquely determine the cause of solar Li depletion. Observations of stellar Li abundances allow one to study the Li depletion problem as a function of age, metallicity and stellar mass. As such, they provide a powerful test for mechanisms which attempt to explain the solar Li depletion. The discovery of a large dip in Li abundances around 6600 K in the Hyades (Boesgaard & Tripicco, 1986) was not predicted by theorists, and remains a major challenge to theoretical stellar evolution models. There is increasing evidence that a dispersion in Li abundances exists among stars with similar ages, metallicities and masses (Soderblom et al. 1993; Boesgaard et al. 1998). Such a dispersion suggests that another stellar property is important in determining the amount of Li which is depleted in stars. There is mounting observational evidence that rotation plays a key role in determining the amount of Li depletion in a star (Barrado y Navascués & Stauffer 1997; Jones et al. 1997). In this review, I will discuss the relationship between mixing, rotation and Li abundances in stars.

2. Solar Observations

The present photospheric abundance of Li in the Sun has been depleted by a factor of 140 ± 40 as compared to the meteoritic value of $\log N(Li) = 3.31 \pm 0.04$ (Anders & Grevesse, 1989)[2]. Solar models which do not allow for transport or mixing in the radiative regions of the Sun predict very little Li depletion (a factor of 2 – 3). This has been a remarkably robust prediction of standard stellar structure theory which has remained unchanged for 40 years (Schwarzschild et al. 1957; Chaboyer et al. 1995a), despite the fact that the opacities and nuclear reaction rates used in stellar structure codes have changed considerably over this time period. Current solar models imply that the region where Li is destroyed in the Sun is $\sim 0.05 \, R_\odot$ below the base of the

---

[1] In this review I will use Li to represent ^{7}Li, the isotope which is produced by big bang nucleosynthesis. ^{7}Li accounts for $\sim 93\%$ of the total Li abundance in meteorites. Observers typically measure the total Li abundance, while theoretical models determine depletion factors for ^{7}Li and ^{6}Li The ^{6}Li isotope is destroyed at much lower temperatures and ^{7}Li. When making the comparison between the observations and theory, it is usually assumed that the ^{6}Li contribution to the stellar Li content is neglible. However, see the discussion on halo stars (§4) where observations of ^{6}Li may be used to elucidate the mixing mechanism operating in these stars.

[2] Using the standard notation where $\log N(Li) \equiv 12 + \log[N(Li)/N(H)]$.

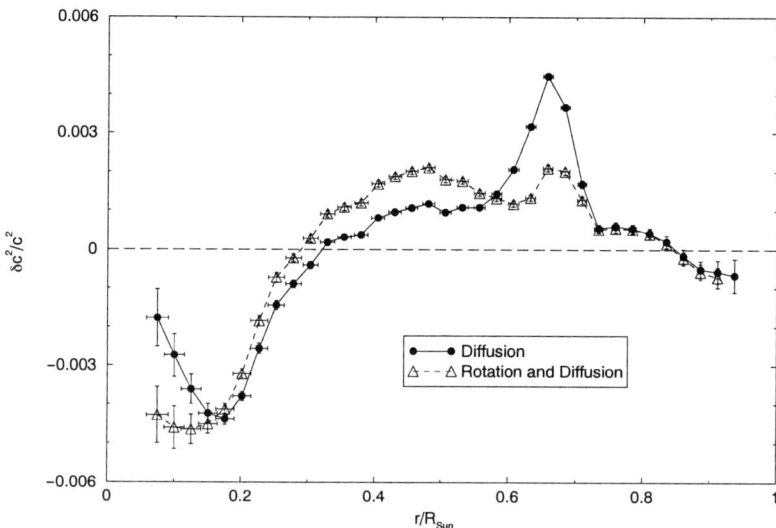

Figure 1. The difference in the square of the sound speed (c^2) between two solar models and the actual Sun, as determined from helioseismology.

solar convection zone. The Li depletion in the Sun suggests that some material from the region where Li is destroyed has been transported to the convection zone, leading to the observed Li depletion at the surface.

The cause of this transport process is still a matter of debate. Observations of Be can constrain the nature of this transport process, as Be is burned at a higher temperature (3.5×10^6 K) than Li, and so probes the deeper interior of the Sun. Observations of the photospheric and meteoritic Be abundances suggest that Be has been depleted at the solar surface by a factor of 1.8 ± 0.5 (Anders & Grevesse, 1989). Thus, at the $3\,\sigma$ level it appears that the Sun has indeed depleted Be at its surface. This is supported by the work of King et al. (1997), who found a minimum surface solar Be depletion level of 12%.

Helioseismic observations can be used to probe the interior structure and rotation of the Sun and provide a strong test for stellar evolution theory. The observed frequencies of the solar p-modes depend primarily on the sound speed c in the solar interior and so it is relatively straightforward to invert the helioseismic observations to probe the variation of sound speed with depth in the Sun. In order to linearize the problem, this inversion is typically done with respect to a theoretical solar model. The data which are now available from the GONG project (Harvey et al., 1996) and SOHO satellite (Rhodes et al., 1997) allow for a very precise determination of the sound speed in the Sun, down to $r \simeq 0.1\,R_\odot$. A typical example of a solar sound speed inversion is shown in Figure 1, which shows that current solar models are able to match the observed value of c^2 to within 0.5% throughout the solar interior. This is a remarkably achievement for stellar evolution theory, made possible by many advances in the last 5 years in the opacities and equation of state used within stellar evolution codes. Another important advance has been the realization that diffusion (whereby the elements heavier than hydrogen sink to the center of the star) must be included in solar models in order to match the helioseismic observations (Christensen-Dalsgaard et al., 1993;

Guenther et al., 1996).

Although the match between the helioseismic observations and theoretical models is impressive, the observations are extremely accurate, and the remaining differences between the models and the Sun can be used to continue to improve the physics used in stellar models. Figure 1 indicates that there is a large error in the diffusion model near the base of the solar convection zone ($r \sim 0.7\,R_\odot$). Detailed studies of this region have suggested that this error is due to the sharp change in the mean molecular weight which occurs at the base of the solar convection zone in models which include diffusion (Gough et al., 1996). The most likely explanation is that some form of slow turbulent mixing is operating below the base of the solar convection zone (Basu, 1997). The model labeled 'rotation and diffusion' in Figure 1 includes slow turbulent mixing generated by rotation (Chaboyer et al., 1995a), and provides a much better match to the observed Sun near the base of the convection zone than the model which only includes diffusion. Helioseismic observations provide strong evidence that slow turbulent mixing occurs below the base of the convection zone in the Sun.

Helioseismology can also probe the interior rotation rate of the Sun. The global modes of oscillation in the Sun can be described by the spherical harmonics, of radial order n, degree ℓ, and horizontal order m (cf. Thompson et al. 1996). If the Sun were spherically symmetric, then the observed oscillations would depend only on n and ℓ. Rotation breaks the spherical symmetry of the Sun, leading to a splitting of the common n, ℓ modes into different m values. The degree of this splitting can be used to infer the solar rotation rate as a function of depth within the Sun. Current data allows for a reliable rotation rate inversions down to $r \sim 0.4\,R_\odot$. Rotation rate rate inversions done by a number of authors, using different data sets have all reached essentially the same conclusion. In the the radiative interior there is quasi-rigid rotation, while a latitude-dependent rotation exists in the entire convection zone (Thompson et al., 1996; Corbard et al., 1997; Kosovichev et al., 1997). The transition between these two regimes occurs in a thin layer below the base of the convection zone and is referred to as the solar tachocline. Recent work suggests that the tachocline is very narrow, with a width estimated to be $r = (0.020 \pm 0.005)\,R_\odot$ (Basu, 1997) or $r = (0.05 \pm 0.03)\,R_\odot$ (Corbard et al., 1998).

3. Young Cluster Observations

Observations of Li in young cluster stars can be used to empirically determine the amount of stellar Li depletion as a function of age, mass and chemical composition. These observations provide very strong constraints for stellar models. A striking result from early observations of Li abundances in cluster stars was the existence of a dip in Li abundances in the F stars (around $T_{\text{eff}} \simeq 6600$ K) in the Hyades cluster (Boesgaard & Tripicco, 1986). Stars around this temperature range have Li abundances which are at least 1.0 dex lower than stars which are hotter or cooler than the dip (see Figure 2). This dip was not predicted by theoretical models. Stars on the hot side of the Li dip do not have convective envelopes, and stars on the cool side of the Li dip have small convective envelopes. The Li dip occurs in stars with very small convective envelopes. The Hyades is a somewhat metal-rich cluster ([Fe/H] = +0.1, (Boesgaard & Friel, 1990)) which is approximately 600 Myr old (Perryman et al., 1998). Observations of Li abundances in other clusters (the Pleiades in particular) have shown that the Li dip does not exist for stars on the zero age main sequence (ZAMS) (Soderblom et al., 1993).

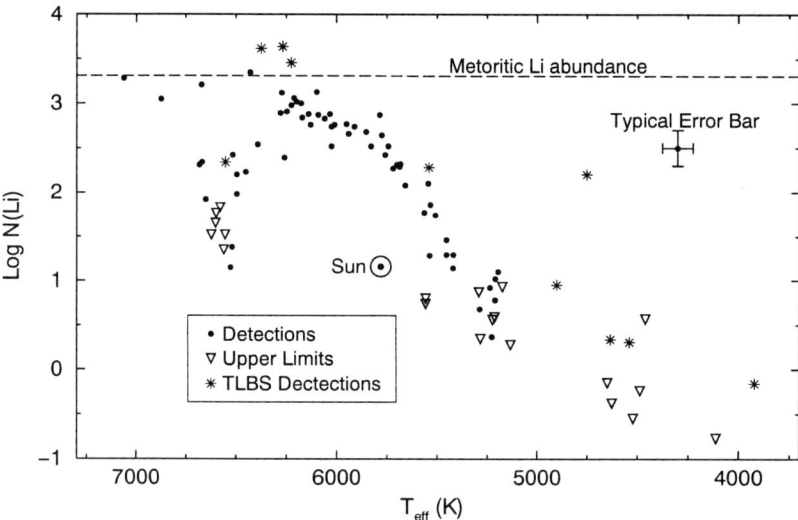

Figure 2. Observations of Li in the Hyades (Balachandran 1995; Barrado y Navascués & Stauffer 1997). The photospheric solar Li abundance is indicated by the solar symbol ⊙, and the meteoritic Li abundance is given by the dashed line.

In addition to the Li dip, there are a number of interesting features in Figure 2. In particular, it appears that the tidally locked binary stars (TLBS) have (on average) a higher Li abundance than single stars or binaries which are not tidally locked (Barrado y Navascués & Stauffer, 1997). This is in good agreement with the theoretical work of Zahn (1994), who postulated that Li depletion is due to rotational mixing and this rotational mixing does not operate in TLBS. The excess Li abundance observed in TLBS is strong evidence that rotation plays a key role in Li depletion.

The Hyades stars around the solar temperature (T_{eff} = 5780 K) have significantly higher Li abundances in the Sun (Figure 2). This is despite the fact that the Hyades is slightly more metal-rich than the Sun and so theoretical models would predict that the Hyades stars should deplete more Li than the Sun on the pre-main sequence. This suggests that Li depletion in these cooler stars ($T_{eff} \lesssim 6000$ K) occurs on the main sequence, a conclusion which is reinforced by the observed Li abundances in the Pleiades (Chaboyer et al., 1995b).

Another interesting feature of Figure 2 is that the highest Li abundances observed are similar to the meteoritic Li abundances, suggesting that their has been no significant Li enhancement over the last 4 Gyr. It is important to note that the only stars which have Li abundances similar to the meteoritic abundances are either TLBS, or are on the hot side of the dip (and hence, do not have a surface convection zone). The question then arises, do all single stars with surface convection zones deplete Li? The stars which have the least amount of Li depletion are those on the cool side of the Li dip, around T_{eff} = 6200 K. In an seminal paper, Boesgaard (1991) demonstrated that, even for the stars around T_{eff} = 6200 K, there was a clear correlation between their Li abundance and age. Older stars have lower Li abundances, implying that all stars with surface convection zones deplete Li on the main sequence. The work of Boesgaard (1991) utilized observations from a number of different observers, and

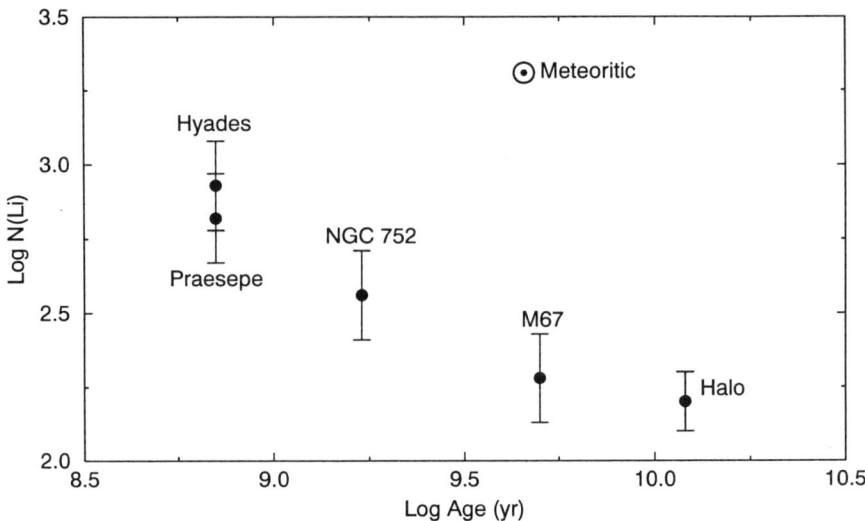

Figure 3. The mean abundance of Li for stars in the temperature range $6300 \leq T_{eff} \leq 6100$ K in four different clusters, as a function of the cluster age (data from Balachandran 1995). The mean abundances for stars in the same temperature range in the halo is taken from the data of Ryan *et al.* (1996). The meteoritic Li abundance is also indicated.

included stars with a rather large range in effective temperatures. In order to see if this had any effect on her conclusions, I have repeated the analysis of Boesgaard (1991) using the uniform data set compiled by Balachandran (1995) and only included stars with $6300 \leq T_{eff} \leq 6100$ K. The results are shown in Figure 3. Even with this restricted data set it is clear that older stars have lower Li abundances, implying that Li depletion has occurred on the main sequence for stars around $T_{eff} = 6200$ K. It is interesting to note that halo stars in the same temperature range appear to match the trend found in open cluster stars (see §4).

The existence of a dispersion in Li abundances at a given effective temperature within a cluster has been the subject of a number of observational papers. If such a dispersion exists, it proves that something besides age, mass and chemical composition must control Li depletion in stars. Thorburn *et al.* (1993) studied in detail the existence of a dispersion among Hyades stars. They found that a such a dispersion does exist for stars with $T_{eff} = 6100$ K, with $\sigma_{N(Li)} \sim 0.15$ dex. The existence of a dispersion among the cooler stars was more difficult to prove, though the data did suggest that a small dispersion of $\sigma_{N(Li)} \sim 0.09$ dex existed among the G stars ($T_{eff} \sim 5500$ K). In contrast, there is clear evidence for a Li dispersion in this temperature range among the Pleiades (a solar metallicity cluster whose stars are on the ZAMS) stars, while there is little evidence for a dispersion among the hotter Pleiades stars (Soderblom *et al.*, 1993). There was also clear evidence for a correlation between rotation rates and Li abundances in the Pleiades, with the fastest rotators having the highest Li abundances. A recent paper by Jones *et al.* (1997) studies in detail the dispersion in Li abundances (at a given effective temperature) among the Pleiades (age 70 Myr), M34 (age 250 Myr) and the Hyades (age 600 Myr). They found that the dispersion among the lower mass stars (M ~ 0.65 to $0.95\,M_\odot$) is greatest on the

ZAMS, and decreases with age. The fastest rotating stars have the highest Li abundances. In addition, they found that Li was depleted on the main sequence, leading Jones et al. to conclude "*high rotation preserves lithium during pre-main sequence evolution and that a high angular momentum loss rate accelerates lithium depletion after the star is on the main sequence*". This is a key observational fact, which implies that rotation must lead to mixing and Li depletion in low mass main sequence stars.

However, the basic assumption used in the above analysis, that a dispersion in Li equivalent widths corresponds to a dispersion in Li abundances has been brought into question by Stuik et al. (1997). These authors suggest that magnetic activity (which leads to the formation of pots and plage on cool stars) may effect the Li line strength. Thus, the observation that fast rotators have high Li equivalent widths could be due to the fact that fast rotators are more active stars than slow rotators. The work of Stuik et al. (1997) is based upon theoretical models (which have known problems), and is a schematic feasibility analysis, rather than a definitive statement that magnetic activity effects Li line strengths. Nevertheless, it raises a key point which clearly requires more attention in the future.

4. Halo Star Observations

Observations of Li in very old stars has the potential to determine the primordial Li abundance. Li is one of only 3 elements (the others being H and He) which are believed to have been produced in significant amounts by big bang nucleosynthesis. Thus, the determination of the primordial Li abundance is a key constraint on theoretical models of big bang nucleosynthesis. In their pioneering work, Spite & Spite (1982) measured Li abundances in 13 metal-poor (halo) stars, and found that all of the stars had nearly identical Li abundances. As there was no relation between Li abundances and metallicity, this suggested that the observed Li abundance was the primordial Li abundance. Since then, there have been numerous studies of Li abundances in halo stars. These studies confirmed that halo stars with $T_{eff} \gtrsim 5600$ K have nearly identical Li abundances, independent of their metallicity or temperature (Spite et al., 1984; Rebolo et al., 1988; Hobbs & Duncan, 1987; Hobbs & Thorburn, 1991).

However, this assumption was brought into question by the discovery of a number of hot halo stars which have abundances significantly lower than the plateau (Hobbs et al., 1991; Spite et al., 1993; Thorburn & Beers, 1993; Thorburn, 1992). About $\sim 5\%$ of the halo stars in the plateau effective temperature range have low Li abundances. These stars demonstrate that at least some hot halo stars do deplete Li, leaving open the possibility that plateau stars may have also depleted Li. The current status of Li abundances measurements in halo stars is shown in Figure 4.

The existence of a dispersion among the Li plateau stars would strongly suggest that even these stars have depleted Li. The question of weather or not a dispersion exists among the plateau stars is an open one. Deliyannis et al. (1993) performed a dispersion analysis in the Li equivalent width–colour data and concluded that an intrinsic Li/H dispersion of 10% existed among the plateau stars. Furthermore, a uniform analysis of ~ 80 plateau halo stars found a correlation between Li abundances, effective temperatures and metallicity (Thorburn, 1994). Hotter, and/or more metal-rich stars were found to have higher Li abundances. This work was confirmed by Ryan et al. (1996). Using a sub-set of the above observations, and adopting a different effective temperature scale Molaro et al. (1995) and Bonifacio & Molaro (1997) found no evidence for a dispersion, or correlation with [Fe/H].

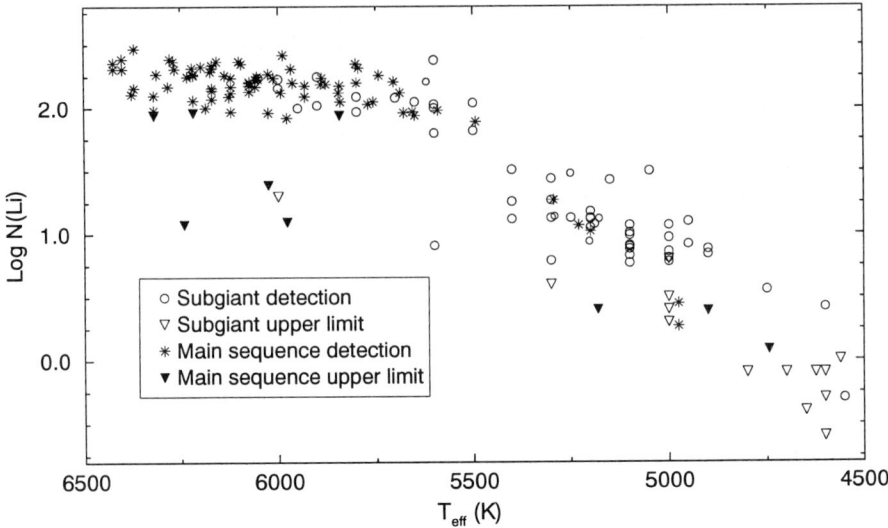

Figure 4. Li abundances in low metallicity ([Fe/H] < −1.0) field stars (halo stars). Data from Pilawchoski *et al.* (1993), Thorburn (1994), and Deliyannis *et al.* (1994).

In order to directly probe for the existence of a Li dispersion among hot halo stars, Boesgaard *et al.* (1998) have determined Li abundances of 7 stars in the globular cluster M92 ([Fe/H] = −2.1). Stars in a given globular cluster all have the same metallicity, age and reddening. In addition, it is possible to observe a number of stars with virtually the same colour, implying that the stars all have the same effective temperature. Boesgaard *et al.* (1998) observed 4 stars with virtually identical effective temperatures, of which 1 star has a significantly higher Li abundance than the others. In all other aspects (including a detailed element by element analysis), the star with the high Li abundance is virtually identical to the other 3 stars. This observation strongly suggests that a dispersion in Li abundances exists among hot halo stars and that this dispersion is due to stellar depletion from a higher primordial Li abundance.

Observations of ^{6}Li can also be used to determine if the plateau stars have depleted Li. ^{6}Li is a destroyed at much lower temperatures than ^{7}Li, so that if enough mixing has occurred to deplete ^{7}Li, virtually all of the ^{6}Li should be destroyed. An observation of ^{6}Li in a hot halo star would suggest that there has not been any significant depletion of ^{7}Li in these stars, and that the observed ^{7}Li abundance is very close to the primordial abundance[3]. The best evidence for a ^{6}Li detection in a hot halo star has been presented by Smith *et al.* (1993) and Hobbs & Thorburn (1997). Both of these groups claim a detection of ^{6}Li in the subgiant HD 84937 ([Fe/H] = −2.4, T_{eff} = 6100 K). The careful analysis of Hobbs & Thorburn (1997) led them to conclude that ^{6}Li/^{7}Li = 0.08 ± 0.04 in HD 84937. Thus, the detection is significant at

---

[3]Standard big bang nucleosynthesis predicts that all of the Li production will be in the form of ^{7}Li. ^{6}Li can be produced by cosmic ray spallation in the interstellar medium. Deliyannis & Malaney (1995) have suggested that ^{6}Li may be produced by stellar flares at the surface of the star, implying that an observation of ^{6}Li in the photosphere of a star does not constrain the depletion of ^{7}Li in the stellar interior.

the 2 σ level. It is premature to conclude that ^{6}Li has definitively been detected in a hot halo star.

Halo stars cooler than the plateau show progressively lower lithium abundances, similar to that seen in open clusters. This is true both for main sequence stars, and subgiant stars (see Figure 4). In both cases, the deepening convective envelopes have brought Li depleted material to the surface of the star. It is interesting to note that stellar evolution models which incorporate the latest available input physics are unable to correctly reproduce the onset of Li depletion observed in main sequence or subgiant halo stars (Chaboyer, 1995). This implies that the models are in need of revision, and casts doubt on the ability of the models to correctly predict the amount of ^{6}Li depletion which occurs.

5. Theoretical Models

The preceding sections have made it clear that Li depletion has occurred in stars with a variety of masses and ages, for which standard stellar evolution models predict no Li depletion. A number of mechanisms have been proposed which deplete Li, in order to explain some or all of the Li observations. These Li depletion mechanisms include: overshoot, mass loss, mixing induced by gravity waves, diffusion, and mixing induced by rotation.

Overshoot at the base of the convection zone has been proposed by a number of authors (most recently by Ahrens *et al.* 1992) to explain the depletion of Li in the Sun. Solar models which include an overshoot of ~ 0.24 pressure scale heights at the base of the convection zone are able to reproduce the observed solar Li depletion. However, this Li depletion occurs on the pre-main sequence, implying very low Li abundances in young, cool ($T_{\rm eff} \sim 5800$ K) stars. This is not seen in any of the open cluster observations (including the Hyades, Figure 2 and the Pleiades). Open cluster observations rule out fast overshoot at the base of surface convection zones as a viable explanation of the observed solar Li depletion (Chaboyer *et al.*, 1995b).

Large amounts of mass loss ($\sim 0.05 \, M_\odot$) sufficient to expose Li depleted material at the surface of a star has been suggested as a cause of the F star Li dip (Schramm *et al.*, 1990). However, a detailed stellar evolution study by Swenson & Faulkner (1992) found that mass loss alone could not explain all open cluster Li observations. Furthermore, a detailed study of Be abundances among Li dip stars has detected moderate Be deficiencies among stars with severe (but detected) Li abundances (Stephens *et al.*, 1997). The mass loss hypothesis requires that all of the Li be depleted before any Be depletion occurs, in contradiction with these observations.

Press (1981) suggested that internal gravity waves generated at the base of the convective envelope of a star may produce weak mixing in the radiative interior. This suggestion was studied in detail by García López & Spruit (1991), who found that mixing induced by internal gravity waves could explain the Li dip observed in F stars if the intensity of the gravity waves was increased by a factor of 15 above mixing length estimates. The degree of Be depletion predicted by these observations appears to be incompatible with the observations of Stephens *et al.* (1997).

The diffusion of Li out of the surface convection zone was suggested by Michaud (1986) to be the cause of the Li gap. The models required a small mass loss rate ($10^{-15} \, M_\odot \, {\rm yr}^{-1}$) to explain the lack of Li depletion on the hot side of the dip. Helioseismology indicates that He diffusion occurs in the Sun (§2), indicating that Li diffusion should also be operating in stars. As the diffusion time scales for Li and Be are similar, Be and Li should be depleted by the same amount if diffusion is the cause of the Li dip. Observations indicate that the degree of Be depletion is much smaller than Li depletion in dip stars (Stephens *et al.*, 1997), ruling out diffusion as the sole cause of the Li dip. Diffusion is also unable to explain the Li depletion observed in the cooler stars ($T_{\text{eff}} \lesssim 6000\,\text{K}$) like the Sun (Chaboyer *et al.*, 1995b).

It has been known for over 70 years that thermal imbalances in rotating stars give rise to large scale flows (von Zeipel, 1924; Eddington, 1925; Sweet, 1950). Thus, it is not surprising that a number of authors have suggested that the Li observations can be best explained by mixing induced by rotation (e.g. Charbonneau & Michaud 1988; Vauclair 1988; Pinsonnealt *et al.* 1990,1992; Charbonnel *et al.* 1992,1994; Zahn 1992; Chaboyer & Demarque 1994; Chaboyer *et al.* 1995a,1995b; Deliyannis & Pinsonneault 1997). The exact physics underlying rotation induced mixing is still not well understood, leading to a variety of approaches in dealing with the mixing. The Yale rotational models (Pinsonnealt *et al.* 1990,1992; Chaboyer & Demarque 1994; Chaboyer *et al.* 1995a,1995b; Deliyannis & Pinsonneault 1997) are perhaps the most ambitious, as they self-consistently include the structural effects of rotation along with the coupled transport of angular momentum and material due to a variety of rotation induced mixing mechanisms. These models, along with those of Charbonnel *et al.* 1994 are able to reproduce a number of features of the observed Li abundances, including the cool side of the Li dip and the main sequence depletion of Li in the Sun and similar temperature cluster stars. Generically, these models also predict a correlation between rotation velocities and Li abundances (observed in young clusters) and that tidally locked binary stars should not deplete Li (Zahn, 1994) as observed in the Hyades (Barrado y Navascués & Stauffer, 1997). The rotation models do a good job of matching the Be observations (Stephens *et al.*, 1997) in the Li dip stars (Deliyannis & Pinsonneault, 1997). These models also predict that a dispersion of Li abundances (due to a dispersion in initial rotation velocities) should exist for stars with equal ages, masses and metallicities, as observed in open clusters.

The small (or non-existent) dispersion among the plateau stars in the halo puts strong constraints on the rotation induced mixing models (Chaboyer & Demarque, 1994), but does not rule them out. Rotation induced mixing models predict some main sequence Li depletion for all stars with surface convection zones, in good agreement with the observations. By extension, these models also predict significant Li depletion among metal-poor halo stars implying that the primordial abundance is higher than the plateau value (Pinsonnealt *et al.*, 1992; Chaboyer & Demarque, 1994). The exact amount of Li depletion depends on the details of the models. The Yale models predict a factor of ~ 10 depletion, implying a primordial Li abundance of $\log N(\text{Li}) = 3.1$, a value which is incompatible with standard big bang nucleosynthesis and current estimates for the primordial abundances of helium and deuterium. However, these models do not match all of the observations, and there are considerable uncertainties associated with these models. Thus, the exact amount of Li depletion which has occurred in the plateau stars is still an open problem.

One of the key problems with the Yale rotational models is that angular momentum transport was inefficient, leading to fast rotating cores. This is clearly in contradiction with helioseismic observations of the solar internal rotation (§2). An easy way to correct this problem has recently been identified by Kumar & Quataert (1997) and Zahn *et al.* (1997). These authors found that low-frequency gravity waves could be excited by convection in the Sun. These gravity waves can transport angular momentum very efficiently. The estimates for the time scale of mixing by these two groups are very similar to each other, and suggest that rigid body rotation would be enforced in the solar radiative region on time scales of 10^7 to 10^8 years. Another difficulty with the Yale models which included the combined effects of rotation and diffusion was that the inclusion of diffusion lead to a differential Li depletion across the effective temperature of the halo Li plateau (Chaboyer & Demarque, 1994). The derived Li abundances did not agree with observations of the most metal-poor stars. This difficulty in reproducing the plateau Li abundances when diffusion is included can be overcome by the addition of a modest stellar wind ($\sim 10^{-12.5}\,M_\odot\,\text{yr}^{-1}$) in the models (Vauclair & Charbonnel, 1995).

6. Summary

There is a wealth of data on Li abundances and rotation velocities in stars with a variety of ages, masses and metallicities. This data clearly indicates that Li depletion occurs on the main sequence for all stars with a surface convection zone. This is in direct contradiction with standard stellar evolution theory. A number of possible mechanisms which lead to extra Li depletion have been put forth. The dispersion in Li abundances at a given age, metallicity and temperature, the correlation between Li abundances and rotation velocities in Pleiades, the fact that tidally locked binary stars in the Hyades have an excess Li abundance as compared to single stars, and the detection of moderate Be deficiencies among Li dip stars with detectable Li abundances, all imply that that rotation induced mixing is leading to Li depletion on the main sequence. Helioseismic observations of the Sun support this hypothesis, for they show that slow form of slow mixing is operating below the base of the solar convection zone (Basu, 1997). Current stellar models which incorporate rotation induced mixing explain many, by not all of the observations. Models which are able to account for all of the data are likely to include diffusion, rotation induced mixing, angular momentum transport by gravity waves and/or magnetic fields and modest stellar winds.

References

Ahrens, B., Stix, M. & Thorn, M. 1992, A&A, 262, 673
Anders, E. & Grevesse, N. 1989, Geochim. Cosmochim. Acta, 56, 197
Balachandran, S. 1995, ApJ, 446, 203
Barrado y Navascués, D. & Stauffer, J.R. 1996, A&A, 310, 879
Basu, S. 1997, MNRAS, 288, 572
Boesgaard, A.M. 1991, ApJ, 370, L95
Boesgaard, A.M., Deliyannis, C.P., Stephens, A. & King, J.R. 1998, ApJ, 493, 206
Boesgaard, A.M. & Friel, E.D. 1990, ApJ, 351, 467
Boesgaard, A.M. & Tripicco, M.J. 1986, ApJ, 302, L49
Bonifacio, P. & Molaro, P. 1997, MNRAS, 285, 847
Chaboyer, B. 1995, in Stellar Evolution: What Should Be Done?, eds. A. Noels, D. Fraipont-Caro, M. Gabriel, N. Grevesse & P. Demarque (Liège: Institut d'Astrophysique), 345 – 358 441, 876
Chaboyer, B. & Demarque, P. 1994, ApJ, 433, 510
Chaboyer, B., Demarque, P. & Pinsonneault, M.H. 1995, ApJ, 441, 865
Chaboyer, B., Demarque, P. & Pinsonneault, M.H. 1995, ApJ, 441, 876
Charbonnel, C., Vauclair, S., Maeder, A., Meynet, G. & Schaller, G. 1994, A&A, 283, 155
Charbonnel, C., Vauclair, S. & Zahn, J.-P. 1992, A&A, 255, 191
Charbonneau, P. & Michaud, G. 1988, ApJ, 334, 746
Christensen-Dalsgaard, J., Proffitt, C.R. & Thompson, M.J. 1993, ApJ, 403, L75
Corbard, T., Berthomieu, G., Provost, J. & Morel, P. 1998, A&A, 330, 1149

Corbard, T., Berthomieu, G., Morel, P., Provost, J., Schou, J. & Tomczyk, S. 1997, A&A, 324, 298
Deliyannis, C.P. & Malaney, R.A. 1995, ApJ, 453, 810
Deliyannis, C.P., Pinsonneault, M.H. & Duncan, D.K. 1993, ApJ, 414, 740
Deliyannis, C.P. & Pinsonneault, M.H. 1997, ApJ, 488, 836
Deliyannis, C.P., Ryan, S.G., Beers, T.C., Thorburn, J.A. 1994, ApJ, 425, L21
Eddington, A.S. 1925, Observatory, 48, 73
García López, R.J. & Spruit, H.C. 1991, ApJ, 377, 268
Gough, D.O., et al. 1996, Science, 272, 1296
Greenstein, J.L. & Richardson, R.S. 1951, ApJ, 113, 536
Guenther, D.B., Kim, Y.-C. & Demarque, P. 1996, ApJ, 463, 382
Harvey, J.W. et al. 1996, Science, 272, 1284
Hobbs, L.M. & Duncan, D.K. 1987, ApJ, 317, 796
Hobbs, L.M. & Thorburn, J.A. 1991, ApJ, 375, 116
Hobbs, L.M. & Thorburn, J.A. 1997, ApJ, 491, 772
Hobbs, L.M., Welty, D.E. & Thorburn, J.A. 1991, ApJ, 373, L47
Jones, B.F., Fischer, D., Shetrone, M. & Soderblom, D.R. 1997, AJ, 114, 352
King, J.R., Deliyannis, C.P. & Moesgaard, A.M. 1997, ApJ, 478, 778
Kumar, P. & Quataert, E.J. 1997, ApJ, 475, L143
Kosovichev, A.G. et al. 1997, Solar Physics, 170, 43
Michaud, G. 1986, ApJ, 302, 650
Molaro, P., Primas, F., Bonifacio, P. 1994, A&A, 295, L47
Perryman, M.A.C. et al. 1998, A&A, 331, 81
Pilachowski, C.A., Sneden, C. & Booth, J. 1993, ApJ, 407, 699
Pinsonneault, M. H., Deliyannis, C. P. & Demarque, P. 1992, ApJS, 78, 179
Pinsonnealt, M. H., Kawaler, S. D. & Demarque, P. 1990, ApJS, 74, 501
Press, W.H. 1981, ApJ, 245, 286
Rebolo, R., Moloaro, P. & Beckman, J.E. 1988, A&A, 192, 192
Rhodes, Jr. E.J., Kosovichev, A.G., Scherrer, P.H., Schou, J. & Reiter, J. 1997, Solar Physics, 175, 208
Ryan, S.G., Beers, T.C., Deliyannis, C.P. & Thorburn, J.A. 1996, ApJ, 458, 543
Schramm, D.N., Steigman, G. & Dearborn, D.S.P. 1990, ApJ, 359, L55
Schwarzschild, M., Howard, R. & Härm, R. 1957, ApJ, 125, 233
Soderblom, D.R., Jones, B.F., Balachandran, S., Stauffer, J.R., Duncan, D.K., Fedele, S.B. & Hudon, J.D. 1993, AJ, 106, 1059
Smith, V.V., Lambert, D.L. & Nissen, P. 1993, ApJ, 408, 262
Spite, F. & Spite, M. 1982, A&A, 115, 357
Spite, M., Molaro, P., Fran cois, P. & Spite, F. 1993, A&A, 271, L1
Spite, M., Maillard, J.P. & Spite, F. 1984, A&A, 141, 56
Stephans, A., Boesgaard, A.M., King, J.R. & Deliyannis, C.P. 1997, ApJ, 491, 339
Stuik, R., Bruls, J.H.M.J. & Rutten, R.J. 1997, A&A, 322, 911
Sweet, P.A. 1950, MNRAS, 110, 548
Swenson, F.J. & Faulkner, J. 1992, ApJ, 395, 654
Thompson, M.J., et al. 1996, Science, 272, 1300
Thorburn, J.A. 1992, ApJ, 399, L83
Thorburn, J.A. 1994, ApJ, 421, 318
Thorburn, J.A. & Beers, T.C. 1993, ApJ, 404, L13
Thorburn, J.A., Hobbs, L.M., Deliyannis, C.P. & Pinsonneault, M.H. 1993, ApJ, 415, 150
Vauclair, S. 1988, ApJ, 335, 971
Vauclair, S. & Charbonnel, C. 1995, A&A, 295, 715
von Zeipel, H. 1924, MNRAS, 84, 665
Zahn, J.-P. 1991, A&A, 265, 115
Zahn, J.-P. 1994, A&A, 288, 829
Zahn, J.-P., Talon, S. & Matias, J. 1997, A&A, 322, 320

LOI/SOHO CONSTRAINTS ON OBLIQUE ROTATION OF THE SOLAR CORE

L. GIZON
W.W. Hansen Experimental Physics Laboratory,
Stanford University, Stanford CA 94305, U.S.A.

T. APPOURCHAUX
ESA / ESTEC,
P.O. Box 299, 2200 AG Noordwijk, The Netherlands

AND

D.O. GOUGH
Institute of Astronomy, and Department of Applied Mathematics
and Theoretical Physics, University of Cambridge,
Madingley Road, Cambridge CB3 0HA, U.K.;
HEPL, Stanford University, Stanford CA 94305, U.S.A.

1. Introduction

The axis of rotation of the Sun's surface is inclined from the normal to the ecliptic by $7°.25$. Is that true also of the rotation of the rest of the Sun? Knowledge of the direction of the angular momentum is pertinent to studies of the formation of the solar system. Moreover, Bai and Sturrock (1993) have recently interpreted temporal variations in the spatial distribution of solar flares as the outcome of the interaction of the Sun's envelope with an obliquely rotating core. We report here an attempt to determine the principal seismic axes of oscillation of the dipole and quadrupole p modes from LOI data obtained as a component of the VIRGO investigation on the spacecraft SOHO. We find that formally their most likely orientation is somewhat closer to being normal to the ecliptic than is the axis of the surface rotation. However, the uncertainty in the determination well encompasses the possibility of them being parallel to the surface rotation axis, yet it does not reject (at a level marginally greater than one standard deviation) the possibility that the Sun's angular momentum is parallel to that of the rest of the solar system.

2. The influence of rotation on oscillations

Rotation splits the degeneracy of oscillation modes in a multiplet (having like radial order n and degree l). If angular velocity is constant on spheres, a scalar p-mode eigenfunction is approximately proportional to $\Re[Y_l^m(\theta, \phi) \exp(-i\omega_m t)]$ with respect to spherical polar coordinates (r, θ, ϕ) about an appropriate axis **P**. The index m denotes the azimuthal order, and Y_l^m is a spherical harmonic function. If rotation is about a unique axis **R**, then **P** coincides with **R**. But if **R** varies with r, **P** is a vector average of $\mathbf{R}(r)$, weighted with the rotational splitting kernel K of the mode (Gough and Kosovichev, 1993). When the magnitude Ω of the angular velocity varies with latitude (defined relative to $\mathbf{R}(r)$), K depends on m and, except in the case $l = 1$, the pulsation axis of the multiplet is not well defined. In the case of the Sun, the latitudinal variation of Ω is small: in the data analysis of quadrupole modes we report below, we assume that there exists a frame in which each azimuthal component can still be described geometrically in terms of a single spherical harmonic, and that the m dependence of the splitting is negligible. If the oscillation axis **P** is not coincident with the surface rotation axis **S** with respect to which one imagines solar oscillations are analysed, each harmonic projection (ℓ, m') with respect to **S** is actually a linear combination of normal modes, and will exhibit all the rotationally split frequencies ω_m (where m is the true azimuthal order, with respect to **P**). The orientation of **P** cannot be determined from frequencies alone, however – the frequencies of an (aspherical) object are independent of the direction from which that object is viewed. To determine **P** it is necessary to measure the eigenfunctions. A first attempt at that (Gough, Kosovichev and Toutain, 1995), made by estimating in the IPHIR data the amplitudes of whole-disc projections of blended components of $l = 1$ multiplets, suggested that **P** might be closer to the normal **n** to the ecliptic than is **S**, but the significance of the measurement is difficult to assess. Here we report a more sophisticated analysis of modes with $l = 1$ and $l = 2$ observed with the LOI (Appourchaux, Andersen, Fröhlich, et al., 1997).

3. Determination of the pulsation axes

The LOI measures radiant intensity $s_p(t)$ integrated over a set of 12 pixels p. The nominal attitude of SOHO is such that the projection of **S** onto the detector should coincide with a principal axis of symmetry **d** of the detector at all times. A multiplet (l, n) is presumed to be composed of $2l + 1$ independently randomly excited modes m, each producing a disturbance of the form $x_m(\theta, \phi, t) \equiv f_m(t; \omega_m) Y_l^m(\theta, \phi)$ about a pulsation axis **P**. We denote by $\beta_d(t)$ the inclination of the axis **P** with respect to **d**, and by $\alpha_d(t)$ the azimuth of **P** relative to the line of sight. During the short selected interval of observation, the variation of $\alpha_d(t)$ and $\beta_d(t)$ is small, and accordingly we ignore it, replacing $\alpha_d(t)$ and $\beta_d(t)$ by their averages $\overline{\alpha}_d$ and $\overline{\beta}_d$. This approximation implies that we neglect the hyperfine structure that results from SOHO's orbit about **n** (see Goode and Thompson, 1992). The sensitivity of a pixel p to a mode m can thus be expressed in terms of the Euler angles $\overline{\alpha}_d$ and $\overline{\beta}_d$ by integrating over the pixel the Fourier amplitude $\hat{x}_m(\theta, \phi, \omega)$ of x_m weighted with the limb-darkening function. Now consider a small frequency range, and construct $2l + 1$ linear combinations $\hat{y}_M(\omega)$, $-l \leq M \leq l$, of the 12 Fourier transforms $\hat{s}_p(\omega)$ in order to isolate each component m. Using a maximum likelihood technique, we fit a parametric model to the $2l + 1$ complex spectra $\hat{y}_M(\omega)$. For each discrete frequency ω, $\hat{f}_m(\omega; \omega_m)$ is presumed to be (a realization of) a centred complex Gaussian random variable whose variance is Lorentzian with amplitude A_m, linewidth Γ, and central frequency $\omega_0 + m\, a_1$, where a_1 is the mean rotational splitting. The $\hat{f}_m$ are assumed to be independent random variables. We add independent Gaussian noise to each pixel with location-dependent

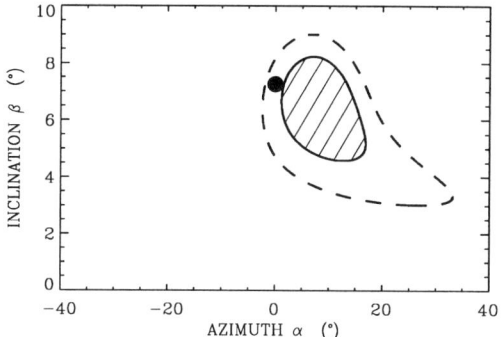

Figure 1. Permitted orientations (α, β) of the mean pulsation axis **P** with respect to the normal **n** to the ecliptic. The hatched area and the dashed boundary refer to the 1-σ and 1.5-σ confidence levels respectively, computed under the assumption that the uncertainties in $\overline{\alpha}_d$ and $\overline{\beta}_d$ are independent. The filled circle indicates the direction of **S**.

variance depending on three undetermined parameters N_j. The parameters ($\overline{\alpha}_d$, $\overline{\beta}_d$, ω_0, a_1, Γ, A_m, N_j) are chosen such that they maximize the joint probability density function over some frequency interval (frequency bins are assumed to be independent of each other). Realistic Monte Carlo simulations have been performed which demonstrate that the estimators are essentially unbiased.

Observations were taken over the four-month interval 1996, July 10 – Nov. 6, centred about Sept. 8 when the inclination $B_0(t)$ of **S** from the axis **d** was the greatest (on average, $\overline{B}_0 = 6°.06$). During this interval the average position of **n** is coincident with **d**. The angles $\overline{\alpha}_d$ and $\overline{\beta}_d$ where determined for each multiplet separately. Means and standard deviations are listed in Table 1. From the knowledge of the satellite's orbit it is also possible to deduce constraints on the fixed azimuth α and inclination β of **P** with respect to the ecliptic normal **n** (Figure 1). The results are not inconsistent with **P** being coincident with **S**.

TABLE 1. Euler angles $\overline{\alpha}_d$ and $\overline{\beta}_d$ which define the directions of the pulsation axes **P** with respect to **d**.

mode set	$\overline{\alpha}_d$	$\overline{\beta}_d$
$l = 1$ and $15 \leq n \leq 24$	$5° \pm 10°$	$4°.9 \pm 2°.5$
$l = 2$ and $14 \leq n \leq 24$	$10° \pm 10°$	$5°.7 \pm 1°.6$
all	$8° \pm 7°$	$5°.5 \pm 1°.4$

4. On the orientation of the core

We adopt a simple two-zone model. The outer zone, $r_c < r \leq R_\odot$ is assumed to rotate about **S** with angular velocity determined from higher-degree modes (Kosovichev, Schou, Scherrer, et al., 1997); beneath the convection zone a uniform rate of 435 nHz

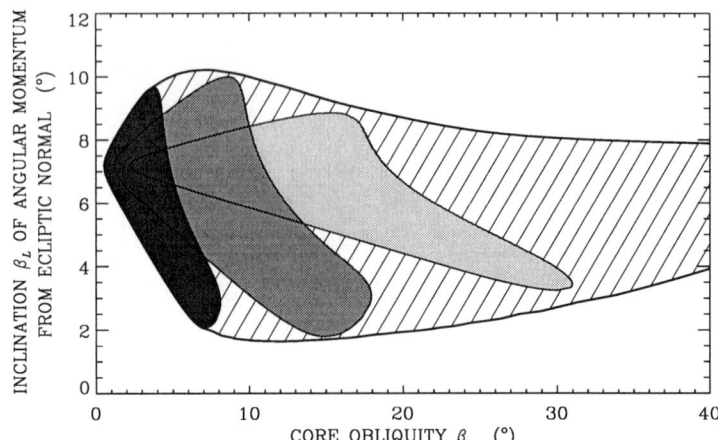

Figure 2. Permitted inclinations β_L of the solar angular momentum when $a_1 = 440$ nHz, for $r_c/R_\odot = 0.25$ (light grey), 0.4 (mid grey) and 0.7 (dark grey). The hatched area is the permitted region for $r_c < 0.7R_\odot$ and 433 nHz $< a_1 <$ 447 nHz.

was adopted. The inner zone, $0 \leq r \leq r_c$, was assumed to rotate uniformly with angular velocity Ω_c about an axis **C**. The constraints on (α, β) can then be used to constrain the obliquity β_c of **C** with respect to **S** for each value of r_c. In this way one can test, for example, the model of Bai and Sturrock (1993) in which the core rotates at a rate $\Omega_c/2\pi = 454$ nHz inclined by $\beta_c = 40°$. In this model, we find that the core radius r_c has to be less than $0.24R_\odot$ in order to be consistent with LOI data.

5. The orientation of the angular momentum

Subject to the same two-zone model, we can also constrain the orientation (α_L, β_L) of the axis **L** of the Sun's angular momentum with respect to the normal **n**. In Figure 2, the hatched region in the (β_c, β_L) plane is the permitted region if α, β and a_1 are all within one standard deviation of their seismically determined values. We measured the averaged rotational splitting over the same set of modes to be $a_1 = 440 \pm 7$ nHz from one year of data. It is evident that at a level of significance only slightly above one standard deviation the possibility of **L** being perpendicular to the plane of the ecliptic is not ruled out.

References

Appourchaux, T., Andersen, B.N., Fröhlich, C., Jiménez, A., Telljohann, U. and Wehrli, C.: 1997, *Solar Phys.* **170**, 27.
Bai, T. and Sturrock, P.A.: 1993, *Astrophys. J.* **409**, 476.
Goode, P.R. and Thompson, M.J.: 1992, *Astrophys. J.* **395**, 307.
Kosovichev, A.G., Schou, J., Scherrer, P.H., Bogart, R.S., Bush, R., Hoeksema, J.T., et al.: 1997, *Solar Phys.* **170**, 43.
Gough, D.O. and Kosovichev, A.G.: 1993, in Weiss, W.W. and Baglin, A. (eds.), *Inside the Stars, ASP Conf. Series* **40**, pp. 566–568.
Gough, D.O., Kosovichev, A.G. and Toutain, T.: 1995, in Ulrich, R.K., Rhodes, E.J. Jr. and Däppen, W. (eds.), *GONG'94, ASP Conf. Series* **76**, pp. 55–58.

HELIOSEISMOLOGY AND THE SOLAR NEUTRINO PROBLEM

H. M. ANTIA AND S. M. CHITRE
Tata Institute of Fundamental Research
Homi Bhabha Road, Mumbai 400005, India

The precisely measured frequencies of solar oscillations provide us with a unique tool to probe the solar interior with sufficient accuracy. These frequencies are principally determined by the dynamical quantities like sound speed, density or the adiabatic index of the solar material and a primary inversion of the observed frequencies yields the sound speed and density profiles inside the Sun (Gough et al. 1996). The equations of thermal equilibrium enable us to determine the temperature and chemical composition profiles, but for this additional prescriptions regarding the input physics (i.e., opacities, equation of state and nuclear energy generation rate) are required (Shibahashi 1993; Antia & Chitre 1995; Shibahashi & Takata 1996; Kosovichev 1996). This information in turn can be used to calculate the neutrino fluxes, and the seismic models can thus be used to explore the possibility of an astrophysical solution to the solar neutrino problem (Roxburgh 1996; Antia & Chitre 1997).

The sound speed and density profiles inside the Sun are inferred from the observed frequencies using a regularized least squares technique (Antia 1996). The primary inversions based on the equations of hydrostatic equilibrium, however, provide us with the ratio T/μ, where μ is the mean molecular weight. In order to separately determine T and μ it is necessary to use the equations of thermal equilibrium. Once the temperature, density and composition are known we can calculate the integrated luminosity and the neutrino fluxes. However, the computed luminosity will not necessarily match the observed solar luminosity because of possible errors in the primary inversions as well as the uncertainties in the nuclear reaction rates. This difference in luminosity will give an estimate of uncertainties in the nuclear reaction rates and also the inversions.

We have used the nuclear reaction rates from Bahcall & Pinnsoneault (1995, hereinafter BP95), except for the pp reaction for which we have used the older reaction rate from Bahcall (1989). These inversions have been obtained using a Z profile including diffusion of heavy elements, with surface $Z = 0.018$. A comparison between the X profile in the Sun obtained using GONG months 4-10 data and that in the Model S of Christensen-Dalsgaard et al. (1996) shows the inverted profile to be distinctly flat for $r > 0.68R_\odot$ (Fig. 1). It thus appears that the region just below the convection zone is probably mixed (Richard et al. 1996; Basu 1997) and this may explain the observed low lithium abundance in the solar envelope.

The helioseismically estimated cross-section for the pp nuclear reaction should be $(4.15 \pm 0.25) \times 10^{-25}$ MeV barns, if we admit 5% error in integrated luminosity that incorporates all systematic errors including those arising from a factor of 2 uncertainty in Z. This value is consistent with 4.07×10^{-25} MeV barns (Bahcall 1989), but slightly

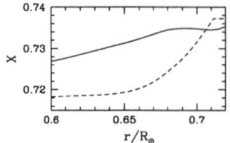

Figure 1. The continuous line shows the hydrogen abundance profile in the Sun as inferred using the GONG months 4–10 data, while the dashed line shows the X profile in Model S of Christensen-Dalsgaard et al. (1996).

higher than the value of 3.89×10^{-25} MeV barns adopted by BP95. However, an increase in the pp nuclear reaction rate by a factor of 2.9 as recently suggested by Ivanov et al. (1997) is certainly unacceptable, because the computed luminosity for seismic models then turns out to be $2.7 L_\odot$. Even if opacities are reduced arbitrarily in an attempt to get the correct luminosity, the temperature is correspondingly lowered and in order to maintain the sound speed the hydrogen abundance X needs to be increased to a value exceeding unity! Clearly, such an increase in the pp reaction rate is totally unphysical and in fact, even an increase by a factor of 1.65 can be ruled out from helioseismic data.

It can be readily seen that the neutrino fluxes in seismic models, computed assuming thermal equilibrium, are not very different from those in the standard solar model and evidently these are inconsistent with the observed fluxes. In order to explore the possibility that a relaxation of thermal equilibrium may alter the neutrino fluxes, we use arbitrary X profiles and even after allowing for arbitrary modifications in opacities (or equivalently in the heavy element abundances Z) it turns out to be difficult to construct seismic models which are simultaneously consistent with any two of the three solar neutrino experiments (Antia & Chitre 1997). This is true even when the Sun is not assumed to be in thermal equilibrium and the deductions are freed from any assumptions about input opacity and specific mode of energy transport in the central regions of the Sun. The only premise that is required for the analysis is that the sound speed and density in the solar interior can be prescribed from primary inversions of the helioseismic data and that the nuclear reaction rates are known and neutrinos have standard properties. These results are consistent with those obtained by Heeger & Robertson (1996) from more general considerations, independent of any solar model. The solution to the solar neutrino problem should therefore be sought in terms of nonstandard neutrino properties (e.g., MSW effect).

References

Antia, H. M. 1996, A&A 307, 609
Antia, H. M. & Chitre, S. M. 1995, ApJ 442, 434
Antia, H. M. & Chitre, S. M. 1997, MNRAS 289, L1
Bahcall, J. N. 1989, *Neutrino Astrophysics*, Cambridge University Press, Cambridge, UK
Bahcall, J. N. & Pinsonneault, M. H. 1995, Rev. Mod. Phys. 67, 781 (BP95)
Basu, S. 1997, MNRAS 288, 572
Christensen-Dalsgaard, J., Däppen, W. et al. 1996, Science 272, 1286
Gough, D. O., Kosovichev, A. G., Toomre, J. et al., 1996, Science 272, 1296
Heeger, Robertson 1196, Phys. Rev. Lett. 77, 3720
Ivanov, A. N., Troitskaya, N. I., Faber, M. & Oberhummer, H., 1997 Nucl. Phys. A 617, 414
Kosovichev, A. G. 1996, Bull. Astron. Soc. India 24, 355
Richard, O., Vauclair, S., Charbonnel, C., Dziembowski, W. A., 1996, A&A 312, 1000
Roxburgh, I. W., 1996, Bull. Astron. Soc. India 24, 89
Shibahashi, H. 1993, in Frontiers of Neutrino Astrophysics, eds. Y. Suzuki & K. Nakamura, Universal Academy Press, Tokyo, p. 93
Shibahashi, H. & Takata, M. 1996, PASJ 48, 377

THE ART OF FITTING P-MODE SPECTRA

T. APPOURCHAUX
ESA/ESTEC
P.O.Box 299, 2200 AG, Noordwijk, Pays-Bas

AND

L. GIZON
W.W. Hansen Experimental Physics Laboratory
Center for Space Science and Astrophysics, Stanford University
Stanford, CA 94305-4085, USA

1. Maximum Likelihood Estimators and p-mode spectra

For deriving p-mode parameters from m, ν diagrammes, one has to treat correctly the statistics of the observation. The correct statistical treatment of these diagrammes was first achieved by Schou (1992) (PhD thesis, Aarhus University). Fitting p-mode spectra requires 4 major steps:

1. Compute the mode leakage matrices
2. Compute mode covariance matrices from the previous matrices
3. Compute the noise covariance matrices
4. Compute and maximize the likelihood of the observation

The leakage matrices link the observed m Fourier spectrum to the modes m'. Most often, the matrices are not diagonal and leads to correlation between the observed m. Nevertheless, we have shown that even if there are correlation between the $2l+1$ Fourier spectra, taking into account correctly the correlation leads implicitly to the removal of these correlations. Unfortunately, what is true for the modes does not apply to the noise mainly because they have different correlation characteristics.

Here we must point out that using only power spectra of each m signal to derive p-mode parameters is an approximation that will lead to underestimating the splitting at low degrees. This is the case of the official GONG splittings that are well known to be severely underestimated. In any case, after having taken into account all the correlations, it can be shown, using for example Monte-Carlo simulations, that the maximization of the likelihood leads to unbiased p-mode parameters. The result we obtained on the GONG data (Appourchaux, Rabello-Soares and Gizon, these proceedings) show that the artificial bias of the GONG splitting can be removed using the steps mentioned above.

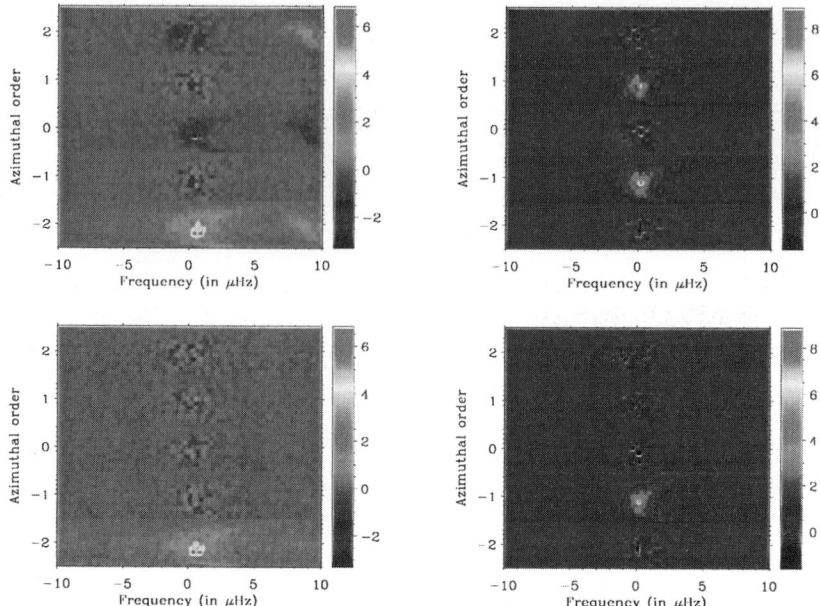

Figure 1. Correlations between the $2l+1$ Fourier spectra of $l=2$ for 1-year of LOI/SOHO data. Two top diagrammes are for spherical harmonic filters: (Top, left) For $l=2, m=2$. (Top, right) For $l=2, m=1$. Two lower diagrammes are for inverted leakage matrix: same as for the previous diagrammes: (Bottom, left) For $l=2, m=2$. (Bottom, right) For $l=2, m=1$. Each panel is composed of 5 $(2l+1)$ echelle diagramme of the correlation between the Fourier spectra, i.e. $\mathrm{Re}(y_m(\nu)y_{m'}^*(\nu))$ or real part of the cross echelle diagrammes. We can note the absence of correlation between the m's for which Δm is odd. We should also point out that the $l=0$ modes have been removed by applying the inverse of the combined leakage matrix of $l=0$ and $l=2$ (Left lower panel).

2. Visualization of correlations

Here we show that even if you do not know anything about the instruments that obtained the observations, it is still possible to measure on the data the leakage and mode covariance matrices. Figure 1 shows a way to visualize the correlations between the $2l+1$ Fourier spectra. These diagrammes are useful for investigating the knowledge of the leakage matrix. For instance, one can make similar diagrammes not only with the original data ($\vec{y_l}$ but with also with $\vec{x_l} = C_l^{-1}\vec{y_l}$, where C_l is the leakage matrix for the degree l. In this case the leakage matrix for $\vec{x_l}$ is purely diagonal for the given l; for other degrees this does not apply. If the assumed leakage matrix is correct then the cross echelle diagrammes will display power only for $m = m'$.

Here we would like to stress that what really matters is the knowledge of the leakage matrices. The main source of bias for the splitting lies in the imperfect knowledge of the leakage matrix elements. We have also shown using Monte-Carlo simulations that underestimating or overestimating leakage matrices elements will result in underestimating the splitting. These effects have also be found using real data such as the LOI or GONG (Appourchaux, Rabello-Soares and Gizon, these proceedings).

A COMPARISON OF LOW-DEGREE SOLAR P-MODE FREQUENCIES FROM BISON AND LOI

T.APPOURCHAUX
ESA/ESTEC
P.O.Box 299, 2200 AG, Noordwijk, Pays-Bas

W.J. CHAPLIN, Y. ELSWORTH, G. R. ISAAK, C. P. MCLEOD AND B. A. MILLER
School of Physics and Astronomy,
University of Birmingham, Edgbaston,
Birmingham B15 2TT, U.K.

AND

R.NEW
School of Science & Mathematics,
Sheffield Hallam University,
Sheffield S1 1WB, UK

1. OBSERVATIONS

Here, we compare the low-degree solar p-mode frequencies returned from the analysis of two, contemporaneous, independent helioseismological data sets collected during 1996. The first comprises Doppler velocity observations of the 770-nm line of potassium, made in integrated sunlight by the six-station, terrestrial Birmingham Solar-Oscillations Network (BiSON). The second consists of irradiance distribution measurements of the solar disc, made at 500 nm, by the Luminosity Oscillations Imager (LOI), which is part of the VIRGO experiment on the ESA/NASA SOHO satellite. The starting date is 27 March 96. The power spectra corresponds to an 8-month time series. Shorter time series of 4 months were also used to verify the starting dates and the temporal behaviour of the p modes.

2. RESULTS

Fitting of p-mode spectra is performed assuming that the statistic is a χ^2 with 2 degrees of freedom. Each degree is a superposition of m for which $l + m$ is even; the relative amplitude of the m's is that of integrated sunlight instrument. The BiSON time series has a duty cycle of 75%, and consequently the first temporal aliases were included for fitting these data. The power spectra of the 2 data sets were analysed by 2

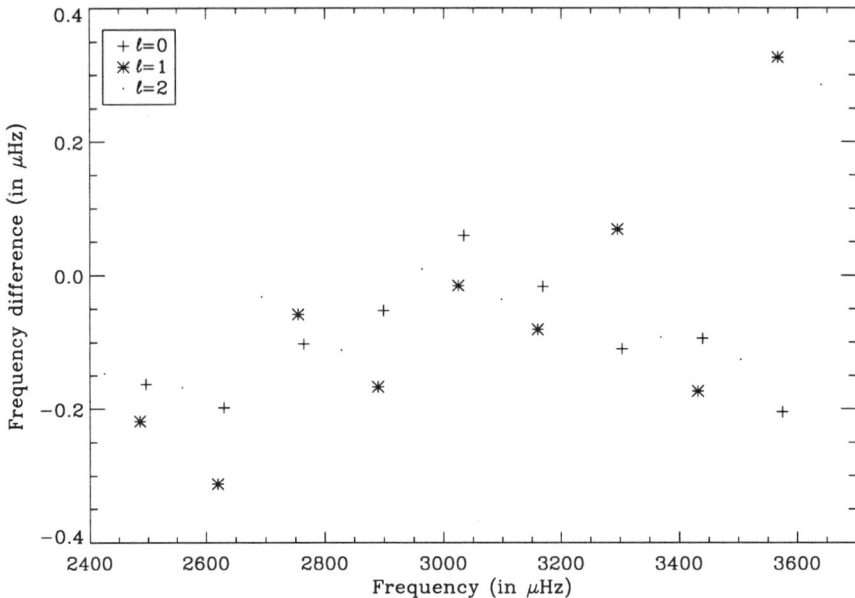

Figure 1. Frequency difference between BiSON and LOI for $l = 0 - 2$ as a function of frequency. The typical error bar on the measurement of a single mode frequency ranges from 0.10 μHz to 0.15 μHz.

different fitting routines (Chaplin's and Appourchaux's). We compared the frequencies returned by the 2 routines on the same data set and found that the differences were about a few nHz over all the degrees. We then compared the frequencies of the LOI and BiSON sets using Appourchaux's routines. Figure 1 shows the result of this comparison. These differences agree with the result of Toutain et al, (1997) (Sol. Phys., in press) showing there are systematic differences between the frequencies of the modes detected in intensity and velocity. Similar frequency differences have been found using GONG and SOHO/LOI data (Appourchaux, 1997, private communication). In addition, Fig. 1 shows that the difference is parabolic with a maximum at about 3000 μHz, and independent of degree.

3. DISCUSSION

The source of the systematic difference is yet to be found. The fact that the difference is independent of degree leads us to conclude that the difference could be due to a surface effect that creates a distortion of the line profile depending on the signal observed. The distortion may create either different asymmetries or different line profiles. The implications for structure inversions are obvious: the inverted internal structure of the Sun will depend upon the signal used for deriving the p-mode frequencies. This was already mentioned by Appourchaux et al (1996) using the VIRGO data (IAU 181, Nice). Further work will be needed to make the intensity and velocity structure inversions agree with each other.

TOUCHING ON THE EFFECTS OF AN IMPERFECT WINDOW FUNCTION

W. J. CHAPLIN, Y. ELSWORTH, G. R. ISAAK, C. P. MCLEOD AND B. A. MILLER
School of Physics & Astronomy, University of Birmingham
Edgbaston, Birmingham B15 2TT UK
E-mail: wjc@star.sr.bham.ac.uk

AND

R. NEW
School of Science & Mathematics, Sheffield Hallam University
Sheffield, S1 1WB

BiSON currently consists of four fully- and two semi-automated observatories, dedicated to the round-the-clock collection of low-degree (low-ℓ) helioseismological data. A complete historical record of the network can be found in Chaplin et al. (1996). The annual duty cycles achieved by BiSON have improved steadily since our sixth site was commissioned in 1992. Over the period 1992 – 1996 inclusive, we have recorded annual duty cycles of: 54, 68, 78, 74 and 78 per cent respectively.

While a quasi-diurnal gap structure persists in combined, 6-station BiSON time series, giving rise to weak sidebands in the frequency domain, they also contain a large number of short gaps. We are currently investigating the feasibility of filling short gaps with estimates of the solar Doppler velocity signal, in an effort to reduce the contamination introduced by the effects of breaks in the data stream. In order to illustrate the impact of an imperfect window function, we have performed a series of simulations with artificial p-mode data.

The plots in Fig. 1 show the effect of imposing a 4-month BiSON window function (1996 Mar 27 – 1996 July 26) on a set of artificial modes generated in the time domain. Some 101 simulated Doppler velocity mode residuals – mapping the range $0 \leq n \leq 3$ and $1185 \leq \nu \leq 4665\,\mu$Hz – were produced by a model based upon a stochastically forced, damped harmonic oscillator (for full details see Chaplin et al. 1997). The mode frequencies, relative powers and lifetimes were selected in order to reflect the characteristics of the observed, low-degree solar p-mode spectrum. The 4-month window function was then applied to the resulting time series, and the power spectrum computed. We also computed: the spectrum from a 100-per-cent fill time series (i.e., no window function applied); and a spectrum where the 4-month window function was applied, but where gaps of duration 1 hr or less were "removed" (i.e., data were restored to the time series, increasing the fill by ≈ 4 per cent). We have also included a plot of the actual BiSON low-degree spectrum for this period.

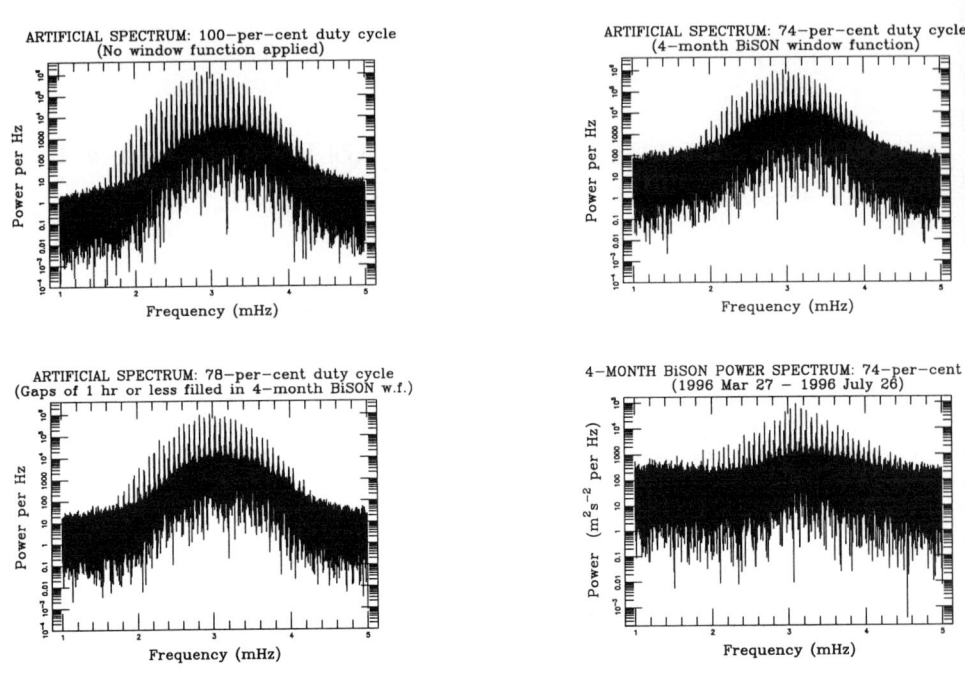

Figure 1. These plots show the results of applying a 4-month BiSON window function from 1996 to a time series of artificial p-mode residuals (see text). Also included is a plot of the actual BiSON low-degree spectrum from this period.

The results are rather striking: the window function gives rise to a factor-of-ten reduction in the signal to noise near the centre of the spectrum. The large increase in the background level also impacts upon the detectability of modes at the extreme ends of the spectrum. By "filling" short gaps in the time series, the deterioration in quality is greatly reduced. Provided one were able to fill such gaps with reliable estimates of the true solar data, these simulations imply that the resulting gains in signal to noise would be quite marked for ground-based networks like BiSON. This is clearly of most importance to those modes with low signal-to-noise ratios – the resulting improvement in the measured precision of the frequencies of the strongest modes would be quite modest.

References

Chaplin W. J., Elsworth Y., Howe R., Isaak G. R., McLeod C. P., Miller B. A., New R., van der Raay H. B. and Wheeler S. J., *Solar Physics*, 1996, **168**, 1

Chaplin W. J., Elsworth Y., Howe R., Isaak G. R., McLeod C. P., Miller B. A. and New R., 1997, *MNRAS*, 1997, **287**, 51

GONG SPECTRA IN THREE OBSERVABLES: WHAT IS A P-MODE FREQUENCY?

J. HARVEY, F. HILL, R. KOMM, J. LEIBACHER, B. POHL
AND THE GONG

National Solar Observatory
Tucson, Arizona, USA

1. Introduction

The Global Oscillation Network Group (GONG) is an international, community-based helioseismology project conducting a detailed study of the internal structure and dynamics of the Sun. GONG is operating a six-station network of full-disk solar imagers exploring the range from $\ell = 0$ to about $\ell = 200$. Full network operation began in October 1995. The observation duty cycle has averaged about 87% over the first 684 days of full operation, and single-site data loss due to instrument down time is less than 2%. The data management group is processing the network data at pace with the collection rate. Here we present a comparison of spectra obtained in Doppler velocity, total spectral intensity, and modulation (a mixture of equivalent width and magnetic field strength).

2. What is a P-Mode Frequency?

A comparison of GONG month 16 (36 days) m-averaged, wavelet-denoised spectra for three different ℓ values and two frequency ranges in the three observables is shown in Figure 1. It is obvious the modes observed in velocity and intensity have the previously known reversal of asymmetry but, more surprisingly, they also show apparently different central frequencies. While this difference is small but non-zero in the five-minute band, it can be as large as 50 μHz close to the acoustic cut-off frequency. Modulation and velocity spectra are more similar than velocity and intensity spectra. However, there are differences between modulation and velocity spectra. For example, there is a visible peak in velocity at $\ell = 20$, $\nu \approx 2.59$ mHz which has no counterpart in modulation.

The large apparent differences in the central frequencies of the p-modes observed in velocity and intensity raise the question of just how to define a p-mode frequency. Clearly, the physics of the oscillation determining the real central frequency in the sun cannot depend on the quantity we observe from the earth. Instead, the observable differences probably arise from at least three effects. First, the details of the source mechanism of the p-modes, particularly the location of the source with respect to the cavity trapping the modes, can change both the apparent acoustic line asymmetry

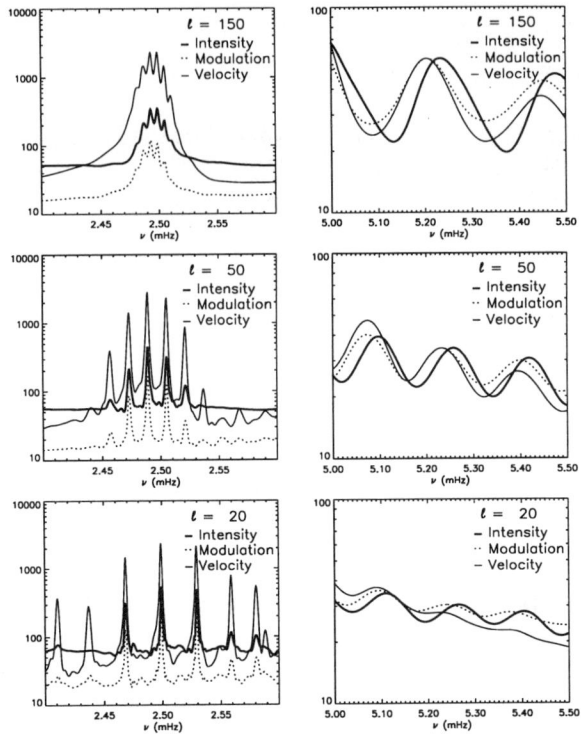

Figure 1. m-averaged GONG spectra observed in velocity, intensity, and modulation.

and frequency (Jefferies 1997). Second, there is a different spatial leakage signature for each of the three observables due to their different center-to-limb behaviors. Finally, there are different levels of solar noise in the three observables. While the comparison between velocity, intensity, and modulation is currently poorly understood, simultaneous modelling of all three types of spectra already in progress will greatly improve our ability to accurately measure the solar oscillation frequencies.

3. Acknowledgements

This work utilizes data obtained by the Global Oscillation Network Group (GONG) project, managed by the National Solar Observatory, a Division of the National Optical Astronomy Observatories, which is operated by AURA, Inc. under a cooperative agreement with the National Science Foundation. The data were acquired by instruments operated by the Big Bear Solar Observatory, High Altitude Observatory, Learmonth Solar Observatory, Udaipur Solar Observatory, Instituto de Astrofisico de Canarias, and Cerro Tololo Interamerican Observatory.

References

Jefferies, S.M.: 1997, these Proceedings.

APLICATION OF A NEW OBSERVATIONAL STRATEGY TO THE STUDY OF GRAVITATIONAL SOLAR MODES.

I. MARTIN MATEOS (IMM@IAC.ES) AND P.L. PALLÉ
Instituto de Astrofísica de Canarias

Abstract.
The aim of the present work is the detection of solar g-modes, making use of their spatial and temporal properties, by means of a new observational strategy. The basic data, gathered at the Observatorio del Teide in 1993, consists on daily solar velocity measurements taken continuous and sequentially at six different and symmetric positions on the solar disk. By correlating the time series obtained from different positions, and considering the geometrical properties of different modes (l, m) on the Sun's surface, some of them can selectively be eliminated or enhanced. In particular, the main spectral features present in the resulting power spectra must have precise phase relations if they correspond to global solar g-modes.

1. Observations and data.

The data available for the analysis were obtained using a resonant spectrophotometer (long stability). This instrument measures the line of sight velocity in the potassium 7699 A line.

The data are obtained with a new strategy using alternative and continuous measurements at 6 different positions, uniformly distributed on the solar disk. The advantage of observing at the solar periphery is to increase the horizontal velocity signal, which theoretical predictions suppose is the highest for g-modes. Since the global modes have different temporal phase at different apertures, we used this property to identify and classify modes, combining measurements from different positions. The combinations let us filter out the signal from some particular modes and to enhance others. The 6 apertures are separated by 60 degrees with the centers at 0.55 $R_\odot$. The measurements are taken alternatively each 20 seconds, with a cycle continued during all the daily observations. Many spectra have been obtained from these series of data, from individual apertures and from the combination of them, by FFT methods.

2. Peak selection.

The work has been performed on the interval 50-250 μHz. The peak selection process seeks for common peaks in independent spectra combinations. First, the method looks for peaks whose amplitude is more than a threshold (2σ, 2.5σ and 3σ; being σ the standard deviation). At this time only peaks which have passed the threshold limit are considered. For the selection of the standard deviation the frequency interval was divided into ten small intervals, with 20 μHz each. Different standard deviations were considered because the noise level is strongly variable in the whole interval. The next step was to study the phase, and several peaks were selected, those which presented a high amplitude and the same phase (within limits of either 10 or 20 degrees) in various combinations of the series.

Finally we checked the time evolution of the selected peaks, and different tests were applied for studying their properties. The following table shows the final results, with the basic properties of the selected peaks. These peaks are the candidates to g-mode frequency peaks.

TABLE 1. Selected peaks. Resolution 0.01 μHz.

Frequency (μHz)	Period (min.)	Amplitude (cm/s)	noise (cm/s)
53.35	312.4	107.69	7.5
58.39	285.43	60.01	4.52
59.28	281.4	67.65	4.52
70.35	236.9	103.94	6.16
70.79	235.43	75.94	4.54
71.08	234.47	96.08	6.16
79.79	208.88	121.76	6.16
105.13	158.53	67.94	5.87
114.03	146.16	62.18	4.77
116.75	142.75	45.89	4.77
129.41	128.78	45.68	4.77
130.54	125.58	28.67	3.57
135.91	122.60	26.65	3.57
164.14	101.53	38.57	2.98
170.65	97.66	31.42	2.96
182.57	91.28	24.81	2.96
194.16	85.83	35.17	2.91
232.41	70.53	18.35	2.46

VAMOS: VELOCITY AND MAGNETIC OBSERVATIONS OF THE SUN

Calibration and Data Analysis

M. OLIVIERO
Dipartimento di Scienze Fisiche dell'Università "Federico II"
Mostra D'Oltremare Pad.19, 80125 Napoli, Italy

G. SEVERINO AND TH. STRAUS
Osservatorio Astronomico di Capodimonte
Via Moiariello 16, 80131 Napoli, Italy

AND

A. CACCIANI, M. DOLCI AND P.F. MORETTI
Dipartimento di Fisica dell'Università "La Sapienza"
P.le A.Moro 2, 00185 Roma, Italy

This poster illustrates the calibration procedure and the analysis pipeline developed for the helioseismology data acquired with the VAMOS (Velocity And Magnetic Observations of the Sun) instrument. The VAMOS, based on the MOF (Magneto-Optical Filter) technology, and its operation are discussed in detail elsewhere (Cacciani et al., 1997 and Moretti et al., 1997).

The data set used in the present work consists of 256 solar Doppler images in the Na I D lines, one per minute, obtained in Napoli on February 20, 1997 starting at 10:38 U.T. The average Sun radius is 226 ± 1 pixels. The spatial resolution is ~ 10 arcsec, i.e. ~ 2 pixels, limited by the 1.5 cm iris diameter.

Calibration has been performed in essentially three steps.
First a theoretical solar image has been constructed containing all the contributions to the Earth-Sun line-of-sight relative velocity for each observing time.
Then the observed Doppler images have been fitted to the simulated velocity images. Because the solar sodium line profiles are center-to-limb dependent, the solar disk has been divided into six concentric rings, and in each of these annuli the observed signal was fitted to the simulated velocity with a second order polynomial.
Finally the Doppler images have been effectively calibrated to produce residual images in absolute velocity units.

The two-dimensional residual images were fitted with the 40401 spherical harmonics having $0 \leq l \leq 200$. The spherical harmonics fit routines, as well as the calibration routines for the VAMOS, have been developed in IDL, version 4.0.1. The resulting fit coefficients for the 256 residual images were Fourier transformed to produce the power spectra plotted in Figure 1.

In the figure, we compare the m-averaged $l - \nu$ diagrams from the VAMOS data and the GONG archive. The GONG data series was obtained on February 19, 1997,

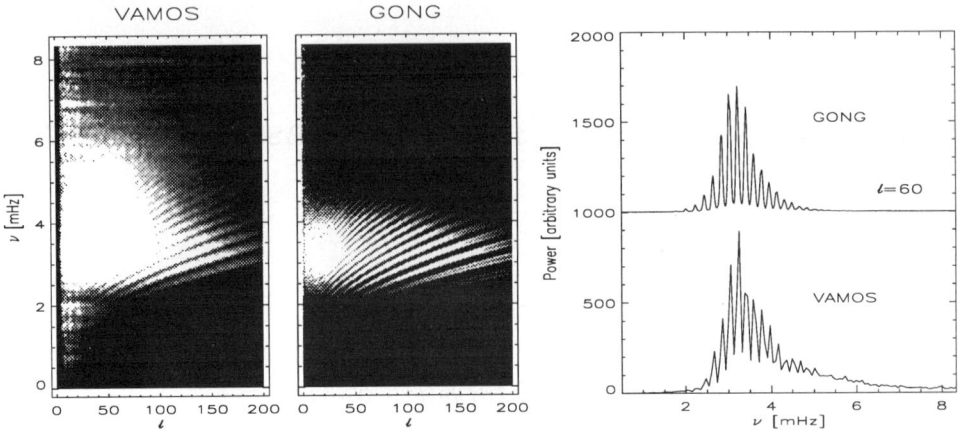

Figure 1. $l-\nu$ diagrams and power vs frequency at $l=60$ for both VAMOS and GONG.

at Tenerife, and lasts 558 minutes. Therefore, we expect that the GONG frequency resolution is about half of our 65μHz frequency resolution.

In the VAMOS diagram, at least 20 ridges are clearly visible, some of which up to $l=200$. The low frequency signal was removed by a high pass filtering before applying the Fourier transform.

In the same figure, power is plotted versus frequency, for the value of $l=60$, again for both VAMOS and GONG. For clarity the GONG data were shifted upward by 1000. The ratio of the highest peak to noise at this l is ~ 30 in the VAMOS data and ~ 140 in the GONG data.

To conclude, on the basis of this data set and with the present, albeit preliminary, setup, the VAMOS is sensitive to p-modes with degree $5 \leq l \leq 200$.

Acknowledgements

We thank L.A. Smaldone and C. Marmolino.
This work was partially supported by ASI through contract ASI 95-RS-69.
This work utilizes data obtained by the Global Oscillation Network Group (GONG) project, managed by the National Solar Observatory, a Division of the National Optical Astronomy Observatories, which is operated by the Association of Universities for Research in Astronomy, Inc., under a cooperative agreement with the National Science Foundation.

References

Cacciani, A., Marmolino, C., Moretti, P.F., Oliviero, M., Severino, G., Smaldone, L.A.: 1997, Mem. SAIt. in press.
Moretti, P.F., Severino, G., Cauzzi, G., Reardon, K., and Straus, T., Cacciani, A., Marmolino, C., Oliviero, M., and Smaldone, L.A.: 1997, in F.P. Pijpers, J. Christensen-Dalsgaard, and C.S. Rosenthal (eds.), "SCORe'96: Solar Convection and Oscillations and their Relationship", Kluwer, Dordrecht, p. 293.

SEARCHING FOR G-MODES AT THE SOLAR LIMB

C.G. TONER AND S.M. JEFFERIES
National Solar Observatory
P.O. Box 26732, Tucson, AZ 85726 USA

Abstract. We have used a differential technique to look for the signature of g-modes at the limb of full-disk intensity images of the Sun. Our spectra show tentative evidence for a set of peaks that are equally spaced in period, as predicted by asymptotic theory. The period spacing that we find is ~ 27.5 minutes. If interpreted as $\ell = 1$ g-modes, this implies $T_0 \sim 39.0$ minutes.

1. Introduction

Unlike p-modes, g-modes are expected to have large amplitudes in the deep interior of the Sun. This makes them significantly more sensitive to the structure of the Sun's central regions than the p-modes, and, potentially, a powerful diagnostic of those parts of the Sun. However, because g-modes are evanescent in the Sun's convective zone, only small amplitude oscillations survive to the surface.

The common way to look for g-modes is to search for peaks that are equally spaced in period as predicted by asymptotic theory. To first order the period $T_{n\ell}$ of a mode ($m=0$) is given by $T_{n\ell} = T_0[n + \ell/2 + \delta][\ell(\ell+1)]^{-1/2}$ where T_0 is the asymptotic period spacing and δ is a constant $\simeq -0.25$. Since the frequencies expected for g-modes are very low, solar and terrestrial atmospheric "noise" seriously hamper any measurement efforts.

A recent calculation predicts that the g-mode signal in intensity should peak very close to the solar limb (T. Toutain, private communication). Therefore, we have developed a technique which differences the signal in a narrow annulus at the limb from the signal in an adjacent (interior) annulus in order to reduce the background "noise" signal and thus enhance the probability of detecting g-modes.

2. Observations and Reductions

The data used for this project was obtained at the South Pole during the austral summer of 1994/95. The full solar disk in the Ca K-line was recorded with ~ 2 arc sec pixels every 42 seconds. In total there are ~ 62 days of observations with a fill factor of $\sim 47\%$.

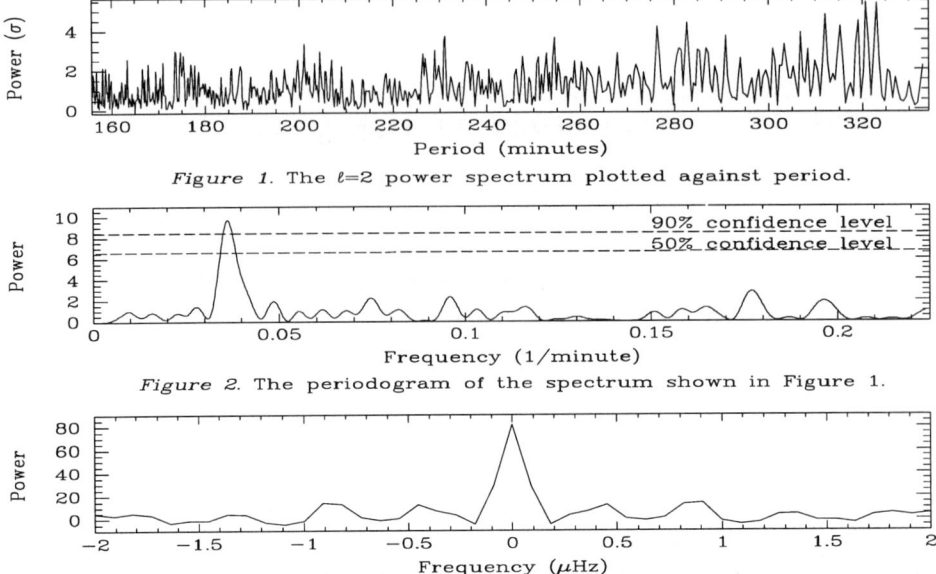

Figure 1. The ℓ=2 power spectrum plotted against period.

Figure 2. The periodogram of the spectrum shown in Figure 1.

Figure 3. The autocorrelation of the spectrum shown in Figure 1.

The images were restored to a common blurring using a modification of the technique described by Toner *et al.* 1997. Each image was registered to a common size and orientation and the ratio $(Observed\ image - Reference\ image) \div Reference\ image$ was binned by azimuthal angle in two narrow annuli at/near the limb, then the annulus signals were differenced. Finally, the power spectrum was obtained using a 2-D FFT of the difference data.

3. Results

The $\ell = 2$ spectrum for $50 \leq \nu \leq 110$ μHz is plotted in Figure 1 as a function of period in minutes. While the spectrum is noisy, there is a suggestion of equally spaced "blobs" of power. We have used the periodogram program of Horne and Baliunas (1986) to see if there a periodic signal in the spectrum. The periodogram shows a strong peak at the ~ 95% confidence level at a frequency corresponding to a peak spacing of 27.6 minutes (Figure 2). If interpreted as $\ell = 1$ g-modes ($\ell = 1$ signal will appear in ℓ =2 and 3 due to spatial leakage), this implies an asymptotic period spacing, T_0, of ~ 39.0 minutes. The $\ell = 1$ and $\ell = 3$ periodograms show similar peaks at 27.4 minutes and 27.2 minutes, respectively. The $\ell = 3$ periodogram also has a peak at 15.9 minutes, which is consistent with the expected spacing for $\ell = 2$ g-modes for the same T_0. Figure 3 shows the autocorrelation of the spectrum shown in Figure 1. The peaks at lags of 0.45, 0.9, and 1.3 μHz may be the signature of rotational splitting. Further work is required to understand the spatial response function of our observations before we can comment on the validity of these signals.

References

Toner, C.G., Jefferies, S.M. and Duvall, T.L.Jr. 1997, *Astrophys. J.*, **478**, 817.
Horne, J.H., and Baliunas, S.L. 1986, *Astrophys. J.* **302**, 757.

AUTOREGRESSIVE MODELING OF *GOLF* VELOCITY TIME SERIES

R. ULRICH, F. VARADI, L. BERTELLO
University of California, Los Angeles

AND THE GOLF TEAM

The frequencies of solar p-modes are usually computed by fitting Lorentzian profiles to the observed signals' Fourier spectra. We offer a new technique to estimate mode parameters: direct modeling by autoregressive stochastic processes, in conjunction with Singular Spectrum Analysis.

A second order autoregressive (AR) stochastic process regresses the signal y_i on its previous values and it is defined by the second order difference equation $y_i = a_1 y_{i-1} + a_2 y_{i-2} + w_i$, where w_i is excitation, assumed to be white noise. This is a forced and damped oscillator in discrete time whose frequency and damping factor are related to the model coefficients a_1 and a_2 by the characteristic equation. It is also a filter with input w_i and output y_i and its spectral response has Lorentzian profile. These and the autoregressive-moving average (ARMA) models are the mainstay of digital signal processing and spectral analysis (Proakis *et al.*, 1992). When additive noise contaminates the signal, the correct model would be ARMA, but it is difficult to fit.

We apply Singular Spectrum Analysis (SSA, Vautard *et al.*, 1992) to extract the AR part of the signal. In SSA, one diagonalizes the autocovariance matrix (considered up to a certain lag, the window length) to obtain its eigenvalues (the singular spectrum) and eigenvectors. The latter are used as filters to isolate various components of the signal. The singular spectrum usually exhibits a noise floor, and the components above it tend to be associated with modes. We use these filtered signals above the noise to fit AR models.

We apply the technique to 370 days of the velocity time series provided by the GOLF experiment (Gabriel *et al.*, 1997). Since SSA cannot deal with thousands of modes, we extract and resample the signal in frequency bands which contain particular modes, yielding a few hundred points in the time domain. Three different widths for these bands and for each band a different SSA window length are used. AR models of orders corresponding to the expected rotational splitting are fitted using both Burg's method and the least-squares modified Yule-Walker equations.

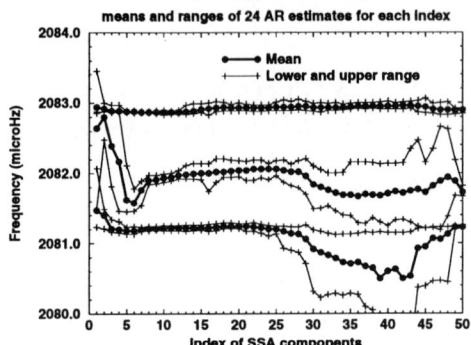

Figure 1. Singular Spectra and AR frequency estimates for the $l = 2$, $n = 13$ p-mode.

The left panel of Fig. 1 shows the singular spectra for a mode. The first 15 or so SSA components appear to be significant and the rest form the noise floor. We reconstructed the signal from the first component up to the 50th and fitted AR models for each reconstruction. The right panel shows the means and ranges of the estimated frequencies. They are fairly stable around the index which marks the beginning of the noise floor — between 10 and 20. We obtained estimates of both frequency and rotational splitting for several $l = 1$ and $l = 2$ modes and for each mode we averaged the stable cases near the beginning of their respective noise floors. These results are in good agreement with maximum likelihood estimates (MLE) of Lorentzian fits (see Table 1).

		MLE+Lorentzian		AutoRegressive	
l	n	Frequency (μHz)	Splitting (nHz)	Frequency (μHz)	Splitting (nHz)
1	10	1612.75 ± 0.03	425 ± 33	1612.71 ± 0.05	400 ± 50
1	11	1749.31 ± 0.03	412 ± 27	1749.31 ± 0.05	410 ± 50
1	12	1885.11 ± 0.03	405 ± 27	1885.12 ± 0.03	425 ± 30
1	13	2020.82 ± 0.04	467 ± 32	2020.85 ± 0.04	485 ± 40
1	14	2156.76 ± 0.06	415 ± 43	2156.76 ± 0.07	445 ± 70
1	15	2292.04 ± 0.07	424 ± 60	2292.02 ± 0.09	400 ± 90
2	12	1945.79 ± 0.03	370 ± 17	1945.81 ± 0.07	395 ± 70
2	13	2082.06 ± 0.05	391 ± 26	2082.08 ± 0.09	425 ± 90

TABLE 1. Mode parameters.

References

Gabriel, A. H. et al.: 1997, 'Performance and early results from the GOLF instrument flown on the SOHO mission', *Solar Physics*, in press

Proakis, J. G., Rader, C. M. Ling, F., and Nikias, C. L.: 1992, *Advanced Digital Signal Processing*, Macmillan Publ. Co.

Vautard, R., Yiou, P. and Ghil, M.: 1992, 'Singular-spectrum analysis: A toolkit for short, noisy chaotic signals,' *Physica D* **58**, 95–126

LARGE SCALE CONVECTION – OBSERVATIONS

How does the energy flow?

R.K. ULRICH
Department of Physics and Astronomy
University of California at Los Angeles

Abstract. Although the rate of production of nuclear energy in the sun's core almost certainly increases smoothly on time scales shorter than and comparable to the solar age, the mechanisms transporting this energy to the photosphere may be subject to more erratic fluctuations. Indeed space observations establish that the total solar irradiance in the direction of the earth is variable on time scales of days to years. I discuss energy transport by convection and magnetic fields with consideration of the possibility of energy storage through changes in the solar radius. Convection is the primary means of energy transport below the solar surface and must be involved in any modulation of energy flow. The large scale and long duration properties of solar convection arising from zones below the solar surface can be well studied using the Michelson-Doppler Imager instrument on SOHO. New evidence of long duration convective structures based on the MDI data is presented. These patterns appear to be related to the torsional oscillations observed at the Mt. Wilson Observatory.

1. Introduction

A major objective of solar physics is to understand the solar cycle. The roughly regular rise and fall of solar surface magnetic fields has a major impact on the earth's environment and could play some role in climate should the pattern undergo a change like that during the Maunder Minimum. While there has been progress in numerical modeling, the observational basis for our understanding of the fundamental mechanism of the solar dynamo has not progressed greatly. I have chosen to focus on this question rather than to prepare a comprehensive review of the literature. Much of the material from the Mt. Wilson synoptic program has not been previously presented.

I suggest that one approach worth exploring is to focus on the energetics of the dynamo. In human detective work a good plan is to follow the money trail. In solar detective work I think a similar good plan is to follow the energy trail. The forms of energy relevant to the solar interior and surface are nuclear, radiative, thermal, gravitational, kinetic and magnetic. The core nuclear reactions are the starting point. The energy moves out to the convective envelope boundary in the form of radiation. These two steps probably do not play a role in the solar cycle of magnetic activity and I will say no more about them. Within the convective envelope, the matter is so opaque that radiative energy transfer can be neglected until the very surface layers are reached. The energy is carried through the convective envelope in the form of correlated thermal heat content of the matter and the vertical velocity of displacement. This process is known as convective energy transport and has been successfully modeled in the near-surface layers but is poorly known over the entire convective envelope. For the largest scales of motion, we can crudely estimate turnover times from mixing length theory (Vitense, 1953, Kim et al. 1996) and find they should be long compared to the rotation period. Consequently, there should be a strong interaction between rotation and convection. This is a case where thermal energy is converted by buoyancy into kinetic energy which then can interact with rotational kinetic energy. In addition, the magnetic energy somehow becomes coupled with the rotation and convection to generate the solar cycle. The entire problem is too much to address as a whole so I will focus on just two parts of the energy flow:

- What can we learn about large-scale and long-lived kinetic energy patterns in the sun's convective envelope?
- Does the variation in localized energy release associated with distinct surface activity cause a correlated change in the thermal or gravitational energy in the immediate subsurface regions?

Such storage would result in a change in the average density of the solar matter (see Spruit 1977, Fox and Sofia 1994). Although these calculations did not estimate the effect on the position of the solar surface, the alteration of the density suggests there should be an observable change in the solar limb figure. While I do not focus on the related magnetic energy questions, it is clear from the large scale velocity measurements presented below that pathways exist for the interconversion of magnetic and kinetic energy.

Our ideas about the sun's convection zone below the solar surface have been largely guided by general theoretical considerations (the mixing length theory of Vitense [1953] is usually a starting point) and numerical modeling experiments (see the recent discussion by Kim et al. [1996] for a comparison between mixing length theory and numerical simulations). Most of the unstable region is close to a polytropic structure. In such a region the temperature is roughly proportional to the distance below the solar surface. The pressure scale height is often taken as providing an estimate of the spatial scale of the convective flows. Because the pressure scale height is proportional to the temperature, it is also proportional to the depth below the solar surface. Thus general theoretical considerations lead us to believe there is a systematic increase in spatial scale of the convective flow as we look deeper below the surface. Similar ideas lead to the expectation that the convective velocity decreases inward below the surface boundary layer. Such simple arguments suggest that the convective cell or eddy lifetimes should range from about 10 minutes at the solar surface to in excess of one month near the inner boundary of the convection zone. At the inner edge of the convective envelope, the pressure scale height is a major fraction of the geometric separation between the inner and outer boundaries of the convection zone so we are led to believe that the largest scale convection currents could have a scale comparable to that of the convection zone as a whole. This general picture was originally introduced by Simon and Weiss (1968).

For the largest scale motions, it is clear that the influence of solar rotation will be dominant. These hypothetical convective flows are known as "giant cells" and have never been identified from solar observations. Theoretically one might expect there to be a wide range of spatial scales extending between the "giant cells" and the surface granulation. The only well accepted convective pattern between these two extremes is the supergranulation. However, the spatial scale of the supergranulation is not precisely defined and probably encompasses a range of sizes from the mesogranulation introduced by November et al. (1981) below 10 Mm to duragranulation discussed by Ulrich et al. (1998, see below) at 50 Mm.

Throughout the convection zone the requirement that the motions carry the energy flux imposes a limitation on the velocities achieved. In the deeper layers, the high density and high heat content per unit mass means that only a small velocity is needed to satisfy the flux requirement. Thus there is an inverse correlation between the spatial scale of the convective flow and the amplitude of its velocities. Unless some filtering technique is used, the smallest scale flows will dominate any velocity map. The larger scale flows can only be revealed by taking advantage of the short lifetime of the smaller scale patterns. Progressively longer temporal averaging intervals should reveal progressively larger scale flows. The granulation may be the dominant spatial scale of convection at the solar surface where we observe the interiors of the hot and cool flows. The spatial and temporal scales immediately larger and longer lived than the granulation cannot be observed easily due to the masking effect of the non-convective phenomenon of the five-minute oscillations. In practice filtering out the oscillations requires averaging over a time period of 20 to 60 minutes. Any convective pattern with a similar lifetime is also removed in this process. The convective pattern which remains is the "mesogranulation". Caution is required in identifying the "mesogranulation" since smaller scale motions are removed by the filtering process needed to suppress the oscillations.

In pushing the study of supergranulation toward larger scales it has been possible to obtain good isolation of these structures by using time sequences as long as possible. From ground-based facilities, it is difficult to obtain observed sequences with high temporal continuity for intervals longer than 14 hours. Supergranulation is the most prominent large scale flow on the solar surface and estimates of its lifetime are about 24 hours. The large amplitude five-minute oscillations impose another requirement on duration of observing sequences – they must be long enough that the oscillations can be averaged out. For convection patterns with lifetimes longer than the supergranulation, solar rotation becomes a major issue. Even a stationary velocity field will undergo apparent evolution due to the changing projection factors as the observed region is carried under the field of view. We do not have an option of observing from a heliostationary platform. This effect seriously compromises any effort to determine the lifetime of the velocity pattern using power spectral techniques. However, the changing projection angles can be used to determine both north-south and east-west velocities for the longest lived flows. The following section describes the implementation of this analysis.

Prior to the development of the methods of helioseismology, the solar surface was the only place where solar velocities could be observed. While the seismic observations can reveal properties of the internal flows, the deduction is difficult to carry out unless observations like those from the Michelson-Doppler Imager experiment on SOHO are available. These have good spatial resolution and during the dynamics program have good temporal continuity.

2. The Observed Velocity

2.1. NATURE OF THE OBSERVATIONS

Direct velocity observations can be made by way of the Doppler shift of spectral lines. Such observations provide only the line-of-sight component of the velocity and measure the apparent shift of some feature of a spectral line. Line shape and offset is influenced by unresolved convective effects as well as radiative transfer effects which depend on the center-to-limb angle ρ of the observed point. Such line shifts are generally referred to as the "limb shift". Differential solar rotation also enters into the apparent line-of-sight velocity as the largest effect and must be removed in order to make any large scale convective patterns apparent. The task of extracting appropriate deviations from the background differential rotation and limb shift functions may have a cross-talk effect in which spurious patterns are added or true patterns are removed. For the Mt. Wilson data the long duration of the observations permit a clean method of making the separation: the background functions are derived from observations prior to 1986 while the deviations are studied for observations after 1986. For the MDI data, Ulrich *et al.* (1998) have restricted the study to spatial scales small enough that the cross-talk problems are minimal.

The interpretation of the line-of-sight velocities in terms of long-lived flows must take into account the gross effect solar rotation has on the angle the line-of-sight makes with the solar surface. The continuously changing viewing angle causes even stationary flows to appear to vary. At each point within the solar atmosphere the stationary or long-lived velocity could have three components: vertical or radial, V_r, horizontal east-west, V_{EW}, and horizontal north-south, V_{NS}. However, since I am primarily interested in the longest-lived and largest scale flows, I can use the small ratio of the vertical density scale height to horizontal structure size in the continuity equation to show that the ratio of horizontal velocity to the vertical velocity must be large. After ignoring the vertical velocity and following the formulation by Howard and Harvey (1970) I can express the line-of-sight velocity V_{los} as:

$$V_{los} = V_{EW} \cos B \cos B_0 \sin L + V_{NS} \sin(B - B_0) \cos L \qquad (1)$$

where L is the central meridian angle of the point in question, B is the solar latitude and B_0 is the tilt of the sun's equatorial plane of rotation relative to our line of sight.

We now fix the background rotation by an external model V_{rot} and calculate a deviation of the line of sight velocity at each point according to:

$$\delta V_{los} = (V_{EW} - V_{rot}) \cos B \cos B_0 \sin L + V_{NS} \sin(B - B_0) \cos L - V_{LS}(\rho) \qquad (2)$$

where V_{LS} is the limb shift velocity which is a function of center-to-limb angle ρ. We use the time dependence of δV_{los} which results from the regula r increase in L to determine both $(V_{EW} - V_{rot})$ and V_{NS} by fitting δV_{los} to the L dependence shown in equation (2). The dependence of δV_{los} on L is nearly linear with the slope giving the EW component and the intercept giving the NS component. Thus I can express the result of the fit in terms of a vector sum:

$$\boldsymbol{V} = (V_{EW} - V_{rot}) \boldsymbol{i}_{EW} + V_{NS} \boldsymbol{i}_{NS} \qquad (3)$$

where $\boldsymbol{i}_{EW}$ and $\boldsymbol{i}_{NS}$ are unit vectors parallel and perpendicular to the sun's equator. This vector $\boldsymbol{V}$ then gives both the direction and magnitude of the velocity deviation from strict differential rotation.

Figure 1. The reference rotation curve subtracted from all observations is shown as the solid line. For comparison the curve derived from the classical $\Omega = A + B \sin^2(\text{Lat}) + C \sin^4(\text{Lat})$ curve is show n as the dashed line. Note that while the curves are similar, there are small but significant deviations between the actual and the classical curve. The top panel gives the change in velocity implied by the difference between the two curves.

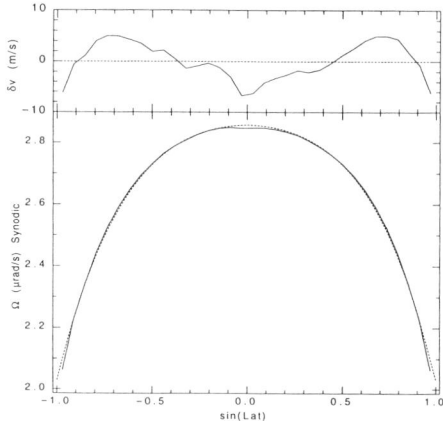

2.2. MT. WILSON VELOCITIES

The synoptic program at Mt. Wilson obtains solar magnetogram observations up to twenty times per day using the Babcock system established at the 150-foot tower. These data are observed using the light of the FeI line at $\lambda 525.02$nm. The spectral sampling of this radiation utilizes fiber-optic image reformattors which select two nearly identical bands on opposite wings of the line. This selection has changed little after 1982. The passbands are characterized by two parameters: the full width of each, w, and the separation of their centroids, s. The detailed velocity and magnetic field values determined by a Babcock Magnetograph depend on these two parameters. There have been two changes since 1982 which have altered w and s: before Nov. 21, 1994 we used a grating having 632 lines/mm in 5th order and after that date we have been using a grating having 367.5 lines/mm in the 9th order. On the date of the change the fiber-optic reformattor was unchanged. Before the grating change the values of w and s were 6.21 nm and 7.76 nm respectively. After the change the values were 5.27 nm and 6.59 nm respectively. Beginning April 13, 1996 we have been using a 24-channel data system which incorporates new fiber-optic reformattors. With these new fiber-optics the values are 5.82 nm and 7.77 nm.

The large scale convective structures we seek may reveal themselves through their velocity variations. Herein comes the problem: a variation is a difference between a particular velocity measurement and some reference state — how do we identify the reference state? There is a temptation to assert that the reference state is defined by some particular combination of terms in a mathematical expansion; but without a physical theory which establishes the appropriate mathematical expansion any variations probably just indicate the inadequacy of the choice of mathematical terms. Our solution is to utilize a temporal average over more than one solar cycle for the reference state. An additional potential problem in the reference state definition is the possibility that any large transient feature could alter the reference state in such a way as to reduce the apparent size of the transient. We avoid this possibility by defining the reference state with data prior to our study period. This procedure gives us the reference rotation curve shown in figure (1) Although we begin our rotation rate analysis with the regular $\Omega = A + B \sin^2(\text{Lat}) + C \sin^4(\text{Lat})$ function, we apply a small correction to this function in order to properly represent the actual rotation rate averaged over the time period 1975 to 1986. The lack of symmetry between north

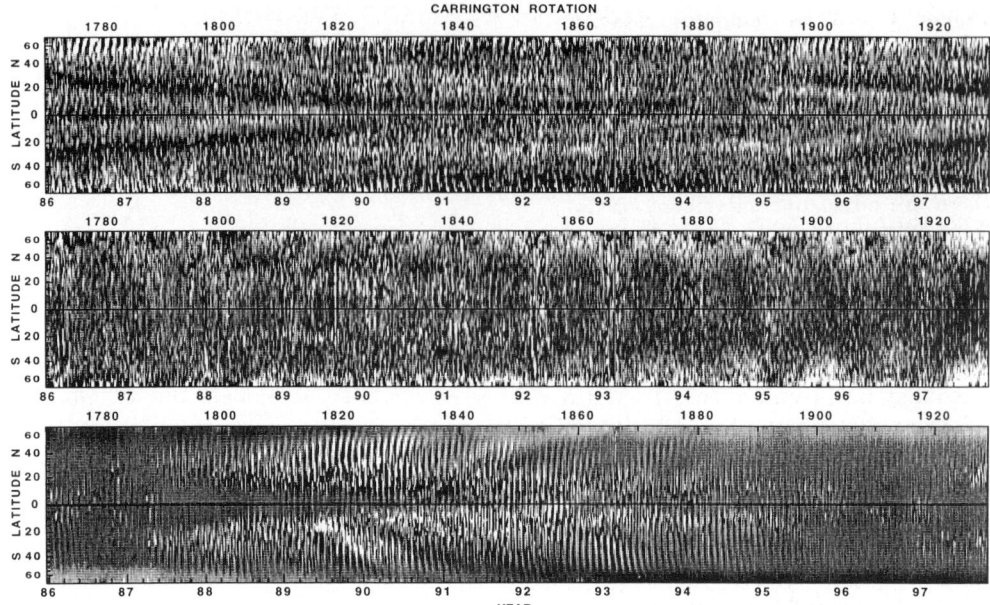

Figure 2. An eleven-year (1986 to 1997) record of the Mt. Wilson velocity and magnetic data showing the development of a full activity cycle. The top figure gives the deviation in rotation rate relative to the curve in figure (1) as a function of time and latitude. The grey-level for this panel is such that black represents 5 m/s faster than average rotation and white represents 5 m/s slower than average rotation. The middle panel gives the results for the north-south component of velocity. This motion mostly represents the meridional flow and is shown as a deviation from a long term average. For this panel white represents recession velocity and black represents approach velocity and the limits are again ±5m/s. The bottom panel gives the magnetic field with a saturation level of 2 gauss.

and south probably results from statistical fluctuations.

We display the magnetic fields and deviations in velocity from the reference state as defined by equation (3) in a format we term supersynoptic charts. A standard synoptic chart displays a full solar surface chart of magnetic field as a function of latitude and Carrington Longitude. The normal convention is to have East on the right and West on the left so that the points crossing the central meridian first are to the right and points crossing later are on the left. We reverse the order of the horizontal axis so the Carrington Longitude decreases from left to right. The advantage is that now time of central meridian passage increases from left to right and we may abut successive synoptic charts into long continuous chart where horizontal axis is the time of central meridian passage. By compressing this chart it can be reduced to a size where a very long time can be displayed compactly. Another format sometimes used to display long data sequences is called a stackplot. In a stackplot typically a range of latitude is selected and then the property as a function of longitude for successive Carrington rotations are plotted as strips above one another. These might be more precisely referred to as a latitude stackplot. One might also consider a longitude stackplot in which a band of longitude is selected and then the variable plotted in strips as a function of latitude for successive carrington rotations. Our supersynoptic chart is a form of longitude stackplot except that all possible longitude strips are included.

In figure (2) the supersynoptic charts for V_{EW}, V_{NS} and B are shown on the top, middle and bottom panels. Owing to the fact that the east-west direction is parallel to the node lines in zonal harmonics and the north-south direction is parallel to node lines in sectoral harmonics, we refer to the top panel as the zonal flow map and the middle panel as the sectoral flow map. The Carrington Rotation number is shown along the top of each panel and this figure includes just over 150 rotations. Each equatorial circumference is divided into eight bins providing minimal resolution in this dimension. The most important results are contained in the top chart which shows the torsional oscillation and related velocity patterns. The middle chart which nominally should show variations in the meridional circulation rate in fact shows a long term trend which is not correlated with the solar cycle in a simple fashion and which may be a result of changes in the data taking system. Both velocity plots include a clear annual effect which is related to the reduced frequency of observation during the months of January to April. The feature which is most evident in the zonal flow map is the torsional oscillation first described by Howard and LaBonte (1980). The bands of faster than average rotation rate begin on the left edge of the map at about $\pm 30°$ latitude and reach near the equator by mid-1990. The analysis method we use provides considerable detail in the structure of this velocity pattern. Apart from a few periods during the southern California rainy season and the long term, large scale trends in the sectoral flow map, the structure shown in the two flow maps is of solar origin. The three successive winters from 1991 to 1993 provide clear examples of the nature of the maps when the data is inadequate.

Keeping the above limitation in mind, the following features can be summarized from the flow maps:

1. During the early part of the eleven-year cycle, the latitude where the zonal velocity has a well defined maximum, this latitude marks a lower boundary below which no substantial magnetic fields are found.
2. During the time around solar maximum, the torsional oscillation pattern is not well defined. A weak pattern at low to middle latitudes is present but it does not drift in latitude.
3. While there are periods when longitudinal structures appear to persist for more than a single rotation, in general such patterns are rare and not prominent. Those longitudinal structures which are present show significant changes from one rotation to the next.
4. Taking times of solar minimum to be 1986.6 and 1996.4, the early phase of the torsional oscillation pattern is much weaker for cycle 23 compared to the same phase of cycle 22. This is illustrated in figure (3) which shows the peak speedup velocity as a function of time for the two cycles. The southern hemisphere was weaker than the northern hemisphere after the pattern began to emerge in mid-1996.
5. The polar regions exhibit substantial changes in their average rotation rate. The south pole in particular for the period 1990.5 to 1994.5 showed this effect with an amplitude reaching 8 m/s.
6. For extended periods the polar regions show a pattern of rising and falling rotation rate synchronized to the polar rotation period. This pattern is largely along lines of constant solar longitude but is slightly tilted in a manner consistent with shear from differential rotation. In the 1986 to 1988 northern pole episode, a

pattern was also seen in the sectoral flow map in a phase consistent with the flow being a polar crossing sheet. In successive quarter rotations such a sheet flow carried with rotation has a sequence of apparent effects:

(a) First it appears to enhance rotation.

(b) One quarter rotation later it appears to be receding.

(c) In the third quarter it appears to diminish rotation.

(d) In the final quarter it appears to be approaching.

The sequence observed is consistent with this pattern.

7. For several periods the torsional oscillation pattern contains structures which persist more than one rotation. These are most evident in 1986 and 1996 prior to the onset of solar activity. These have sizes comparable to the solar circumference and would be described by low degree spherical harmonics.

8. Although the polar speedup and the remnants of the equatorial faster- than-average zone are present in 1991 to 1994.5, the torsional oscillation pattern does not seem to be connected as a single and unified phenomenon. Rather the torsional oscillation seems to precede the onset of activity at the beginning of the cycle and govern its initial migration toward the equator. During the declining part of the cycle the pattern does not seem to be related to the activity.

9. The magnetic field reversal occurs at the time when the polar speedup is most evident. During this period the fluxtubes associated with the decaying active regions are being swept toward the poles by the meridional circulation. Such fluxtubes could be aligned radially and provide a coupling between the more rapidly rotating interior at lower latitudes and the more slowly rotating polar regions. Thus the polar speedup may be a consequence of this coupling. The polar crossing flows are also evident during these periods and could be a consequence of unbalanced torques produced by the fluxtubes.

The representation given in figure (2) contains more information than is necessary to illustrate the torsional oscillations. If, instead of dividing each Carrington rotation into eight latitude bins, we use just one, the structures are smoothed out leaving only the longitude dependence at a lower velocity amplitude. The resulting figure (3) shows the migration of the torsional oscillation pattern very clearly. While the equatorward drift of the latitude of rotation rate increase is easily seen during the early years of the cycle, the pattern reaches the equator after only about 5 years of the cycle and then remains at a nearly fixed latitude while gradually weakening. During the final three years of cycle 22, the pattern was nearly absent. The relationship between the polar speedup noted above and the later torsional oscillation is unclear. If the speedup had drifted equatorward following the appearance at the polar regions to join the subsequent mid-latitude torsional pattern, it would have been easy to associate the two phenomena as part of an extended solar cycle. Instead with the several-year gap, the connection is less clear.

The grey-level plots of figures (3) and (2) provide a good representation of the evolution in two dimensions but are difficult to compare quantitatively. To facilitate such comparisons, I have averaged the rotation rate over three successive rotation periods. The results of this procedure are shown in figure (4). Each average deviation curve is displaced vertically by an increment of 5 m/s. The central time for the average is given to the right of each line. The left-hand panel shows the result for solar cycle 22 and the right for the early part of solar cycle 23. The migration of the maximum is quite evident.

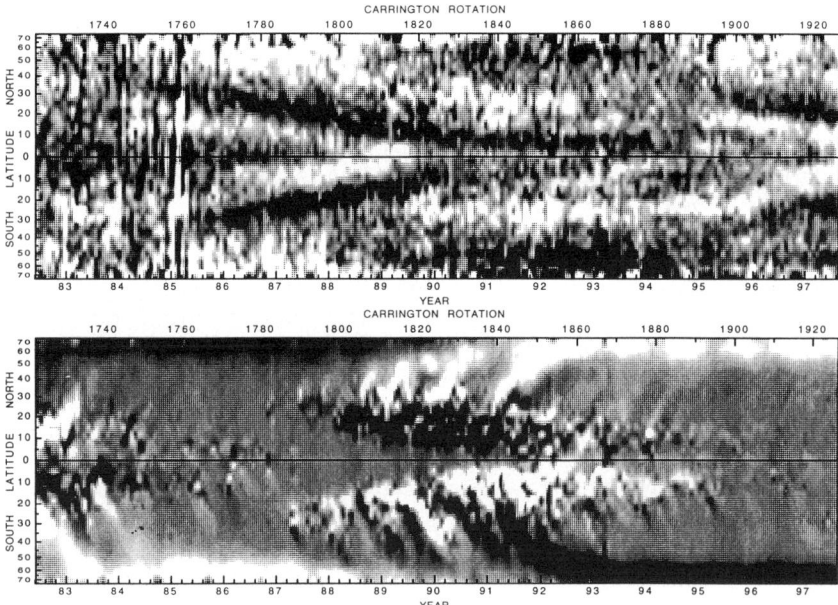

Figure 3. The torsional oscillations averaged over the solar circumference. The top panel gives the zonal velocity and the bottom panel gives the average magnetic field. The zonal velocity is coded in grey level such that white represents 3 m/s less than average rotational velocity and black represents 3 m/s faster than average rotational velocity. The magnetic field, also averaged over the circumference is indicated by black/white saturation of ±0.75 gauss.

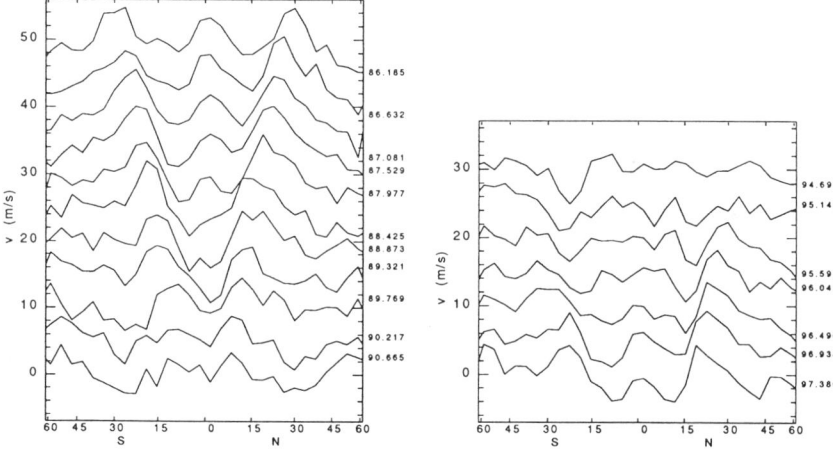

Figure 4. The change in rotation rate averaged over three successive Carrington Rotations for specified time intervals near the beginnings of solar cycles 22 and 23. Each line gives this average and is offset from zero velocity by an integer multiple of 5 m/s. The time in years at the center of the averaging interval is indicated to the right of each line. These results are only shown near the beginnings of each cycle where the pattern is well defined.

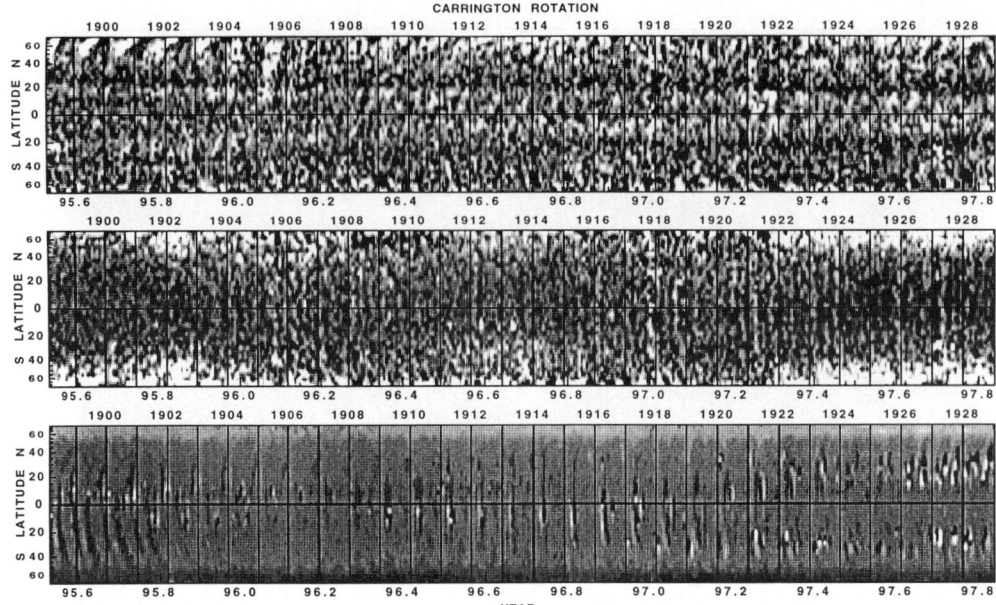

Figure 5. The zonal flow map for the past two years together with the sectoral flow map and the magnetic field map. In this figure each Carrington Rotation boundary is indicated by a vertical line.

The helioseismic analysis of observations from the MDI instrument on SOHO now makes it possible to obtain regular observations of the torsional oscillation pattern with another system (see Schou et al. in these proceedings). To facilitate comparison of these new results with the evolution of the torsional oscillation pattern, I include here as figure (5) a magnified version of figure (2) for just the past two years.

An important question to be applied to data such as shown in figures (2) and (5) is whether there are velocity structures of very large scale which persist over more than one solar rotation. The statistics of the flow maps are difficult to define because of the substantial noise which is present each winter. There are however hints of repeating patterns. The maps shown in figure (5) are best suited for this examination. Here as in figure (2) the polar crossing flows are most evident after mid 1995 in the north polar region. On the sun's equator, there is a zone of faster than average rotation rate which coincides with Carrington longitude 0° lasting from CR1919 to CR1922. At the latitude of the northern faster-than-average zone, there is a longitude about 330° (recall that the Carrington Longitude at the central meridian decreases with time) of faster-than- average rotation rate which is evident from CR1912 to CR1913. These features produce peaks at frequencies $\nu = m/\text{Rot. Per.}$ which are evident for $m \leq 8$.

2.3. MDI VELOCITIES

The data from the MDI instrument on SOHO provides an opportunity to study the solar velocities without any temporal gaps. During the dynamics program period (see Scherrer et al. 1995 for a description of the MDI instrument and programs) full disk

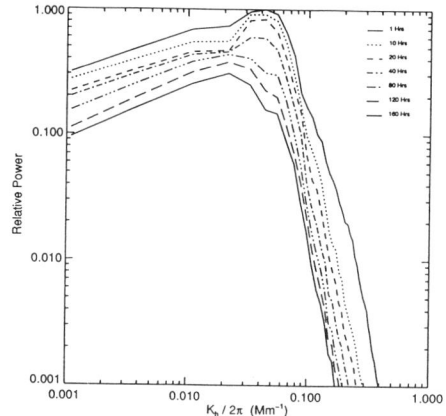

Figure 6. The spatial power spectra averaged over all azimuthal directions as a function of the inverse of the horizontal wavelength $K_h/2\pi$. Averaging periods for the curves from top to bottom are 1, 10, 20, 40, 80, 120 and 160 hours. The power is normalized relative to the maximum for the 1 hour averaging period. These curves are the average of the results for all eight study areas.

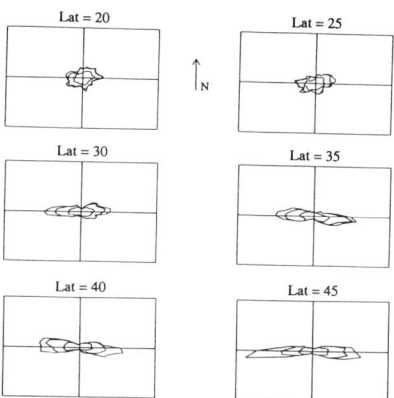

Figure 7. Windsock diagrams for six of the eight study areas from Ulrich et al. (1998). These represent the asymmetry of the velocity distribution as a function of the azimuthal angle relative to the local NS direction. Each contour is the average $|V|$ in the direction relative to the crossing point of the vertical and horizontal lines. The three contours are for all velocities less than 200 m/s, 300 m/s and no upper limit. The vertical axis represents ±2 m/s.

velocity images made up of 2 arcsec square pixels are available for periods up to 60 days. Solar rotation prevents the continuous observation of any part of the solar surface for longer than about 180 hours. Although a part of the surface can be seen longer than this, the foreshortening outside of the 180-hour window is too severe to permit the kind of analysis necessary to study surface dynamics. Beck (1996) and Ulrich et al. (1998) have extracted from the MDI dynamics program images a sequence of 176 hours and carried through an analysis similar to that used for the Mt. Wilson velocities. Due to the high duty cycle for the MDI instrument and the absence of atmospheric disturbances, this data set yields a much cleaner set of velocity maps for both the NS and EW components than is available from the ground-based observations. A full discussion of these results is given in Ulrich et al. (1998). Here I summarize the principal results.

An important question is the temporal decay of the velocity field. Solar rotation makes this question difficult to analyze due to the changing projection angle. We extract the NS velocity component from the full images by averaging together images which are symmetrically placed relative to the central meridian thereby canceling the EW component. By extending the total time interval included in the average, we isolate the time decay of the NS component alone. The spatial power spectra as a function of this averaging interval is shown in figure (6). Note the initial rapid decay followed by a more gradual shift toward longer wavelengths. The decay at long averaging times is sufficiently slow that we believe there is a good chance the features persist for more than one rotation.

The full maps include both components of velocity and allow Ulrich et al. (1998) to study the isotropy of the flows. Each horizontal velocity vector defines an amplitude and an azimuth angle relative to the local North direction. The isotropy can be quantified in terms of these azimuth angles by dividing the possible 360° into bins and calculating the average of the absolute values of the velocity vectors which fall into each bin. This produces what Ulrich et al. (1998) referred to as a windsock diagram after the patterns given for airplane pilots to evaluate the prevailing winds near an airport. Windsock diagrams have been calculated by Ulrich et al. (1998) for each of the eight study areas and are shown in figure (7) for six of these. The two areas nearest the equator are not reliable due to a contribution from vertical velocities. Most striking in figure (7) is the progressive domination of the EW velocities. In addition the peak of asymmetry lies along a line which deviates slightly from the EW direction. The sense of the deviation is that the faster than average vectors systematically move toward the equator and the slower than average vectors systematically move toward the pole. This effect has been found in the motion of sunspots (Ward 1965, Gilman and Howard 1984) but never before in the kinematics of velocity vectors for the gas. This tilt of the windsock diagrams at mid-latitude could be a direct indication of the process which maintains the solar differential rotation.

3. Energetics

The rise and decline in the strength of the sun's magnetic field is the major component in the modulation of the flow of energy from the interior to space. Throughout most of the interior, the magnetic energy density is far less than the kinetic energy of the convecting and rotating gases. Consequently, the dynamics of the convective envelope ultimately govern the dynamo process. However, even though the magnetic field is weak overall, it must be capable of modifying the dynamic flow through the weak effects of the magnetic forces. Based on these ideas, it is commonly assumed that the differential rotation is a consequence of the occurrence of a convecting fluid in a rotating, self-gravitating body. Numerical models (see Gilman 1980, 1992 for reviews) support this interpretation. The Ward (1965) correlation between NS and EW velocities would support the existence of differential rotation, and the determination from the MDI observations that this correlation is present in non-magnetic regions indicates that it can indeed lead to the existence of the differential rotation. Gilman (1980) suggests that the kinetic energy in the differential rotation is comparable to the convective energy. A critical question is the balance between the rotational and convective influences. When rotation dominates, the rotation tends to be constant on cylinders. The results from helioseismology show (Korzennik 1990, Kosovichev et al. 1997) that the rotation rate increases inward near the surface as if the radially moving currents were preserving angular momentum, then the rotation rate decreases inward as is required by the constant rotation rate on cylinders model. Perhaps this composite behavior is a result of a variable balance between rotational energy and convective kinetic energy with the former dominating in the deepest parts of the convective envelope and the latter dominating near the surface.

Superimposed on the differential rotation pattern are the solar cycle correlated velocity variations of which the torsional oscillation discussed above is the only phenomenon identified. While it has been suggested that the motion is generated by surface Lorentz forces (Schüssler [1981], Yoshimura [1981]), rough calculations based on this model (Rüdiger et al. [1986]) yield phasing inconsistent with the observations. The faster-than average band is observed to be closer to the equator than ±35° latitude and reaches the equator about at the time of maximum activity. For most of the cycle, there is no clear evidence of two bands of faster-than-average rotation co-existing in either the north or south hemisphere. None of the Rüdiger et al. (1986) models show this single mode behavior and the one which comes closest has the faster-than-average band reaching the equator at sunspot minimum. Furthermore, these models require an interaction between a weak poloidal field and electrical currents generated by the deep toroidal field which is the ultimate source of sunspots and active regions. Current ideas about the solar dynamo place the toroidal field below the convective/ radiative boundary and it is unclear that the generated currents can persist to the solar surface.

The torsional oscillations occur both with and without any surface magnetic field. The pattern was clearly present through the minimum between cycles 21 and 22 but underwent a period of absence between cycles 22 and 23. The close association between the latitude of maximum speedup of rotation and the equatormost location of the magnetic fields indicates that the torsional oscillations are intimately connected with the dynamo. As the most evident large-scale velocity feature associated with the solar cycle, they represent the most significant kinetic energy component which is part of the solar cycle. The recent finding by Schou (these proceedings) that the torsional oscillations can be detected in horizontal velocity maps deduced from solar p-modes indicates that they are not confined to the solar surface. Snodgrass (1988) and Gilman (1992) suggest that they could be a consequence of Parker's (1987) "thermal shadow" effect. In this model the toroidal fields serving as the source of the activity block the flow of energy from the interior and produce a cool overlying layer which systematically induces downflows over a broad area. The coriolis force on matter flowing into the downflow region causes the poleward slower-than-average rotation and the equatorward faster-than-average rotation. Such behavior is seen in figure (3) in the northern hemisphere during the interval 1986 to 1992. The most recent pattern has the slower-than-average band equatorward of the faster-than-average band and only a weak indication of a poleward component.

The original title of this paper (assigned by the organizing committee) included the solar radius. I have been silent on this question because there is almost no data which can be brought to bear on it. It is a difficult observational task, probably requiring a space-based instrument. Brown and Christensen-Dalsgaard (1998) report tight limits on the possible variation of the sun's radius as defined by a parameter dependent primarily on the hydrostatic structure. Ulrich and Bertello (1996) using radiation emitted near the temperature minimum found a correlation in radius change and the solar cycle with an amplitude of 400 milli-arcsec (approximately two pressure scale heights). This latter result may due to a change in the average structure of the solar atmosphere. The energetic implications of these radius variations are unclear. Blocking or enhancement of the local rate of energy flow could produce a variation in the thermal energy of the underlying convective region and cause a change in the average volume and radius. Such an effect would be limited to a portion of the solar surface where the blocking or enhancement has occurred. Since the Brown and Christensen-Dalsgaard (1998) observations are derived from EW drift of the solar image across the detector, any high latitude bumps or valleys would not have been detected.

If Parker's "thermal shadow" effect is associated with the torsional oscillations, it might produce a zone of cooler-than-average material which could influence the overlying solar figure. A careful study of the time dependence of the solar figure might reveal such a depressed zone. However, the underlying zone of enhanced magnetic field could have the opposite effect of increasing the average radius. In addition, atmospheric effects associated with surface activity on the overlying solar surface can also increase the apparent radius. This last source of confusion could be removed through the use of observations in the solar continuum and in spectral lines since the effect should be enhanced for the lines.

As a potential means of probing the energy flow through the convective envelope, observations of the solar figure carry great promise. Unfortunately, such observations are extremely difficult through the earth's atmosphere and are only now becoming available from space. Consequently, this portion of the paper remains a "wish list" rather than a report of results.

Acknowledgments: The long-term synoptic Mt. Wilson program has received support over the time interval shown in figure (3) from a variety of agencies including: US Government agencies NASA, ONR and NSF through a number of grants programs and by the Carnegie Institute of Washington. The acquisition and analysis of the MDI data has been made possible by support of the SOI Team at Stanford University which operates the MDI instrument, a NASA sponsored experiment on the ESA/NASA SOHO mission.

4. References

Beck, J.G. 1997, *Velocity Fields on the Solar Surface with Time Scales up to 30-Days*, PhD Thesis UCLA.
Brown, T. and Christensen-Dalsgaard, J.: 1998, *Astrophys. J.* Submitted.
Fox, P.A. and Sofia, S.: 1994, in J.M. Pap, C. Fröhlich, H.S. Hudson, S.K. Solanki (eds.), *The Sun as a Variable Star Solar and Stellar Irradiance Variations*, Cambridge Univ. Press, p. 280.
Gilman, P.A.: 1980, in D.F. Gray and J.L. Linsky (eds.), *Stellar Turbulence*, Lecture Notes in Physics, 114, Springer-Verlag, p. 19.
Gilman, P.A.: 1992, in K.L. Harvey (ed.), *The Solar Cycle*, ASP Conf. Vol. 27, p. 241.
Gilman, P.A. and Howard, R.F.: 1984, *Solar Phys.* **93**, 171.
Howard, R. and Harvey, J.: 1970, *Solar Phys.* **12**, 23.
Howard, R. and LaBonte, B.: 1980, *Astrophys. J.* **239**, L33.
Kim, Y.-C., Fox, P.A., Demarque, P. and Sofia, S.: 1996, *Astrophys. J.* **461**, 499.
Korzennik, S. 1990, *Helioseismic Observations from Mt. Wilson*, PhD Thesis, UCLA.
Kosovichev, A.G., *et al.* : 1997, *Solar Phys.* **170**, 43.
November, L., Toomre, J., Gebbie, K. and Simon G.: 1981, *Astrophys. J.* **245**, L123.
Parker, E.N.: 1987, *Astrophys. J.* **321**, 984.
Rüdiger, G., Tuominen, I., Krause, F. and Virtanen, H.: 1986, *Astron. Astrophys.* **166**, 306.
Scherrer, P.H. *et al.* : 1995, *Solar Phys.* **162**, 129.
Schussler, M.: 1981, *Astron. Astrophys.* **94**, L17.
Simon, G.W. and Weiss, N.O.: 1968, *Z. Astrophys.* **69**, 435.
Snodgrass, H.B.: 1988, in P.F. Schewe (ed.), *Phys. News in 1987*, AIP Special Report, p. S-11.
Spruit, H.C.: 1977, *Solar Phys.* **55**, 3.
Ulrich, R.K., Beck, J., Hill, F. and Bogart, R.: 1998, *Astrophys. J. Letters* Submitted.
Ulrich, R.K. and Bertello, L.: 1995, *Nature* **377**, 214;.
Vitense, E.: 1953, *Z. Astrophys.* **32**, 135.
Ward, F.: 1965, *Astrophys. J.* **141**, 534.
Yoshimura, H.: 1981, *Astrophys. J.* **247**, 1102.

STELLAR CONVECTIVE CORES

I.W. ROXBURGH

Astronomy Unit, Queen Mary College, University of London
Mile End Road, London E1 4NS, UK

1. Abstract and Introduction

The internal structure of stars is governed by hydrostatic support, the distribution of the chemical elements, the transport of energy by radiation and convection, and the liberation of energy by nuclear reactions. The evolution of stars is primarily determined by the changing composition due to the nuclear burning of elements in the central parts of the star, and the redistribution of the products of these reactions by mixing processes. The dominant mixing process is convection: it governs the extent of the mixed cores in moderate and large mass main sequence stars and their subsequent evolution, it mixes nuclear processed material into the envelopes of giants affecting the composition of material ejected into the interstellar medium, thereby affecting the chemical (and luminosity) evolution of galaxies. Understanding convection is essential if one is to understand the evolution of stars. Here I am concerned with convection in stellar cores and in particular with the extension of these cores by the penetration of convective motions into the surrounding stable layers affecting the internal structure and enlarging the chemically mixed region, which in turn affects the subsequent evolution. I briefly discuss a number of approaches to this problem: isochrone fitting of clusters and binary stars; simple theoretical models, the integral constraint, numerical simulation and what we can hope to get from asteroseismological observations of individual stars and of clusters and stellar groups.

2. Isochrone fitting to binary stars and stellar clusters

Since the structure and evolution of a star is affected by the extent of its convective core so too will be the isochrone (or locus at constant age) of a group of stars in the H-R diagram. One way to seek to quantify overshooting is therefore to seek the best fit to both the H-R diagram of individual clusters and the width of the main sequence band for a collection of clusters with differnt assumptions on convective overshooting.

In studies by Maeder and Mermillod (1992) and Meynet et al. (1993), overshooting was parametrised by taking the radius of core to be enlarged by a multiple α of the local pressure scale height $H_p = P/|dP/dr|$. These authors found that the best fit to isochrones of a large number of clusters indicated that $\alpha \approx 0.2 - 0.3 H_p$. In a more recent study Dowler and VandenBerg (1995, private communication) found that cluster isochrones could be well fitted if core overshooting was as given by the integral condition I derived some years ago (Roxburgh 1978,89, see section 5). Figure 1 shows their results for the Cluster IC 4651. Similar results were found for the range of clusters they analysed. These authors also found that the same model of core overshooting improved the fit of theoretical evolutionary models with the observations of binary stars.

3. The classical mixing length model of convection

The usual approach to modelling convection is to use the mixing length model, or some presumed refinement thereof, but at best such models only give an indication of the magnitude of the effects being studied, and their quantitative predictions should not be taken seriously. Unfortunately not all astronomers exercise such restraint! In this "classical" mixing length model of convection, turbulent eddies are imagined to rise or sink under the action of bouyancy, travelling a distance ℓ, the mixing length, conserving entropy and in pressure equilibrium with their surroundings, and then to mix into their surroundings. The mixing length ℓ is taken as αH_p where H_p is the pressure scale height and α is an unknown parameter adjusted so that a solar model has the observed solar radius, or by some other empirical fitting condition.

In this simple model the temperature excess δT, density excess $\delta \rho$, velocity v and fluxes F_{rad}, F_{conv} are given by

$$\frac{d}{dz}\left(\frac{\delta T}{T}\right) = \Delta \nabla \frac{1}{H_p}, \quad v\frac{dv}{dz} = g\frac{\delta T}{T}, \quad \Delta \nabla = \nabla - \nabla_{ad}, \quad \nabla = \frac{dLogT}{dLogP} \quad (1)$$

$$F_{rad} = \chi(\nabla_{ad} + \Delta\nabla), \quad F_{conv} = c_p \rho T v \left(\frac{\delta T}{T}\right), \quad F_{rad} + F_{conv} = F, \quad \chi = \frac{4acT^4}{3\kappa\rho H_p} \quad (2)$$

where $\nabla_{ad} = (\Gamma - 1)/\Gamma$ with $\Gamma = c_p/c_v$ the ratio of specific heats. The kinetic energy flux is neglected on the (weak) argument that upward and downward moving eddies occupy equal areas and the viscous energy flux is asssumed negligibly small. Taking $\Delta\nabla, g, H_p, \rho, T$ constant and integrating over the mixing length $\ell = \alpha H_p$ gives

$$\delta T \approx \alpha T \Delta\nabla, \quad v^2 \approx \alpha^2 c_p T \Delta\nabla, \quad F_{conv} \approx \alpha^2 \rho (c_p T)^{3/2} (\Delta\nabla)^{3/2}. \quad (3)$$

Deep inside the convective zone with $F_{conv} \approx F$, $\Delta\nabla \approx 10^{-8}$. At the boundary $(z=0), v=0, \Delta\nabla=0, F_{rad} = F = \chi_0 \nabla_{ad}, F_{conv} = 0$. On the unstable side of the boundary $(z < 0)$, $F_{rad} \approx (1-x)F$, $F_{conv} \approx xF, \Delta\nabla \approx x^{2/3} \Delta\nabla_0$ where $x = -z/H_\chi$ and $\Delta\nabla_0$ is the value well inside the stable layer In the stable region $F_{conv} = 0$, $F_{rad} = F = (\chi_0 + \chi' z)(\nabla_{ad} + \Delta\nabla)$ and $\Delta\nabla \approx -(z/H_\chi)\nabla_{ad}$ decreases to a value of about $-\nabla_{ad}/2$ in a distance of order $H_\chi/2$. There is no convective penetration into the surrounding stable region in this simple model.

However since $\Delta\nabla \geq 0$ in the unstable region eddies are accelerated up to the boundary and therefore continue into the stable layers. With $\Delta\nabla \approx (-z/H_\chi)^{2/3} \Delta\nabla_0$ an eddy starting from $z = -h$ with $v = 0, \delta T = 0$ accelerates under gravity and arrives at the boundary $z = 0$ with

$$v_s^2 = 2\int_{-h}^{0} g\frac{\delta T}{T}dz = \frac{3}{4}\left(\frac{g\Delta\nabla_0}{H_p H_\chi^{2/3}}\right)h^{8/3}, \quad \frac{\delta T_s}{T} = \frac{3}{5}\left(\frac{\Delta\nabla_0}{H_p H_\chi^{2/3}}\right)h^{5/3}. \quad (4)$$

The eddy continues into the stable layer where $\Delta\nabla \approx -(z/H_\chi)\nabla_{ad}$ and for $z > 0$ has a temperature excess δT and velocity v given by

$$\frac{\delta T}{T} = \frac{3}{5}\Delta\nabla_0 - \frac{1}{2}\nabla_{ad}x^2, \quad \frac{v^2}{v_0^2} = \frac{3}{8} + \frac{3}{5}x - \left(\frac{1}{6}\frac{\nabla_{ad}}{\Delta\nabla_0}\right)x^3 \quad (5)$$

where $v_0^2 = 0.8c_p T\Delta\nabla_0$, $x = z/H_p$ and we have taken $H_\chi = H_p$. The penetration distance d is given by $v = 0$ which with $\Delta\nabla_0 \approx 10^{-8}$ gives $x = d/H_p \approx 0.004$, a very small distance.

This calculation is essentially the same as that in Roxburgh (1965), which was likewise based on an eddy picture, and closely related to that of Saslaw and Schwarzschild (1965) in which they calculated the eigenfunction of the lowest unstable eigenmode. In both cases the variation of $\Delta\nabla$ was taken as given by a model with no overshooting as in the above analysis.

4. Non-local models of convective overshooting

The error in these analyses was to ignore the feedback of the overshooting on the thermal structure. Even a small amount of penetration disturbs the structure of the thin overshooting region, mixing matter from the convective layer and therefore sharing entropy and making the layer slightly less stable. With a slightly less stable layer the next eddy or convective cell can penetrate that little bit further, increasing slightly the region that is mixed with the unstable layer. Since the convective turnover times are very short compared with the thermal relaxation time of a star these successive mixings change the entropy gradient in the overshoot layer making it almost adiabatic. It is necessary therefore to determine the equilibrium structure after many such mixings.

A number of non-local models have been developed to seek to incorporate this feedback; Shaviv and Salpeter (1973), Maeder (1975), Roxburgh (1978, 1985), Schmitt et al. (1984) and Zahn (1991). Whilst these models incorporate some measure of feed back they still consider convection as an extended local phenomenon. It is the authors view that since the "mixing time" in a convective region is small compared with the thermal relaxation time the mean equilibrium structure of a convecting region is a global rather than a local property. Many years ago I developed (but never published) a global model in which the structure of a convective core and surrounding layers was determined by the condition that the mean convective flux, calculated with the properties of the lowest unstable eigenmode, exactly balanced the difference between the total and radiative fluxes (cf. Roxburgh 1994). This model gave similar results to the integral constraint on convective overshooting discussed in section 5, not surprisingly since these models, and indeed the non-local models referred to above, all neglect viscous dissipation.

In a non-local model we take the boundary of the classical convective core to be at the place where $F_{rad} = F$. Neglecting the kinetic energy flux this requires F_{conv} to be zero so we consider eddies arriving at the boundary $z = 0$ with $\delta T = 0, v_0 > 0$. For simplicity we also take $\Delta\nabla = 0$ at the boundary but there is no difficulty in extending the analysis to non zero values.

The equilibrium of the overshoot region is given by exactly the same equations (1), (2), as the standard mixing length model. These equations can be integrated (numerically) to give $v(z), \delta T(z), \Delta\nabla(z)$ in the overshoot region given $\delta T(0) = 0, v(0), \Delta\nabla(0)$ at $z = 0$ Since $dv^2/dz = 0$ at $z = 0$ and $v(z)$ is a slowly varying function of z, with $\Delta\nabla(0) = 0$, equations (1) and (2) can be integrated to give

$$\Delta\nabla(x) \approx -\frac{\nabla_{ad}}{1+\Lambda}\left[1 - e^{-(1+\Lambda)x}\right], \quad where \quad \Lambda = \frac{c_p \rho T v \nabla_{ad}}{F}, \quad x = z/H_\chi \quad (6)$$

$\Delta\nabla$ remains small in the overshoot region as long as $\Lambda > 1$ or $v > v_0\, \Delta\nabla_0/\nabla_{ad}$ where $v_0, \Delta\nabla_0$, are typical values inside the convectively unstable region. The variation of v and the penetration distance d where $v = 0$ can be estimated by multiplying the second of equations (1) by v, using $F_{conv} = c_p\rho T v \delta T$ to eliminate $v\delta T$, noting that $F_{conv} = F - F_{rad} \approx -Fz/H_\chi$ and integrating to give (cf. Zahn 1991)

$$v^3 = v_0^3 - \frac{3}{5}\frac{F}{\rho}\frac{z^2}{H_\chi H_p} = 0 \quad when \quad d = \left(\frac{5\rho v_0^3}{3F}\right)^{1/2} (H_\chi H_p)^{1/2} \quad (7)$$

These results should not be taken too literally, but they show that inclusion of the feedback of the convective overshooting on the thermal structure gives a large slightly subadiabatic region out to a distance of the order of (say) 0.3 of a scale height. The mass of the convectively mixed region is then substantially greater than that of the unstable convective core, with considerable consequences for the structure and evolution of the star.

5. The Integral Constraint on convective overshooting

An alternative approach is to seek analytical constraints on the maximum extent of convective overshooting (Roxburgh 1978,89,92). On using the equation of continuity $\partial\rho/\partial t + div(\rho\mathbf{v}) = 0$ the thermal energy equation ($dS = dQ/T$) can be expressed as:

$$\frac{\partial}{\partial t}(\rho S) + div(\rho S\mathbf{v}) = \frac{\Phi + \epsilon\rho - div\mathbf{F}_r}{T} \quad (8)$$

where S is the entropy per unit mass, ρ the density, $\mathbf{v}$ the velocity, $\mathbf{F_r}$ the radiative flux, ϵ the energy generation per unit mass and Φ the viscous dissipation per unit volume.

On integrating this equation over a volume V containing both the convective core and any overshoot region such that the convective velocites $\mathbf{v} = 0$ outside V then on using Gauss's theorem this can be expressed as

$$\frac{\partial}{\partial t}\int_V \rho S\, dV + \int_\Sigma \rho S\mathbf{v}.d\Sigma = \int_V \frac{(\Phi + \epsilon\rho - div\mathbf{F}_r)}{T} dV \quad (9)$$

where Σ is the surface of V. For a stationary or statistically steady state the first term on the left hand side is zero, and with $\mathbf{v} = 0$ on Σ the second term also vanishes so that the right hand side is zero. The total flux $\mathbf{F}$ is given in terms of the energy generation by $div\mathbf{F} = \epsilon\rho$, and on integrating by parts, using Gauss's theorem to convert to a surface integral, and noting that $\mathbf{F_r} = \mathbf{F}$ on Σ, we find

$$\int_V (\mathbf{F}_r - \mathbf{F}).\nabla \left(\frac{1}{T}\right) dV = \int_V \frac{\Phi}{T} dV > 0 \qquad (10)$$

where the last inequality comes from the fact that the viscous dissipation Φ is positive definite. If the viscous dissipation is neglected, (as is done in the simple non-local models described above) then since $F_r < F$ inside the unstable layer it follows that there must be a penetration region where $F_r > F$ the integral condition allowing one to determine the extent of this region, quantifying the earlier estimates derived using the eddy/plume mixing length models. But there is a major conceptual difference between the integral condition and other analyses which, in the author's opinion, is of major importance. The integral condition is a global condition on the whole convective region not just a condition in the neighbourhood of the boundary, that is the whole convecting region has to adjust to satisfy this constraint (including viscous dissipation). This is entirely reasonable since we impose the condition of statistical stationarity.

It should be stressed that the viscous dissipation is not zero, indeed is necessarily positive. In Roxburgh (1978) Φ was set zero to give an estimate of the upper limit on the extent of convective penetration, in contrast to the classical model where the kinetic energy flux is set to zero and there is no overshooting. Simple stellar models using this condition give an enhancement of the core mass of the order of 50% (Roxburgh 1978). It is this condition (with Φ neglected) that was used in the work of Dowler and VandenBerg described in section 2 and Figure 1.

However 2-D and 3-D resolved numerical simulations of convection in an unstable layer surrounded by stable layers (Roxburgh and Simmons 1993, Roxburgh 1998) demonstrated that, within their assumptions, the contribution of viscous dissipation to the integral condition decreased as the Prandtl number was decreased, and for Prandtl numbers less than 0.1 the mean properties of the solution were adequately described by the integral condition with $\Phi = 0$ (Figure 2).

6. Numerical simulation of convection

There are several groups undertaking 2-dimensional and 3-dimensional numerical simulation of convection in the sun and stars, mostly for plane parallel geometries rather than for convective cores. I am currently developing my 3-D code for studing overshooting from cores but unfortunately this has not yet been completed. Whilst such simulations can advance our understanding of convection it is important to remember that they do not simulate the real astrophysical conditions. These calculations fall into two categories: laminar calculations in conditions of very high viscosity, "convection in treacle", and large eddy simulations with some sub-grid scale modelling of the turbulent flow. Sub-grid modelling approximations is an area that requires much detailed study, Canuto (1996) has recently claimed that many schemes in current use are inconsistent as they do not satisfy Galilean invariance - this criticism (fortunately!) does not apply to the Smagorinski scheme used by myself and colleagues Kwing Chan and Harinder Singh.

3-D simulations by Nordlund and Stein (eg 1997) of the solar convective envelope are most impressive in that they include radiative losses at the upper boundary. Their results on overshooting at the base of the layer do not give a sharp transition between the convective region - which includes a large marginally stable zone - and the radiative interior, the sharp transition being smoothed out by averaging over many penetrat-

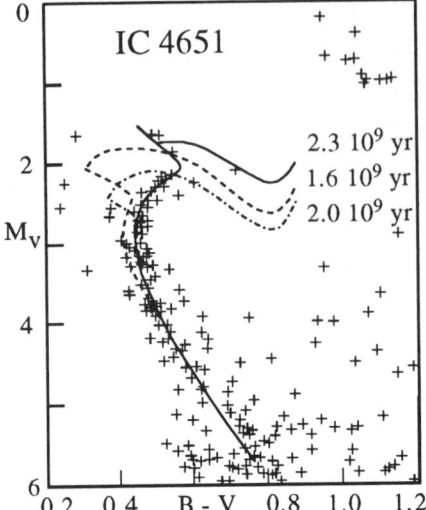

Figure 1. The best isochrone fit to Cluster IC 4651 is with core overshooting using the Integral Constraint and t = 2.3 10^9 yrs. Standard models give a worse fit and a younger age (Dowler and VandenBerg)

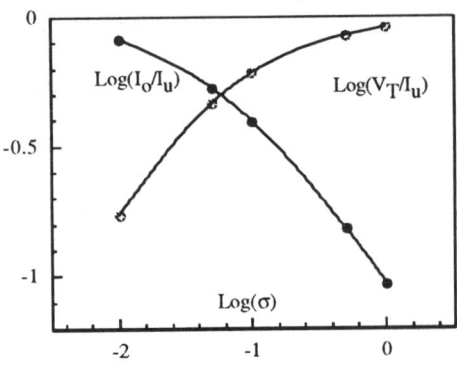

Figure 2. Variation with Prandtl number σ of the ratio of the contributions to the integral constraint from the overshoot region I_o and the total viscous dissipation V_T, to the contribution from the unstable region I_u

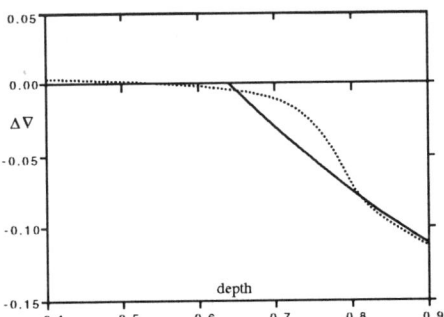

Figure 3. Variation of the superadiabatic gradient $\Delta \nabla$ at the boundary of a convective layer: numerical solution with $\sigma = 0$ (dotted line) and mixing length model (solid line).

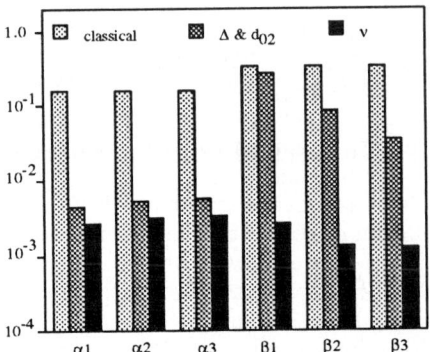

Figure 4. Error estimation on the determination of core overhooting parameters $\beta_i(M)$ and mixing length parameters $\alpha_i(M)$ for a set of 6 stars in a model Hyades cluster.

ing plumes. 2-D resolved laminar simulations by Roxburgh and Simmons (1993) do however find such a penetration layer that is at least similar to that predicted by simple analytical models (Figure 3). Likewise 3-D laminar simulations by Roxburgh (1998) and turbulent simulations by Singh, Roxburgh and Chan (1994,95,97) using Smagorinski sub-grid modelling, find such a transition. Much remains to be done!

7. Asteroseismology and convection

A major advance in our understanding of convective cores should be achieved through asteroseismology, especially through high precision space observations such as those planned for the French satellite mission COROT (cf. Baglin et al. 1997), and possible larger missions such as STARS (Badiala et al. 1996). Advances should also be achieved through the development and application of diagnostic and inversion techniques to coordinated ground based observations of (for example) δ Scuti and β Cephei stars. The boundary of convective cores generates a periodic signal in the oscillation frequencies produced by the steep change in sound speed gradients and composition as the star evolves with mixing in the core and associated overshoot region (Roxburgh and Vorontsov 1994). Some oscillation modes can be exceedingly sensitive to the region around the core, giving a valuable diagnostic tool with which to probe the internal core structure. The problem with such stars is that of mode identification and fitting the observed spectrum of frequencies to a model of the star. At the moment this is still an uncertain process but promises rich rewards. The method used is essentially model fitting, that is to produce a set of models of different mass and with different assumptions on the internal physics, in particular on convective overshooting, that satisfy such classical observational constraints as exist (M_V, T_{eff}, abundances), computing the oscillation frequencies of these models and seeking to find a fit to the observations. This is not easy since only a subset of modes may be observed and because the predicted values of the frequencies depends on the structure the surface layers where there is considerable uncertainty in our understanding of the physics. A new "differential response" technique which gives a way round this problem has recently been developed by Vorontsov (1997) and gives some hope that progress can be made in this area.

For a group of stars in a cluster, or in a binary system, asteroseismology is a potentially powerful tool for probing both convective overshooting in the core and the properties of convection in outer envelopes. For such stars which may be assumed to have the same age and initial composition (and in the case of binaries possibly good constraints on the masses) we can parametrise the unknowns in the models of stellar evolution and seek to determine these by a simultaneous fit to the classical observables (M_V, T_{eff}, and possibly some constraints on composition X, Y, Z) and the measured oscillation frequencies. Several such analyses have been undertaken to quantify the accuracy with which one can determine these parameters by such cluster/group fitting (cf. Audard et al. 1996). Figure 4 shows the results of an analysis by Audard and Roxburgh (1997) for a group of 6 stars with properties similar to those in the Hyades, in which the unknown stellar parameters are β_i, the fractional increase in core mass due to overshooting (assumed to vary with core mass), the parameter α_i in the mixing length model of convection, (assumed to vary with stellar surface properties g, T_{eff}), the masses of the stars M_i, the age t, the initial composition Y, Z, and the distance to the cluster D. The classical observables and their assumed errors are magnitudes m_V (0.01mag), $B - V$ (0.01mag), $LogZ$ (0.1), π (0.004"). The asteroseismological observables are assumed to be frequencies with $\ell = 0, 1, 2$ either just enough to give an average value (over n, ℓ) for the large separation $\Delta = \nu_{n,\ell} - \nu_{n-1,\ell}$ ($0.05\mu Hz$) and small separation $d = \nu_{n,\ell} - \nu_{n-1,\ell+2}$ ($0.3\mu Hz$), or individual frequencies $\nu_{n,\ell}$ ($0.3\mu Hz$) with the assumption that for large mass stars only half the frequencies with $n = 1, 10$ are measured, and for smaller mass stars half the frequencies with $n = 17, 29$. This "cluster fitting" procedure needs to be refined both for binary systems and for clusters to demonstrate (with artificial data) that it is possible to reproduce the input physics that went into producing the artificial data, and in particular to use Vorontsov's "differential response" technique when individual frequencies are assumed measured.

8. Conclusions

Convection and convective overshooting is one of the most important and least understood processes that determine the structure and evolution of the sun and stars. There is evidence from observations of clusters and binary systems that penetration above convective cores is significant. A major advance can be expected from seismic observations of other stars, from the ground and from space, providing data that can be used to diagnose the properties of convective core overshooting and the efficiency of surface layer convection. Numerical simulations, whilst restricted in the range of parameters they can study, can also be used to address problems in this field and to guide theorteical developments. A major effort however needs to be directed towards understanding sub-grid scale modelling and eddy viscosity.

References

Audard et al. 1996, *Bull. Astr. Soc. India*, **24**, 305
Audard N, Roxburgh I W, 1997, *Proc IAU 181*, ed F-X Schmider, in press
Badiala, M, et al., 1996. STARS. *ESA D/ScI 96/4*, Paris.
Balgin A et al., 1997 COROT, Convection and Rotation, Phase A study, CNES, France.
Canuto V., 1997, *Solar Convection Oscillations and their Relationship*, eds Pijpers F, Christensen-Dalsgaard J and Rosenthal C, Kluwer, Dordecht, 1997.
Dowler P, 1995, Thesis, University of Vancouver.
Dowler P., VandenBergh D., 1995, (private communication).
Maeder A, 1975, *Astron. Astrophys*, **40**, 303
Maeder A, Mermilliod J-C,1992, *Astron. Astrophys. Supp.* **98**, 477
Meynet G., Mermilliod J-C., Maeder A., 1993, *Astron & Astrophys Supp*, **98**, 477.
Nordlund Å. Stein R.F., 1997, this volume
Roxburgh I W, 1965, *Mon. Not. R. astr. Soc.*, **130**, 223
Roxburgh I W, 1978, *Astron. Astrophys* **65**, 281
Roxburgh I W, 1985, *Solar Physics*, **100**, 21-51
Roxburgh I W, 1989, *Astron. Astrophys* **211**, 361, 1992, ibid **266**, 291
Roxburgh I W, 1998, to be published
Roxburgh I W, Simmons J., 1993, *Astron. Astrophys*, **277**, 93
Roxburgh I W, Vorontsov S V., 1994, *Mon. Not. R. astr. Soc.*, **267**, 297
Schmitt J H M M, Rosner R, Bohm H U, 1984, *Astrophys J.*, **282**, 316.
Shaviv G and Salpeter E, 1973, *Astrophys J.*, **184**, 191
Singh H P, Roxburgh I W, Chan K L, 1994, *Astron. Astrophys*, **281**, L73, 1995, ibid, **295**, 703, 1997, this volume.
Vorontsov S V, 1997, *Sounding Solar and Stellar Interiors*, ed F-X Schmider, in press
Zahn J-P, 1991, *Astron. Astrophys.*, **252**, 179

A SEISMIC MODEL OF THE SOLAR CONVECTIVE ENVELOPE

H. SHIBAHASHI[1], K. M. HIREMATH[1,2], AND M. TAKATA[1]
[1]*Department of Astronomy, University of Tokyo, Japan*
[2]*Indian Institute of Astrophysics, Bangalore 560034, India*

Abstract. We determine the structure of the solar convective envelope by solving the basic equations for mass conservation and hydrostatic equilibrium with the imposition of the sound-speed profile determined from helioseismology and the equation of state. The solution is required to match with the structure of the radiative core, which is also determined with the imposition of the sound speed profile. The helium abundance is obtained as a part of the solutions.

1. Introduction

One major success of helioseismology is the determination of the sound speed profile in the solar interior from the observed eigenfrequencies. The inversion has so far been carried out based on either the variational principle or the asymptotic formula. In the former method, where we deal with a small deviation of the true sun from a model, a reference model is always needed. The asymptotic inversion does not necessarily require a reference model, but in order to determine the sound speed profile with a good accuracy, it is better to carry out the linearized or calibrated inversion by introducing a reference model. Usually an evolutionary solar model is used as the reference model. Since the inversion in these cases is a linearized problem, the solution, the deviation of the sound speed profile in the sun from the reference model, becomes more accurate with the introduction of a better reference model. The relative difference between the inverted sound speed profile and the recent evolutionary models is of the order of 1%. If the results from helioseismology can be incorporated in solar modeling, we can improve the solar model and, in turn, we will be able to more accurately determine the sound speed profile of the sun by iteration. Takata and Shibahashi (1997a, b) have succeeded in constructing a seismic model of the radiative core of the sun from the observed eigenfrequencies. It is obvious that we also need to construct a seismic model of the convective envelope to compute the theoretical eigenfrequencies of the seismic solar model. Note that the theory of convection is not yet complete. Evolutionary models of the sun are constructed by adjusting the efficiency of the convective energy transport as a free parameter, which is not determined uniquely in the current theory of convection. It would be nice if the structure of the solar convective envelope can be helioseismically determined. In this paper, we describe our attempt to construct a convective envelope model from helioseismic data.

2. Methodology of Making a Seismic Solar Model

We try to construct a snapshot model of the present-day sun by imposing the sound-speed distribution $c(r)$ obtained from helioseismology (Shibahashi 1993, Shibahashi and Takata 1996a, b, Takata and Shibahashi 1997a, b). We assume that the sun is in hydrostatic equilibrium and in thermal balance. The model is spherically symmetric and we ignore the effects of rotation and the magnetic field. The basic equations for constructing a model with the above assumptions are the same as those used in the theory of stellar structure:

$$dM_r/dr = 4\pi r^2 \rho, \quad (1)$$

$$dp/dr = -GM_r\rho/r^2, \quad (2)$$

$$dL_r/dr = 4\pi r^2 \rho \varepsilon, \quad (3)$$

$$dT/dr = \begin{cases} -\frac{3}{4ac}\frac{\kappa\rho}{T^3}\frac{L_r}{4\pi r^2} & \text{if radiative} \\ (dT/dr)_{\text{conv}} & \text{if convective} \end{cases}, \quad (4)$$

along with the equation of state and equations for κ and ε. Since the sound speed, which we regard as a known function of r, is a thermodynamically determined quantity, it is a function of two other thermodynamic quantities, such as p and T, along with the mass fraction of each chemical composition X_i. If we consider hydrogen ^{1}H and helium ^{4}He separately as X and Y, respectively, and treat all the other elements collectively as heavy elements Z, only two of them should be treated independently because $X + Y + Z = 1$. Furthermore, if we know the distribution of $Z(r)$, then only X or Y is regarded as an independent variable; $c = c(p, T, X)$. Inversely,

$$X = X(p, T, c). \quad (5)$$

We assume for simplicity that Z is homogeneously distributed. We can then express the density as a function of c, p and T:

$$\rho = \rho(p, T, X) = \rho(p, T, c). \quad (6)$$

Similarly, $Y = Y(p, T, c)$, $\kappa = \kappa(p, T, c)$, and $\varepsilon = \varepsilon(p, T, c)$. Then equations (1)–(4) are a set of equations for M_r, p, L_r, and T for the given $c(r)$ and the given Z.

3. Seismic Model of the Radiative Core

We divide the solar interior into two parts, the radiative core and the convective envelope, and treat them separately. As for the radiative core, for a given value of Z, equations (1)–(4) form a boundary value problem with the following boundary conditions: $M_r = 0$ and $L_r = 0$ at $r = 0$, and $L_r = L_\odot$ and $\nabla_{\text{rad}} = \nabla_{\text{ad}}$ at $r = r_{\text{conv}}$ (Takata and Shibahashi 1997a, b). The extent of the convection zone is helioseismologically determined from the kink of $c(r)$ with a reasonable accuracy. We temporarily adopt the thus-determined extent: $r_{\text{conv}}/R_\odot \simeq 0.709$. We adopt the OPAL equation of state (Rogers et al. 1996), the OPAL opacity (Iglesias and Rogers 1996), and the nuclear reaction rates compiled by Bahcall and Pinsonneault (1995). The profiles of $X(r)$ and $Y(r)$ are obtained as a part of the solutions. The Z-value is adjusted so that the ratio $Z/X(r_{\rm conv})$ matches with the spectroscopically obtained value near to the solar surface, since the convection zone is assumed to be chemically homogeneous; $Z/X(R_\odot) = 0.0245$ (Grevesse and Noels 1993). The solutions of the radiative core are seen in Takata and Shibahashi (1997a).

4. Making a Seismic Model of the Convective Envelope

As for the convective envelope, any physical quantities have already been given at the base of the convection zone from the solution of the radiative core. Since the chemical composition is uniform in the convective envelope and X has been fixed at the base of the convection zone, equation (5) can be regarded as an equation giving T in terms of p and $c(r)$; $T = T(p, c)$. Substitution of this into equation (6) allows us to express the density in terms of only p and $c(r)$; $\rho = \rho(p, c)$. Then the equations governing the hydrostatic balance (1)–(2) are decoupled from equations (3) and (4). We can determine the density and pressure by solving only equations (1) and (2) by integrating them outward from the base of the convection zone as an initial value problem with the help of equation (6). As for the equations governing the thermal structure, equations (3) and (4), since there is no energy source in the convective envelope, we do not need to consider equation (3). The temperature T is determined from $T = T(p, c)$ and we do not need to solve equation (4), either. In order for the model to be regarded as a solar model, the mass at the solar radius must be equal to the solar mass; $\Delta M = 0$, where ΔM is defined by

$$\Delta M \equiv M_r(R_\odot) - M_\odot \tag{7}$$

and it is dependent on $r_{\rm conv}$, Z/X, the sound speed profile, and the other microphysics. There is no guarantee that this condition is satisfied. Indeed, for the most likely sound speed profile inverted by Takata and Shibahashi (1997a) from the observational data together with $r_{\rm conv}/R_\odot = 0.709$ and $Z/X = 0.0245$, $\Delta M/M_\odot$ is $\simeq -2.74 \times 10^{-3}$. Since the uncertainty in the measurement of the mass and the radius of the sun is of the order of 10^{-4}, $|\Delta M/M_\odot| \simeq 10^{-3}$ is a problem. This discrepancy cannot be compensated by changing $r_{\rm conv}/R_\odot$ or Z/X in their reasonable ranges or by taking account of the uncertainties in the opacity and the nuclear reactions. We have numerically evaluated the dependence of ΔM upon the uncertainties of the sound speed profile, the depth of the convection zone, and the metal abundance Z/X. As a consequence, it was found that ΔM is highly sensitive to the uncertainty in the sound speed profile, while it is not so sensitive to the uncertainties in the depth of the convection zone and in the metal abundance. Hence we have to adopt some other sound speed profiles rather than the most likely profile. Both the statistical error and the systematic error in the sound speed profile are as small as 0.1% for $r/R_\odot > 0.3$, while they are as large as 0.3% near the center. In order to seek for the sound speed profile with which the seismic model satisfies the boundary condition (7), we modified only the sound speed profile for $r/R_\odot \leq 0.35$ by parametrizing how much it deviates at the center from model S of Christensen-Dalsgaard et al. (1996) (see figure 1), while we kept $r_{\rm conv}/R_\odot = 0.709$ and $Z/X = 0.0245$. We found that the boundary condition is satisfied for the profile which deviates at the center by $\sim 0.22\%$ from model S. Once we choose this sound speed profile and adopt the above values for $r_{\rm conv}/R_\odot$ and Z/X, any physical quantities of both the radiative core and the convective envelope are uniquely determined. The

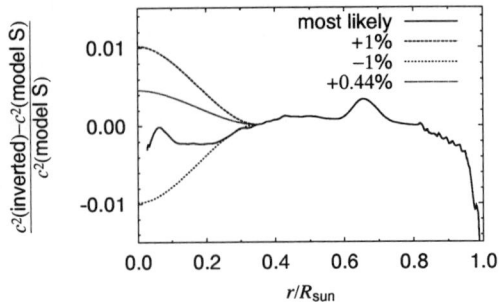

Figure 1. The inverted squared sound speed used as the imposition in solving equations (1)–(4). We modified the most likely profile in $r/R_\odot < 0.35$, and found that the curve labeled "+0.44%" satisfies the outer boundary condition $\Delta M = 0$.

resultant chemical abundances are determined as $X = 0.755$, $Y = 0.226$, and $Z = 0.0185$. This is a new method of determining the chemical abundances. The helium abundance has so far been estimated from the local profile of the sound speed at the HeII ionization zone (see e.g., Basu 1997). This method depends on the fact that the function $W(r) \equiv r^2/(GM_r)dc^2/dr$ has a peak at the He II ionization zone and the peak height is sensitive to the helium abundance. It is then expected that the helium abundance can be well estimated from the sound speed profile at the He II ionization zone. Practically, however, the uncertainty in the equation of state and the quality of the oscillation data produce fairly large errors in the estimation of the helium abundance. On the other hand, the present method of determining the helium abundance is based on the global functional form of $c(r)$. Hence, we think that these two methods should be complementary to each other. We had to modify the sound speed profile within the error bar so that the resultant seismic model satisfies the boundary condition (7). Note that the modification is not unique, and we will have to evaluate more carefully the influence of the error in the sound speed profile upon the seismic model.

References

Bahcall, J. N. and Pinsonneault, M. H. (1995), *Rev. Mod. Phys.*, **67**, pp. 781-808.
Basu, S. (1997), in *Proc. IAU Symp. No. 181, Sounding Solar and Stellar Interiors*, ed. J. Provost and F.-X. Schmider (Dordrecht: Kluwer), in press.
Christensen-Dalsgaard, J. et al. (1996), *Science*, **272**, pp. 1286-1292.
Grevesse, N. and Noels, A. (1993), in *Origin and Evolution of the Elements*, ed. N. Prantzos, E. Vangioni-Flam, and M. Cassé (Cambridge: Cambridge Univ. Press), pp. 15-25.
Iglesias, C. A. and Rogers, F. J. (1996), *Astrophys. J.*, **464**, pp. 943-953.
Rogers, F. J., Swenson, F. J. and Iglesias, C. A. (1996), *Astrophys. J.*, **456**, pp. 902-908.
Shibahashi, H. (1993), in *Frontiers of Neutrino Astrophysics*, ed. Y. Suzuki and K. Nakamura (Tokyo: Universal Academy Press), pp. 93-103.
Shibahashi, H. and Takata, M. (1996a), *Publ. Astron. Soc. Japan*, **48**, pp. 377-387.
Shibahashi, H. and Takata, M. (1996b), *Bull. Astron. Soc. India*, **24**, pp. 301-304.
Takata, M. and Shibahashi, H. (1997a), *Astrophys. J.*, submitted.
Takata, M. and Shibahashi, H. (1997b), in these proceedings.

THE CONVECTION ZONE AND OSCILLATIONS

V.A.BATURIN, I.V.MIRONOVA AND S.V.AYUKOV
Sternberg Astronomical Institute,
Universitetsky pr., 13, Moscow, Russia

1. Introduction

We study the influence of the structure and physics of the uppermost superadiabatic layers on the oscillation spectrum of p-modes. These modes are trapped inside the Sun by reflecting at the upper layers of the Sun, so these layers are important for the oscillation spectrum. Indeed, the main difference between the computed and observed frequencies seems to come from these layers (Christensen-Dalsgaard, 1985; Christensen-Dalsgaard and Thompson, 1997).

We discuss three sources of uncertainties: atmospheric opacity, convection description and the dynamic terms in pressure, i.e., the turbulent pressure.

2. Atmospheric opacity

The density profile in the solar radiative atmosphere is known with a significant uncertainty chiefly originating from the assumed opacity tables. An increase of the atmospheric opacity leads to quite substantial changes of the oscillation spectrum which can be interpreted as an effective increase of the acoustic radius of the Sun (Christensen-Dalsgaard, 1990; Baturin and Mironova, 1995). The resulting spectrum modifications may remove a significant part of the differences between the observed and computed frequencies. This approach has several limitations. Firstly, the opacity can hardly be amplified to arbitrary values. Secondly, it cannot explain the frequency differences for the modes between 2 and 3 mHz (not to mention the high-frequency end of the spectrum). Nevertheless, opacity changes in the atmosphere provide an example of a prominent effect on the oscillation spectrum.

3. Theory of convection

The next potential cause for the frequency differences is the structure of the uppermost superadiabatic region of the solar convection zone. Given the fixed entropy at the bottom of the convection zone the problem is reduced to an adequate description of the structure of the superadiabatic layers. The possible influence of the convective description on the spectrum of oscillations was studied in a series of papers (e.g. Baturin and Mironova, 1995; Monteiro et al., 1995; Rosenthal, 1997). We restrict our

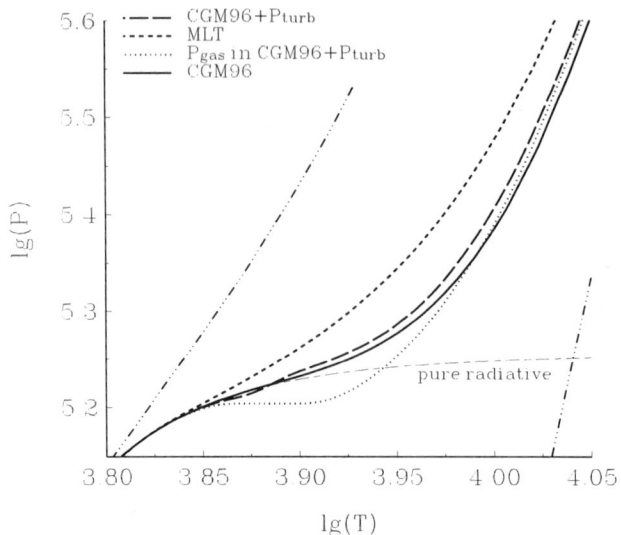

Figure 1. Different convection predictions in the temperature-pressure plane. Only superadiabatic part of the convection zone is shown. MLT—mixing length theory; CGM96—Canuto, Goldman and Mazzitelli (1996) theory. Models with turbulent pressure are plotted as well. The limiting adiabats are dot-dashed.

considerations to convection theories with horizontal averaging and local convective formulations. A convection theory should predict the temperature gradient in the superadiabatic layers. The region of superadiabatic convection is very shallow, only a couple of hundred kilometers in height. Fig. 1 illustrates the role of the convection theory in describing the transition from the top boundary of the convection zone (minimum of entropy) to the deep adiabatic layers. MLT predicts a fairly smooth transition between those two adiabats, whereas CGM96 theory (Canuto et al., 1996) corresponds to a sharper transition. It is also clear that there is not much room left for other completely different convective descriptions, and the CGM96 description is a physically reasonable example of the sharpest transition in the superadiabatic layers.

The influence on the frequency spectrum is shown in Fig. 2. Sharpening the convective transition (i.e., increasing the temperature gradient) leads to a desirable effect on the frequencies, but the amplitude is not as large as in case of the opacity correction. Another variant of the Canuto and Mazzitelli theory has been plotted in Fig. 2 to demonstrate the possibility of amplifying the effect (CMz91). However, this modification corresponds to extremely large variations of the temperature gradient which seem to be unrealistic. But it is worth noticing that changes of the description of convection are able to alter the frequency behaviour in 2–3 mHz range.

4. Turbulent pressure

The most intriguing effect is the dynamic features of convection (Antia and Basu, 1997; Kosovichev, 1995; Mironova and Baturin, 1997; Nordlund and Stein, 1996). More exactly, we consider the additional pressure term in the hydrostatic equation (turbulent pressure). We do not attempt a detailed description of the exact nature

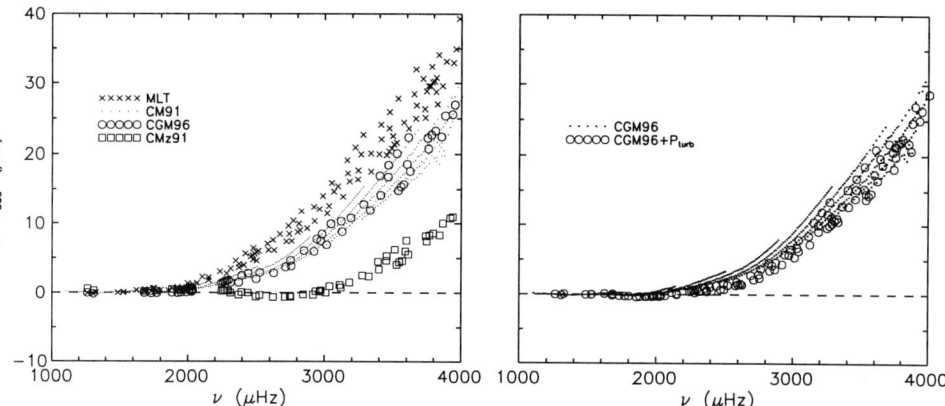

Figure 2. (*left*) Oscillation spectrum of models with different convection descriptions. CM91 and CMz91 —convection theory by Canuto and Mazzitelli (1991). See also caption to Fig. 1 for abbreviations.

Figure 3. (*right*) Effect of inclusion of turbulent pressure on model frequencies.

of this term or of the properties of turbulent convection. Instead of this we estimate the possible effect of the additional pressure term on the oscillational spectrum. To achieve this we have computed a model with a turbulent pressure term according to Canuto, Goldman and Mazzitelli (1996).

The structural changes are easily predictable. The gas component of pressure is reduced, as is density, but the total pressure is roughly the same. The effect is rather strong (turbulent pressure can contribute as much as 12 per cent to the total pressure), but strictly localized—the stratum where turbulent pressure appears is five times shallower than the superadiabatic region. This explains the rather surprising result—the frequencies of the model modified by turbulent pressure are close to the "basic" model frequencies (Fig. 3.)

This result was obtained using Canuto, Goldman and Mazzitelli formulation. We did not investigate the question whether other turbulent pressure descriptions are similarly localized.

Turbulent-pressure terms also occur in oscillation equations used for the frequency calculations. This complex effect can be simulated by changing the adiabatic compressibility Γ_1. We have recalculated the theoretical frequencies with the so-called reduced Γ_1 in the model with turbulent pressure (following a recipe by Rosenthal, 1997) and found that an influence on frequencies is quite small. This might be explained by the fact that the region where turbulent pressure has some effect is very narrow; also, the acoustic waves are predominantly evanescent in these layers.

5. Summary

The possibilities for improving the theoretical p-mode spectrum within the framework of local convective models and adiabatic oscillations are mostly exhausted by the effects considered. A combination of these effects can reproduce the observational frequencies up to 3 mHz, but fails to do so in the high-frequency range. For further improvement it may be feasible to consider non-adiabatic effects and horizontally inhomogeneous convection theories. The influence of the dynamic features of convection seems to be less significant than expected.

The authors thank J. Christensen-Dalsgaard for very useful comments.

References

Antia H.M., Basu S.: 1997, in F.P.Pijpers, J.Christensen-Dalsgaard, and C.S.Rosenthal (eds), *Proceedings of Solar Convection and Oscillations and their Relationship*, Kluwer, p. 51.

Christensen-Dalsgaard J.: 1985, in D.O.Gough (ed.), *Seismology of the Sun and the Distant Stars*, Reidel Publ. Co., Dordrecht, Holland, **169**, p. 23.

Christensen-Dalsgaard J.: 1990, in G.Berthomieu, M.Cribier (eds), *Inside the Sun*, Kluwer, Dordrecht, p. 305.

Christensen-Dalsgaard J., Thompson M.J.: 1997, *Monthly Notices Roy. Astron Soc.* **284**, 527.

Baturin V.A., Mironova I.V.: 1995, *Astronomy reports* **39**, 107.

Canuto V.M., Goldman I., Mazzitelli I.: 1996, *Astrophys.J.* **473**, 550.

Canuto V.M., Mazzitelli I.: 1991, *Astrophys.J.* **370**, 295.

Kosovichev A.G.: 1995, in J.T.Hoeksema, V.Domingo, B.Fleck, B.Battrick (eds), *Helioseismology: Proceedings of the 4th SOHO Workshop*, **1**, ESA SP-376, ESTEC, Noordwijk, p. 165.

Mironova I.V., Baturin V.A.: 1997, in F.-X.Schmider and J.Provost (eds), *Proceedings of the IAU Symp. No.181, Sounding Solar and Stellar Interiors*, Nice Observatory (in press).

Monteiro M.J.P.F.G., Christensen-Dalsgaard J., Thompson M.J.: 1996, *Astron. Astrophys.* **307**, 624.

Nordlund A., Stein R.F.: 1996 in A. Noels et. al. (eds), *Proceedings of the 32d Liege Colloquium*, Liege: Univ. Liege, p.75.

Rosenthal C.S.: 1997, in F.P.Pijpers, J.Christensen-Dalsgaard, and C.S.Rosenthal (eds), *Proceedings of Solar Convection and Oscillations and their Relationship*, Kluwer, p.145.

TOTAL SOLAR IRRADIANCE VARIATIONS

The Construction of a Composite and its Comparison with Models

CLAUS FRÖHLICH
Physikalisch-Meteorologisches Observatorium Davos
World Radiation Center, CH–7260 Davos Dorf, Switzerland

AND

JUDITH LEAN
E.O.Hulburt Center for Space Research
Naval Research Laboratory, Washington, D.C.20375-5320, U.S.A.

Abstract. Measurements of the total solar irradiance (TSI) during the last 18 years from spacecraft are reviewed. Corrections are determined for the early measurements made by the HF radiometer within the ERB experiment on NIMBUS7 and the factor to refer ACRIM II to the ACRIM I irradiance scale. With these corrections a composite TSI is constructed for the period from 1978–1997. This time series is compared with a model that combines a magnetic brightness proxy with observed sunspot darkening and explains nearly 90% of the observed short and longterm variance. Possible, but still unverified degradation of the radiometers hampers conclusions about irradiance changes on decadal time scales and longer.

1. Introduction

The radiation from the Sun at the mean Sun-Earth distance (1 AU), integrated over all wavelengths, hence total solar irradiance (TSI), is traditionally called the "solar constant" although it has been shown to vary on time scales from minutes to decades. The largest amplitude variance of up to a few tenths percent occurs on time scales from days to several months and is related to the photospheric features of solar activity: decreases in the irradiance during the appearance of sunspots, and increases when bright faculae are present. This modulation of TSI is conceptually well understood, but a detailed understanding is still missing. While the long-term modulation of TSI by the 11-year activity cycle is well documented and generally accepted, understanding of the mechanisms governing the variations is incomplete, and limits conclusions about longer term changes of TSI, such as those which may have occured during the Maunder Minimum associated with the Little Ice Age in Europe. Knowledge of such changes is important for the understanding of possible solar forcing of climate change during the past, present and future.

The radiometric accuracy of irradiance measurements made by individual instruments, of the order of 0.2%, is insufficient to determine long term changes of only about 0.1% that occur during the 11-year modulation. While the instrument repeatablility is adequate to monitor short term changes, the long term behaviour can only be retrieved by careful tracing of one experiment database to the other, incorporating good knowledge of the degradation of radiometers operating in space. Fortunately several time series of TSI exist, made from different spacebased platforms by different radiometers. This allows the construction of a composite time series having improved long term precision, thus yielding an unbiased estimate of TSI variability during the past 18 years. In the following a critical review of TSI measurements is presented and some new corrections for the early periods of the ERB experiment on NIMBUS-7 introduced. A composite TSI is then constructed by combining overlapping time series. Comparing this composite with empirical models deduced from proxies for TSI variability sources permits an independent 'adjustment' of the extrapolated degradation behaviour of the ERB experiment. We use our newly constructed composite time series to compare the behaviour of the TSI during the last two solar cycles which in turn improves our understanding of this type of modulation.

2. Review of Total Solar Irradiance Monitoring Programmes

The TSI measurements from satellites discussed here have been performed by the following experiments (ordered chronologically) and are shown (already corrected, as discussed below) in Fig. 1:

- Hickey-Frieden radiometer (HF) of the Earth Radiation Budget (ERB) experiment on the NIMBUS-7 satellite from November 16, 1978 until January 24, 1993 (Hoyt et al., 1992 and references therein). The measurements are performed during the passage of the Sun through the angle of view at the southern terminator of the satellite and last for a few minutes. Only daily values with more than 5 readings (orbits with measurements) per day are included. Thus no interpolated values, as listed in the published time series are used. It is important to note that for the periods of the '3-days-on-1-day-off' operation, the ERB experiment yields interrupted lines in Fig. 1. Data prior to the end of 1980 have been adjusted downward corresponding to a slip in the NIMBUS7 orientation relative to the sun. The need for this correction is explained below together with the correction for degradation. Moreover, the data after October 1, 1989 have been decreased by 0.31 Wm^{-2} and after May 8, 1990 by another 0.37 Wm^{-2}. The need for these latter adjustments are indicated from comparison with ERBE data (Lee et al., 1995) and models deduced from photospheric observations (Chapman et al., 1996).
- ACRIM I (Active Cavity Radiometer for Irradiance Monitoring) on the Solar Maximum Mission Satellite (SMM) from February 14, 1980 until June 1, 1989 (Willson & Hudson, 1991 and references therein). Only those data are taken which have a standard deviation of < 0.2 Wm^{-2}. Moreover, a correction has been added for unamended degradation during 1980 as described below.
- Solar Monitor of the Earth Radiation Budget Satellite (ERBS) which is similar to the radiometer ACRIM (Lee et al., 1987) since October 25, 1984 with a 4 months gap in 1993 due to battery problems of the satellite and consequent switch-off of the experiment (Lee et al., 1995). As for HF on NIMBUS7 the

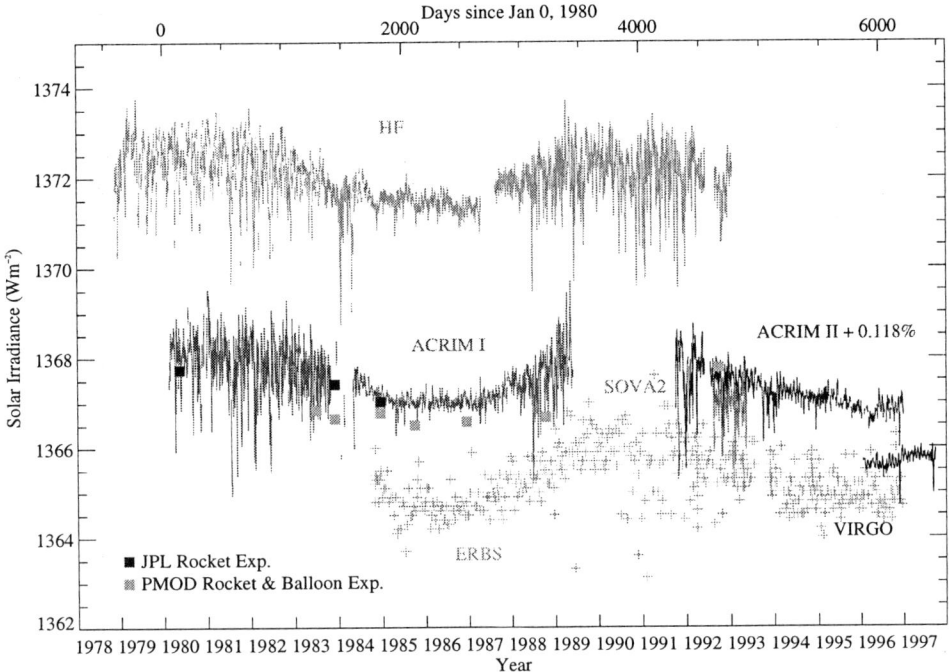

Figure 1. Time series of daily values of total solar irradiance in Wm^{-2} as observed by HF/ERB on NIMBUS7, ACRIM I & II on SMM and UARS, the solar monitor on ERBS, SOVA2 on EURECA and VIRGO on SOHO. For corrections applied to the original time series see text. The results of the rocket and balloon experiments of JPL and PMOD/WRC are also plotted for comparison.

measurements are performed during the drift of the Sun through the angle of view of the instrument, but only once in two weeks. No corrections are applied.
- ACRIM II on the Upper Atmospheric Research Satellite (UARS) since October 1991 (Willson, 1994). As for ACRIM I only those data are taken which have a standard deviation < 0.2 Wm^{-2}. No corrections are applied.
- PMO6 of SOVA (Crommelynck *et al.*, 1993) on the European Retrievable Carrier (EURECA) from August 11, 1992 until May 14, 1993 (Romero *et al.*, 1994)
- DIARAD and PMO6-V of VIRGO (Variability of solar IRradiance and Gravity Oscillations, Fröhlich *et al.*, 1995, 1997a) on SOHO (SOlar and Heliospheric Observatory) since January 18, 1996. The VIRGO values used here are composed of the PMO6-VA values 'adjusted' to the long-term behaviour of the DIARAD-L as described in Fröhlich *et al.*(1997b).

For comparison, Fig. 1 also includes the rocket-based irradiance measurements made by JPL and the rocket and balloon-borne measurements made by PMOD/WRC; these experiments were used in 1986 to provide additional evidence that the downward trend of solar irradiance during the declining phase of the cycle 21 was real and not of instrumental origin (Willson et al., 1986).

The differences among the absolute values of the various spacebased datasets in Fig. 1 are due to differences in their absolute calibrations which are accurate to 'only' about ±0.2% in absolute terms. The repeatability of the daily measurements within each dataset is generally much better than 0.01% per year. The HF time series is essentially uninterrupted and can be used (with adjustments noted above and described below) as an overlapping reference for over 14 years. It cannot be proven, however, that the sensitivity of HF has no long term trend or that the changes in operational procedures have not influenced the absolute values more than that accounted for here (see also comments and corrections by Lee et al., 1995 and Chapman et al., 1996). Moreover, sensitivity changes early in the NIMBUS7 mission have never been assessed. Similarily, the ERBE data are uninterrupted, starting 6 years after NIMBUS7 and continuing still. The low duty cycle of the ERBS measurements (once every 2 weeks) hampers their usefulness as a reference data set; as a cross check they are, however, valuable.

To utilize the early HF/ERB data (before 1982), corrections for degradation of this radiometer have to be made. Before proceeding, however, we have must address the issue of early degradation of ACRIM I first raised by Foukal & Lean (1988) and extensively discussed by Willson & Hudson (1991). The issue was that during 1980 the ACRIM I data decreased more rapidly than did proxy variability models which could be interpreted as an uncompensated degradation. Willson & Hudson (1991) justified their determination of the ACRIM I data by the fact that they were based on measurements with two back-up radiometers, and that the result was also in agreement with recently evaluated HF data. However, they overlooked the fact that the degradation as measured after the repair in 1984 happened effectively in 1980. Since the exposure time during the spin mode phase was about hundred times less than during normal operation, and equivalent to only about 15 days, no significant degradation happened during the spin mode phase of more than 3 years. This factor corresponds to the ratio of available data points during spin mode compared to normal operation as stated in Willson & Hudson (1991). Thus, from Fig. 3 of Willson & Hudson (1991) the degradation at the end of the normal operation in 1980 is about 300 ppm, corresponding to the value measured in 1984. If we account for an early degradation (before day 62) of about 40 ppm, the total amounts to 340 ppm compared with 180 ppm used by Willson & Hudson (1991). Two corrections are needed to account for this: the data have to be reduced by $340 - 180 = 160$ ppm to refer them to the 1984 value and the downward trend has to be corrected by $340 - 180 = 160$ ppm over the period of normal operation in 1980. Thus, the '300-ppm excess' can be compensated for by taking the full degradation into account and the corrected irradiance in Fig.2 of Willson & Hudson (1991) fits the proxy model as well as it does later on.

With the corrected ACRIM I the HF degradation can now be assessed. The original analysis of the HF could not correct for this because of lack of information from e.g. a backup instrument. Thus, any correction has to be based on assumptions about the instrument performance. The HF has the same geometry of the cavity as does the PMO6 type instruments (inverted 60° cone) and uses the same specular black paint; so these radiometers may give guidelines. The analysis of the PMO6V radiometers within VIRGO on SOHO has shown that the overall degradation can be modelled by an exponential function with a time constant $\tau \approx 280$ days exposure time (Anklin et al., 1997). Moreover, these radiometers show at the very beginning of exposure to the sun an increase in sensitivity which is very fast and can be modelled with an exponential function with $\tau = 4\ldots 10$ days exposure time. This effect is not readily understood, but observed in both PMO6V radiometers consistently in amplitude and τ with exposure-to-mission times differing by a factor of ≈ 50. The degradation and initial increase are evident in Fig. 2b. Fig. 2a shows the correction for the slip at the end of 1980 together with a comparison to the corrected ACRIM I after the shift has been removed. The lower 81-day running mean of the ACRIM I comparison is prior to correcting HF for degradation; the curve above is after. The exposure-to-mission time is about 1:25 for the HF during the phase of the 3 days operation out of 4. The time constant of the increase is about $\tau_{incr} = 70$ mission days and similar to that observed by the PMO6V, the amplitude (145 ppm) is, however, much higher. The HF degradation corresponds to a time constant of $\tau_{degrad} = 385$ mission days and an amplitude of 510 ppm, which is also larger than observed by PMO6V. This may be due to the different environment in a Earth orbit relative to the one at SOHO orbiting around the Langrange point $L1$.

The time series of the two ACRIM instruments without adjustment of their respective irradiance scales demonstrate the difficulty in bridging gaps in data if no other overlapping irradiance experiments are available. The ACRIM I and ACRIM II absolute scales differ by more than 0.1% as shown below, which is larger than the peak-to-peak modulation amplitude of the 11-year cycle. In relation to the state-of-the-art of room temperature radiometry this is a very good result, but quite insufficient for the long-term monitoring of TSI variability. Fortunately the time-series of NIMBUS7 (corrected for degradation) and ERBE both overlap the ACRIM I and II and can be used for intercomparison. The intercomparisons are performed for each pair of TSI observations for the days when both time series have valid data. For ACRIM I only the data after the repair in 1984 are used in this comparison as the earlier data need some correction, as described above, and the spin mode data are less reliable and should not be used. The calculations, as first presented in Fröhlich (1997), have been refined by minimizing $\chi^2 = \sum((y_i - \alpha x_i)/\sigma_i)^2$ with y_i being the mean ratios of ACRIM I to HF and ERBS and x_i the corresponding ones of ACRIM II, σ_i is the geometrical mean of the standard deviations of the means to ACRIM I and II mean respectively and α is the scaling factor to refer ACRIM II to ACRIM I. The results of the comparisons are shown in Fig. 3 and listed in Table 1. The averages used for the determination of α are indicated as horizontal lines and the periods are listed in Table 1. The correction factor for ACRIM II amounts to $\alpha = 1.001180 \pm 0.000153$. The unweighted fit gives a slightly higher value of 1.001222 ± 0.000165 which means that the ERBE data pull the ratio up. The indicated uncertainty is a formal 1σ error and the change relative to the formerly reported value of 1.001245 is mainly due to the fact that the averages are weighted by their standard deviations which gives less weight to the ERBE data and that the period for the fit has been reduced to the period after the repair in 1984. This is justified in order to avoid influence of the early corrections. The individual values of the comparisons are the listed in Table 1. For the corrected ACRIM II ratios a regression line is calculated primarily to check for possible trends; their slopes are not significantly different from zero; the trend against VIRGO is due to low values of ACRIM II in January to March 1996 which are believed to be less reliable because of adverse operational conditions as indicated by the many gaps in the ACRIM II data.

This cross calibration exercise demonstrates the possibility of increasing the relative precision of the TSI temporal data. But it shows also the limitations of this

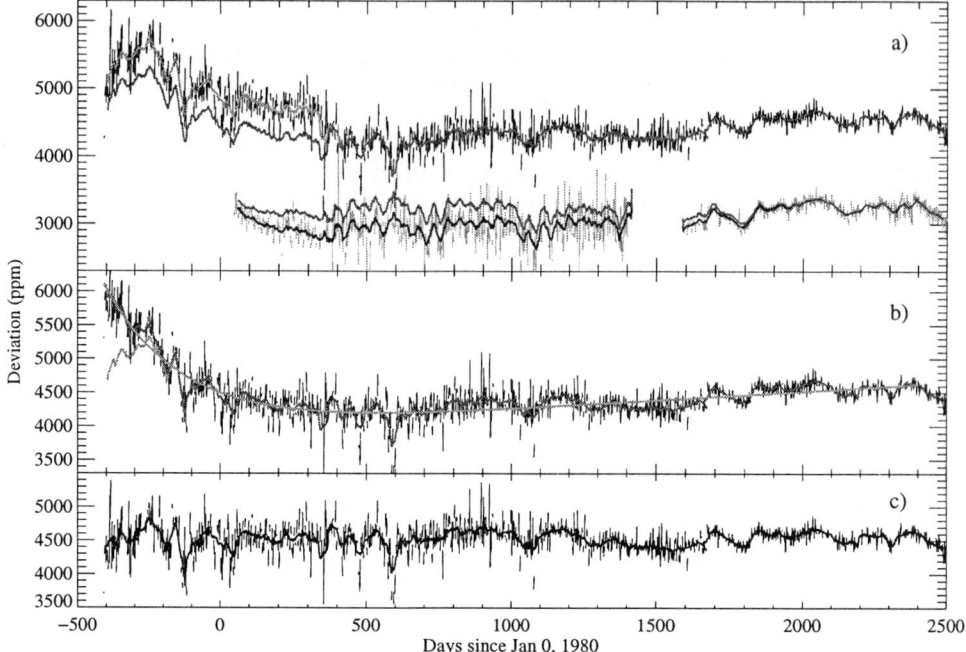

Figure 2. Determination of the degradation and initial increase of the HF radiometer sensitivity and the correction for the slip at the end of 1980. The thin lines are running means through daily values, the thick ones are 81-day running means. *a)* The upper curve corresponds to the comparison with the model and the lower two curves to ACRIM I, before and after the correction for degradation. The correction around day 350 is indicated by the downward shifted line in the HF data. Both comparison show a steady increase of about 0.3 ppm/day which must be due to the performance of HF. *b)* Comparison to the model with the fitted exponential and linear increase for the degradation and the correction for the early increase in sensitivity. *c)* Comparison with the model after all corrections.

approach: Since one time series covers the maxima of cycle 21 and 22 and the minimum in between (HF/NIMBUS7), and another covers both minima and the maximum in between (ERBE), the comparability of both cycles in terms of their absolute level and amplitude may still be questionable. The ratios in Fig. 3 illustrate another problem: the different instruments (e.g. ACRIM I and HF) can record the short- and medium-term variations quite differently. Part of these differences may be related to the way the radiometers are operated: continuous measurements during the sunlit part of an orbit as for ACRIM I and SOVA, or only short periods with the sun sweeping through the field of view as for NIMBUS7 and ERBE, or an intermediate way as for UARS where a solar pointing mechanism allows for measurements during an extended period of time, but still not during the full sunlit part of an orbit. The different standard deviations of the ratios as listed in Table 1 may reflect such effects. The smaller standard deviations of the ACRIM II comparison with VIRGO and SOVA are due to

TABLE 1. Comparison of the different TSI results.

Instruments	Mean Ratio	Std.Dev.	Number of Data Points	Period
ACRIM I/HF	0.996790	0.000142	1622	18-May-84 – 1-Jun-96
ACRIM I/ERBE	1.001596	0.000260	132	25-Oct-84 – 24-May-89
ACRIM I/PMOD	0.999574	0.000319		
ACRIM II_{corr}/HF	0.996861	0.000240	360	5-Oct-91 – 24-Jan-93
ACRIM II_{corr}/ERBE	1.001441	0.000286	215	9-Oct-91 – 18-Dec-96
ACRIM II_{corr}/SOVA2	1.000400	0.000088	269	11-Aug-93 – 15-May-94
ACRIM II_{corr}/VIRGO	1.000882	0.000115	282	29-Jan-96 – 30-Dec-96

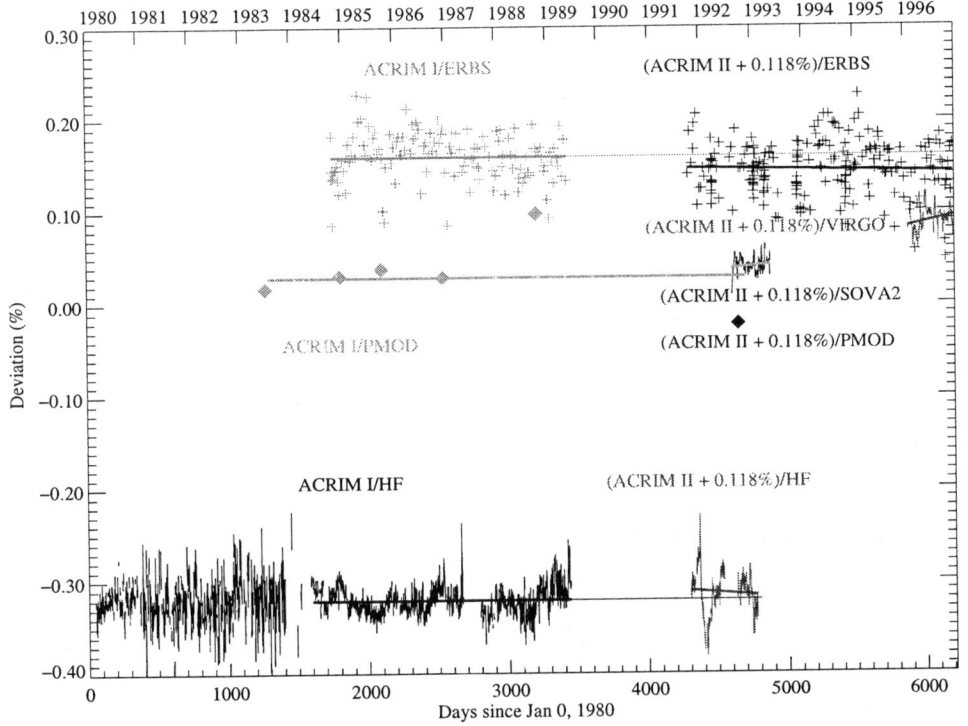

Figure 3. Comparison of ratios of different TSI time series for the determination of the scaling factor for ACRIM II in order to refer it to ACRIM I.

the fact that these results are the only comparison with similar sampling strategies. This result emphasizes how important it is to define the 'mean' value to which the observation corresponds. The comparisons with the results from the rocket and balloon experiments show the difficulties of using single point measurements to deduce longterm variability; but the reproducibility is still better than 0.1% and, perhaps fortuitously, the SOVA2 results are very close to those. Significant differences exist between the amplitudes of medium term variations evident in the 6 months modulation of TSI during the solar minimum between cycle 21 and 22 where HF shows twice the amplitude of ACRIM I (Fig. 1 and 3). Within VIRGO a similar difference in behaviour has been observed between the readings of PMO6-V and DIARAD (Fröhlich et al., 1997b). As the sensitivity of all radiometers to sunspot blocking is very similar, the observed difference may be explained by a difference in their spectral sensitivity to UV radiation which is much stronger relative to the total during enhancement of TSI, as observed during minimum. Definitive conclusion about this effect requires a thorough investigation of the spectral sensitivity of the different radiometers which is under way.

TABLE 2. Time series used for the composite TSI and correction factors F_1 at the beginning of the period (relative to the time series before) and F_2 at the end (relative to the time series after) determined for a period of overlap of 80 days. The data are scaled with F_{aver} listed in the last column.

Period	Instrument	F_1	F_2	F_{aver}
16-Nov-80 – 6-Mar-80	HF		0.996783	0.996783
7-Mar-80 – 22-Nov-80	ACRIM I			1.000000
23-Nov-80 – 3-May-84	HF	0.996869	0.996912	0.996890
4-May-84 – 2-Jun-89	ACRIM I			1.000000
3-Jun-89 – 4-Oct-91	HF	0.996924	0.997069	0.996996
5-Oct-91 – 17-Jan-96	ACRIM II$_{corr}$			1.000000
18-Jan-96 – 31-Dec-96	VIRGO	1.000923		1.000923

3. Construction of a Homogenious Composite TSI Time Series

For constructing a composite TSI time series we need a reference irradiance scale. As the absolute accuracy is still insufficient compared to the solar variability, any choice of a reference instrument is acceptable. Until the advent of SOHO, ACRIM was the only experiment which secured a reliable determination of the degradation of the operational radiometers by virtue of its redundant receivers. Thus, ACRIM I and the corrected ACRIM II values are used as an initial reference scale for the composite time series. Before the repair of SMM in spring 1984 and in the gap between ACRIM I and II values from the HF instrument are used. After the start of the observations by VIRGO (18 January 1996) those data are used. VIRGO's continuous sampling yields reliable time series and improved determination of the degradation because two completely different radiometers PMO6V and DIARAD and their backups are used. The analysis of the degradation by Anklin et al.(1997) emphasizes the difficulty in attaining levels of 10 ppm repeatability. Table 2 summarizes the time series used

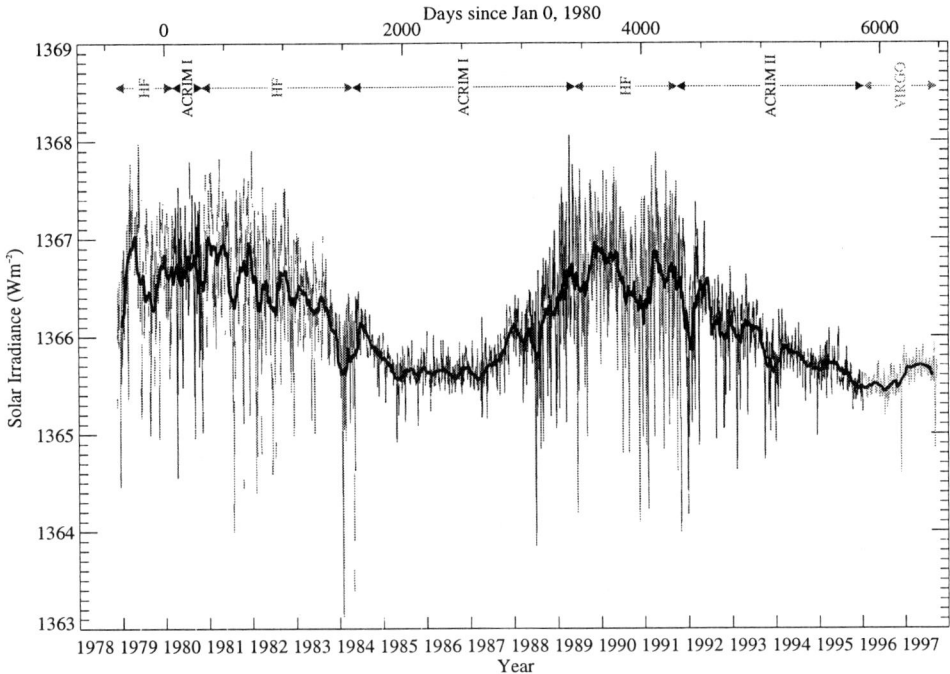

Figure 4. Composite total solar irradiance for 1978-1997. Its absolute value is referred to SARR. Note the '3-day-on–1-day-off' periods of the HF operation which is seen as 'dashed line' e.g. before 1984.

and the corresponding correction factors for the different periods. The four values ACRIM I/HF show an overall upward drift of about 290 ppm indicating a possible degradation of HF not yet compensated. This underscores the care that is needed for the interpretation of individual time series, and the importance of redundant instruments on the same platform to assess the degradation of the operational data by comparison with measurements by less often exposed radiometers. The ratios at the end of ACRIM I and at the beginning of ACRIM II are very close, a result of the adjustment of ACRIM II as described in the previous section. The difference of 145 ppm is due to the fact that only 80 days at the end of ACRIM I and at the beginning of ACRIM II are used to determine the ratios. After the adjustment of the HF and VIRGO values to fill the gaps within and between the ACRIMs the whole time series is adjusted to the Space Absolute Radiometer Reference (SARR) defined by Crommelynck *et al.*(1995) by applying the factor for ACRIM II$_{corr}$ of 0.998996. This does not improve the absolute accuracy, but allows comparison of repeated space experiments with the same radiometer.

The composite TSI is shown in Fig. 4. Willson (1997) suggested that the 1996/97 minimum is about 0.5 Wm^{-2} higher than the one in 1986/87; our composite would rather indicate a decrease of about 0.13 Wm^{-2}. The difference between these two

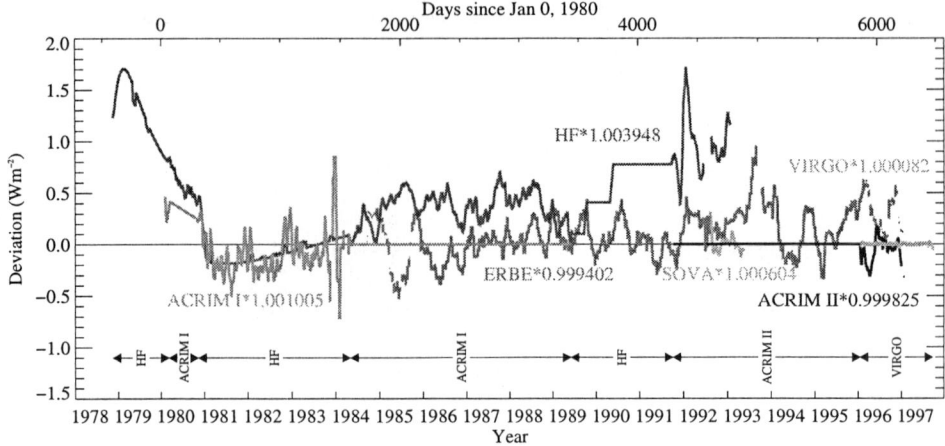

Figure 5. Comparison of the original, uncorrected time series with the composite TSI. The factors listed are the ratios to scale the original data to the composite referred to SARR.

estimates of decadal TSI variability corresponds almost exactly to the sum of the two corrections applied to the HF data in October 1989 and May 1990 of 0.68 Wm^{-2} which were not taken into account by Willson (1997). The significat uncertainties in the knowledge of the degradation indicate that our decrease of 0.13 Wm^{-2} is probably not significant (see Fig. 4). As to the differences between the behaviour of the two cycles the following can be noted: Although no significant activitiy of the new cycle has been observed before mid 1997 it seems that – at least for the irradiance – the minimum after cycle 22 is shorter than the former one. A slight increase after the sunspot in November 1996 – one of the old cycle – is observed; the first important sunspot, however, of the new cycle arrived only in September 1997. Comparison of the maxima is more difficult as the variance is greater. In order to illustrate the effect of the corrections made to the different time series Fig. 5 shows the comparison of the uncorrected values with the composite TSI record.

4. Comparison of the Composite with a Proxy Model

Models have been developed to reconstruct solar irradiance from parameterizations of independently measured proxies of magnetic variability sources, specifically dark sunspots and the bright faculae and network. White light images made most recently from the US Airforce SOON sites and archived by the NOAA World Data Center (WDC) provide information about the areas and locations of dark sunspots from which to quantitatively determine the net irradiance reduction on a daily basis. Summing the irradiance deficit due to all sunspots present on the disk yields a parameterization variously termed the bolometric sunspot blocking function, P_S, (Foukal, 1981) or the photometric sunspot index, PSI, (Hudson et al., 1982; Chapman and Meyer, 1986). We calculate a time series of daily sunspot blocking following Foukal

(1981) and Fröhlich et al.(1994), incorporating results from recent studies of the dependence of sunspot residual intensity contrast on sunspot area, in the sense that larger sunspots are darker than smaller spots (Brandt et al., 1994).

An analogous approach is in principle applicable for the estimation of facular brightening, also called the photometric facular index PFI. But uncertainties in observational determinations of the area, contrast and center-to-limb functions of faculae are much larger than for sunspots. In lieu of a reliable facular brightness specification directly from solar imagery, use is made of full disk measurements of solar emissions whose variations predominantly arise from magnetic sources associated with photospheric faculae. The index of Ca K core emission relative to the nearby continuum (Livingston et al., 1988), an analogous Mg II core-to-wing ratio (de Toma et al., 1997), and the He I (1083 nm) equivalent width (Harvey & Livingston, 1992) all vary in response to enhanced emission from chromospheric plages that overlay photospheric faculae, and from the surrounding chromospheric network. We have constructed a composite facular brightening time series using the NOAA Mg index obtained from Solar Backscatter Ultraviolet data from 1978 to 1992, the Solar Stellar Intercomparison Experiment Mg II index thereafter, and the He EW from 1976 to 1978. The composite is placed on the scale of the NOAA Mg index and extended by linear relationships determined from the data in the period of overlap. Although chromospheric proxies track bolometric brightness changes relatively well over solar cycle and active region time scales (Foukal and Lean, 1988), differences in center-to-limb variations and filling factors of the sources of variability in the chromospheric proxies limit their ability to track photospheric faculae brightness changes on shorter times scales of days to months.

We first obtain an empirical representation of TSI variability from multiple regression of our composite TSI time series with parameterizations of sunspot blocking, as calculated above, and the Mg index proxy of facular brightening. This two-component empirical model accounts for 80.0% of the variance in the composite TSI and its components are shown in Fig. 6a and for the sunspots in c. Recognizing that the chromospheric brightness sources may relate somewhat differently to photospheric facular brightness sources over shorter (rotational) versus longer (solar cycle) time scales, we separate the Mg II index proxy into a smoothly varying longer term component and a shorter term component associated with rotational modulation, as shown in Fig. 6b and c. Multiple linear regression of the sunspot darkening, slowly varying Mg II index and short term Mg II index facular proxies yields an empirical model that now accounts for 82.9% of the TSI variability, and, if the shortterm variability and the TSI are smoothed, for 88.0%. Besides the fact that the three-component model yields quite different factors for the slow (130.2) and the fast (78.2) Mg II index, the factor for the PSI is much closer to one (0.9913) than in the two-component model, where it is 1.082, indicating that part of the faculae compensate the sunspots. As Fig. 6d indicates, the differences of the composite and the empirical model show only a small long term trend of the same order as the downward trend seen in the composite. This could be taken as confirmation that indeed the trend is due to unrecognized degradation. Although some non neglectible differences exist during high activity, the overall agreement is quite good, implying that this representation captures the predominant solar cycle variability sources.

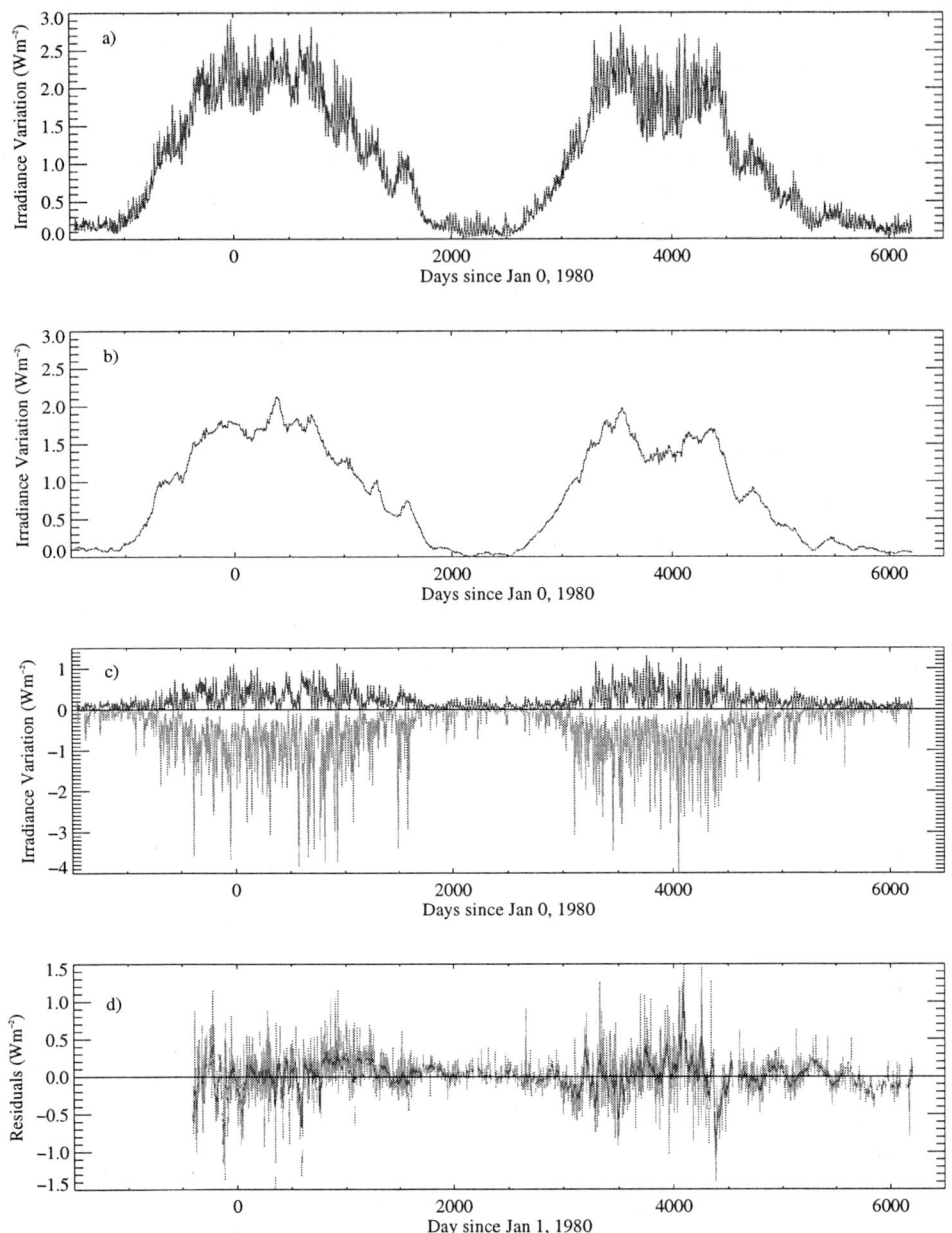

Figure 6. a) Facular brightening from Mg index, b) Longterm enhancement of the irradiance, c) Shortterm variability due to faculae and sunspots, and d) the residuals between the model and the composite TSI without and with smoothing the short term variation over 21 days. The parameters are plotted as a result of the multiple regression in Wm^{-2}: two-component model for a), and three-component for b)–d).

5. Conclusions

ACRIM II can be reliably traced back to ACRIM I only with both the HF and ERBE data sets and the recognition of instrumental changes in HF. This underlines the importance for having even more than two experiments simultaneously in space to measure TSI. With only one radiometer even the medium term variability would be questionable. The results from the comparison of these four data sets – ACRIM I, II, HF and ERBE – permit the construction of a reliable composite TSI. As demonstrated by the god agreement with a proxy model, this composite promises to be very valuable for the study of solar irradiance variability for time scales up to the solar cycle modulation. The assessment of the intercycle variability, however, is still hampered by the fact that undetected long term trends of the radiometers can not be conclusively excluded. However, some conclusion about the longterm variation of the Sun may still be drawn from the understanding of the difference between the behaviour of the two cycles which is now well documented by the composite TSI.

Acknowledgements

The authors thanks Dr. R.C.Willson for many helpful discussions for the interpretation of ACRIM data, Dr. R.D.Lee III for providing recent ERBE irradiance data and the VIRGO team for their data and the many contributions to their interpretation. VIRGO is an investigation on the Solar and Heliospheric Observatory, SOHO, which is a mission of international cooperation between ESA and NASA. Part of this work (CF) is supported by the Swiss National Science Foundation, which is gratefully acknowledged, as is a NASA/UARS Guest investigators grant (JL).

References

Anklin, M., Fröhlich, C., Finsterle, W., Wehrli, Ch., Dewitte, S., Crommelynck, D.: 1997, in Fleck, B.& Wilson, A., ed(s)., *31. ESLAB Symposium: Correlated Phenomena at the Sun, in the Heliosphere and in Geospace*, ESA SP-415, Noordwjick, NL., in press

Brandt, P. N., Stix, M., and Weinhardt, H.: 1994, *Solar Phys.* **152**, 119

Chapman, G. A., and Meyer A. D.: 1986, *Solar Phys.* **103**, 21

Chapman, G.A., Cookson, A.M., Dobias, J.J.: 1996, *J.Geophys.Res.* **101**, 1354

Crommelynck,D.A., Domingo, V., Fichot, A., Fröhlich, C., Penelle, B., Romero J., Wehrli, Ch.: 1993, *Metrologia* **30**, 372

Crommelynck, D., Fichot, A., Lee III, R.B., Romero, J.: 1995, *Adv. Space Res.* **16**, (8)17

de Toma, G., White, O. R., Knapp, B. G., Rottman, G. J., Woods, T.N.: 1997, *J. Geophys. Res.* **102**, 2597

Foukal, P.: 1981, in L. E. Cram, J. H. Thomas, ed(s)., *The Physics of Sunspots*, Sacramento Peak Observatory, New Mexico, 391

Foukal, P., Lean, J.: 1988, *Astro.Phys.J.* **328**, 347

Fröhlich, C.: 1997, in Pap, J., Fröhlich, C., & Ulrich, R., ed(s)., *Proceedings of the SOLERS22 Workshop, Sacramento Peak, June 1996*, Kluwer Academic Publ., Dordrecht, The Netherlands, in press

Fröhlich, C., Pap J. M., and Hudson, H. S.: 1994, *Solar. Phys.* **152**, 111

Fröhlich, C., Romero, J., Roth, H., Wehrli, C., Andersen, B.N., Appourchaux, T., Domingo, V., Telljohann, U., Berthomieu, B., Delache, P., Provost, J., Toutain, T., Crommelynck, D., Chevalier, A., Fichot, A., Däppen, W., Gough, D.O., Hoeksema, T., Jiménez, Gómez, M., Herreros, J., Roca-Cortés, T., Jones, A.R., Pap, J. and Willson, R.C.: 1995, *Solar Phys.* **162**, 101

Fröhlich, C., Andersen, B., Appourchaux, T., Berthomieu, G., Crommelynck, D.A., Domingo, V., Fichot, A., Finsterle, W., Gómez, M.F., Gough, D.O., Jiménez, A., Leifsen, T., Lombaerts, M., Pap, J.M., Provost, J., Roca Cortés, T., Romero, J., Roth, H., Sekii, T., Telljohann, U., Toutain, T., Wehrli, C.: 1997a, *Solar Phys.* **170**, 1

Fröhlich, C., Crommelynck, D., Wehrli, C., Anklin, M., Dewitte, S., Fichot, A., Finsterle, W., Jiménez, A., Chevalier, A., Roth, H.J.: 1997b, *Solar Phys.* **176**, in press

Hoyt, D.V., Kyle, H.L., Hickey, J.R., Maschhoff, R.H.: 1992, *J.Geoph.Res.* **97**, 51

Harvey, J. W., and Livingston, W. C.: 1994, in D. M.Rabin, J. T. Jefferies and C. Lindsey, ed(s)., *International Astronomical Union Symposium 154: Infrared Solar Physics*, Kluwer Academic Publ., Dordrecht, The Netherlands, 59

Hudson, H.S., Silva, S., Woodard, M., and Willson, R.C.: 1982, *Solar Phys.* **76**, 211

Lee, III., R. B., Barkstrom, B. R. , Cess, R. D.: 1987, *Appl. Optics* **26**, 3090

Lee, III., R. B., Gibson, M. A., Wilson, R. S., Thomas, S.: 1995, *J. Geoph.Res.* **100**, 1667

Livingston, W.C., L. Wallace, and O.R. White: 1988, *Science* **240**, 1765

Romero, J., Wehrli, C., Fröhlich, C.: 1994, *Solar Phys.* **152**, 23

Willson, R.C.: 1994, in J.M. Pap, C. Fröhlich, H.S. Hudson & S.K. Solanki, ed(s)., *The Sun as a Variable Star: Solar and Stellar Irradiance Variations*, Cambridge Univ. Press, 54

Willson, R.C.: 1997, *Science* **277**, 1963

Willson, R.C., Gulkis,S.,Janssen,M., Hudson, H.S., Chapman.G.A.: 1981, *Science* **211**, 700

Willson, R.C., Hudson, H.S., Fröhlich, C., Brusa, R.W.: 1986, *Science* **234**, 1114

Willson, R.C. Hudson, H.S.: 1991, *Nature* **351**, 42

SOLAR IRRADIANCE VARIATIONS: THEORY

H.C. SPRUIT

Max-Planck-Institut für Astrophysik
Postfach 1523
D-85740 Garching, Germany

1. Introduction

The following is a somewhat condensed version of discussions previously given elsewhere (Spruit, 1991, 1992). Some new developments not discussed there are presented in sections 4 and 5.

Since the observed irradiance variations are so clearly associated with manifestations of the solar magnetic field, I focus here on magnetic causes. Much of the physics of irradiance variations, however, is governed by the thermal response of the convective envelope and this response is similar for other possible causes of irradiance variations.

Conceptually, one can separate the thermal effects of magnetic fields into two or three types:

1. 'Sources and sinks'. The generation of a magnetic field involves the conversion of energy of motion into magnetic energy. Since the motions in the solar envelope are thermally driven, this ultimately means conversion of thermal into magnetic energy: building up a magnetic field produces a thermal sink somewhere. The opposite happens when the field decays: magnetic energy is converted into heat. These thermal effects exist only during *changes* in the magnetic energy content of the envelope.

2. Changes in thermal transport coefficients. Magnetic fields interfere with convection, causing a reduction in the efficiency of heat transport in the envelope. In contrast to (1), these changes last as long as the magnetic field itself is present.

Both these sources and sinks and changes in the transport coefficient cause thermal perturbations, varying with magnetic activity, which propagate through the envelope and cause variations in surface energy flux. Related to the second class of perturbations are:

3. The effects of magnetic fields at the surface of the star. Sunspots, being dark, radiate less than the surrounding photosphere, while the small elements that make up plages and the network have an excess emission. In addition, it is conceivable that the magnetic elements have an indirect effect by modifying the convective flow in their surroundings slightly (for which there is some observational evidence, see section 4).

2. Time scales

The response of the Sun to thermal perturbations is not governed by a single time scale, but by a wide range of time scales. The longest of these is the thermal time scale of the Sun as a whole, called the Kelvin-Helmholtz time scale, about 10^7 years. If the central heat source of the Sun were switched off, the internal structure and the luminosity would start to change on this time scale. More generally, we can define the thermal time τ_t scale as a function of depth:

$$\tau_t \equiv U(z)/L(z) \approx \frac{1}{L}\int_{R-z}^{R} 4\pi r^2 u\, dr, \tag{1}$$

where L is the luminosity at depth z, U the thermal energy of the envelope down to a depth z, and u the thermal energy per unit volume, approximately (for an ideal gas of constant γ) given by $u = P/(\gamma - 1)$. This is the time scale on which the structure of the envelope, and the observed luminosity, would start changing when the heat flux in the star were interrupted, by some magical means, at depth z. Some rough values for this quantity are $\tau_t \sim 10^5$yr at $z = 2\,10^5$km (depth of the convection zone), 10 yr at 20 000km (the size of a supergranule), 10 hrs at 2000km (size of a granule). This shows that the thermal time scale is a very sensitive function of depth in the Sun. As a result, the thermal response of the Sun also depends critically on the location of the disturbance. A fairly good approximation to the stratification of the convective envelope of the Sun is a polytrope of index $n = 2$ (this is better than the standard 'convective' value n=1.5 because $\gamma < 5/3$ due to partial ionization). Hence the gas pressure and thermal time scale vary roughly as $P \sim z^3$, $\tau_t \sim z^4$, where the depth z is counted from a level 3 scale heights above the surface $\tau = 1$.

A second kind of time scale involved in thermal readjustments is the *diffusive* time scale. In the mixing length approximation, transport processes in the convection zone can be computed with a turbulent diffusion coefficient $\kappa_t \approx \frac{1}{3}l_c v_c$ where l_c and v_c are the convective length scale and velocity. In the solar convective envelope, this quantity varies only weakly, at a value of the order 10^{13} cm^2/s. Thermal inhomogeneities (more precisely: entropy inhomogeneities) of length scale d are smoothed by turbulent diffusion on a time scale

$$\tau_d = d^2/\kappa_t. \tag{2}$$

For $d = 2\,10^5$km this is about 1 yr, for $d = 2000$km about 1 hr. Comparing τ_t and τ_d, we see that they are of similar magnitude close to the surface (to be precise: in the surface boundary layer where convection is not efficient enough to keep the stratification close to adiabatic). In deeper layers, the thermal time scale is much *longer* than the diffusive time scale, by a factor of up to 10^5.

The two time scales measure different types of thermal adjustment process. These same processes appear in the thermal behavior of, say, a chunk of metal, aluminum say, heated from the inside and suspended in space. The thermal time scale is the time scale on which its temperature adjusts to a change in the heat input, such that the heat radiated from the surface into space balances the heat input again. It is determined by the heat capacity [U in eq. (1)] and the power level (L). The time scale on which different parts of the chunk equilibrate to the same temperature is governed by a different process, namely the thermal conduction (the equivalent of the turbulent diffusion in the Sun). The diffusion (conduction) time scale is much shorter than the thermal time scale, because of the large heat conductivity of Al. In the Sun, it is the very large turbulent diffusivity in the bulk of the convection zone that causes the very short diffusive time scale compared with the thermal time scale.

How do these different time scales come into play when the convection zone is thermally perturbed by, say, the storage of energy in a growing magnetic field? Such perturbations can be computed in detail, either by numerical methods (Endal et al.1985, Gilliland, 1988) or more analytically. We can, for example, consider the initial value problem in which a perturbation is allowed to evolve in time by thermal transport in the convection zone. In general this evolution has components on all the time scales of the problem, including the very long thermal time scale. Formal aspects of this problem have been discussed elsewhere (Spruit, 1982ab, 1991, Arendt 1992). In the following, the basic conclusions of these analyses are summarized.

3. Expected level of luminosity variations

For quantitative estimates, the strength of the field and its filling factor in the convection zone have to be specified. Assume that we have a layer of field with strength of the order of 10 000G (equipartition with the convective flows as estimated by a mixing length model), one scale height deep, near the base of the convection zone (where most of the magnetic flux is probably located). I summarize here some results, dicsussed in greater detail elsewhere (Spruit 1991).

3.1. SOURCES-AND-SINKS

If the energy needed to build up this field during one half of the solar cycle is taken out of the thermal energy near the base of the convection zone, the calculations show that a surface luminosity variation of only 10^{-7} results. This is due to the very large heat capacity of the lower convection zone. The effect is stronger if the source of the magnetic field is assumed to be closer to the surface, but is still much smaller than the observed effect. Recent models for the emergence of magnetic flux from the base of the convection zone (D'Silva and Choudhuri 1993, D'Silva and Howard 1993, Caligari et al. 1995) indicate that the actual field strength at the base of the convection zone is probably about 10 times higher than the equipartition estimate. The magnetic energy per unit of magnetic flux is then also 10 times higher, but this still does not lead to a significant luminosity effect.

3.2. SHADOWS

If magnetic fields interfere with convection, a thermal perturbation develops as well. This effect depends crucially on the 'covering factor'. If a reduction of convective efficiency is assumed that uniformly covers a horizontal surface at some depth z, magnetic fields comparable to equipartition with convective flows can have stronger effects than the source-and-sink perturbations (Gilliland, 1988), though a measurable effect is predicted only if the field is located close below the surface. The effect is reduced, however, if there are 'holes' in this cover. This is because the turbulent heat conductivity in the convection zone is so high that the heat flux is easily 'shunted' past blocking objects below the surface (Spruit, 1977).

3.3. SURFACE EFFECTS

By far the most effective way in which a magnetic field influences the irradiance is by its effect on the *net surface emissivity*. The reduced emission from a sunspot area shows up directly in the irradiance records as a dip tracking the passage of the area across the disk. The only complication is that one might expect that part of this reduction could be compensated by a brightening elsewhere, for example in the form of a 'bright ring' surrounding the spot. Evidence of such bright rings is absent for most spots that have been studied for this effect. Any brightening in the surroundings can usually be attributed to facular emission, which is present at the same levels in active regions with or without sunspots. The absence of such bright rings is understood in terms of a turbulent diffusion model for the heat flux in the convection zone (Spruit, 1977, 1982b, Foukal et al. 1983). The 'blocked heat flux', for the most part, does not reappear elsewhere on the surface, but stays inside the convection zone, being stored/released on the very long thermal time scale of the convective envelope. This conclusion holds, in the diffusion model, as long as the blocking effect of the spot extends to a depth of at least 1000 km, a mild requirement given that the observed Wilson depression of the umbra of a spot is already of the order of 500km.

Changes in heat flux through modified granulation will affect the convective envelope in the same way as the excess emission from the small scale magnetic field, since it is also an effective change of the emissivity of the solar surface. In particular, one does not expect these changes to be 'compensated' by opposite changes elsewhere on the surface (except, as before, on the 10^5yr thermal time scale of the envelope).

4. 'Magnetically modified' granulation

In 3.3 above, I have focused on the direct effects of magnetic fields on the net emissivity of the solar surface. More indirect effects may play a role as well. The shape of granules appears to be different in magnetic regions (Macris and Roesch 1983, Muller and Roudier 1984, Muller 1986, Title et al. 1992). They are smaller, more irregular, and the measured flow speeds are lower. These effects are present *in addition* to the magnetic elements themselves, which change the appearance of granulation by filling in the intergranular lanes. The changes may be the result of the geometric constraints the magnetic tubes put on the convective flow outside them.

Since the flow appears to be different, it would seem possible that the heat flux it carries is also different. This might contribute to the observed solar cycle variations of irradiance (Muller 1986, Kuhn and Libbrecht 1988). Limits on this contribution can be put by observations of the *colors* of the solar cycle variation signal. Solanki (1997) shows that the variation as seen in the near UV follows the behavior expected from the small magnetic elements themselves rather than the weak color signal expected from a small variation in the surface temperature over a larger area, but he does not quote quantitative limits. In order to contribute in the right sense to the solar cycle variation, the constraints imposed on the flow by the magnetic flux elements would have to lead to an *increase* in the heat flux carried by granulation. Direct measurements of a temperature change in granulation in magnetic regions are probably difficult, since the effect would be small and hard to separate from the enhanced emission from the magnetic elements. Numerical simulations (Nordlund, private communication) indicate, however, that an effect of the right sign may be present.

5. Improving on the turbulent diffusion picture

In the results quoted a diffusion model for convection was used. It assumes that convection can be modeled by a turbulent viscosity for momentum and a turbulent diffusivity for the transport of heat. While this was a simple and somewhat justifiable model of convection in the absence of detailed knowledge of the convective flows in a stellar envelope, we now know that it does not represent stellar convective envelopes very well. Numerical simulations (Nordlund 1982, 1985ab, 1986, 1990, Nordlund and Dravins 1990, Nordlund and Stein 1990, 1991, 1996, Stein and Nordlund 1989, 1991, Steffen et al. 1989, Steffen 1993) show an extremely *nonlocal* picture. They show that convective flows are driven almost exclusively by cooling at the surface, with narrow fast moving downdrafts between slow almost isentropic upflows. For detailed descriptions of these results, see the references given (cf also Spruit 1997). In the present context, the most important property is that the convective flow at all depths is driven by cooling at the surface rather than by a local overturning process. The material in the cool downdrafts survives to large depths below the surface with little mixing into the upflows.

5.1. SPOT BLOCKING

Given this extremely nonlocal picture, it is appropriate to ask how the sunspot blocking problem can be explained without appealing to a turbulent diffusion model.

Below the spot (modeled as a region of reduced heat transport efficiency extending to some depth below the surface) the upflows have exactly the same temperature as upflows in the unspotted surroundings at the same level, namely that given by the entropy at the base of the upflows. In this sense, there is no 'pile up of heat below the spot'. Because of the reduced heat loss at the surface, however, the downflows below the spot will be less vigorous. The unspotted surface notices nothing of the spot's presence (except for an extremely narrow ring where lateral radiative exchange takes place, and except for the presence of a moat flow, see below). It continues to cool upwellings into downdrafts as before, since the entropy in the upflows has not changed. Thus, we expect again that bright rings will be absent around spots, but the reason is even simpler than in the diffusion model. The spot is a region at the surface where less heat is radiated away, and this is now independent of the depth of the spot below the surface (in contrast to the diffusion model, where the spot has to extend to a minimum depth of 1000 km for the explanation to work).

Still, one may wonder what happens to the amount of heat generated in the solar interior that now fails to be emitted at the surface. This part of the problem is the same as in the diffusion model: the imbalance causes a secular increase of the entropy in the entire convection zone until a new thermal equilibrium is reached. Because of the very long thermal time scale of the convection zone (10^5 yr), the effect is negligible on observable time scales. In a steady state, when the average number of spots does not change, the convection zone does not heat up, because its mean temperature is higher than it would be without spots (Spruit and Weiss, 1986). Episodes of larger than average spot coverage cause heating, those of less than average spot coverage cause cooling on this time scale.

5.2. MOAT CIRCULATION

In the diffusion model, the horizontal flow away from a spot observed as the 'moat' has a simple explanation, since the convection zone is hotter than average below the spot in this model, causing an upflow by thermal buoyancy. In our new view of the convection zone, the explanation does not work any more in this simple form, since there is no upward buoyancy anywhere. A circulation still results, however, due to the buoyancy associated with a reduced density below the spot. Downflows are cooler and denser than the surrounding upflows, so that the *mean density* on a horizontal surface is greater than the density in the upflowing fluid. Below the spot this effect is less strong than outside the spot, since less cooling has taken place at the surface. This difference in mean buoyancy of the fluid below the spotted and unspotted areas drives a circulation as before. The difference in explanation may sound a bit pedantic, since one could also have said that the mean temperature below the spot is higher than outside because of the weaker downflows. The new explanation is more satisfactory since it appeals more directly to the physical cause of the circulation, namely an imbalance in the fluid density. More importantly, this explanation shows that the moat circulation *does not carry extra heat* to the surface: on a given horizontal level below the surface the mean temperature is higher in the circulation, but this increase is entirely caused by fluid which is moving downward, and therefore not visible at the surface, rather than upward as it is in the diffusion picture! This difference again is due to the extremely nonlocal nature of convection in stellar envelopes.

5.3. THERMAL SHADOWS

The thermal shodow expected from a blocking object below the surface (Spruit 1977, Goode and Kuhn, 1990), which is a small effect already in the turbulent diffusion model, is likely to be even smaller in the new picture of stellar envelope convection. Since all upward flowing gas has nearly the same entropy, the only effect of a blocking object is to temporarily halt the downward motion of gas cooled at the surface. As in the case of the moat circulation discussed above, this has no consequence for the surface temperature above the blocking object, since only a small fraction of the downward moving gas is carried back up to the surface. As before, this is a consequence of the non-local nature of the convective flow.

6. Acknowledgement

It is a pleasure to thank Alan Title for a discussion on the effects of surface magnetic fields on granulation, which led to the inclusion of section 4.

References

Arendt, S., 1992, *Astrophys. J* **389**, 421
Caligari, P., Moreno-Insertis, F., Schuessler, M., 1995, *Astrophys., J.* **441**, 886
Chiang, W. H. & Foukal, P. V., 1984, *Solar Phys.* **97**, 9
D'Silva, S. & Choudhuri, A. R., 1993 *Astron. Astrophys.* **272**, 621
D'Silva & Howard, R. A. 1993, *Solar Phys.* **148**, 1
Endal, A. S., Sofia, S. & Twigg L. W., 1985, *Astrophys. J.* **290**, 748
Foukal, P. V., Fowler, P., & Livshits, M., 1983, *Astrophys. J.* **267**, 863

Gilliland, R. L., 1988, in *Solar radiative output variation*, P. Foukal, ed., Cambridge Research & Instrumentation Inc., 21 Erie st. Cambridge, MA 02139, p239
Goode, P.R., Kuhn, J.R., 1990, *Astrophys. J.* **356**, 310
Kuhn, J.R., Libbrecht, K.G., Dicke, R.H., 1988, *Science* **242**, 908
Muller, R., 1986 *Solar Phys.*, **119**, 229
Muller, R., Roudier, T., 1984, in *The Hydromagnetics of the Sun*, ESA SP xx, p. 51
Macris, C.J., and Roesch, J., 1983, *Comptes Rendus, ser II*, **296**, 265
Nordlund, Å., 1982, *Astron. Astrophys.* **107**, 1
Nordlund, Å., 1985a, in *Small scale dynamical processes in quiet stallar atmospheres* ed. W. Keil, Sacramento Peak Observatory, Sunspot, NM 88349, USA
Nordlund, Å., 1985b, in *Progress in stellar spectral line formation theory*, eds. J.E. Beckman and L. Crivellari (NATO ASI series **152**), Reidel, Dordrecht, p. 215
Nordlund, Å., 1986, *Solar Phys.*, **100**, 209
Nordlund, Å., 1991, in *Stellar atmopheres: Beyond classical models*, eds. L. Crivellari, I. Hubeny and D.G. Hummer (NATO ASI series **341**), Kluwer, Dordrecht, p. 61
Nordlund, Å. & Dravins, D., 1990, *Astron. Astrophys.* **228**, 155
Nordlund, Å. & Stein, R.F., 1990, *Comp. Phys. Comm.* **59**, 119
Nordlund, Å. & Stein, R.F., 1991, in *Stellar atmospheres: beyond classical models*, eds. L. Crivellari, I. Hubeny and D.G. Hummer (NATO ASI series **341**), Kluwer, Dordrecht, p. 263
Nordlund, Å. & Stein, R.F., 1996, in *Proceedings of the 32nd Liège Int. Astrophys. Colloquium 'Stellar Evolution: What should be done'*, eds. A. Noels et al. p.75
Solanki, S., 1997, submitted to *Astron. Astrophys.*
Spruit, H.C., 1977, *Solar Phys.* **55**, 3
Spruit, H.C. 1982a, *Astron. Astrophys.* **108**, 348
Spruit, H.C. 1982b, *Astron. Astrophys.* **108**, 356
Spruit, H.C. 1991, in *The Sun in Time*, eds. C. Sonett, M. Giampapa & M.S. Matthews, University of Arizona Press, Tucson, p118
Spruit, H.C., 1992, in *Sunspots: Theory and Observations*, eds. J.H. Thomas and N.O. Weiss, Cambridge: CUP, p.163
Spruit, H.C., 1997, *Mem. Soc. Astron. It.*, in press
Spruit, H.C. & Weiss, A., 1986, *Astron. Astrophys.* **166**, 167
Steffen, M., Ludwig, H.-G., Krüss, A.: 1989 *Astron. Astrophys.* **213**, 317
Steffen, M.: 1993, in *Inside the stars* (IAU Coll 137), eds. W. Weiss & A. Baglin, Astron. Soc. Pac. Conference series **40**, p300
Stein, R.F. & Nordlund, Å, 1989, *Astrophys. J.* **342**, L95
Stein, R.F. & Nordlund, Å, 1991, in *Challenges to Theories of the Structure of Moderate-Mass Stars*, eds. D.O. Gough and J. Toomre, Lecture Notes in Physics **388**, Springer, Berlin, p195
Title, A.M., Topka K.P., Tarbell T.D., Schmidt, W., Balke C., Scharmer G., 1992, *Astrophys. J.* **393**, 782

OBSERVATIONS OF THE LATITUDINAL VARIATION OF THE SOLAR RADIANCE OF NON ACTIVE REGIONS OF THE SUN

V. DOMINGO AND L. SANCHEZ
*ESA at NASA/GSFC, code 682, Greenbelt, MD 20771, USA
(vdomingo@esa.nascom.nasa.gov, lsanchez@esa.nascom.nasa.gov)*

AND

T. APPOURCHAUX
*Space Science Department of ESA, ESTEC, Postbus 299,
AG 2200 Noordwijk, The Netherlands (thierrya@so.estec.esa.nl)*

Kuhn et al. (1988) have found that there are variations in the photospheric temperature with the solar cycle that depend on solar latitude. This would be an independent mechanism, other than the effects of sunspots and faculae, contributing to the change in solar irradiance (Kuhn 1991). It is important to pursue this investigation as such variations would be related to the transport of energy through the convection zone, and thus give a good indication of its structure and evolution.

Aboard SOHO, the VIRGO instrument (Fröhlich et al. 1995), measures the total solar irradiance and its variations, and also, with the Luminosity Oscillations Imager (LOI) (Appourchaux et al., 1997), measures the radiance of the Sun at 500 nm in an image of the Sun in 16 pixels, the outer narrow four being used, with a servo system, to achieve fine pointing and the other 12 distributed in four latitudinal bands. This instrument gives us the opportunity to try to find a photometric equivalent of the temperature measurements by observing the variation of the radiance at four different latitudinal bands. SOHO, by being operated at the transition between two solar cycles, may provide a particularly good occasion to measure changes in the structure of the energy flow in the convection zone.

To select observations of the Sun with "no active regions" during the period April 1996-August 1997 we have chosen intervals of time when no region of the solar disk had a magnetic field above the arbitrary level of 500 microtesla, in the Wilcox Observatory synoptic charts (Solar Geophysical Data, 1996 and 1997, and University of Stanford web site http://solar.stanford.edu/SolarData/). Figure 1 shows the result of comparing the normalized intensities of the pixels that observe the Sun above about 30 degrees heliolatitude (called "polar" thereafter) and those that observe the latitudes below 30 degrees("equatorial"). On December 1996 one observes (Fröhlich 1997) a step increase of the solar irradiance of about 1.3×10^{-4}. If the irradiance change is due to change in the equatorial band, as could be expected following the temperature changes reported by Kuhn et al. (1988) for the previous solar cycle, the

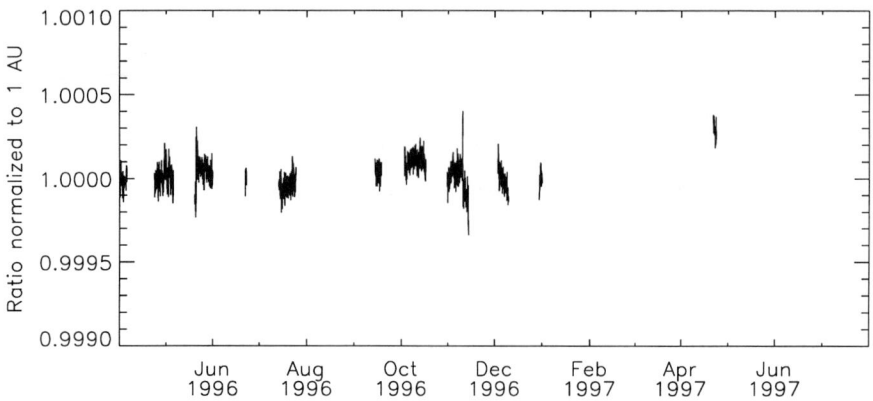

Figure 1. Ratio of the solar radiance measured in two bands, one (polar), comprising the heliolatitude above 30 degrees north and south and the other (equatorial) combining the regions below 30 degrees. The ratio has been normalized to 1 AU. Periods of "quiet" sun have been selected

ratio of polar to equatorial radiance would change about twice as much or about 3×10^{-4}. We have only two "quiet" intervals after the solar irradiance change. The observation during the first one, on 30-31 December 1996, would indicate no change or a change smaller than 10^{-4}, thus suggesting that the change in solar irradiance seen at the start of the 23rd solar cycle is not due to a latitudinal change of the temperature. On 22-23 April 1997, the ratio "polar"/"equatorial" radiance is higher than in 1996, but we consider, pending further analysis of higher resolution data, that this may not be a "quiet" sun period: the sunspot number is higher in this interval than in the others.

In conclusion, at first glance this result would contradict what we would expect by extrapolating the observations of Kuhn et al. (1988) to the present solar cycle. But it is important to realize that this is a very preliminary analysis of the VIRGO data and more detailed analysis taking into account the effect of solar active regions, including higher resolution data from the Michelson Doppler Imager (MDI), is underway.

References

Appourchaux, T., Andersen, B. N., Fröhlich, C., Jimenez, A., Telljohann, U., Wehrli, C.: 1997, In-flight Performance of the Luminosity Oscillation Imager of VIRGO Aboard SOHO, *Solar Phys.* **170**, 27

Fröhlich, C. et al.: 1995, Virgo: Experiment for Helioseismology and Solar Irradiance Monitoring, *Solar Phys.* **162**, 101

Fröhlich, C.: 1997, Solar Irradiance Variations, *This issue.*

Kuhn, J. R., Libbrecht, K. G., Dicke, R. H.: 1988, The Surface Temperature of the Sun and Changes in the Solar Constant, *Science* **242**, 908-910

Kuhn, J. R.: 1991, Inferring Solar Structure Variations from Photometric and Helioseismic Observations, *Adv. Space Res.* **11**, No. 4, pp. 171-184

AMPLITUDE MODULATION OF RADIAL P-MODES FROM VIRGO

T. LEIFSEN
Institute of Theoretical Astrophysics, PO Box 1029, Blindern, N-0315 Oslo, Norway

B.N. ANDERSEN
Norwegian Space Centre, PO Box 113 Skøyen, N-0212 Oslo, Norway

AND

THE VIRGO TEAM

Abstract. The radial order solar p-modes show an amplitude modulation at the equatorial siderial rotation frequency of the Sun. The peak is clearly separated from the peak at the synodic rotation visible in the irradiance data.

1. Data and analysis

In this study a continuous dataset of solar irradiance sampled every 60 seconds with the VIRGO SPM blue channel on the SOHO spacecraft for the time period 29-01-96 to 23-07-97 has been used to study the time and frequency variation of the solar $l=0$ p-modes. Also the Stanford index of solar global magnetic field from 29-01-96 to 30-06-97 was used. Wavelet analysis with a modified Morlét wavelet as basis function was used to study the time variation of the amplitudes of the $l=0$ p-modes with radial orders of 12-32. The different modes have been studied using a variety of combination of temporal and frequency resolution. The amplitude modulation was studied using temporal resolution of about half a day. The $l=0$ data have been filtered in the frequency domain to remove the effects of the nearby $l=2$ modes. Supercomputing time was provided by The Research Council of Norway.

2. Results and Conclusions

The power spectra of the SPM data and magnetic index values show a clear peak at the solar synodic rotation frequency. The power spectrum of the radial order averaged (n=14-26) $l=0$ mode amplitudes show a clear peak at the solar siderial rotation frequency. Correcting for the siderial/synodic effect in the sampling of the mode amplitudes makes the rotation peak coincident with the peaks of the SPM data and the magnetic index. In addition the power around 1 μHz is reduced due to removal of the phasedrift caused by the orbital motion of the Earth. The autocorrelation function of the SPM data has twice the width of the magnetic index and the mode amplitudes. This is due to the opposite center to limb variation of sunspots and faculae. The

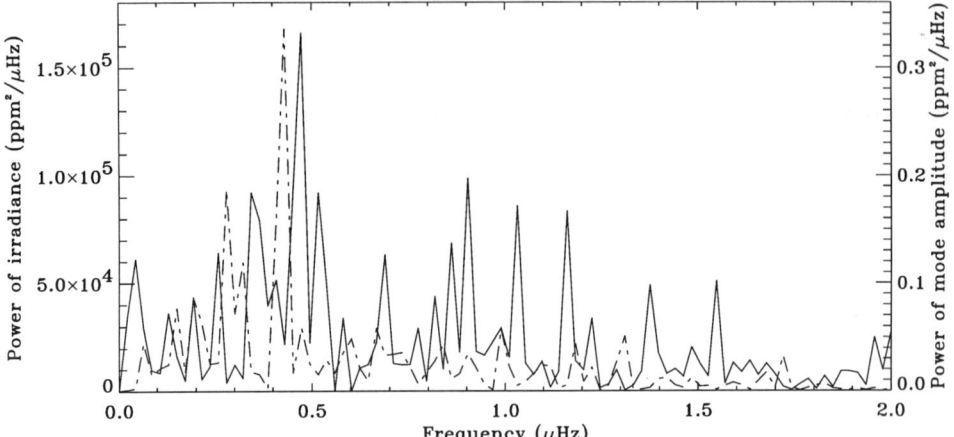

Figure 1. The power spectrum of the amplitude (averaged over $n=$ 16-26) of $l=0$ (solid line) and the power spectrum of the intensity in the SPM blue channel (dashed line).

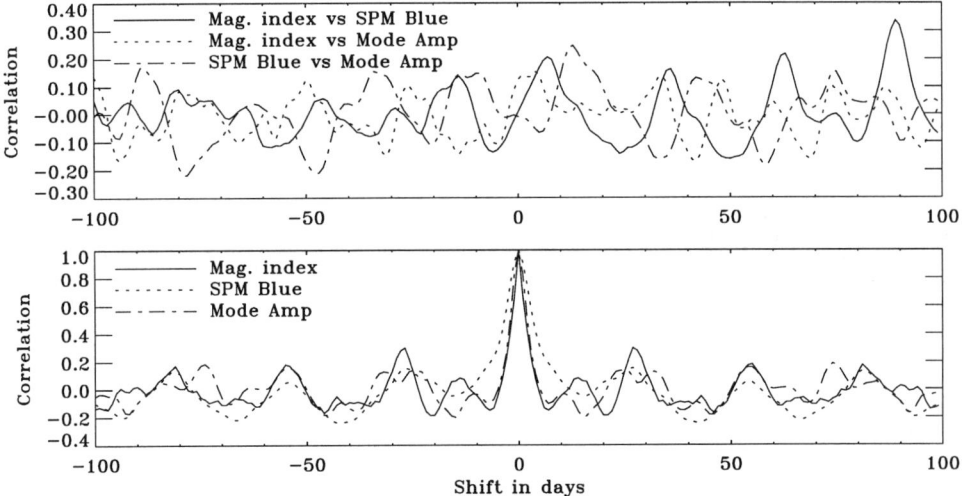

Figure 2. The panels show the cross- (upper) and autocorrelations of the timestrings as given in the legends. Mode amplitude timestrings are resampled synodically.

synodically sampled mode amplitudes averaged for $n=14$-26 are nearly in phase with the magnetic index, while there is a weak anticorrelation with the SPM irradiance. The temporal and frequency behaviour of the $l=0$ modes are very structured with accumulated pulses down to the time resolution of half a day.

The mode amplitudes of the $l=0$ p-modes are strongly modulated by the siderial rotation of the Sun. This modulation is not associated with the meridian passing of the active regions/magnetic fields. This is clearly seen in the direct power spectra where the peaks are separated by a frequency difference corresponding to the siderial/synodic effect. However, the amplitude of the rotational modulation is clearly dependent on the amount of solar activity on the visible hemisphere.

AN IMPROVED CALIBRATION OF THE MIXING-LENGTH BASED ON SIMULATIONS OF SOLAR-TYPE CONVECTION

H.-G. LUDWIG
*Astronomical Observatory, Niels Bohr Institute,
DK-2100 Copenhagen, Denmark, [hgl@astro.ku.dk]*

B. FREYTAG
*Institut für Astronomie und Astrophysik der Universität Kiel,
D-24098 Kiel, F.R.G., [Freytag@astrophysik.uni-kiel.de]*

AND

M. STEFFEN
*Astrophysikalisches Institut Potsdam, D-14473 Potsdam, F.R.G.,
[MSteffen@aip.de]*

Based on detailed 2D numerical radiation hydrodynamics (RHD) calculations of time-dependent compressible convection, we have studied the dynamics and thermal structure of the convective surface layers of stars in the range of effective temperatures and gravities between $4500\,\mathrm{K} \leq T_\mathrm{eff} \leq 7100\,\mathrm{K}$ and $2.54 \leq \log g \leq 4.74$. Although our hydrodynamical models describe only the shallow, strongly superadiabatic layers at the top of the convective stellar envelope, they provide information about the value of the entropy s^* of the deeper, adiabatically stratified regions. E.g. in the solar case the helioseismically measured entropy jump is predicted within 9% of its actual value.

Despite the complex interplay of hydrodynamics and radiative transfer we find a rather simple functional dependence $s^*(T_\mathrm{eff}, \log g)$ across the HR diagram. s^* can be translated into an effective mixing-length parameter α_MLT suitable for constructing standard stellar structure models by matching s^* in envelope models basing on mixing-length theory (MLT) (see Fig. 1). The α_MLT's are derived adopting the formulation of MLT by Böhm-Vitense (Zs. Ap. 46, p.108, 1958). The atmospheric $T(\tau)$ relation in the envelope models was chosen to mimic closely the average structure of the RHD atmospheres in order to eliminate its influence on the calibration of α_MLT. We find a moderate, nevertheless significant variation of α_MLT — primarily with effective temperature — over the range studied. A similar calibration of the convection theory by Canuto & Mazzitelli (Ap.J. 370, p.295, 1991) extended by including a variable amount of overshoot does not lead to a smaller variation of the controlling parameter α_CM (Fig. 2). The presented calibrations were derived for solar metallicity and helium content. Preliminary results of RHD models for lower metallicity show a complex dependence of α_MLT on T_eff and $\log g$. The MLT calibration is improved with respect to an earlier version (cf. Ludwig/Freytag/Steffen, in: *Solar Convection and Oscillations*, eds. F.P. Pijpers, J. Christensen-Dalsgaard, & C. Rosenthal, in press).

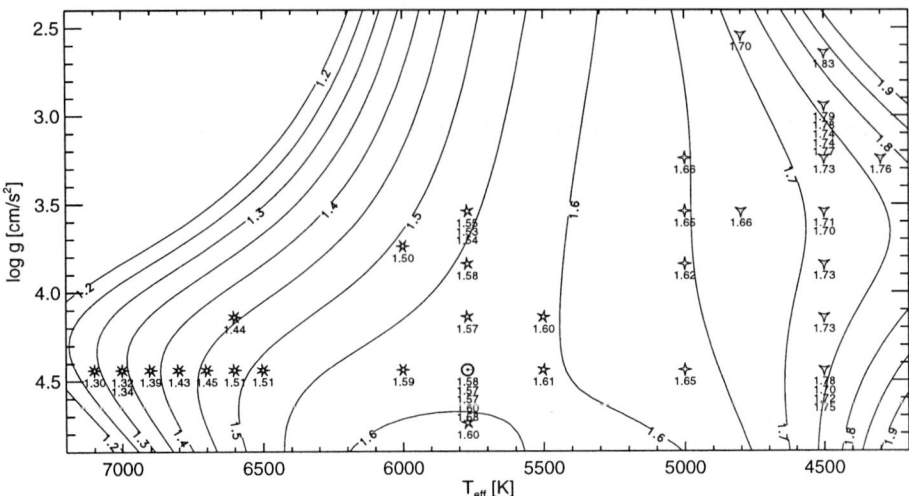

Figure 1. $\alpha_{\rm MLT}$ for standard mixing-length theory with $l_{\rm mix} = \alpha_{\rm MLT} H_{\rm p}$ ($l_{\rm mix}$: mixing-length, $H_{\rm p}$: local pressure scale height) derived from radiation-hydrodynamics (RHD) simulations as a function of effective temperature and gravitational acceleration. Symbols indicate RHD models. Attached to the symbols the actual data values are given, the contour lines present a polynomial fit to them. For some parameters several RHD models were computed. They differ in numerical details to provide an estimate of the internal uncertainty of the calibration.

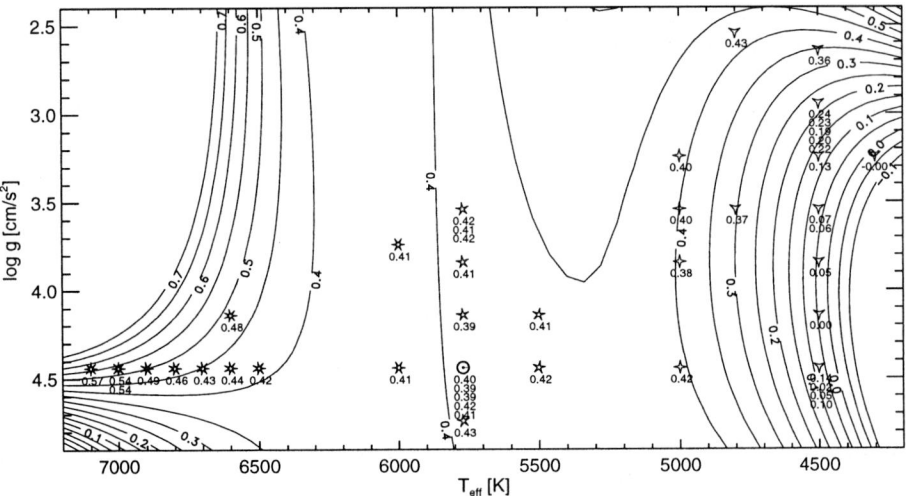

Figure 2. $\alpha_{\rm CM}$ for the Canuto & Mazzitelli convection theory with $l_{\rm mix} = \Lambda + \alpha_{\rm CM} H_{\rm p,top}$ (Λ: distance to the upper (Schwarzschild) boundary of the convective zone, $H_{\rm p,top}$: pressure scale height at the upper boundary). The $\alpha_{\rm CM}$-calibration reproduces the same underlying s^*-distibution as the $\alpha_{\rm MLT}$-calibration shown in Fig. 1.

NONLINEAR INVERSION FOR THE HYDROSTATIC STRUCTURE OF THE SOLAR INTERIOR

K.I. MARCHENKOV, I.W. ROXBURGH AND S.V. VORONTSOV
Astronomy Unit, Queen Mary College, University of London
Mile End Road, London E1 4NS, UK

The technique which we use is based on the "quasi-asymptotic" description of solar p-modes developed by Roxburgh & Vorontsov (1996). As a significant improvement of the standard asymptotic analysis, this description incorporates a Born approximation to allow for a detailed treatment of the regions of rapid variation of seismic parameters with depth (base of the convection zone) and strong influence of gravity perturbations (low-degree modes which penetrate deep into the solar core).

Essentially the eigenfrequency equation used is approximated by a second-order inhomogeneous Airy equation with a frequency dependent "acoustic potential", which is solved by a Born-type perturbation analysis. Matching the solutions in the interior with "exact", non-asymptotic solutions in the surface layers, where asymptotic analysis becomes locally invalid, leads to the eigenfrequency equation

$$\int_{r_1}^{R} s\, dr \simeq \frac{\pi}{\omega}\left(n + \frac{5}{72}\frac{1}{\pi^2 n} + \alpha^{int} + \alpha^{out}\right), \quad s^2 = \frac{1}{c^2} - \frac{\tilde{w}^2}{r^2}, \quad \tilde{w} = \frac{\ell + 1/2}{\omega}. \quad (1)$$

The outer layers contribute to the eigenfrequency equation with a surface phase shift α^{out}, similar to that of the standard asymptotic analysis. α^{int} is the frequency and degree dependent internal phase shift, given in the first order by an integral over the interior of the acoustic potential weighted with Airy function squared, r_1 denotes the position of the turning point $s^2(r_1) = 0$, c^2 is squared sound speed, ω is frequency.

To develop an inversion technique based on this analysis we proceed as follows. First of all, we infer the leading-order approximation to $F(\tilde{w}) = \int_{r_1}^{R} s\, dr$ by putting $\alpha^{int} = 0$. In this stage we separate the input data $(\pi n/\omega)$ in the right-hand side of equation (1) into three parts: a) the leading-order asymptotic term $F(\tilde{w})$, b) the frequency dependent surface phase shift for low-degree modes $\alpha_0^{out}(\omega)$, c) the first-order correction to the surface phase shift for intermediate and high degree modes $\tilde{w}^2 \alpha_2^{out}(\omega)$. Secondly, we invert $F(\tilde{w})$ using standard Abel integral transform to get first approximation for the sound-speed profile. After that we use the equations of hydrostatic support to determine the Brunt-Vaisala frequency profile and calculate the internal acoustic Born potential and corresponding Born phase shift for each mode used in the analysis. With the surface phase shift determined in the first step and with the internal phase shift determined in the previous step we infer an improved value of $F(\tilde{w})$ from equation (1) and an improved estimate of the sound speed-profile.

We repeat the above steps until there is no change in the sound speed profile between successive iterations.

The accuracy, stability and convergence of the iteration scheme have been investigated in considerable detail using artificial data (frequencies) sets computed for solar models. The procedure is stable and converges in few iterations.

The numerical resuls are presented in the following figures.

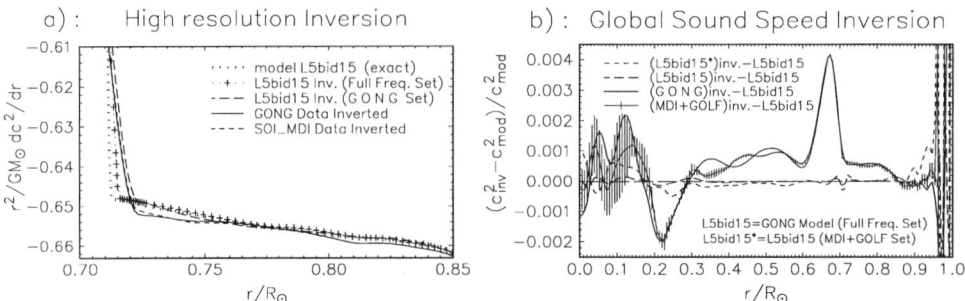

Figure 1. a) Sound speed gradient inversion. b) Global sound speed inversion. Vertical bars mark the SOHO curve and show the envelope of the solutions, when the white noise of the reported observational error amplitudes was added to the oscillation frequencies, in 100 random realizations.

The GONG data set used consists of 1803 modes with frequencies in the range $1 \leq \omega/(2\pi) \leq 4.5$ mHz. There may be further information in the data at higher and lower frequencies, but the quality of GONG data at those frequencies is not good enough. In Figure 1a we compare the results of the inversions for the sound speed gradient dc^2/dr obtained with GONG data, SOI/MDI data (144 days data set, 1802 modes used), and GONG reference solar model (Model S, l5bi.d.15, Christensen-Dalsgaard et al., 1996) with the exact values derived from the model. The primary target of our analysis is to achive the highest possible resolution of the region around the base of the solar convection zone, for searching possible signatures of penetrative convection, element diffusion and/or strong magnetic fields. Hopefully, it should allow the direct calibration of equation of state in the Sun, given high quality data of intermediate degree range.

Also the Born approximation technique successfully incorporates the effects of gravity perturbations in the solar core and the inversion procedure is now being used to diagnose the structure of the deep interior, given high quality data on modes with low ℓ values. In Figure 1b we present preliminary results for global sound speed inversion obtained with GONG data and combined data set of SOI/MDI (144 days) and low ℓ GOLF (110 modes used) frequencies. The difference in results in the deeper regions between GONG and SOHO is clearly seen. Unfortunately MDI and GOLF data sets contain almost no frequencies lower than 1.5 mHz and higher than 4.0 mHz, which restricts the resolution of our inversion.

The iterative inversion technique described here is fully nonlinear, uses only oscillation frequencies as the input data and does not use any reference solar model.

References

Christensen-Dalsgaard, J., et al.,: 1996, *Science* **272**, 1286.
Roxburgh, I.W., and Vorontsov, S.V.: 1996, *Monthly Notices Roy. Astron. Soc.* **278**, 940.

QUASI-10-DAY AND 4-DAY PERIODICITIES IN SOLAR IRRADIANCE

M.V. NIKONOVA, N.V. KLOCHEK AND L.E. PALAMARCHUK
Institute of Solar-Terrestrial Physics
P.O.Box 4026 Irkutsk 664033 Russia

There are many investigations of the total solar irradiance (TSI) variability over the time scales from days to years. Particular emphasis has been placed on the sources of this variability and on the problems concerning stable equipment operation. The investigations made aboard SMM/ACRIM showed a relation between solar variability and various solar activity manifestations. Even then the TSI power spectrum revealed some peculiar periodicity in the range from 9 to 11 days that was not explained and also a well-defined quasi-4-day periodicity (Froelich and Pap, 1989). Their distinctive feature is that they manifest themselves in many processes both on Sun and on Earth. An understanding of their nature is of importance to solar-terrestrial physics. The earlier investigations of Nimbus-7 data pointed out the possible effect of solar sources of unknown nature and also some residual instrumental effects on the behaviour of TSI variations. As a further development of these studies the spectral analysis was made of the TSI variations using the final data set derived over the range from November 1978 to April 1992 aboard Nimbus-7. In our study the data gaps were filled with an artificial noise signal. The autocorrelation function (ACF) was used for detrending and filtering of the data.

Figure 1 shows the power spectrum of ACF of TSI variations over the short-period range. The spectral power within the frequency range corresponding to the solar rotation period does not exceed the confidence level $p = 0.99$; other spectral peaks (SPs) are concentrated within two frequency ranges. Within the first range there is the SP (marked as 1 on fig.1) corresponding to well-known 13-day periodicity observed in various solar parameters including Nimbus-7 data. The next five SPs (2-6) form the quasi-10-day range with the mean value being equal to 10.54 days. Within the second range three SPs (7-9) form the quasi-4-day range. Taking into consideration the fact that the gaps were regular at the beginning of Nimbus-7 mission the 4-day periodicity might be considered as artificial one. However, filling the gaps with white noise and the

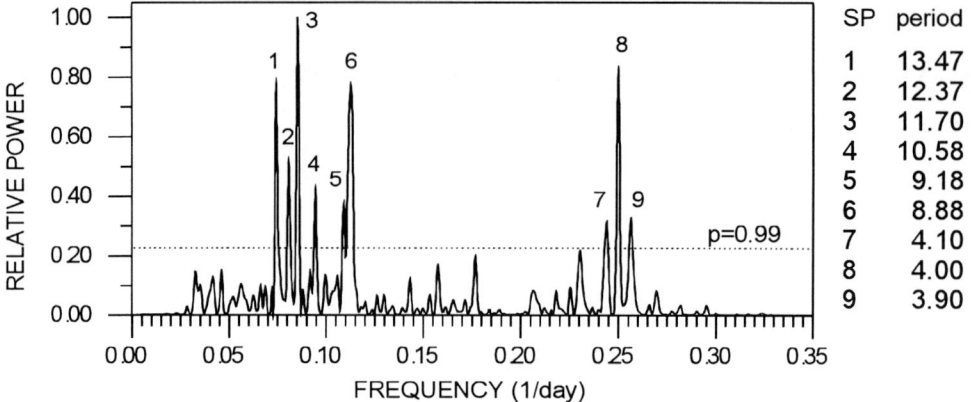

Figure 1. Power spectrum of TSI variations with basic peaks marked

following filtering of data suggest that 4-day periodicity is related to the gaps to a lesser extent. It might be assumed that the gaps are related to the regular interruptions in the normal operation of Nimbus-7. These upsets might be associated with the wave disturbances in the Earth's atmosphere that reveal 10- and 4-day periodicities with the last one being the most pronounced (Philbrick and Chen, 1992). These two periodicities are often revealed simultaneously. In geophysics they are typical of so-called planetary waves. 10- and 4-day periodicities are also present in the Earth's seismic noise, in the ionospheric variations and in variations of the solar wind density.

Thus, the 10- and 4-day periodicities are suggested to be global. As to the quasi-10-day periodicity it was shown (Klochek and Nikonova, 1989) that this periodicity revealed in the power spectrum of X-ray solar radiation may be caused by explosive processes of stellar pulsation type. It is interesting to note that variable stars of classical cepheid type that may be grouped with the stars of G2 spectral class, at the epoch of the middle light of cepheids, reveal the basic periods of the fundamental radial pulsations close to this periodicity.

References

Froelich, C. and Pap, J. (1989) Multi-Spectral Analysis of Total Solar Irradiance Variations, *Astron. and Astrophys.* **Vol. 220, no. 1/2,** pp.272-280

Klochek, N.V. and Nikonova, M.V. (1989) Some Results of the Cepstral Analysis of the Solar Activity Phenomena, *Solar Magnetic Fields and Corona* Novosibirsk, Nauka, **Vol. 1,** ss.398-400 (in Russian)

Philbrick, C.R. and Chen, B. (1992) Transmission of Gravity Waves and Planetary Waves in the Middle Atmosphere Based on the Lidar and Rocket Measurements, *Advances Space Res.* **Vol. 12, no. 10,** pp.303-306

A k–ϵ MODEL OF THE LOWER OVERSHOOT LAYER

K. PETROVAY
Instituto de Astrofísica de Canarias
E-38200 La Laguna, Tenerife, Spain

One region where discrepancies between the seismic and standard solar models continue to exist is the overshoot layer below the convective zone (Gough et al. 1996). A main reason for this discrepancy is clearly the lack of a realistic physical model for astrophysical overshoot. Beside numerical experiments, the most promising approach to the construction of such a model is based on the Reynolds-stress formalism of turbulence theory (Speziale 1991). As a first step toward the development of a more complete Reynolds stress model of the overshoot, here we present a simple k–ϵ model, valid under the following conditions:

(a). *Incompressibility*: $\partial_i v_i = 0$. Owing to the short length scale (cf. Fig. 1), this assumption is somewhat better founded here than in other models.

(b). *Closure*: Let $\overline{wv_i^2}/2 = C_{wk} k^{3/2}$ and $\overline{w\epsilon} = C_{w\epsilon} k^{1/2}\bar\epsilon$. ($w = v_z$ vertical velocity; $k \equiv V^2/2$: kinetic energy; ϵ: dissipation rate). Now *assume* C_{wk} =const., $C_{w\epsilon}$ =const. This is warranted if $l/H_V \simeq$constant (H_V: scale height of V^2), which will be seen indeed to be the case in our model, cf. eqs. (3)–(4).

(c). *No correlation* with thermodynamic fluctuations: $\overline{v_i\rho} = \overline{v_i P} = \overline{v_i T} = 0$. The assumption implies that these correlations vanish in a very thin ($\ll H_P$) transition layer below $\Delta\nabla = 0$. No change in the geometrical flow properties is expected in such a thin layer, so l_0, C_{wk}, and $C_{w\epsilon}$ should be the same as in the lower part of the unstable layer. On the basis of numerical experiments here we use $l_0 = H_P$, $C_{wk} = 0.6$, $C_{w\epsilon} = 3$, $q = 2.4$. (The index '0' refers to values at $z = 0$ i.e. at $\Delta\nabla = 0$.) Note however that V may be significantly reduced in the transition layer, so V_0 (and thus $\bar\epsilon_0$) is undetermined in the present model.

The fluctuating part of the equation of motion now reads

$$\partial_t v_i + v_j\partial_j v_i = -\partial_i P'/\rho + \rho' g_i/\rho + \nu\partial_j^2 u_i. \tag{1}$$

With the above assumptions, the first moment of (1) with v_i yields:

$$\partial_j(k^{3/2}) = -\bar\epsilon/C_{wk}.$$

In all previous models of the overshoot layer the length scale $l = V^3/\bar\epsilon$ was set arbitrarily. Here in contrast we perform the operation $2\nu\overline{(\partial_k v_i)\partial_k}$ on eq. (1):

$$\partial_j(k^{1/2}\bar\epsilon) = 2\nu^2\overline{(\partial_j\partial_k v_i)^2}/C_{w\epsilon} = -(C_\epsilon/C_{w\epsilon})\bar\epsilon^2/k$$

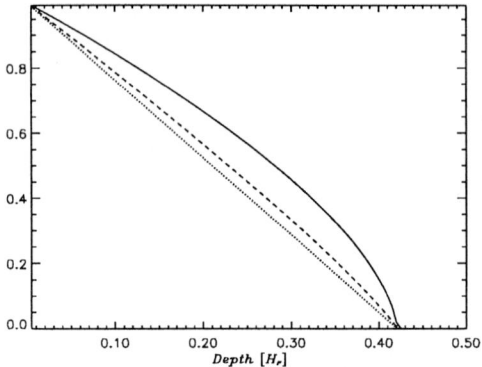

Figure 1. Normalized values of the turbulent velocity V (solid), dissipation rate $\bar{\epsilon}$ (dashed) and length scale l (dotted) as functions of z in a plane parallel model

where $C_\epsilon \simeq 1.8$ from experiments and RNG theory. In the plane-parallel limit then

$$d_z(V^3) = -2\sqrt{2\bar{\epsilon}}/C_{wk} \qquad d_z(V\bar{\epsilon}) = -2\sqrt{2}\,(C_\epsilon/C_{w\epsilon})\,\bar{\epsilon}^2/V^2, \qquad (2)$$

and this system admits the analytic solution (Fig. 1)

$$\bar{\epsilon}/\bar{\epsilon}_0 = (V/V_0)^{q-1} \qquad (3)$$

$$V^{4-q} = V_0^{1-q}\left(V_0^3 - \frac{2^{3/2}}{3}\frac{4-q}{C_{wk}}\epsilon_0 z\right) \qquad q = 3C_\epsilon\frac{C_{wk}}{C_{w\epsilon}}. \qquad (4)$$

It is apparent that $l = V^3/\bar{\epsilon}$ decreases linearly with depth z, i.e. the widely used assumption $l \sim H_P$ is indeed very unrealistic. The overshoot distance in the kinematic (i.e. $V \neq 0$) sense is of order H_P, its exact value being rather sensitive to C_{wk} and $C_{w\epsilon}$: $z_{ov}/l_0 = 3/2^{3/2}C_{wk}/(4-q)$.

In conclusion, this illustrative model shows that the length parameter can be determined self-consistently. Fitting the model to the results of numerical experiments on deep turbulent convection yields $z_{ov} \sim H_P$ for the kinematic extent of overshoot. On the other hand, the overshoot *in thermal sense* (i.e. the quasiadiabatic part of the subadiabatic layer) is negligible in this model (as a direct consequence of assumption *(c)* above). This could help in explaining the remaining discrepancies between observed and calculated oscillation frequencies, and it would also be beneficial for the dynamo.

The practical value of the model hinges upon assumption *(c)*, which should be tested in a more complete model including thermodynamic correlations.

This work was funded by the DGES project no. 95-0028, by the OTKA grant no. 016196, and by the FKFP grant no. 0201/97.

References

Gough D. O. *et al.* 1996, Science, 272, 1296
Speziale, C. G. 1991, Ann. Rev. Fluid Mech., 23, 107

NUMERICAL SIMULATION OF PENETRATIVE CONVECTION: A PARAMETRIC STUDY

HARINDER P. SINGH
Department of Physics
Sri Venkateswara College, University of Delhi

IAN W. ROXBURGH
Astronomy Unit
Queen Mary & Westfield College, University of London

AND

KWING L. CHAN
Department of Mathematics
The Hong Kong University of Science & Technology

Abstract. In continuation of our earlier studies (Singh, Roxburgh & Chan 1997a,b; hereafter SRC97a,b) we perform some further tests to study the general behaviour of penetrative convection and its scaling with rms vertical velocity by varying a number of input parameters like the aspect ratio and the positioning of the interface between the unstable-lower stable layer.

1. Introduction

In a recent study (SRC97a,b), we performed three-dimensional simulations of convection in a three-layer configuration with a view to study the penetration of flows into the lower stable layer and to examine the scaling relationship between the penetration distance (Δ_d) and the vertical velocity given earlier by Schmitt et al. (1984) and Zahn (1991) which may be stated as $\Delta_d \propto V_{zo}^{3/2}$, where V_{zo} is the vertical velocity at the bottom of the middle convective layer. The total domain of computation consists of ~ 7.5 pressure scale heights with the middle unstable region having 5.4 p.s.h. The lower stable layer extends to a height of 0.4 from the bottom and contains 1.2 p.s.h. The aspect ratio of the numerical box is 1.5. Four different models corresponding to four values of the input flux F_b were computed and the following equation was evaluated to examine the relationship described earlier by taking pair of cases

TABLE 1. Relationship between penetration distance and rms vertical velocity

Model No.	Aspect Ratio	$(F_b(1), F_b(2))$	LHS Eqn.(1)	RHS Eqn.(1)
1	1.5	(0.1875,0.125)	1.18	1.24
2	1.5	(0.1875,0.0625)	1.72	1.78
3	1.5	(0.1875,0.03125)	2.46	2.66
4	1.5	(0.125,0.0625)	1.46	1.43
5	1.5	(0.125,0.03125)	2.08	2.14
6	1.5	(0.0625,0.03125)	1.43	1.49
7	1.5	(0.125,0.0625)	1.40	1.41
8	2.0	(0.125,0.0625)	1.26	1.41
9	3.0	(0.125,0.0625)	1.44	1.54

corresponding to two flux values:

$$\frac{\Delta_d(F_b(1))}{\Delta_d(F_b(2))} = \frac{V_{zo}^{3/2}(F_b(1))}{V_{zo}^{3/2}(F_b(2))}. \tag{1}$$

We obtained six values for the left and the right hand sides of above equation from our four cases. The results are given in Table 1 (Models 1-6). It may be seen that $\Delta_d \propto V_{zo}^{3/2}$.

We do some further tests by shifting the location of the interface between the unstable and the lower stable region from a height of 0.4 from the bottom to a height of 0.5. This provided a larger (2 p.s.h.) lower stable layer at the bottom. Two cases were studied with two different values of the input flux F_b. The ratios of the penetration distances in the two cases were equated with the right hand side of eq.(1). The result for Model 7 confirms the scaling relationship between the penetration distance and the vertical velocity.

The aspect ratio of the box was changed from 1.5 to 2.0 and also by a factor of two from 1.5 to 3.0. Two models (with different F_b values) were computed for each of these aspact ratios. As may be seen from Table 1 (Models 8-9), the scaling relationship implied by eq.(1) has deteriorated slightly and could be attributed to the fact that the number of grid points in the two horizontal directions (35×35) were not changed for the larger aspact ratios. The calculations with improved horizontal resolution for larger aspect ratios are currently under way.

References

Schmitt, J. H. M. M., Rosner, R., Bohm, H. U. 1984, ApJ, 282, 316
Singh, H. P., Roxburgh, I. W., Chan, K. L. 1997a, in Stellar Astrophysics, ASP Proc. 1997
    Pacific Rim Conf., Eds. K. L. Chan, K. S. Cheng, & H. P. Singh (in Press)
Singh, H. P., Roxburgh, I. W., Chan, K. L. 1997b, A&A (submitted)
Zahn, J. P. 1991, A&A, 252, 179

INVERSION METHODS

M.J. THOMPSON
Astronomy Unit, School of Mathematical Sciences,
Queen Mary & Westfield College, London, E1 4NS, England

1. Introduction

I want to start by addressing the question, 'What is inversion?' My answer would be that inversion is the process of going from data to making inferences about the object under study. In the case of helioseismology, the data at the present time are principally the mode frequencies, and the object under study is the solar interior.

For example, suppose we found that observed frequencies of p modes confined to the convection zone agreed well with the frequencies of a theoretical model, but that the observed frequencies of modes penetrating just beneath the convection zone were higher than those in the model. We might reasonably infer that there was something wrong about our model just beneath the base of the convection zone. Moreover, since for acoustic modes the frequency $\nu \sim A/\int c^{-1} ds$, where c is the sound speed, the integral is with respect to distance s along a ray path, and A is just some constant of proportionality, we might infer that the sound speed was too small in this region in the model. Since we know that increasing the opacity in the model in this region would have the effect of increasing the sound speed there, we might infer therefore that the opacity in the model might need to be increased in this region.

In my view, this hypothetical scenario (based though on the actual findings of Christensen-Dalsgaard et al. 1985) is already an example of inversion. It illustrates how knowledge about the details of the mode physics – the depth of penetration of different modes, and the role of the sound speed – is used to make localized inferences about possible errors in the model. In a real case, one would also make some quantitative estimate of the shortfall in the model sound speed, and give proper consideration too to the uncertainties on that inference.

The frequencies of the Sun's global modes depend on conditions in the interior through adiabatic exponent Γ_1, pressure p and density ρ – all functions of position $\mathbf{r}$. The observed p modes depend on these three quantities principally in the combination $c^2 = \Gamma_1 p/\rho$. They also depend on the interior velocity field $\mathbf{v}(\mathbf{r})$ (in particular the rotation) and magnetic field $\mathbf{B}(\mathbf{r})$. These – and functions that can be derived just from a knowledge of these quantities – are the primary seismic variables about which we can make helioseismic inferences. Inferences about secondary quantities, such as the opacity in the example above, which affect the modes only through influencing the primary seismic variables, are at a different level, since there may be other physical causes that may change lead to the same change in the primary variables.

The simple example above also illustrates that it is necessary to have a framework for solving the forward problem in order to perform an inverse analysis. Only if one can predict (albeit perhaps in some approximation) what data would be produced by a particular physical model can one hope to use the data to make inferences about that physics.

The dependence of the mode frequencies on the structure and dynamics of the solar interior is intrinsically nonlinear. Most helioseismic inversion is based on some form of linearization, and for that reason I shall concentrate on the linearized problem for most of this review. However, I should at least note that there are other approaches. One such might be described as inversion by doing the forward problem, namely changing some aspect of the model, seeing what effect this has on the frequencies and thus finding which model amongst some set best fits the observational data. This approach is known as model calibration. Recently there have been several attempts at what is in effect an implementation of this approach with an automated search through some set (or space) of models. Specifically Paul Charbonneau at HAO and Barry Lapthorn, a student of mine, have both been looking at using genetic algorithms for inverting for the solar internal rotation (see, e.g. Charbonneau et al. 1997) . The approach looks promising, at least when only a few dozen independent parameters are used to parametrise the space of models to be explored. Another set of nonlinear methods is based on using an asymptotic description of the forward problem, usually leading to having to invert an Abel integral (e.g. Gough 1984, 1986; Brodsky & Vorontsov 1988; Shibahashi 1988).

In the context of inversions for radial hydrostatic structure, linearization is generally performed by formally perturbing a variational principle for the frequencies. Retaining only first-order perturbation quantities, one obtains a linear functional equation that relates differences in global frequencies between the Sun and a model to differences in their primary variables (e.g. Gough 1985). Further assumptions and manipulations can be used to relate the frequency differences instead to differences in secondary quantities such as opacity, chemical abundances or temperature.

It is rather natural also to handle the effects on the mode frequencies of the internal magnetic and velocity fields, and in particular the internal rotation, in a linearized approximation. For example, the rotation period of roughly one month is very much longer than the typical period of a mode, which is around five minutes, so that including the effects of rotation on the mode as a linearized perturbation is a very good approximation.

As a very simple example of a set of linear constraints to be inverted, I introduce the archetypal equation

$$d_i = \int_0^1 K_i(x)\Omega(x)\,dx + \epsilon_i \qquad (i = 1, 2, \ldots, M) . \tag{1}$$

Here d_i represent the M pieces of observational data, $\Omega(x)$ the unknown function about which we wish to make inferences, ϵ_i the errors in the data, and $K_i(x)$ are known functions. For simplicity I take the errors to be independent normally distributed with zero mean and with variance σ_i^2. The linearized problem for inferring the Sun's internal rotation is very similar to this, except that the function Ω depends on radius and latitude rather than on just one variable x: the data are the so-called frequency splittings between modes with different azimuthal degrees m. The linearized problem for inferring the differences in radial structure between Sun and reference model is also of this form, except that the structural differences generally need to be represented by more than one unknown scalar function $\Omega(x)$: the data are then differences (or relative differences) between observed frequencies and those of the reference model.

2. Data fitting

An obvious way to solve the observational constraints represented by equation (1) is to use the least-squares method. Thus one might seek a representation of Ω of the form

$$\bar{\Omega}(x) = \sum_{j=1}^{N} \bar{\Omega}_j \phi_j(x)$$

where $\phi_j(x)$ ($j = 1, \ldots, N$) are chosen basis functions, and the constants $\bar{\Omega}_j$ are determined by minimizing the data mismatch

$$\sum_{i=1}^{M} \left[d_i - \int_0^1 K_i \bar{\Omega} \, \mathrm{d}x \right]^2 / \sigma_i^2 \,. \tag{2}$$

In practice this is not very successful unless N is chosen to be very small. The reason is that the data depend on the unknown Ω through $\int K_i \Omega \mathrm{d}x$: this tends to smooth variations in Ω, just as straightforward integration of a function tends to produce a result that is smoother than the original function itself. Inversion – inferring Ω from the data d_i – is analogous to the inverse of integration, namely differentiation; and as is well known, differentiating noisy data means that the noise tends to get magnified. The same happens with a naïve application of least squares to the helioseismic inversion problem: if the base functions permit it, the solution will generally be dominated by rapidly varying oscillations caused by the solution attempting to fit the noise in the data.

A commonly employed resolution of this problem is to introduce some regularization into the least-squares method. Instead of minimizing expression (2) one instead minimizes an expression such as

$$\sum_{i=1}^{M} \left[d_i - \int_0^1 K_i \bar{\Omega} \, \mathrm{d}x \right]^2 / \sigma_i^2 + \lambda^2 \int_0^1 (\bar{\Omega}'')^2 \, \mathrm{d}x \,. \tag{3}$$

Here the double prime denotes second derivative with respect to x, and the inclusion of the new term penalizes solutions which have large second derivatives. The value of the constant λ is chosen so as to trade-off the essentially opposing aims of minimizing the data mismatch and minimizing the penalty term. The choice of the exact form of the penalty term is rather ad hoc, but the advantage of making it quadratic in $\bar{\Omega}$ is that one is still led to a set of linear equations to be solved to find the coefficients $\bar{\Omega}_j$.

It is illuminating to see how a Bayesian would analyse the difficulty we encounter with unregularized least-squares and our ad hoc solution of regularized least squares. Bayes's Theorem says that

$$P(\Omega|DX) \propto P(D|\Omega X) P(\Omega|X) \,,$$

The expression on the left is the *a posteriori* probability of Ω, given the data D and any prior information (or prejudice) X. The first term on the right is the probability of observing the data D given Ω and any prior information X, and the remaining term is the prior probability of Ω before the data is obtained. Given our assumption about the data errors being independent and normal,

$$P(D|\Omega X) \propto \prod_{i=1}^{M} e^{-\left(d_i - \int K_i \Omega dx\right)^2 / 2\sigma^2},$$

so $-2 \ln P(D|\Omega X)$ is precisely the expression (2). Thus minimizing (2) is equivalent to maximizing the likelihood of the data. If we accept that what we should be doing in fact is finding the function Ω that maximizes $P(\Omega|DX)$, then we see that unregularized least squares is equivalent to assuming that the prior probability $P(\Omega|X)$ is the same for all Ω. Maximizing $P(\Omega|DX)$ is equivalent to minimizing $-2 \ln P(\Omega|DX)$ which, apart from an additive constant, is equal to the usual least-squares term plus $-2 \ln P(\Omega|X)$. By equating this with expression (3), it is evident that the regularized least squares is equivalent to assuming a prior probability satisfying $-2 \ln P(\Omega|X) = \lambda^2 \int (\Omega'')^2 dx$ plus constant, i.e.

$$P(\Omega|X) \propto e^{-\frac{1}{2}\lambda^2 \int (\Omega'')^2 dx}.$$

This then is our prior probability for Ω. It embodies a prejudice that functions Ω with large second derivatives are less probable (prior to getting any data to the contrary). However, its form has no firm justification. Could we instead introduce a prior which had physical motivation, perhaps updated in the light of previous helioseismic inference?

3. Characterizing all functions that are consistent with the data

The regularized least-squares method at best provides just one example of a function that is consistent with the data. (I say 'at best' because if the value of the chi-squared per degree of freedom is not close to unity then the method fails even to do that.) Because the data provide only a finite number of linear constraints (1), there must be an infinite number of other solutions that are equally consistent with the data.

An alternative approach to the inversion problem is to try to characterize all functions that are consistent with the data. It is almost tautological to say that the set of original constraints (1) provides such a characterization. And each of the original constraints individually provides a partial characterization, since all solutions must satisfy each such constraint. It follows then that any linear combination of the original constraints also provides a partial characterization: any function consistent with the original data must satisfy

$$\sum_{i=1}^{M} c_i d_i = \int_0^1 \mathcal{K} \Omega \, dx + \sum_{i=1}^{M} c_i \epsilon_i, \qquad (4)$$

where

$$\mathcal{K} = \sum_{i=1}^{M} c_i K_i(x), \qquad (5)$$

for any constants c_i we choose. Why is this progress? We have still just got an integral constraint on Ω. But in helioseismology the original kernels K_i are oscillatory functions with substantial amplitude over perhaps a large part of the solar interior. Suppose that with the freedom to choose the constants c_i one could arrange to make the averaging kernel $\mathcal{K}$ localized close to some point $x = x_0$ and moreover nicely peaked there, with unit area when integrated from $x = 0$ to $x = 1$. Then the localized average of any solution Ω, weighted by the averaging kernel $\mathcal{K}$, would have to satisfy constraint (4), the left-hand side of which is determined by the data. It turns out that in the various helioseismic applications and over a large part of the Sun it is indeed possible to construct beautifully localized averaging kernels. Each of these corresponds to a constraint of the form (4) on the set of all possible solutions, which I at least find easier to comprehend than the original data constraints. Each of the constraints says that although there may be an infinite set of functions Ω consistent with the data, in the vicinity of x_0 where one has succeeded in constructing a well-localized averaging kernel any solution Ω must have a local average (made precise by the weighting $\mathcal{K}$) equal to the left-hand side of (4), modulo the error term $\sum c_i \epsilon_i$ whose statistical properties can be established if the statistics of the original data errors ϵ_i are known.

There are various ways in which the coefficients c_i can be found in order to construct averaging kernels with desired degree of localization. Such optimally localized averages methods are based on ideas by Backus & Gilbert (1968), and are described in the helioseismic context by e.g. Gough (1985) and Pijpers & Thompson (1992). As with regularized least squares, there is a trade-off to be made, since trying to localize the averaging kernel too much would increase the variance of the error term $\sum c_i \epsilon_i$ and would eventually render the constraint (4) practically useless.

4. Resolution

If it is possible to localize an averaging kernel in the vicinity of x_0, with a substantial positive peak there and with little amplitude elsewhere, then one can define the resolution in the x-direction attained by the inversion in terms of some measure of the width of the averaging kernel peak, e.g. its full width at half maximum (FWHM). In the case of a 2-D inversion, this can be done in both the radial and latitudinal directions.

With a quadratic penalty term, as in eq. (3), the regularized least-squares (RLS) method is linear, in the sense that the solution at any point x_0 is a linear combination of the data d_i. Thus implicitly coefficients $c_i(x_0)$ exist, and so the RLS solution at x_0 is actually also an average of the true function Ω weighted by an averaging kernel of the form (5). Likewise, the solution from any linear data-fitting method can be interpreted point-by-point in terms of averaging kernels and linear functionals characterising the set of all functions that are consistent with the data. Of course the RLS averaging kernels may not be so nicely localized – after all, one did not explicitly seek to localize them. However, in practice they often are, albeit perhaps with negative sidelobes adjacent to the main positive peak, and so for the RLS too one can define attained resolution in terms of the width of the averaging kernel's main peak in regions where the kernel is well-localized. And if over a range of different target locations x_0 it is possible using the explicit averaging kernel construction methods to produce well-localized kernels, why not interpret the solutions $\sum c_i(x_0) d_i$ as forming a continuous curve – a solution – which probably does not fit the data but which can rather be interpreted as a smoothed-out approximation to the true $\Omega(x_0)$.

The standard deviation of the error $\sum c_i(x_0) \epsilon_i$ in the solution is $(\sum c_i^2 \sigma_i^2)^{1/2}$. The coefficients c_i depend on any adjustable parameters in the method, and therefore so too do the error standard deviation and the attained resolution as defined above. One can then define at a given location x_0 a trade-off curve of error standard deviation against resolution, for any given method. As one tries to squeeze the coefficients to get better resolution, the error increases; and, conversely, as one tries to reduce the error, the resolution gets poorer. Not only do the trade-off curves for different linear helioseismic inversion methods have the same qualitative behaviour, Christensen-Dalsgaard et al. (1990) found that they were very similar quantitatively, in regions where well-localized kernels could be constructed. The resolution, in the sense defined above, cannot be reduced to arbitrarily fine values. In fact there is a fundamental limit on how fine the resolution can be, which is determined by the shortest scales of variation of the kernels K_i in the vicinity of x_0. For a primary variable such as rotation or sound-speed, this scale is directly related to local wavenumber of the mode eigenfunctions. Within the propagation region of p modes the wavenumber is related to c/ω, where c is adiabatic sound speed evaluated at location x_0 and ω is the angular frequency of the mode. Thus the finest attainable resolution at x_0 when inverting p-mode data for rotation or sound speed is $c/\nu_{\max}$ times a number of order unity (to allow for the conversion from eigenfunction to kernel), $\nu_{\max}$ being the top limit on the range of frequencies in the data set. The practical resolution limit may be poorer than this because of the noise in the data, which prevents the averaging kernel from being so well localized while keeping the error standard deviation within acceptable bounds.

The best resolution in the convection zone is probably now achieved with the SOI-MDI dataset. Just to give some idea of the currently attainable resolution, I quote some values for the typical resolution achieved with the 144-day MDI set of splitting coefficients in the region where averaging kernels can be reasonably well localized (values are for the RLS inversion of Schou et al. 1997a; but values for 2-D optimally localized averages would be similar). At 30° latitude, for example, the FWHM radial widths of the averaging kernels (i.e. radial resolution) are $0.02R$ at $r = 0.965R$ (R being the solar radius), $0.05R$ at $r = 0.82R$, $0.08R$ at $r = 0.72R$ and $0.11R$ at $r = 0.52R$; the corresponding FWHM widths measured in the latitudinal direction (i.e. latitudinal resolution) are $0.13R$, $0.14R$, $0.20R$ and $0.29R$. These results are summarised in Table 1. Widths vary less with latitude than with radius: at radius $0.72R$, for example, the FWHM radial widths at different latitudes are $0.07R$ at 15° latitude, $0.08R$ at 45° and $0.09R$ at 60°. The latitudinal widths hardly vary with latitude at fixed radius, in the region where the averaging kernels are reasonably well localized. The latitudinal resolution is poorer than the radial resolution, probably because the splittings are only available as so-called a-coefficients rather than individual splittings. Nonetheless, with a-coefficients up to a_{36} (see Schou et al. 1997b), the latitudinal resolution is the best yet achieved.

Table 1. Radial and latitudinal resolution attained with a 2-D RLS inversion of the 144-day MDI set of splitting coefficients, as measured by averaging kernel widths at selected target locations.

Target radius	Target latitude	Radial width(FWHM)	Latitudinal width(FWHM)
$0.965R$	30°	$0.02R$	$0.13R$
$0.82R$	30°	$0.05R$	$0.14R$
$0.72R$	30°	$0.08R$	$0.20R$
$0.52R$	30°	$0.11R$	$0.29R$

Note that if one quotes interquartile widths instead of FWHM (e.g. Christensen-Dalsgaard et al. 1990) the interquartile widths are smaller by a factor of about 0.57 when the shapes of the kernels are approximately Gaussian.

Before leaving the subject of resolution, I would like to point out that what one might call super-resolution can be achieved, but only by the imposition of additional assumptions in the inversion. If for example in a hypothetical problem the underlying function Ω were assumed to be a delta function, and if the averaging kernels were (for simplicity) nice Gaussians whose properties varied only slowly with target location x_0, then the position of the delta function could be inferred from where the solution (4) – viz the convolution of the kernels with the delta function – attained its maximum. Provided the data noise were sufficiently small, the position of this maximum could be located much more precisely than one FWHM of the Gaussian kernels. I call this super-resolution. But features have not really been resolved beneath the lengthscale given by the width of the kernels. If for example, the true Ω had in fact contained two closely-spaced delta functions, these would not have been resolved. A similar example, of more relevance to our applications, is if the true Ω were assumed to have some simple form locally like a step function, then the position at which the step occurs could be localized more precisely than one kernel width. Helioseismic examples where specific forms have been assumed include the attempts to locate the base of the convection zone (Christensen-Dalsgaard et al. 1991) and to measure the position and thickness of the tachocline (e.g. Kosovichev 1996). In these cases, a particular form of the radial structure or rotation profile were assumed, and the data then allowed the location of the assumed form to be determined rather precisely. However, these regions might have a more complex structure not represented by the simple assumed form and unresolved by the inversions.

5. On interpreting results of inversions

When presented with the result of a linear helioseismic inversion, there are several questions one would be wise to ask, and also several issues to be borne in mind. The first question is, what is the resolution? In 1-D inversions, this might be represented by horizontal bars on a solution plot, denoting the FWHM of the averaging kernels. Fuller information, at selected points, can be provided by seeing the averaging kernels themselves. In 2-D, it is again feasible only to show kernels at selected targets; or one might display radial width and latitudinal width of kernels as functions of position as plots in their own right. At least in the (one hopes) substantial region where the kernels can be well localized, such measures are an adequate substitute for seeing the kernels themselves. Of course, one also wants to ask, what is the uncertainty ('error bar') on the solution? The formal error at a point, which is a measure of how much the solution at that point could change if the data errors ϵ_i had a different realization, is just $(\sum c_i^2 \sigma_i^2)^{1/2}$ (under the assumption of independent errors) and this can be included as error bars or bands on 1-D plots or can be plotted in its own right. But it should be borne in mind that the assumptions about the size or independence of the data errors may be wrong (see below). Also the error bars are only formal: because the data are integrals, the true Ω could take arbitrarily large (positive or negative) values over a sufficiently small region and not affect the data perceptibly, so that the true solution at a single point has in a sense infinitely great uncertainty (e.g. Genovese et al. 1995). This problem is resolved if one thinks of the solution as a localized average, which is really what the inversion provides. Also, even if the original data errors are independent, the solution at different points x_0 and x_1 say are different linear combinations but of the same data, so that the errors $\sum c_i \epsilon_i$ at the two locations are correlated (e.g. Howe & Thompson 1996). Howe & Thompson found that, for least-squares and optimally localized averages inversions for rotation, the lengthscale over which errors are positively correlated is similar to the resolution length as defined by the width of the averaging kernel's main peak. Thus it is difficult to know whether a feature in the solution at the limit of resolution is (a) a true feature with that scale; (b) a smaller-scale feature blurred out by the finite resolution of the inversion; or (c) an artifact of the data noise, because the error at neighbouring points in the solution are correlated. (For inversions for structural differences, e.g. for density differences, errors can be correlated over essentially the whole of the solar interior, which introduces further problems of interpretation.) The possibility also exists that features in the solution are due to systematic errors in the data, which we have explicitly ignored in our assumptions about the statistics of the ϵ_i.

6. Sources of uncertainty in current inversions

A quick survey through the literature of recent helioseismic results will reveal that, although the results agree in broad terms, when one looks at fine details there are discrepancies (e.g., Tomczyk et al. 1995; Thompson et al. 1996; Schou et al., 1997b). What has been achieved in making deductions about the solar interior is remarkable, but as the data become more and more precise we want to use that improvement by making more ever more subtle inferences. Thus it is pertinent to ask what is the reason that inversion results are discrepant, in whatever respect. Korzennik et al. (1997) identified three particular sources of uncertainty in current inversions: (a) the inversion results are obtained by different methods and/or different choices of trade-off parameter values; (b) the inversions use data from different instruments and/or observations; and (c) the inversions use data that has been reduced by different peak-bagging pipelines (viz the reduction from observations to finally identifying the modes and quantifying the frequencies and/or frequency splittings). Let us consider each of these briefly.

That applying different inversion methods to the same data yields different looking results should not be a surprise, in the light of the discussion in the preceding section. The data-fitting methods at best yield but one example of a solution that fits the data, so there is no surprise if two such solutions differ, provided the difference between the two solutions lies in the annihilator (see Gough 1985; Gough & Thompson 1991). But even if this is not the case, and of course methods like optimally localized averages don't even purport to find a solution that 'fits' the data, we may interpret the two solutions at a given point as averages of the true solution weighted by the averaging kernels for the respective methods. Because the averaging kernels inevitably differ, the solutions are taking different averages of the true underlying function and if the solutions at a given point differ it does not necessarily mean that the solutions are inconsistent. The solutions are only truly inconsistent if there is no possible underlying Ω which, when convolved with the averaging kernels for the two methods, would give the respective solutions. This is hard to assess, but one possible test when comparing the solution of a data-fitting method with the result of an optimally localized averages approach would perhaps be to see whether the former solution, when convolved with the optimally localized averaging kernels, produced a result that was similar to the optimally localized averages solution. I am not aware that this has ever been done.

Of course the way in which the two methods handle data errors might also produce different results, but if the different averaging kernels are again taken into account, then the difference in the two solutions at a given point should be consistent with the error budgets, $\sum c_i \epsilon_i$ from each of the two methods. Of course, the two errors are correlated and this must be taken into account when making the comparison, but that is easily done if the coefficients c_i are known and the statistical properties of the data errors are as we have assumed. These kinds of intercomparisons have at best only been done to a very limited extent, but Hare and Hounds exercises – where different inversion methods are applied to the same artificial data – are also useful in helping understand the differences between results of different inversion methods, and these have been used in several studies.

The second source of uncertainty is that data come from different instruments or observations. Different instrumental characteristics might lead to systematic differences in the time series. Although a comparison of the time series from the three SOHO experiments shows that they are indubitably looking at the same Sun (Toutain et al. 1997), it is possible that the differences that do exist might lead to systematic errors in the frequency determinations and in the inversions. The actual solar frequencies also change with time. Thus observations from different epochs should only be combined in a single inversion with great caution. In terms of our eq. (1), these sources of uncertainty can be viewed as translating into the errors ϵ_i having a systematic component. (Alternatively, one could view the fact that Ω is time-varying as being an instance of a systematic error in the solution arising from the inadequacy of the underlying model inherent in eq. 1.) One way in which progress can be made on eliminating this source of uncertainty is a careful study of the systematic differences between frequencies and splittings from different instruments or observations.

The third source of uncertainty is that the data is reduced through different peak-bagging pipelines. The peak-bagging is itself a form of inversion, and has similar potential susceptibilities to incorrect assumptions about the errors on the data put into the pipeline. It also in general uses a model of the line shape which is now thought to be deficient in that it does not take into account the asymmetry of the line profile. And the peak-bagging is a nonlinear optimisation problem in which a given algorithm might fall into a local minimum or otherwise be sensitive to an initial guess. Assuming that all the internal checks one can think of have been done, it is highly desirable that systematic comparisons be made of the frequencies and splittings coming out of different peakbagging pipelines, particularly if they can be applied to exactly the same input data. Such comparisons are being undertaken.

I should not end, though, on the problems with inversion! This paper has been a review of some aspects of the inversion methods, and it is a fact that the methods have been developed in helioseismology to an impressive degree. More importantly, the inversions of solar data are revealing the structure and dynamics of the solar interior in ways that a few years ago would have seemed unimaginable. It falls however to other authors in this volume of proceedings to discuss those exciting results.

Acknowlegements

I gratefully acknowledge the co-operation of the SOI Internal Rotation Team, specifically the authors of Schou et al. 1997a and in particular Rachel Howe and Jørgen Christensen-Dalsgaard, for helpful comments and for allowing me to quote the resolution values in Section 4, which are based on their work.

REFERENCES

Brodsky, M. A. & Vorontsov, S. V., 1988. In *Proc. IAU Symposium No 123, Advances in helio- and asteroseismology*, p. 137 – 140, eds Christensen-Dalsgaard, J. & Frandsen, S., Reidel, Dordrecht.

Charbonneau, P., Tomczyk, S., Schou, J. & Thompson, M. J., 1997. *Astrophys. J.*, in press.

Christensen-Dalsgaard, J., Gough, D. O. & Thompson, M. J., 1991. *Astrophys. J.*, **378**, 413 – 437.

Christensen-Dalsgaard, J., Schou, J. & Thompson, M. J., 1990. *Mon. Not. R. astr. Soc.*, **242**, 353 – 369.

Christensen-Dalsgaard, J., Duvall, T. L., Gough, D. O., Harvey, J. W. & Rhodes Jr, E. J., 1985. *Nature*, **315**, 378 – 382.

Genovese, C. R., Stark, P. B. & Thompson, M. J., 1995. *Astrophys. J.*, **443**, 843 – 854.

Gough, D. O., 1984. *Phil. Trans. R. Soc. London, Ser. A*, **313**, 27 – 38.

Gough, D. O., 1985. *Solar Phys.*, **100**, 65 – 99.

Gough, D. O., 1986. In *Seismology of the Sun and the distant Stars*, p. 125 – 140, ed. Gough, D. O., Reidel, Dordrecht.

Gough, D. O. & Thompson, M. J., 1991. In *Solar interior and atmosphere*, eds Cox, A. N., Livingston, W. C. & Matthews, M., p. 519 – 561, Space Science Series, University of Arizona Press.

Howe, R. & Thompson, M. J., 1996. *Mon. Not. R. astr. Soc.*, **281**, 1385 – 1392.

Korzennik, S. G., Thompson, M. J., Toomre, J. & the GONG Internal Rotation Team, 1997. To appear in *Proc. IAU Symposium 181 'Sounding solar and stellar interiors'*, eds Provost, J. & Schmider, F. X. (Kluwer, Dordrecht).

Kosovichev, A. G., 1996. *Astrophys. J.*, **469**, L61 – L64.

Pijpers, F. P. & Thompson, M. J., 1992. *Astron. Astrophys.*, **262**, L33 – L36.

Schou, J., Antia, H. M., Basu, S., et al., 1997a. *Astrophys. J.*, submitted.

Schou, J. & the SOI Internal Rotation Team, 1997b, these proceedings.

Shibahashi H., 1988. In *Proc. IAU Symposium No 123, Advances in helio- and asteroseismology*, p. 133 – 136, eds Christensen-Dalsgaard, J. & Frandsen, S., Reidel, Dordrecht.

Thompson, M. J., Toomre, J., et al., 1996. *Science*, **272**, 1300 – 1305.

Tomczyk, S., Schou, J. & Thompson, M. J., 1995. *Astrophys. J. Lett.*, **448**, L57 – L60.

Toutain, T., Appourchaux, T., Baudin, F., et al., 1997. *Solar Phys.*, in press.

HELIOSEISMIC CONSTRAINTS ON THE SOLAR STRUCTURE

Evidences of Element Segregation and Mild Mixing below the Convection Zone

S. VAUCLAIR
Laboratoire d'Astrophysique de Toulouse
14, avenue Edouard-Belin, 31400-Toulouse, France

1. Introduction

The study of the internal structure of the Sun entered a new age with helioseismology. Several ground based networks, as well as the space mission SoHO, continuously observe the solar oscillations. In particular GONG, the "Global Oscillation Network Project", gathers six sites around the world with six identical Doppler instruments. These instruments observe the phase shift of the Ni 676.8nm line with 3 images every minute, 1.8 sites observing simultaneously. Millions of solar p-modes have been detected. The inversion of the measured frequencies yields accurate and detailed information about the sound velocity in the Sun's interior, which in turn leads to constraints on the equation of state, opacities, chemical composition. Precise informations on the differential rotation inside the Sun have also been obtained. (See the special Science issue on GONG, vol 272, 31 May 1996).

The Sun is the best known of all the stars. Its mass, radius, luminosity and age are known with a high degree of precision (see values and references in Richard et al. 1996, herafter RVCD). The photospheric abundances have also been precisely determined (Grevesse 1991). For the light elements, the abundance determinations show that lithium has been depleted by a factor of about 140 compared to the protosolar value while beryllium is generally believed to be depleted by a factor 2. These values have widely been used to constrain the solar models (e.g. RVCD). However, while the lithium depletion factor seems well established, the beryllium value is still subject to caution. Balachandran (1997) argues that the beryllium depletion is not real because of insufficient inclusion of continuous opacity in the abundance determination. Her new treatment leads to a solar value identical to the meteoritic value.

Observations of the ^{3}He/^{4}He ratio in the solar wind and in the lunar rocks (Geiss 1993, Gloecker and Geiss 1997) show that this ratio may not have increased by more than $\cong$ 10% during these last 3 Gyr in the Sun. While the occurence of some mild mixing below the solar convection zone is needed to explain the lithium depletion and, as we will see below, is consistent with helioseismology, the ^{3}He/^{4}He observations put a strict constraint on its efficiency.

In the present paper we discuss the constraints on the element segregation and mixing processes occuring in the Sun, obtained from helioseismology and abundance measurements.

2. Element segregation induced by microscopic diffusion in stars

Element diffusion inside the stars represents a basic physical process which cannot be ignored in the computations of stellar structure. When stars form out of gas clouds, they built density, pressure and temperature gradients which force the various chemical species present in the stellar gas to move with respect to one another. As a consequence, the abundances observed at the surface of stars are not always representative of their protosolar values, even if they did not suffer any nuclear processing.

The resulting abundance variations depend on several effects. What we use to call "microscopic" diffusion of the chemical elements represents in fact a competition between two kinds of processes. First the atoms move under the influence of external forces (due to gravity, radiation, etc.), second they collide with other atoms and share the acquired momentum with them in a random way, which slows down their motion. This competition leads to a process of element segregation with a time scale decreasing with increasing density.

Although recognized by the pioneers of the study of stellar structure, this fundamental physical process was long forgotten in the computations of stellar models, except for white dwarfs (Schatzman, 1945). Only with the discovery of large abundance anomalies in main-sequence type stars (the so-called Ap and Am stars), which present characteristic variations with the effective temperature, was element diffusion brought into light fifty years later (Michaud, 1970, see other references in Vauclair and Vauclair, 1982).

At that time, the effects of element diffusion were supposed to be important only when the diffusion time scale was smaller than the stellar age. In the Vauclair and Vauclair (1982) review paper, Fig. 1 shows the regions in the HR diagram where diffusion could lead to "observable" abundance variations. The Sun was excluded, as diffusion could not lead in it to abundance variations larger than some ten percent. At that time evidence of abundance variations could be obtained through spectroscopic observations only, and there was no hope to be able to detect differences of order ten percent.

In the present days, due to helioseismology, we know the internal structure of the Sun with a high precision, and evidence for the occurence of element diffusion is obtained with a high degree of accuracy. Abundance variations of the order of a few percent now become indirectly detectable, by comparisons of the theoretical computations with the results of the inversion of pulsating modes. The confirmation by helioseismology of the predictions concerning element diffusion in the Sun represents a great success for the theory of stellar structure.

3. Evidence of Element segregation and mixing in the Sun from helioseismology

Many authors have computed the gravitational and thermal diffusion of helium and heavier elements in the Sun with various approximations (Cox, Guzik and Kidman, 1989; Bahcall and Pinsonneault, 1992; Proffitt, 1994; Thoul, Bahcall and Loeb, 1994; Christensen-Dalsgaard, Proffitt and Thompson 1993; Richard et al., 1996; etc.). Most models, obtained with various codes and numerical techniques, are consistent with each other, which is quite encouraging.

In RVCD, computations of the solar internal structure have been done with the Toulouse code, a version of the Geneva stellar evolution code in which we have precisely included microscopic diffusion processes and possibilities of testing the effects of macroscopic motions like turbulence and stellar winds. Microscopic diffusion was treated as in Charbonnel, Vauclair and Zahn (1992), and the diffusion coefficients were computed using the Paquette et al. (1986) approximation. The abundance variations were followed separately for helium and 14 heavy elements, and they were iterated so that the final abundances correspond to those given by Grevesse (1991).

These models have been compared to the results of helioseismology in collaboration with the Warsaw group (Dziembowski et al., 1994). The values of the function $u = P/\rho$ as obtained from the models are compared to those of the "seismic Sun" (Figure 1). The best solar model is obtained by taking into account both the element segregation induced by diffusion and a mild mixing necessary to account for the lithium depletion. Then the comparison between the models and the "seismic sun" is very good below the convection zone. However in this model the ^{3}He increase with time is too large compared to the observations.

More recent solar models have been computed with a smaller mildly mixed zone below the convection zone in order to account for the lithium depletion and the constraint on the ^{3}He enhancement (Vauclair and Richard 1997a). In these models the critical μ-gradient able to stabilize the mixing processes was varied as a parameter, with $(\nabla \ln \mu)_{crit}$ between 5 and 1.5×10^{-13} cm^{-1}. The variation of the ^{3}He/^{4}He ratio was found small enough during these last 3 Gyrs as soon as $(\nabla \ln \mu)_{crit} \leq 2 \times 10^{-13}$.

If the constraint on the beryllium depletion is relaxed, it allows a still thinner mixed zone. It is interesting in this case to compute the minimum enhancement of the ^{3}He/^{4}He implied by the lithium observed depletion. Vauclair and Richard (1997b) show that it is possible to deplete lithium by a factor larger than 100 as observed and not increase ^{3}He/^{4}He by more than 5 percent since the solar origin. In this case beryllium is only depleted by about 10 percent.

4. Helioseismology and the Solar Neutrino Problem

The comparison between the computed solar models and the "seismic Sun" (Figures 1 and 2) is also able to rule out the core mixing processes which have been invoked to account for the solar neutrino deficiency (e.g. Morel and Schatzman 1996, Cumming and Haxton 1996). Such a mixing, which could be induced by internal waves, nuclear instabilities or any other process, remains an a priori possibility to decrease the solar neutrino flux. The basic reason is that it brings ^{3}He down towards the solar center and increases the rate of the ^{3}He (^{3}He, 2p) ^{4}He nuclear reaction yield, while the ^{3}He (^{4}He, α) ^{7}Be reaction is reduced.

In the present computations (Richard and Vauclair 1997), we have introduced in the best solar model of RVCD a parametrized mixing region located at the edge of the nuclear burning core. We have introduced this extra-mixing in the form of a gaussian centered at a radius r/R$_\odot$=.2, with a maximum turbulent diffusion coefficient of 1000 cm^2.s^{-1}, and a width δ/R$_\odot$=.04, similar to the parametrization introduced by Morel and Schatzman (1996).

Such a core mixing can indeed reduce the neutrino fluxes. Table 1 presents the luminosity, radius, and neutrino production for the two models (RVCD : unmixed in the core; cmix : core mixed). The neutrino fluxes are smaller in the second model, close to the detection values. However this solar model is inconsistent with helioseismology (Figure 1).

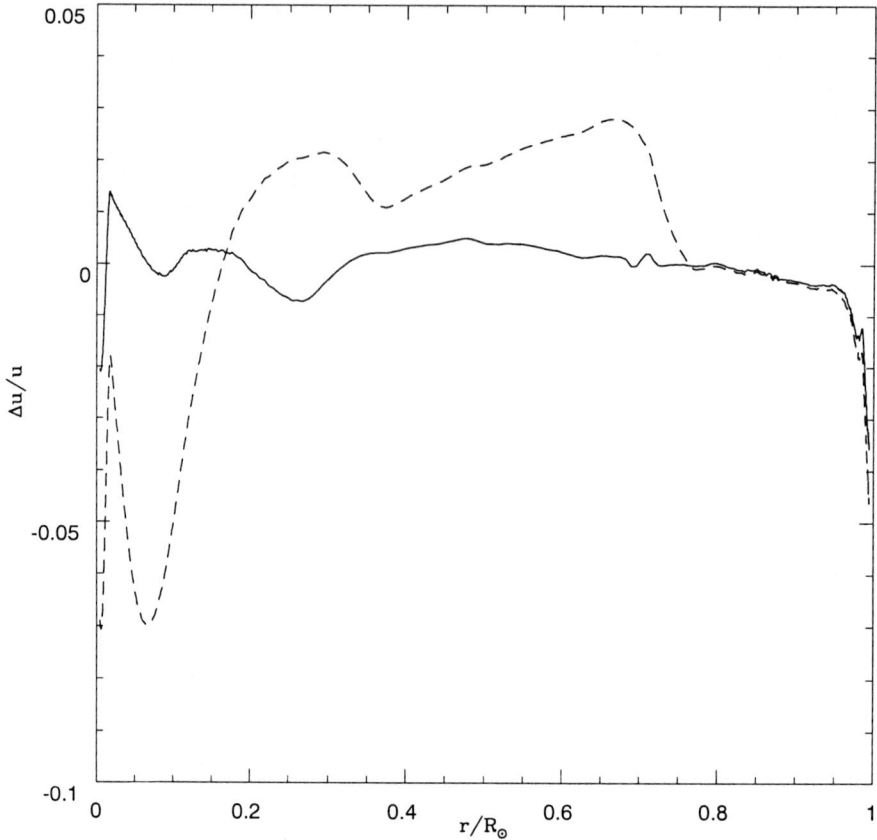

Figure 1. Differences between the $u = P/\rho$ function in the seismic Sun and the best model of RVCD (solid line), and a model including core mixing (dashed line). Mixing near the solar core leads to a modification of the solar structure which is incompatible with the seismic Sun (after Vauclair and Richard 1997a).

5. Conclusion

In summary, the constraints implied by both the helioseismic inversions and abundance determinations in the Sun converge towards the existence of a small mild mixing region below the convection zone, which would extend down to a depth of the order of one scale height. The implied mixing region must be very mild, with diffusion coefficients of 10^3 - 10^4 only. This is quite different from a traditional adiabatically stratified overshooting zone, which is excluded from helioseismology (Christensen-Dalsgaard et al 1995, Basu 1997). It could be related to the differential rotation which occurs below the convection zone.

Such a mild mixing zone can lead to a lithium depletion by a factor 140 as observed, without increasing too much the ^{3}He/^{4}He ratio. In this case beryllium is not

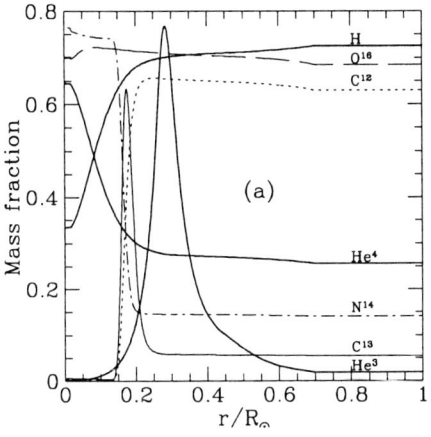 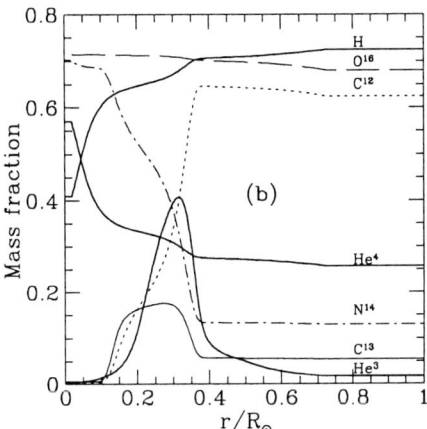

Figure 2. Chemical composition inside solar models (the conventions are the same as in RVCD). Graph (a) represents the case of the best solar model in Richard et al., with no core mixing. The influence of the mixing layer below the convection zone on the ^{3}He abundance is clearly seen on the right side of the ^{3}He peak. Graph (b) represents the model with a partially mixed core. In this model the neutrino fluxes are decreased, but it is not compatible with helioseismology.

TABLE 1.

model	L (10^{34} erg.s^{-1})	R (10^{11} cm)	ϕ (^{8}B) (10^6 cm^{2}s^{-1})	$(\phi\sigma)$ Cl (SNU$_S$)	$(\phi\sigma)$ Ga (SNU$_S$)
RVCD	0.3851	0.6960	6.06	8.14	130.84
cmix	0.3857	0.6959	2.60	3.90	108.75
Observed	0.3851±0.0005	0.6959±0.0001	2.80±0.19±0.33	2.54±0.14±0.14	Sage: 72^{+12+5}_{-10-7} Gallex: $69.7 \pm 6.7^{+3.9}_{-4.5}$

depleted, which is consistent with Balachandran (1997) suggestion. The computed sound velocity is very close to that derived from inversion procedures. The helium profiles directly obtained from helioseismology (Basu 1997, Antia and Chitre 1997) show indeed a helium gradient below the convection zone which is smoother than the gradient computed from pure element segregation. On the other hand, this gradient is compatible with the one obtained when the mild mixing region is added.

In any case this localized mixing region must be completely disconnected from the solar core. The μ-gradient may play the role of a cut-off in this respect. No mixing can indeed be allowed in the nuclear energy production region as it would lead to a sound velocity incompatible with helioseismology. In particular the mixing processes invoked to decrease the neutrino fluxes are excluded.

References

Antia, H.M., Chitre, S.M..: 1997, submitted to A&A
Bahcall, J.N., Pinsonneault, M.H.: 1992, *Reviews of Modern Physics* **64**, 885
Balachandran, S.: 1997, preprint
Basu, S.: 1997, *Mon. Not. R. astr. Soc.*, **288**, 572
Charbonnel, C., Vauclair, S., Zahn, J.P.: 1992, A&A **255**, 191 (CVZ)
Cumming, A., Haxton, W.C. : 1996 *Phys. Rev. Lett.* **77**, 4286
Christensen-Dalsgaard, J., Monteiro, M.J.P.F.G., Thompson, M.J.: 1995, *MNRAS* **276**, 283
Christensen-Dalsgaard, J., Proffitt, C.R., Thompson, M.J.: 1993, *ApJ* **408**, L75
Cox, A.N., Guzik, J.A., Kidman, R.B.: 1989, *ApJ* **342**, 1187
Dziembowski, W.A., Goode, P.R., Pamyatnikh, A.A., Sienkiewicz, R.: 1994, *ApJ* **432**, 417
Geiss, J.: 1993, *Origin and Evolution of the Elements*, ed. Prantzos, Vangioni-Flam & Cassé (Cambridge Univ. Press), **90**
Gloecker, G., Geiss, J.: 1997 preprint
Grevesse, N.: 1991, *A&A* **242**, 488
Michaud, G.: 1970, *ApJ* **160**, 641
Morel, P., Schatzman, E.: 1996, *A&A* **310**, 982
Paquette, C., Pelletier, C., Fontaine, G., Michaud, G.: 1986, *ApJS* **61**, 177
Proffitt, C.R.: 1994, *ApJ* **425**, 849
Richard, O., Vauclair, S., Charbonnel, C., Dziembowski, W.A.: 1996, *A&A* **312**, 1000
Richard, O., Vauclair, S.: 1997, *A&A* **322**, 671
Schatzman, E.: 1945, *Ann. d'Astr.* **8**, 143
Thoul, A.A., Bahcall, J.N., Loeb, A.: 1994, *ApJ* **421**, 828
Vauclair, S., Vauclair, G.: 1982, *ARA&A* **20**, 37
Vauclair, S., Richard, O.: 1997a, Proceedings of the Los Alamos meeting "a half century of stellar pulsation interpretations : a tribute to Arthur N. Cox", June 16-20 , 1997
Vauclair, S., Richard, O.: 1997b, preprint

SOLAR INTERNAL ROTATION

JESPER SCHOU AND THE SOI INTERNAL ROTATION TEAM
HEPL Annex A201
Stanford, CA 94305-4085
USA

Abstract.
With the flood of high quality helioseismic data from the instruments on the SOHO spacecraft (MDI/VIRGO/GOLF) and ground based instruments (e.g. GONG, LOWL, BiSON, IRIS, MWO-CrAO and TON) we have been able to get increasingly detailed information on the rotation and other large scale flows in the solar interior.

Here we will discuss some of the highlights of what we have learned so far and what we may expect to learn in the near future. Among the recent advances have been tighter constraints on the tachocline at the bottom of the convection zone and detection of details in the surface rotation rate similar to the torsional oscillations found in the surface Doppler shift.

1. Introduction

The determination of the solar internal rotation has been one of the central problems in helioseismology and several significant results have been obtained (see e.g. Duvall *et al.*, 1984; Christensen-Dalsgaard & Schou, 1988; Tomczyk, Schou & Thompson, 1995; Thompson *et al.*, 1996 and references therein). While substantial theoretical efforts have been made to understand the origin of the solar differential rotation, no clear understanding of underlying mechanisms has yet emerged. It is widely believed that an understanding of the differential rotation is a key element in understanding the solar cycle. For these and other reasons it is important to continue to constrain the solar internal rotation further.

In this review we will briefly discuss some of the recent helioseismic results concerning solar rotation. We shall not go into any technical details but rather refer the interested reader to the latest papers describing the results, in particular Schou *et al.* (1998), Kosovichev & Schou (1997) and Basu (1997). Details of the various inversion methods can be found in Thompson (1998).

Global flows other than rotation (e.g. meridional flows) will not be discussed here. Giles et al. (1997) and Giles & Duvall (1998) discuss some findings on meridional flows from time-distance helioseismology.

2. Global Picture

Most of the results presented here are from 144 days of MDI Medium-l data from May 9, 1996 to September 29, 1996. The Medium-l program is designed to cover the range $0 \leq l \leq 300$, where l is the spherical harmonic degree. For this study only p modes with degree up to ≈ 200 and f modes up to 250 have been used, corresponding to the modes that are well separated in the power spectra.

The leading-order effect of solar rotation is to split the $(2l+1)$ modes in a single (n,l) (with n being the radial order of the modes) multiplet by an amount (Schou et al., 1994)

$$\Delta\omega_{nlm} \equiv \omega_{nlm} - \omega_{nl} = \int_0^R \int_0^\pi K_{nlm}(r,\theta)\Omega(r,\theta)r\,dr\,d\theta \tag{1}$$

where m is the azimuthal order of the mode, R is the solar radius, K_{nlm} is a mode kernel (that can be calculated from a solar model), Ω is the rotation rate and ω_{nl} is the (m-independent) unperturbed mode frequency.

Unfortunately it is difficult to measure the splittings for individual modes and the m-dependence of the splittings is generally expanded using so-called a-coefficients in order to stabilize the fit:

$$\Delta\omega_{nlm}/2\pi \equiv \Delta\nu_{nlm} = \sum_{i=1}^{i_{max}} a_i(n,l) P_i^{(l)}(m) \tag{2}$$

where the $P_i^{(l)}$ are suitably chosen polynomials of degree i and the a_i are the a-coefficients (Schou et al., 1994). For the present dataset the multiplets were fitted with $i_{max} = 6$, $i_{max} = 18$ and $i_{max} = 36$. For each multiplet one of the three fits was chosen based on l and the convergence of the fits. In most cases $i_{max} = 36$ was used.

Figure 1 shows the rotation rate in the solar interior inferred from the 144d dataset using a Regularized Least Squares (RLS) inversion method (Schou et al., 1994). Other methods, such as optimally localized averages methods and so-called 1.5d and $1d \times 1d$ methods, have also been applied to this dataset, but for lack of space, only the results from the RLS method are presented in this review. The RLS method is currently the one for which most experience exists, but most of the other methods show very similar rotation profiles in the regions for which the methods produce reliable results. For more results and a discussion of the different methods see Schou et al. (1998) and Thompson (1998).

Several features stand out in Fig. 1: the differential rotation in the convection zone, the transition to (near) solid body rotation at the bottom of the convection zone, the steep gradient close to the surface and the jet in the convection zone near the pole. Some of these features have been seen before (e.g. Christensen-Dalsgaard & Schou, 1988; Tomczyk, Schou & Thompson, 1995; Thompson et al., 1996), but some are new and others have been substantially better resolved.

The large scale differential rotation in the convection zone has, of course, been seen before in numerous studies, as mentioned above. Again it appears that the rotation rate is closer to being constant on cones of constant latitude rather than constant

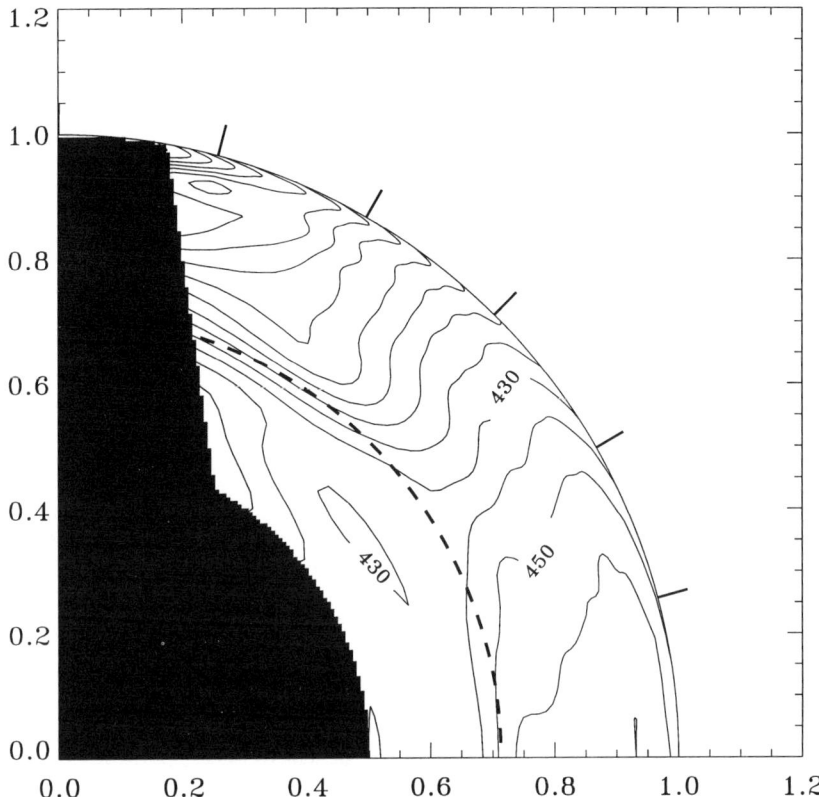

Figure 1. Contour plot of the solar rotation ($\Omega/2\pi(\text{nHz})$). The input data were a-coefficients (up to a_{35}) from 144 days of MDI Medium-l data. The rotation rate in the region close to the rotation axis and close to the center could not be reliably determined from this dataset. This inversion was done using a regularized least squares method. Adapted from Schou *et al.* (1998).

on cylinders, contrary to earlier theoretical predictions. There are however significant deviations from this general behaviour, as outlined below.

The steep transition (tachocline) between the radiative interior and the convection zone has been seen before (Basu, 1997; Kosovichev, 1996; Charbonneau *et al.*, 1997). Unfortunately inversions, such as the one shown in Fig. 1, are not able to resolve this region. To constrain the properties (e.g. width and position) of this region it is thus necessary to make further assumptions. One popular assumption is that the transition is well represented by a function with a few adjustable parameters. Kosovichev (1996) and Charbonneau *et al.* (1997) used an error function with adjustable position, height and width. Basu (1997) used a different parametrization of the transition. In all the cases shown here a width of zero would correspond to an abrupt step, but otherwise the numerical values of the widths are not directly comparable. Typically the fit was

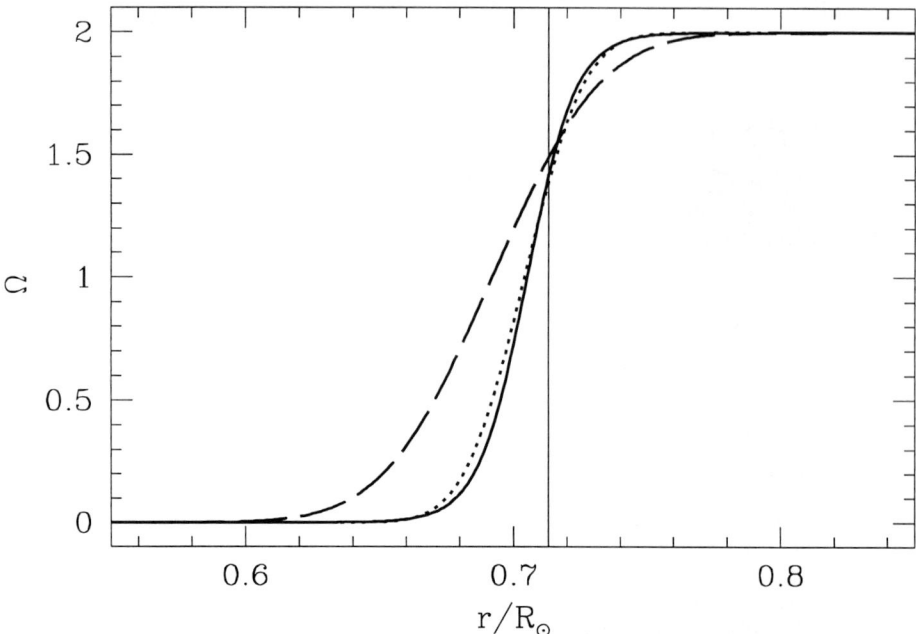

Figure 2. Rotation rate across the tachocline. Solid line from Basu (1997) using BBSO/GONG data. Dashed line from Kosovichev (1996) using BBSO data. Dotted line from Charbonneau *et al.* (1997) using LOWL data. The vertical solid line shows the bottom of the convection zone. The functions fitted were not the same in all cases. All studies used the a_3 coefficient in the fitting. The rotation profiles were rescaled such as to make the asymptotic value 0 in the deep interior and 2 in the convection zone.

carried out only for one measured quantity, in the cases shown here to the a_3 term in the expansion (1) which is related to the first non-constant term in the expansion of Ω with latitude.

A number of such fits are shown in Fig. 2 (adapted from Basu (1997)). The Charbonneau et al. and Basu fits are quite similar with both quoting widths different from zero by about 4σ. The Kosovichev result for the width is different from zero by just over 2σ, and is thus only marginally different from the others. It does, however, appear that the transition is not a step function and that most of the gradient is within the radiative interior.

A quite striking but possibly less reliable feature is the jet seen at 75 degrees latitude about 5% below the solar surface. This is more clearly seen in Figure 3 which also shows that it is quite statistically significant. This jet has a magnitude of around 20nHz (corresponding to about 20m/s), a width of a few percent in radius and an extent of a few degrees in latitude. It should, however, be remarked that this feature is close to the edge of the reliable region and that other inversion methods show it less clearly. Also the resolution, especially in the latitude direction, is comparable to the extent of the feature, and therefore the jet may be narrower (and correspondingly faster) than the inversion seems to show. More a-coefficients or individual splittings would be needed to resolve this and other features better.

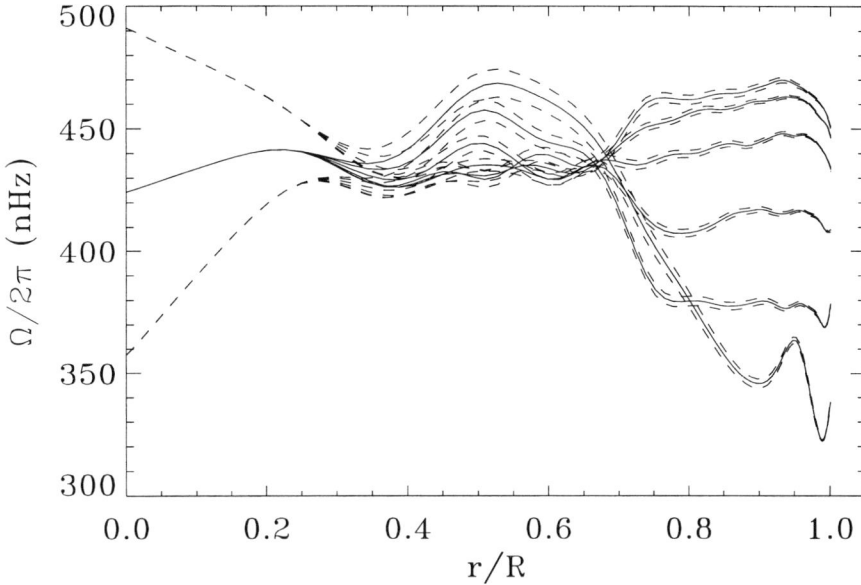

Figure 3. Plot of the rotation rate as a function of radius for a number of latitudes. Latitudes are from top to bottom near the solar surface: 0°, 15°, 30°, 45°, 60° and 75°. One σ error bars have been indicated by dashed lines. Results are from the inversion shown in Fig. 1.

Also seen in Fig. 3 is the rapid increase in the rotation rate as one goes inwards close to the equator. Partly because of the many f modes available in the dataset, we have been able to obtain significantly improved results close to the surface. This increase near the surface may explain the difference between the rotation rates observed using surface Doppler shift and by following features such as sunspots and supergranulation. It is interesting to note that the depth dependence changes as latitude varies, and may even become a decrease at the higher latitudes. The cause of this is not currently clear.

3. Near-Surface Rotation

The inferred rotation rate close to the solar surface is dominated by the contributions of the a_1, a_3 and a_5 terms. Higher order terms are also quite significant, as illustrated by the inversion shown in Fig. 4, where a_1, a_3 and a_5 were artificially set to zero. There are small scale variations in the rotation rate reminiscent of the so-called torsional oscillation (Howard & LaBonte, 1980; Hathaway et al., 1996) seen in surface Doppler shift data. These patterns seem to exist to a significant depth (at least 20Mm) as inferred from the weak l dependence of the high-order splittings and from detailed studies of the global inversion results. The results shown in Fig. 4 are most sensitive to the rotation rate between about 2Mm and 9Mm below the solar surface. Some of these results were discussed in Kosovichev & Schou (1997).

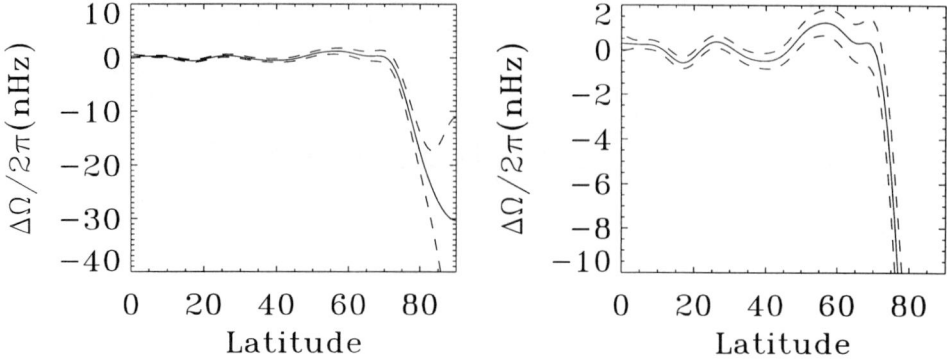

Figure 4. The near-surface rotation rate obtained by setting $a_1 = a_3 = a_5 = 0$ and inverting for the latitude dependence (Kosovichev & Schou, 1997). Only splittings from f-modes with $88 \leq l \leq 250$ were used. 10σ error bars are indicated by dashed lines. The results above $75 - 80°$ latitude are unlikely to be reliable.

Another feature in Fig. 4 is the very slow rotation of the polar regions. This can also be seen in the global inversion in Figs. 1 and 3. This "polar vortex" seems to extend to a significant depth. Again this is close to the edge of the reliable region and it is unclear whether it is related to the 'jet'.

Since the surface Doppler shift signal seems to indicate that the zonal flows migrate to lower latitudes during the solar cycle, five 72 day time-series were analyzed individually. Preliminary results of this analysis, shown in Fig. 5, suggest that the patterns drift towards lower latitudes in a manner similar to the drift of the torsional oscillation signal and the drift in the latitude of emergence of active regions during the solar cycle (as seen in the butterfly diagram).

It is important to note that this signal is an average down to a substantial depth, not simply a surface measurement. Earlier suggestions that the observed surface Doppler shift signal could be due to some sort of systematic error caused by the magnetic fields would thus seem to be ruled out.

4. Conclusion

Recently available datasets from a variety of instruments have allowed us to probe the solar internal rotation in considerably more detail than previously possible. We have succeeded in constraining the tachocline at the bottom of the solar convection zone; we have been able to measure small scale features in the solar rotation rate near the surface; we have detected variations in the near surface rotation rate and we have seen hints of a jet in the convection zone at high latitude.

It is, however, also clear that we have not fully exploited the new datasets and that further discoveries are likely to be made in the near future. In particular we hope to be able to detect other solar cycle variations in addition to the drift of the small scale zonal flows, obtain better constraints on the core rotation and further constrain the properties of the tachocline. It has been suggested that the position and width

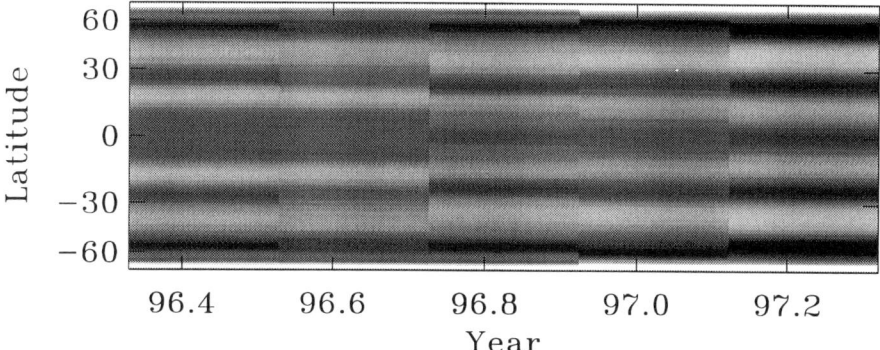

Figure 5. The near surface rotation rate, inferred as in Fig. 4, for five different 72 day time-series. The results for each time period have been shown as a grey scale plot with white being 5m/s slower rotation and black 5m/s faster. Due to the shorter duration of the individual time-series and the desire to use the same set of modes for each period, the set of modes used for this plot is smaller than that used for Fig. 4.

of the tachocline may vary with latitude and that the variation might be detectable by current datasets or those available in the near future. Such a latitude dependence could be detectable both in the rotation rate and in the solar sound speed using the even a-coefficients. One might also hope that it will be possible to detect the small-scale features found close to the surface at greater depths and to detect the possible motion of the jet.

To achieve some of these goals it is clear that we will need observations over longer periods of time, preferably at least one solar activity cycle. Fortunately several instruments, including MDI, GONG and LOWL, are likely to continue operations for several years. Each project requires, of course, continued funding and, especially in the case of space based instruments, the absence of fatal technical problems. Due to the often subtle nature of the data analysis and interpretation, it is also highly desirable to check results using different instruments and analysis techniques, when feasible.

5. Acknowledgements

The 'SOI internal rotation team' co-authors are the co-authors of Schou *et al.* (1998), viz, H.M. Antia, S. Basu, R.S. Bogart, R.I. Bush, S.M. Chitre, J. Christensen-Dalsgaard, A. Eff-Darwich, D.O. Gough, J.T. Hoeksema, R. Howe, S.G. Korzennik, A.G. Kosovichev, R.M. Larsen, F.P. Pijpers, P.H. Scherrer, T. Sekii, T.D. Tarbell, A.M. Title, M.J. Thompson and J. Toomre.

The MDI project is supported by NASA contract NAG5-3077 at Stanford University. SOHO is a mission of international cooperation between the ESA and NASA.

References

Basu, S. (1997) *Monthly Notices Roy. Astron. Soc.* **288,** 572.
Charbonneau, P. *et al.* (1997) in J. Provost and J. Schmider (eds.), IAU Symp. 181: *Sounding Solar and Stellar Interiors*, Kluwer, Dordrecht, Holland, in press.
Duvall, T. L., Jr. *et al.* (1984) *Nature* **310**, 22.
Christensen-Dalsgaard, J. & Schou, J. (1988) in E. J. Rolfe (ed.), ESA SP-286: *Seismology of the Sun & Sun-Like Stars*, p. 149.
Giles, P. M. & Duvall, T. L., Jr. (1998) *these proceedings*, in press.
Giles, P. M., Duvall, T. L. Jr., Scherrer, P. H., and Bogart, R. S. (1997) *Nature*, in press.
Hathaway, D. H. *et al.* (1996), *Science* **272**, 1306.
Howard, R. & LaBonte, B. J. (1980), *Astrophys. J.* **239**, L33.
Kosovichev, A. G. (1996) *Astrophys. J.* **469** L61.
Kosovichev, A. G. & Schou, J. (1997) *Astrophys. J.* **482**, L207.
Schou, J., *et al.* (1998) *Astrophys. J.* in preparation.
Schou, J., Christensen-Dalsgaard, J. and Thompson, M. J. (1994) *Astrophys. J.* **433**, 389.
Thompson, M. J. (1998) *these proceedings*, in press.
Thompson, *et al.* (1996) *Science* **272**, 1300.
Tomczyk, S., Schou, J. and Thompson, M. J. (1995) *Astrophys. J.* **448**, L57.

SOLAR ROTATION AND LARGE-SCALE FLOWS MEASURED BY TIME-DISTANCE HELIOSEISMOLOGY FROM MDI

P. M. GILES
Dept of Applied Physics, Stanford University
Stanford, CA 94305-4085 USA

AND

T. L. DUVALL, JR.
Laboratory for Astronomy and Solar Physics,
NASA/Goddard Space Flight Center, Greenbelt, MD 20771 USA

Abstract. The technique of time-distance helioseismology provides a new tool for examining large-scale flows beneath the sun's surface. We have used this technique to study the meridional flow and have found that the poleward flow observed at the surface appears to persist to a depth of at least 26,000 km. We have also examined the differential rotation in these uppermost layers, and find a slight asymmetry between the northern and southern hemispheres.

1. Introduction

Time-distance helioseismology relies on the measurement of wave travel times between points on the solar surface. These times can then be used to determine the properties of the solar interior. In the simplest approach, a wave is considered to travel along a narrow ray path between reflection points at the photosphere. Any material motion $\mathbf{v}$ will cause a difference in travel time, $\delta\tau$, between a wave travelling in one direction along the ray path, and one travelling in the opposite direction:

$$\delta\tau \simeq -2 \int_\Gamma \frac{\mathbf{v} \cdot \hat{\mathbf{n}}}{c^2} ds. \qquad (1)$$

Here s is the distance along the ray path, $\hat{\mathbf{n}}$ is the tangent to the path, and c is the sound speed. The ray path Γ is determined by the distance Δ between the end points; waves which travel a larger distance penetrate more deeply into the interior.

The travel times, therefore, can be used to measure flows beneath the solar surface. The most prominent solar flow, namely the rotation, has already been accurately measured from analysis of the normal mode frequencies (Thompson *et al.*, 1996). However, the modal approach only measures the mean rotation in the northern and southern hemispheres. The time-distance method should be able to detect any asymmetry that might exist (Duvall et al., 1997a). In addition, a time-distance analysis can be used to measure the flow in meridian planes, known as the meridional circulation (Giles et al., 1997).

2. Results

We used observations covering one solar rotation, during the month of June, 1996. The data used were full-disk doppler images of the sun from the Michelson Doppler Imager (MDI) on SOHO. For each image, a region of the solar surface spanning 100 degrees in longitude and extending to 80 degrees latitude in each hemisphere was remapped onto a rectangular grid. The one-month time series of remapped images was divided into eight-hour segments, for the purpose of computing temporal cross correlations.

In order to study flows in the north-south direction, the cross correlations were computed for pairs of points having a common longitude. Similarly, to study flows in the east-west direction, we chose pairs of points with the same latitude. In either case, the cross correlations for all eight-hour segments during the month were averaged together, to determine the mean flow during one solar rotation.

From the cross correlations, the travel times τ were determined, as described in Duvall et al. (1997b), for each distance Δ and latitude λ. These were then used to compute the reciprocal travel time differences $\delta\tau$. An average of the flow velocity $\mathbf{v}$ in the plane of the ray path was computed using equation (1), by assuming that the flow was entirely horizontal and that it was constant along the ray path.

2.1. MERIDIONAL CIRCULATION

In figure 1 we show the results of the measurement of the north-south flow. The observed values of $\delta\tau$ have been averaged over the entire available distance range. The deepest rays included have photospheric reflection points separated by 73 Mm, and reach a maximum depth of 26.4 Mm.

A positive value of $\delta\tau$ indicates a northward flow, so figure 1a clearly indicates a poleward flow in both hemispheres. This is consistent with previous measurements of the meridional flow at the surface (LaBonte and Howard, 1982; Hathaway et al., 1996; Komm et al., 1993). The solid curve represents a least-squares fit to the measurements of the form

$$\delta\tau = a_1 \cos \lambda + a_2 \sin 2\lambda. \qquad (2)$$

For the average over all the available distances, the coefficients are found to be $a_1 = (-0.40 \pm 0.04)$ s and $a_2 = (1.94 \pm 0.05)$ s. The coefficient a_2 corresponds to a peak poleward velocity of (23.5 ± 0.6) m/s, which is somewhat larger than the value previously reported from surface measurements. This might be due to a real time variation in the flow. The term in $\cos \lambda$ is included to account for the southward velocity observed at the equator. This flow may result from an error in the orientation of the image on the MDI camera, but we do not have an independent method of determining the misalignment. We have therefore treated this cross-equator flow as an instrumental error, with the caveat that it could be a real solar feature.

We next divided the observations according to travel distance, in order to obtain information about the depth dependence of the flow. Four distance ranges were chosen, and in each case a function in the form of equation (2) was fitted to the data.

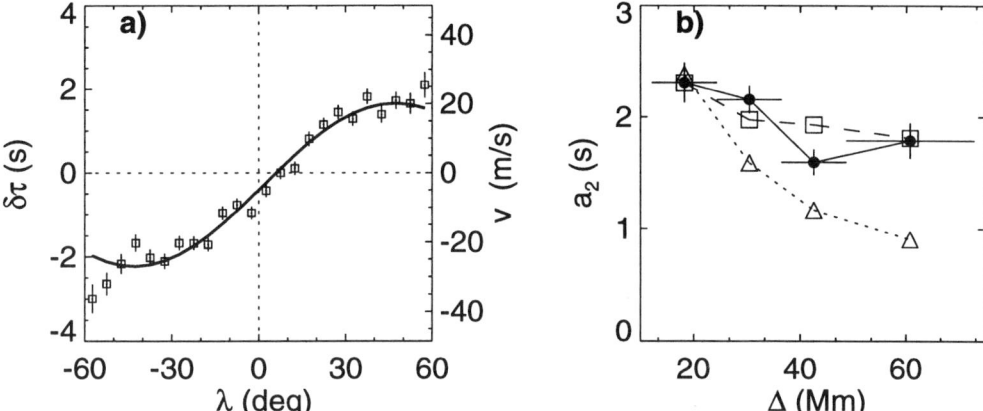

Figure 1. (a) Reciprocal travel-time differences for north-south propagating waves, as a function of latitude, and the inferred flow velocity (see text). (b) Observed travel-time differences as a function of distance travelled. The observed values are shown as the filled circles, with errors. The horizontal error bars indicate the range of distances included in the measurement for each point. Also shown are the calculated coefficients for two different flow models, both of which vary with latitude according to $v = v_0 \sin 2\lambda$. In one model (open squares) the velocity is constant with depth and the magnitude $v_0 = 23.5$ m/s. In the second (open triangles), the flow is constant to a depth of 9.3 Mm and zero in the deeper layers.

In figure 1b the values of the coefficients a_2 are shown for the selected distances. We have compared the observational results to two simple models, described in the figure caption. For each model, the expected travel time differences were calculated using equation (1), and the function in equation (2) was fitted to the result. It seems clear that the model which confines the flow to a thin layer near the surface cannot reproduce the observations as well as the model which has a flow present throughout the region of propagation. Of course, this simple analysis does not rule out some more complicated intermediate behaviour.

2.2. ROTATION

Figure 2a shows a rotation profile computed by the same method as the meridional velocity in figure 1a. For comparison, we have overplotted the surface rotation velocity (dashed curve) as obtained by Ulrich et al. (1988). Clearly, even with the simple approximations used, the measurements agree quite well. The discrepancy at the equator is consistent with helioseismic inversions which indicate that the rotation rate increases with depth very near the surface (Thompson et al., 1996).

It is also clear from figure 2a that the rotation in the southern hemisphere is faster than in the northern. In figure 2b we show the velocity difference at 60 degrees latitude, as a function of the travel distance. This asymmetry also appears to persist to the maximum depth of the ray penetration.

3. Conclusions

We believe that this is the first measurement of the global meridional flow beneath the solar surface, although previous measurements have been made over a smaller area

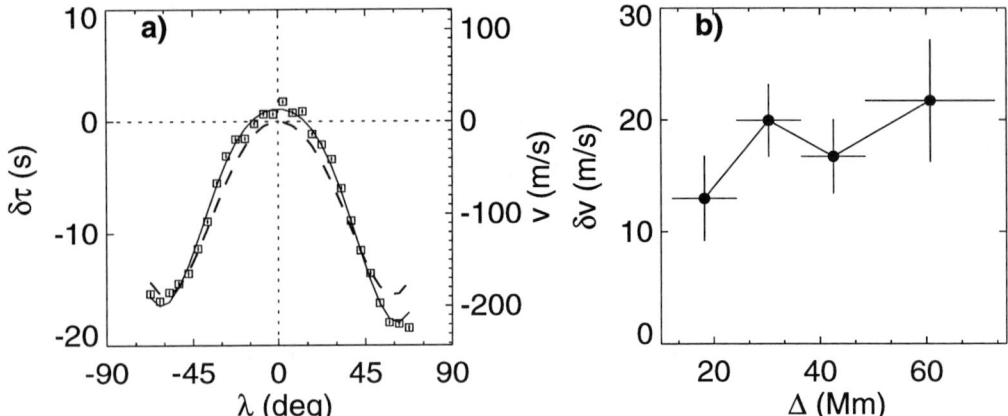

Figure 2. (a) Reciprocal travel-time differences for east-west propagating waves, as a function of latitude, and the inferred flow velocity (see text). The solid curve is a fit to the observations, and the dashed curve is the measured surface rotation rate. Note that the solid-body component of the equatorial surface rate has been subtracted from all the velocities. (b) Inferred velocity difference between northern and southern hemispheres, as a function of wave travel distance. The velocities were taken at ±60° latitude. The sense of the asymmetry is that the southern hemisphere is rotating more rapidly.

using a different technique (Patrón et al., 1995). Our results indicate that the poleward flow previously observed is not confined to a thin layer near the surface, but persists to a depth of at least 26 Mm. We estimate the velocity of this flow to have a maximum value of (23.5 ± 0.6) m/s. We have also shown that the time-distance technique can be used to measure the differential rotation, and we find that the southern hemisphere is rotating slightly faster than the northern, by 10-20 m/s at 60 degrees latitude.

The preliminary results presented here will be extended in the future to cover a longer time period, and a larger part of the solar surface. Particularly interesting will be to extend the measurements to greater depths, and to examine the time variation of the flows during the solar cycle.

We would like to acknowledge support for this research by NASA contract NAG5-3077 at Stanford University.

References

Duvall, T. L. Jr., Kosovichev, A. G., and Scherrer, P. H. (1997a) in *Sounding Solar and Stellar Interiors*, in press.
Duvall, T. L. Jr., et al. (1997b) *Solar Physics* **170**, pp. 63-73.
Giles, P. M., Duvall, T. L. Jr., Scherrer, P. H., and Bogart, R. S. (1997) *Nature*, in press.
Hathaway, D. H. et al. (1996) *Science* **272**, pp. 1306-1309.
Komm, R. W., Howard, R. F., and Harvey, J. W. (1993) *Solar Physics* **147**, pp. 207-223.
LaBonte, B. J. and Howard, R. F. (1982) *Solar Physics* **80**, pp. 361-372.
Patrón, J., Hill, F., Rhodes, E. J. Jr., Korzennik, S. G., and Cacciani, A. (1995) *Ap. J.* **455**, pp. 746-757.
Thompson, M. J. et al. (1996) *Science* **272**, pp. 1300-1305.
Ulrich, R. K., Boyden, J. E., Webster, L., Padilla, S. P., Gilman, P., and Sheiber, T. (1988) *Solar Physics* **117**, p. 291.

LINE PROFILES AND ROTATIONAL SPLITTING OF INDIVIDUAL P-MODES

V.G. GAVRYUSEV
CAISMI CNR, Largo E.Fermi 5, 50125, Florence, Italy

AND

E.A. GAVRYUSEVA
*Osservatorio Astronomico di Capodimonte, via Moiariello 16,
I-80131, Naples, Italy
Institute for Nuclear Research RAN, Moscow, Russia*

1. Data and their Evaluation

We used GONG time series for $\ell = 0, 1, 2, 3$ from June 10, 1995 to January 7, 1997, 578 days in total. The duty cycle varies significantly during this period of time but on average it is sufficiently high, and the remaining gaps are irregular. Because of this we do not see daily side lobes in the spectra obtained.

The time series for $\ell = 0$ (**ts0**) is nearly the reduction of the complex GONG image to the "Sun as a star" measurement. Radial, dipole and quadrupole solar p-modes are visible in this time series. The $\ell = 1$ doublets have the highest power among the other p-modes, in the spectrum of this time series. Then the $\ell = 0$ modes and the triplets of $\ell = 2$ modes follow. The $\ell = 3$ multiplets are poorly visible in **ts0**.

The series for $\ell = 1$ (**ts1**), $\ell = 2$ (**ts2**) and $\ell = 3$ (**ts3**) are regular GONG time series obtained by the complex procedure of spatial filtering of the complete image. This filtering procedure is not ideal for several reasons, and the resulting time series are contaminated by the spatial leakage of the surrounding p-modes of other ℓ. It leads to serious difficulties in the interpretation of the power spectra obtained in many cases.

As it is known, the power of the solar p-modes varies with time. As a result, the power spectra of long time series, obtained by standard Fourier transform, have erratic structures at the location of the p-mode profiles and multiplets. We have applied the FPA technique, described by Gavryusev & Gavryuseva (1996), to reveal the mode line profiles and their parameters. This simple modification of standard Fourier transform permits to eliminate the main contribution of the natural window function into the resulting power spectra and provides an adequate approximation to the **current** line shapes. The existence of strong variations of the p-mode power on time scales of the order from several days to several weeks and months limits the accuracy of the line profile parameters obtained from a time series of a given duration. Each time span of a given duration provides only some approximation to the limit line profile. Due to statistical fluctuations the deviation of the current line shape parameters from the limit parameters can vary drastically depending on the inner statistics of the given time span. Evidently, the strong variation of the mode power on a certain time scale means that the statistics can be significant only when the time span is **much longer** than this time scale.

2. $\ell = 0$ Mean Line Profiles

From the analysis of the **ts0** series we have obtained the line profiles for radial modes of radial order from 11 to 26. While it was claimed that due to spatial leakage other time series can contain the surrounding modes, satisfying the condition $|\Delta\ell + \Delta m| = even$, we have not found the presence of the radial modes in **ts1**, **ts2** and **ts3**, except perhaps for some traces. We investigated the frequency spread for time spans of different duration (Gavryusev & Gavryuseva, 1997). For the 600-day time span we have deduced an uncertainty of the frequency of the order of $30 - 40\ nHz$. We determined all parameters of the modes, i.e., frequency, peak power density and half-line width, but in this short note we present only the frequencies (Table 1).

The statistics of the big pulses, which give the strongest impact on the mode line profile, seems already sufficiently high for 600 days. All current line shapes for the modes with radial order less than 23 seem sufficiently good approximations to the expected symmetrical Lorentzians. In some cases there is a small asymmetry between left and right half-line widths, which could still be caused by statistical fluctuations or even by possible trends of the mean frequencies in time. We cannot yet distinguish one reason from another. But there is certainly no visible asymmetry between the left and the right line wings, as it is seen for high ℓ.

3. $\ell = 1$ and $\ell = 2$ Mean Line Profiles

The dipole solar oscillations ($\ell = 1$) are visible in all GONG time series we have analyzed. In the spectra of **ts0** they are visible as a doublet. In the spectra of **ts1** series for $m = -1, 0, +1$ there should be visible only one component in each series. But it is not exactly like this. While the $m = 0$ series contains always only a single corresponding component, the two other series manifest usually the doublets. Additional components of different azimuthal order m are suppressed, but visible. Hence the parameters of the $m = \pm 1$ components are still affected by one another. For $\ell = 1$ there is an even more serious problem which strongly complicates the possibility of evaluating the multiplet parameters for many radial orders. Due to the spatial leakage of the modes with ℓ probably as high as 10, the precise determination of all parameters in many multiplets is impossible. We used the information, obtained from all (16) time series analyzed, supposing that the effect of leakage is different for different time series and discarding all ambiguous situations. Finally it was possible to evaluate the line parameters for practically all m-components of the dipole modes of radial orders $n = 11 - 23$. Their frequencies and the corresponding splittings are presented in Table 1.

In the **ts2** series only the corresponding m components are visible. The contamination by spatial leakage of other modes is also less severe for $\ell = 2$ modes. All that makes the results obtained for the $\ell = 2$ multiplet parameters more satisfactory. We evaluated the line parameters for all m-components of the quadrupole modes of

TABLE 1. The $\ell = 0$ and $\ell = 1$ modes frequencies, μHz. The splitting is sideral; σ is the standard deviation. The last line gives the average splitting.

ℓ	0	1				
n / m		-1	0	+1	splitting	σ
10			1612.734			
11	1686.523	1748.888	1749.350	1749.677	0.493	0.025
12	1822.124	1884.632	1885.019	1885.468	0.459	0.040
13	1957.451	2020.373	2020.771	2021.092	0.392	0.073
14	2093.506	2156.361	2156.767	2156.991	0.338	0.096
15	2228.693	2291.644	2291.960	2292.253	0.334	0.109
16	2362.806	2425.235	2425.629	2425.834	0.340	0.099
17	2496.181	2558.616	2559.055	2559.527	0.489	0.073
18	2629.674	2692.660	2693.329	2693.757	0.567	0.114
19	2764.060	2827.633	2828.261	2828.518	0.473	0.035
20	2898.967	2962.930	2963.290	2963.564	0.339	0.056
21	3033.675	3097.890	3098.286	3098.818	0.494	0.095
22	3168.596	3232.570	3233.040	3233.653	0.581	0.082
23	3303.488	3367.925	3368.343	3369.014	0.574	0.162
24	3438.752					
25	3574.738					
26	3710.313					
					0.458	0.137

$n = 10 - 25$. The m-component frequencies and the splittings are shown in Table 2. For each individual mode, including the components of the rotational multiplets, we have determined the parameters directly from the mode profile revealed by the FPA method. The presence of neighboring modes disturbs the profile. The joint fitting of neighboring modes have been done to reconstruct the characteristics of the oscillations. It is very important to stress that components of different azimuthal order m as a rule have different amplitudes. For example, only the $n = 13$ and $n = 22$ modes amongst the dipole oscillations presented in the Table 1 have about the same amplitude. The line width is also different. Such asymmetry could be due to the temporal variation of the mode power. But our study (Gavryusev & Gavryuseva, 1997) clearly shows that the power of modes of different azimuthal order is stably different during approximately 800 days. Such asymmetry must be taken into account when spectra are fitted with Lorentzians.

4. Rotational splitting

The averages of the splittings for all radial orders of dipole and quadrupole multiplets are shown in the bottom lines in Tables 1 ($\ell = 1$) and 2 ($\ell = 2$).

The individual values of the splitting obtained from GONG for different radial

TABLE 2. The $\ell = 2$ modes frequencies, μHz. The splitting is sidereal; σ is the standard deviation. The last line gives the average splitting.

ℓ			2				
n / m	-2	-1	0	+1	+2	splitting	σ
10	1673.730	1674.170	1674.510	1674.900	1675.305	0.424	0.030
11	1809.433	1809.840	1810.188	1810.660	1811.171	0.464	0.065
12	1944.956	1945.423	1945.681	1946.164	1946.654	0.454	0.066
13	2081.132	2081.740	2082.202	2082.512	2082.985	0.493	0.084
14	2216.743	2217.323	2217.680	2218.020	2218.459	0.459	0.082
15	2351.323	2351.898	2352.217	2352.790	2353.190	0.497	0.099
16	2484.993	2485.455	2485.907	2486.193	2486.700	0.457	0.074
17	2618.803	2619.275	2619.625	2619.980	2620.544	0.465	0.086
18	2753.330	2753.990	2754.603	2754.647	2755.152	0.486	0.162
19	2888.848	2889.076	2889.490	2889.792	2890.153	0.356	0.057
20	3023.834	3024.515	3024.790	3025.145	3025.448	0.433	0.115
21	3158.867	3159.561	3159.813	3160.360	3160.861	0.528	0.089
22	3294.436	3294.860	3295.155	3295.460	3296.062	0.436	0.094
23	3429.794	3430.220	3430.520	3430.854	3431.547	0.468	0.173
24	3565.664	3566.624	3566.574	3566.949	3567.371	0.457	0.273
25	3702.423	3702.767	3702.912	3703.740	3704.295	0.498	0.163
						0.461	0.038

order n have large error bars, corresponding to the high errors still obtained for the frequencies of the components in the multiplets. The distances between the components vary significantly from one another and from the mean splitting value. For $\ell = 1$ this happens much more frequently than for $\ell = 2$. It seems that this effect is mainly due to the contamination by the spatial leakage of the other modes.

Acknowledgements

This work utilizes data obtained by the Global Oscillation Network Group (GONG) project, managed by the National Solar Observatory, a Division of the National Optical Astronomy Observatories, which is operated by AURA, Inc. under a cooperative agreement with the National Science Foundation. The data were acquired by instruments operated by the Big Bear Solar Observatory, High Altitude Observatory, Learmonth Solar Observatory, Udaipur Solar Observatory, Instituto de Astrofísico de Canarias, and Cerro Tololo Interamerican Observatory.

References

Gavryusev V.G., Gavryuseva E.A., 1996, A&A, **310**, 651
Gavryusev V.G., Gavryuseva E.A., 1997, submitted to ApJ

SPHERICAL AND ASPHERICAL STRUCTURE OF THE SUN: FIRST YEAR OF SOHO/MDI OBSERVATIONS

A.G. KOSOVICHEV, R.NIGAM, P.H. SCHERRER AND J. SCHOU
W.W.Hansen Experimental Physics Lab., Stanford University, U.S.A.

J. CHRISTENSEN-DALSGAARD
Teoretisk Astrofysik Center, Danmarks Grundforskningsfond, and Institut for Fysik og Astronomi, Aarhus Universitet, Denmark

W.A. DZIEMBOWSKI
Copernicus Astronomical Center, Poland

P.H. GOODE
Big Bear Solar Observatory, New Jersey Institute of Technology, U.S.A

D.O. GOUGH
Institute of Astronomy and Department of Applied Mathematics and Theoretical Physics, University of Cambridge, U.K.

J. REITER
Technische Universität München, Germany

AND

E.J. RHODES, JR.
University of Southern California, U.S.A.

Abstract. We report the initial results of one year of continuous observations of the Sun's internal structure from the Michelson Doppler Imager (MDI) on board SOHO. The results have been obtained by inverting frequencies of p and f modes determined with two different methods of averaging over split multiplets. Small systematic differences between the two frequency sets depend primarily on mode frequencies, and, thus, did not significantly affect the inversions. A preliminary study of the systematic effects resulting from asymmetry of oscillation power peaks has also shown no significant influence on the inversion results. The inferred sound-speed profile is in general agreement with the previous data from MDI and ground-based networks. In the energy-generating core, the resolution is substantially improved, and the inversion results indicate a sharp negative perturbation of the sound speed in the core, tending to a positive value near the center. High-precision measurements of the f-mode frequencies have been used to determine the seismic radius of the Sun. The global asphericity estimated from frequency variation across the split mode multiplets has been found to be small, and is consistent with the asphericity during the previous activity minimum. Variations of the solar frequencies during the first year of MDI observations have also been detected.

1. Observations

The Medium-ℓ Program of the Michelson Doppler Imager instrument on board SOHO provides continuous observations of oscillation modes of angular degree, l, from 0 to ~ 300 (Scherrer et al., 1995). The data for the program are partly processed on board because only about 3% of MDI observations can be transmitted continuously to the ground. The on-board data processing, the main component of which is Gaussian-weighted binning, has been optimized to reduce the negative influence of spatial aliasing of high-degree oscillation modes (Kosovichev et al., 1997). The first 360-day observing run, between May 1, 1996 and April 25, 1997, was nearly continuous with a duty cycle of 95%. Figure 1 shows the m-averaged oscillation power spectrum. The mode ridges (the lowest and weakest of which corresponds to the f mode) cover most of the $l - \nu$-plane extending to the upper limits of the observing program, $l = 300$ and $\nu = 8.333$ mHz.

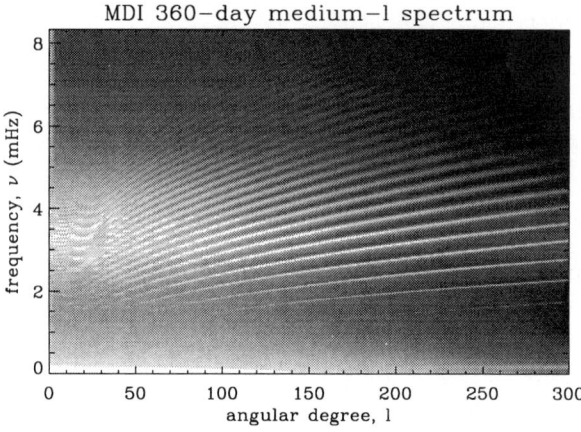

Figure 1. Oscillation power spectrum obtained from 360 days of the medium-ℓ data.

The noise in the medium-ℓ oscillation power spectrum from MDI is substantially lower than in ground-based measurements. This enables us to detect lower-amplitude modes and, thus, to extend the range of measured mode frequencies, which is important for inferring the Sun's internal structure and rotation.

2. Frequency Estimations

Two different methods have been used to estimate the frequencies of the solar normal modes from the oscillation power spectra (Rhodes et al., 1997). In the first, so-called "mean-multiplet" method (Schou, 1992), the power spectral peaks are assumed to have a symmetric Lorentzian shape, and a maximum likelihood method is employed to determine the parameters of Lorentzian profiles. The peaks are fit simultaneously in all of the $2l + 1$ individual power spectra for each multiplet so that the effects of overlapping peaks can be included in the fits. These $2l + 1$ frequencies are effectively averaged to yield a single mean frequency, ν_{nl}, for that multiplet. In addition, an associated set of frequency splitting coefficients, $a_{k,nl}$, for which k runs from 1 to 36, is obtained for the same multiplet.

The second frequency estimation technique (Korzennik, 1990; Reiter and Rhodes, 1997) is becoming known as the "averaged-spectrum" method because it employs the m-averaged power spectra rather than the $2l + 1$ individual power spectra which the mean-multiplet method employs. The advantage of this method is in greater stability of the peak fits. However, it relies on the multiplet splitting coefficients obtained by the first method and used for averaging the individual spectra. The results of the frequency measurements are shown in Fig. 2. The comparison between the two

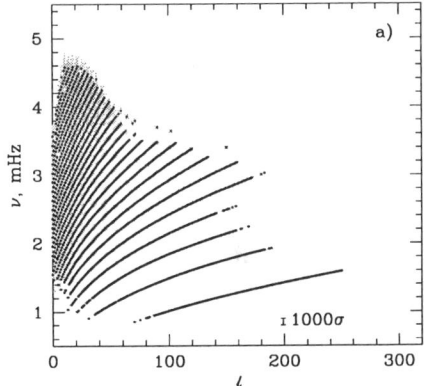
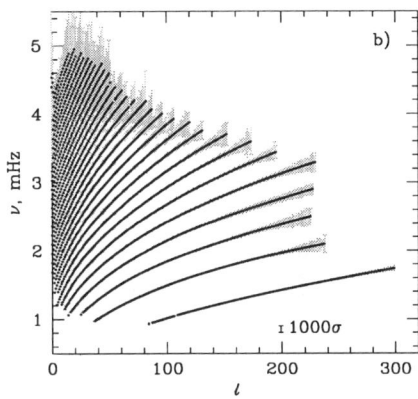

Figure 2. Frequencies of solar modes determined from the 360-day medium-ℓ spectra using (a) the mean-multiplet method (Schou, 1992); (b) the averaged-spectrum method (Rhodes & Reiter, 1997). The error bars (shown in grey) correspond to 1000 standard deviations of the measurements.

sets of mode frequencies (Fig. 3a) reveals the systematic difference of $\simeq 0.05\mu$Hz at frequencies around 2000 μHz, and $\simeq 0.1\mu$Hz above 3500μHz. The systematic difference has been noticed from the 144-day MDI series (Rhodes et al., 1997). The 360-day data indicate that the systematic effect is essentially a function of frequency, and, therefore, most likely related to near-surface perturbations of solar oscillations.

3. Effects of Asymmetric Power Peaks

The medium-ℓ spectra from MDI have revealed asymmetry of the line profiles, which has the opposite sense in Doppler velocity and intensity spectra (Fig.4). The asymmetry has been noticed in the ground-based data (Duvall et al., 1993). Several authors (e.g. Gabriel, 1992; Roxbourgh & Vorontsov, 1995; Abrams & Kumar, 1996) have studied this problem theoretically and have found that there is an inherent asymmetry whenever the waves are excited by a localized source. The reversal of the asymmetry has been recently explained (Nigam, et al., 1998) by a higher level of the solar

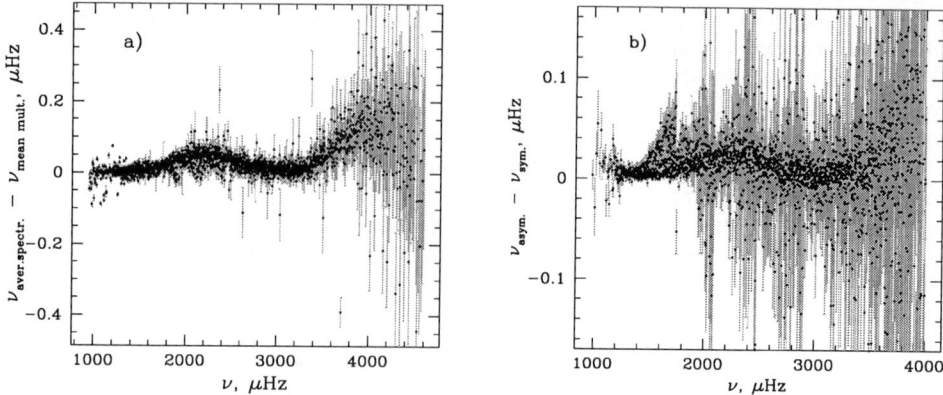

Figure 3. (a) The difference between the frequencies obtained with the averaged spectrum and mean-multiplet methods. (b) The difference between the frequencies determined with asymmetrical ($p4 \neq p2$ in Eq.1) and symmetrical ($p4 \equiv p2$) line profiles using the averaged spectrum method.

noise correlated to the excitation sources in the intensity spectra. The asymmetry effect leads to systematic errors in the determination of frequencies if the mode peaks are fitted with a Lorentzian profile (Hill et al., 1996; Abrams & Kumar, 1996). To

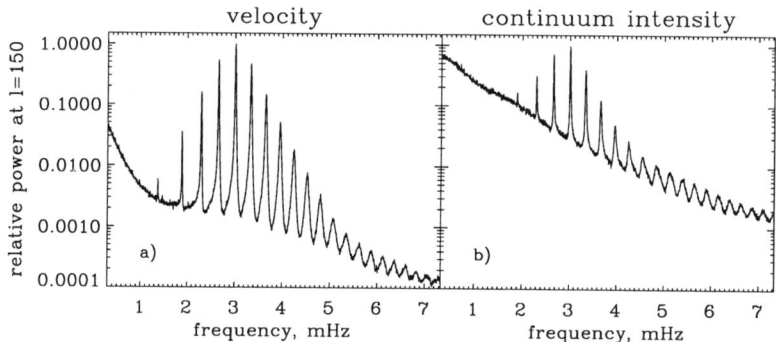

Figure 4. Power spectra of $l = 200$ mode observed in (a) Doppler velocity and (b) continuum intensity.

estimate the effects of the asymmetry on the frequency estimates, we modified the numerical algorithm used in the average-spectrum frequency estimation program to fit asymmetrical profiles to all of the observed peaks in our 144-day spectra:

$$A(\nu) = p_1 \frac{1 - (\nu - p_4)^2/p_3^2}{1 + (\nu - p_2)^2/p_3^2} + p_5 + p_6\nu, \qquad (1)$$

where parameters $p_1 - p_6$ represent the amplitude, the eigenfrequency, the linewidth, the asymmetry, and the background noise respectively. The asymmetry is measured via deviation of p_4 from p_2. This formula is an approximation to the theoretical power spectra.

Preliminary estimates of the asymmetry effect are shown in Fig.3b. The similarity with the frequency dependence of the systematic errors in Fig.3a is intriguing, and suggests that the systematic frequency shift between the two frequency estimates presented in the previous section may result partly from line asymmetry.

4. Radial Stratification

The spherically symmetric structure of the Sun was determined by using a version of the optimally localized averaging method (e.g. Gough & Kosovichev, 1988). Figure 5 shows the relative difference between the square of the sound speed in the Sun and a reference solar model. The reference model is model S of Christensen-Dalsgaard et al. (1996).

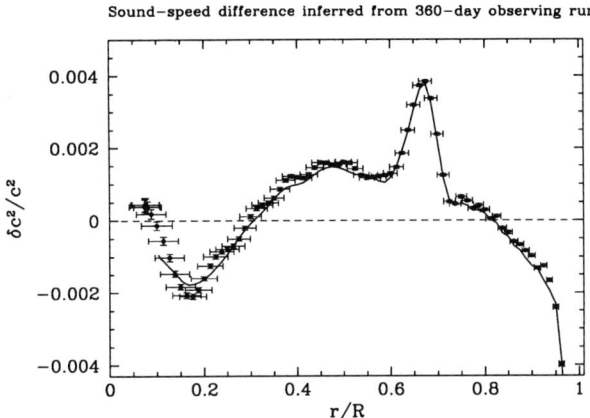

Figure 5. Sound-speed difference between the Sun and solar model S inferred from the averaged spectrum frequencies (points with crosses) and the mean-multiplet frequencies (solid curve) using the optimal averaging inversion method. The vertical error bars show the formal error estimates of the localized averages of the sound speed, and the horizontal bars show the characteristic width of the localized averages.

The results of the inversion of the two frequency sets described in Sec.2 agree with each other within the estimated errors. The results confirmed the previous conclusions about a sharp peak at the base of the convection zone ($r \approx 0.7R$), which is, probably, related to some kind of mixing (e.g. Gough et al., 1996), and rapid variation near the surface, which can be partly explained by correcting the solar radius (Schou et al., 1997). In addition, the new data provide substantially better resolution in the central region where the sound-speed perturbation profile is consistent with partial mixing as suggested from the initial inversion results by Gough & Kosovichev (1988). There is evidence from the averaged-spectrum data set of a sharp negative peak of the sound-speed perturbation between 0.15 and 0.2 R, which may indicate the boundary of such mixing. Of course, at this stage, other possible explanations, e.g. variations of opacity and nuclear reaction rates, are not ruled out. Also, our inversion results contradict the recent results from GOLF (Turck-Chiéze et al., 1997). Obviously, it will take longer time series to understand the structure of the core.

5. Seismic radius

In Figure 6, we compare the observed f-mode frequencies with the theoretical frequencies of solar model S. The main component of the relative frequency difference is a constant offset. This shift suggests that the theoretical frequencies are incorrectly scaled. The f-mode frequencies depend primarily on the gravitational acceleration at the solar surface and the horizontal wavenumber, and, therefore, are proportional to $R^{-3/2}$. This leads to the idea that the value of the solar photospheric radius, R, used to calibrate the model is somewhat different from the actual radius. A detailed analysis of the f-mode frequencies

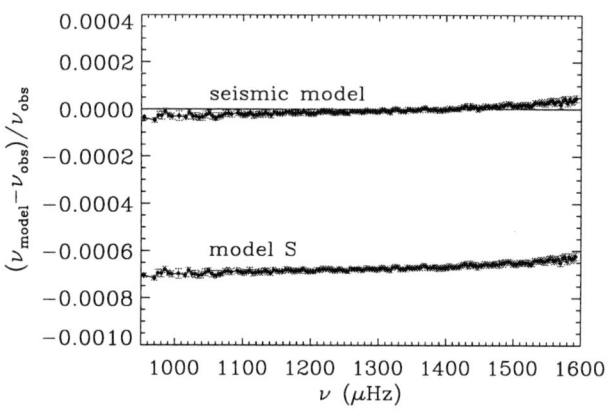

Figure 6. Relative differencies between the f-mode frequencies of $l = 88 - 250$ computed for solar model S of Christensen-Dalsgaard *et al.* (1996) and the observed frequencies. The 'seismic model' frequencies are obtained by scaling the frequencies of model S with the factor 1.00067 which corresponds to reducing the model radius by the factor $(1.00067)^{2/3} \approx 1.00044$. The error bars are 3σ error estimates of the observed frequencies.

(Schou *et al.*, 1997) has shown that the value of the solar photospheric radius used to calibrate the standard solar model has to be reduced by approximately 300 km in order to match the model frequencies with the observed frequencies. The f-mode frequencies provide a strict constraint on the density profile $4-10$ Mm beneath the surface, but the precise correction to the calibration radius of a solar model depends on the description of the superadiabatic layer in the model. For the model S of Christensen-Dalsgaard et al. (1996), originally calibrated to 695.99 Mm, the new calibration radius is approximately 695.68 Mm. The uncertainty due to the statistical errors in the frequency measurements is only 0.008 Mm. However, the systematic error estimated from the deviation of the points of the seismic model in Fig.6 from the zero line could be about 0.03 Mm.

6. Asphericity

For these inversions, we represent the even-a coefficients by

$$a_{2k,nl} = a_{2k,nl,\text{rot}} + (-1)^k \frac{(2k-1)!!}{(2k)!!} \frac{\gamma_k(\nu_{nl})}{\sqrt{l(l+1)}I_{nl}}, \qquad (2)$$

where I_{nl} is the mode inertia, $a_{2k,nl,\text{rot}}$ describe the contribution from the second-order effect of rotation which we have calculated using the rotation law following from the odd-a MDI data, and $\gamma_k(\nu_{nl})$ is a characteristic of asphericity. The functions $\gamma_k(\nu_{nl})$ determined by inversion separately for the p and f modes are shown in Fig. 7. For the p modes, we have found that γ_2

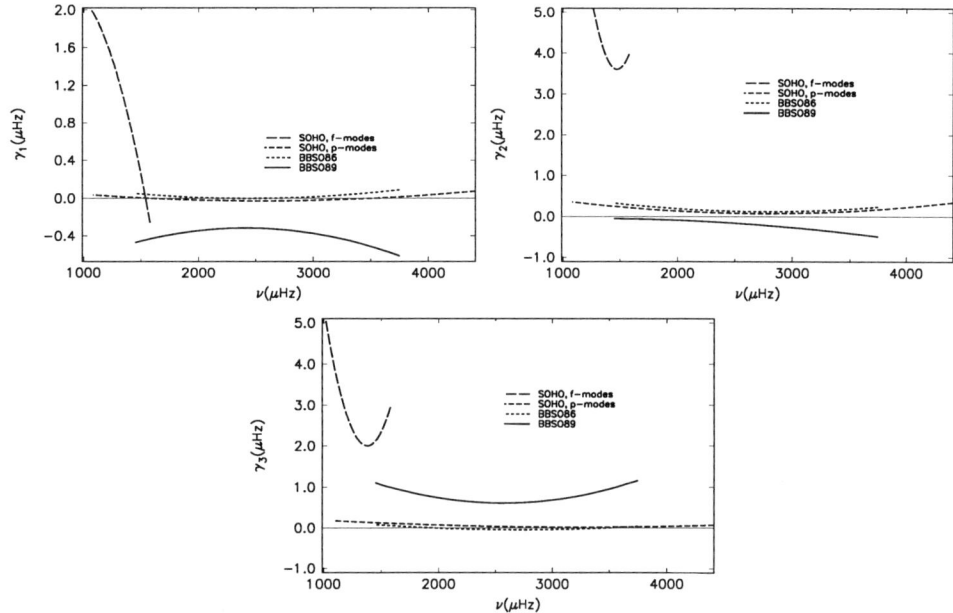

Figure 7. The asphericity coefficients, γ_1, γ_2 and γ_3 (see Eq 2) from the SOHO/MDI data and from two years of the BBSO data, 1986 and 1989, corresponding to the periods of low and high solar activity during the previous solar cycle.

dominates the results from the MDI data. This parameter corresponds to an aspherical structure described by the Legendre polynomial $P_4(\theta)$. This type of asymmetry is consistent with the results from the last activity minimum (compare with BBSO 86 data in Fig. 7). However, this P_4-asphericity is smaller than that from the LOWL data which cover a period earlier in the current activity minimum. During the rising phase of the last activity cycle, this term changed sign, and by 1989 was strongly negative (see Fig. 7). For the f modes, we have found that the P_4 asymmetry also dominates. It will be important to track the behavior of the asphericity, especially for the f modes, during the rising phase of the coming cycle.

7. Temporal variations

To detect temporal variations in the radial solar structure, the mode frequencies have been estimated for five consecutive 72-day intervals using the mean-multiplet method. Variations of the frequencies averaged over l and over 0.2 mHz frequency intervals are shown in Fig. 8 relative to the first 72-day observing interval. The results show variations of the order of 0.02 μHz for the modes

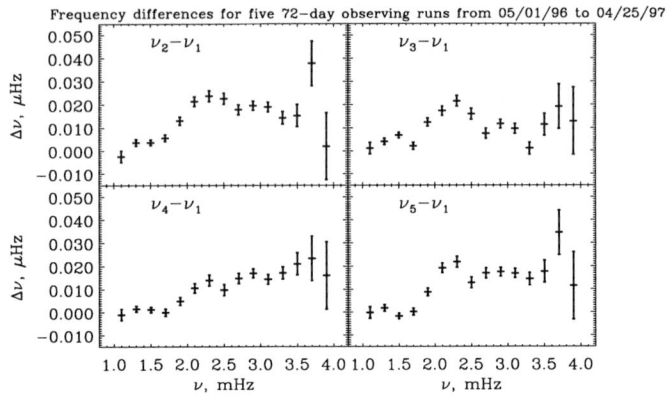

Figure 8. Averaged (over l) frequency differences for five 72-day observing intervals during the 360-day run. These intervals are: (1) 05/01/96-07/11/96, (2) 07/12/96-09/21/96, (3) 09/22/96-12/02/96, (4) 12/03/96-02/12/97, (5) 02/13/97-04/25/97.

above 2 mHz. These changes are likely to be related to near-surface variations resulting from solar activity. However, no apparent solar cycle variations of the internal structure have been detected yet.

References

Abrams, D., and Kumar, P.: 1996, *ApJ*, **472**, 882.
Christensen-Dalsgaard, J., Däppen, W., et al.: 1996, *Science*, **272**, 1286.
Duvall,T.L.,Jr, Jefferies,S.M., Harvey,J.W., Osaki,Y. & Pomerantz,M.A.:1993, *ApJ*, **410**, 829
Dziembowski.W.A., Goode,P.R., Pamyatnykh,A.A. & Sienkiewicz R. 1994, *ApJ*, **432**, 417.
Hill, F., Stark, P.B., Stebbins, R.T., et al.: 1996, *Science*, **272**, 1292.
Gabriel, M.: 1992, *Astron. Astrophys.*, **265**, 771.
Gough, D.O. & Kosovichev, A.G.: 1988, in *Seismology of the Sun and Sun-like Stars*, ed. E.J.Rolfe, ESA SP-286, Noordwijk, 195
Gough, D.O., Kosovichev, A.G., et al.: 1996, *Science*, **272**, 1296.
Guzik, J.A., & Swenson, F.J.: 1997, ApJ, in press.
Korzennik, S.G.: 1990, Ph.D. Dissertation, Univ. Calif. Los Angeles.
Kosovichev, A.G.,Schou, J., Scherrer, P.H., et al.: 1997, *Solar Phys.*, **170**, 63.
Nigam, R., Kosovichev, A.G., Schou, J. & Scherrer, P.H., 1998, these Proceedings.
Reiter, J. and Rhodes, E.J.,Jr.:1997, SOI Technical Note, in preparation.
Rhodes,E.J.,Jr., Kosovichev,A.G., Schou,J, Scherrer,P.H., Reiter,J.:1997, *Solar Phys.*, **175**.
Roxburgh, I.W., and Vorontsov, S.V.: 1995, *MNRAS*, **272**, 850.
Schou, J.: 1992, *On the Analysis of Helioseismic Data*, Thesis, Aarhus University .
Scherrer, P.H., Bogart, R.S., Bush, R.I. et al.: 1995, *Solar Phys.*, **162**, 129.
Schou, J., Kosovichev, A.G., Goode, P.R., Dziembowski, W.A.: 1997, *ApJL*, **489**, L197.
Turck-Chiéze, S. et al.: 1997, *Solar Phys.*, **175**.

WHAT DO SOLAR F-MODE FREQUENCIES TELL US?

H. M. ANTIA AND S. M. CHITRE
Tata Institute of Fundamental Research
Homi Bhabha Road, Mumbai 400005, India

The fundamental mode (f-mode) which is essentially a surface mode can be expected to provide a diagnostic of flows and magnetic fields etc. present in the surface regions (Ghosh, Chitre & Antia 1995; Rosenthal & Christensen-Dalsgaard 1995). It turns out that the observed f-mode frequencies can also provide an accurate measure of solar radius (Antia 1997; Schou et al. 1997).

The frequencies of f-modes are asymptotically expected to satisfy the simple dispersion relation, $\omega^2 = gk$, where g is the acceleration due to gravity at the surface and $k = \sqrt{\ell(\ell+1)}/r$ is the horizontal wave number. At the moderate values of degree ℓ for which the frequencies have been measured, the f-modes are effectively localized somewhat below the solar surface, where gk would be larger than its value at the solar surface. As a result, there is a small departure from this simple dispersion relation. In order to account for this difference we take the ratio of the observed frequencies and those computed for a solar model, and the results are shown in Figure 1. It is clear that this ratio is more or less constant within the expected errors even when ℓ varies by more than a factor of two. In order to explain the observed discrepancy the solar radius will need to be decreased by about 0.03% or about 210 km, which is perhaps somewhat larger than the quoted uncertainty of 70 km in the solar radius.

The frequencies of f-modes will also be affected by other factors such as the density stratification in the outer layers of the Sun, the surface hydrogen abundance, atmospheric opacities and treatment of convection. For estimating this effects we have constructed different solar models with the revised solar radius estimate of 695.78 Mm. Thus Model M1 uses the CM prescription (Canuto & Mazzitelli 1991), while Model M2 is based on the usual mixing-length prescription, both using the atmospheric opacity tables of Kurucz (1991). Model M3 is based on the OPAL opacities (Iglesias & Rogers 1996) and Model M4 which does not include diffusion of elements has a lower hydrogen abundance X and a shallower convection zone. Figure 2 shows the relative difference in frequencies between observed (MDI) values and those for different solar models. We expect the difference $\sim 10^{-4}$ between frequencies of various solar models to be reflected in the estimated solar radius. Thus, if model M2 with MLT treatment of convection is accepted the radius will need to be decreased by about 0.038% (265 km), while if model M3 is taken as standard then the radius will need to be reduced by about 0.041% (285 km) from the standard value. In general, depending on the treatment of surface layers in the model, the radius needs to be decreased by 200-300 km over the standard value to match the f-mode frequencies.

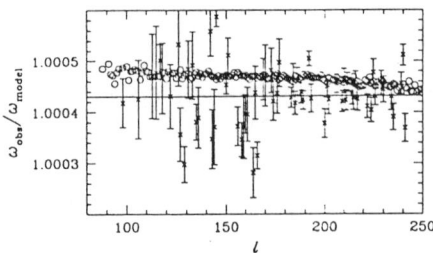 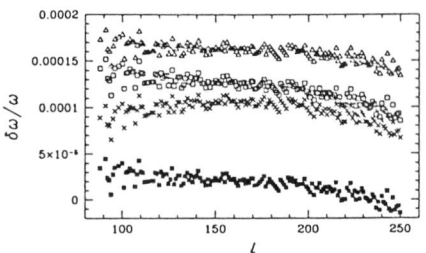

Figure 1. The ratio of observed and model frequencies for f-modes. The crosses with error bars represent the GONG data, while circles represent the MDI data. The horizontal line defines the average over all modes for the GONG data.

Figure 2. The relative difference in frequencies of f-modes between the observed (MDI) values and those of different solar models, with a radius of 695.78 Mm. The filled squares, open squares, triangles and crosses respectively, mark the difference for Models M1, M2, M3 and M4.

Further, the lower hydrogen abundance in model M4 also gives rise to a small dip at the lower frequency end in the frequency differences.

When better data on the f-mode frequencies at higher ℓ become available, it may be possible to separate out contributions from various sources and estimate the value of solar radius more accurately. Since the frequencies of these modes can be determined to a relative accuracy of 10^{-5}, in principle, it would be possible to determine the solar radius much more accurately. The effects arising from different treatment of surface layers are unlikely to change with solar cycle, and any possible temporal variation in solar radius may be reliably determined via the variation in f-mode frequencies with time.

It may be noted that most of the current standard solar models (e.g., Christensen-Dalsgaard et al. 1996) use the value of solar radius as 695.99 Mm with the surface defined at a level where the optical depth is between 1 and 1/3. Clearly, there is a need to revise these models with a possible reduction in the value of solar radius by about 200-300 km. Likewise, most helioseismic inversions that assume a similar definition of solar radius will also need to be revised. In order to estimate the possible errors arising from uncertainty in the value of solar radius, we have attempted helioseismic inversions for the sound speed using the GONG months 4–10 data with different estimates of radius. It appears that the difference caused due to a change of solar radius by 210 km, is much more than the estimated errors in helioseismic inversions over most of the solar interior (Antia 1997). Clearly, we need a reliable measure of the solar radius for inferring conditions in the solar interior with sufficient accuracy.

References

Antia, H. M. 1997, A&A (in press) astro-ph/9707226
Canuto, V. M., & Mazzitelli, I. 1991, ApJ 370, 295
Christensen-Dalsgaard, J., Däppen, W. et al. 1996, Science 272, 1286.
Ghosh, P., Chitre, S. M. & Antia, H. M. 1995, ApJ 451, 851
Iglesias, C. A. & Rogers, F. J. 1996, ApJ 464, 943
Kurucz, R. L., 1991, In Stellar atmospheres: beyond classical models, ed. L. Crivellari, I. Hubeny, D. G. Hummer, NATO ASI Series (Dordrecht: Kluwer), 441
Rosenthal, C. S. & Christensen-Dalsgaard, J. 1995, MNRAS 276, 1003
Schou J., Kosovichev A. G., Goode P. R., Dziembowski W. A., 1997, ApJL (in press)

LOI AND GONG LOW-DEGREE ROTATIONAL SPLITTINGS

T.APPOURCHAUX, M.C.RABELLO-SOARES
ESA/ESTEC
P.O.Box 299, 2200 AG, Noordwijk, Pays-Bas

AND

L.GIZON
W.W.Hansen Experimental Physics Laboratory
Center for Space Science and Astrophysics, Stanford University
Stanford, CA 94305-4085, USA

1. Observations

Two different data sets have been used to derive low-degree rotational splittings. One data set comes from the Luminosity Oscillations Imager of VIRGO on board SOHO; the observation starts on 27 March 96 and ends on 26 March 97, and are made of intensity time series of 12 pixels (Appourchaux et al, 1997, Sol. Phys., 170, 27). The other data set was kindly made available by the GONG project; the observation starts on 26 August 1995 and ends on 21 August 1996, and are made of complex Fourier spectra of velocity time series for $l = 0 - 9$. For the GONG data, the contamination of $l = 1$ from the spatial aliases of $l = 6$ and $l = 9$ required some cleaning. To achieve this, we applied the inverse of the leakage matrix of $l = 1, 6$ and 9 to the original Fourier spectra of the same degrees; cleaning of all 3 degrees was achieved simultaneously (Appourchaux and Gizon, 1997, these proceedings).

2. p-mode fitting

The Fourier spectra were fitted using the technique described by Appourchaux and Gizon, (1997) (these proceedings) after the work of Schou (1992) (PhD thesis, Aarhus University). The leakage matrix of LOI and GONG were derived from theoretical computations for spherical harmonics filters. The noise covariance matrix of the LOI was also computed theoretically, while that of GONG was measured directly on the data using cross-spectra. For each n, l mode we fitted a central frequency, a common linewidth to the $2l + 1$ components, a_i splitting components (using Clebsh-Gordan polynomials), $2l + 1$ amplitudes, and various noise parameters (3 for the LOI and $l + 1$ for GONG).

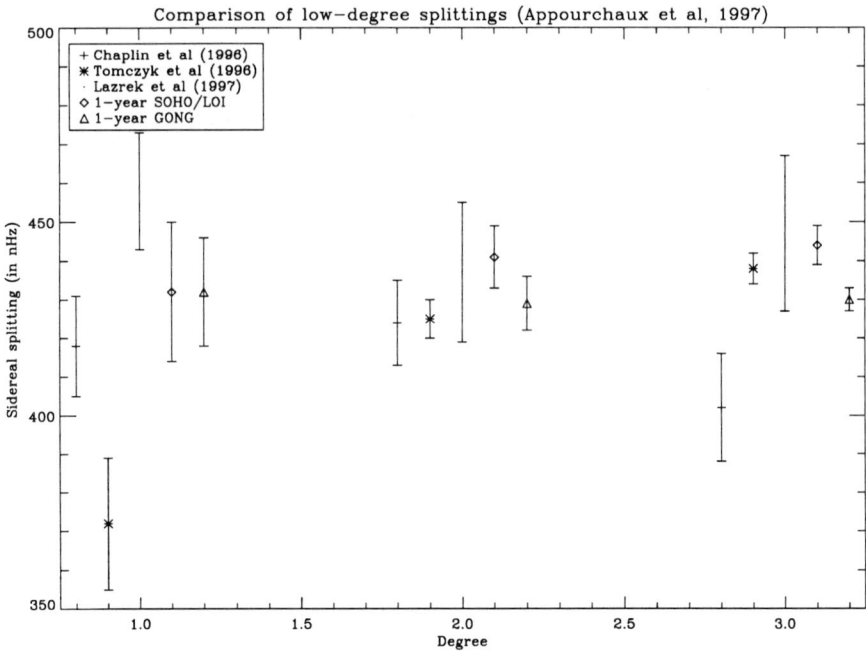

Figure 1. Comparison of low-degree splittings from different instruments. For all the instruments but GOLF (Lazrek et al, 1997, Sol. Phys., in press) the splittings were computed using a weighted mean over n using the formal errors returned by the fit. All the instruments but GOLF displays a_1; GOLF displays a weighted average over the odd a_i. The LOWL data are known to be biased for $l = 1$ due to the poor signal-to-noise ratio and the presence of the temporal alias of the $l = 3$. The LOI and GONG splittings result from this analysis. Note that 'our' GONG splittings do not have the large systematic errors that of the GONG project; these latter were obtained by fitting power spectra. (Chaplin et al, 1996, MNRAS, 280, 849; Tomczyk et al, 1996, ApJ, 488, L57)

3. Results

Figure 1 shows the results of our measurements compared to other previous results. We noticed that the LOI data for $l = 2-3$ are systematically higher than the other imaging velocity instruments. The preliminary forward modeling, that we made, showed that the shell $0 - 0.3 R_\odot$ was rotating about 30 ± 18 % faster with the LOI than with the other instruments. We have shown that an error in one of the leakage element will lead to underestimating the a_1 coefficient. For example, for GONG and $l = 1$ we have $a_1 = 430 - 160(1 - \alpha)^2$ where α is the matrix element relative to its nominal value. For a 10 % error in the leakage element, we underestimate the splitting by less than 2 nHz. This is a general feature of any fitted a_1, either GONG or LOI (Appourchaux et al, 1997, submitted to A&A). If the LOI splittings are overestimated, we have no reason to believe that it is directly due to the fitting procedure. Other source of bias such as leakage of other degrees or of solar origin will be investigated.

A SEARCH FOR $\ell = 2$ ASYMMETRIES IN BISON DATA

W. J. CHAPLIN, Y. ELSWORTH, G. R. ISAAK, C. P. MCLEOD AND
B. A. MILLER
School of Physics & Astronomy, University of Birmingham
Edgbaston, Birmingham B15 2TT UK
E-mail: wjc@star.sr.bham.ac.uk

AND

R. NEW
School of Science & Mathematics, Sheffield Hallam University
Sheffield, S1 1WB

An in-depth discussion of the analysis presented here can be found in an up-coming paper (Chaplin *et al.*, 1997).

1. Overview

The presence of a magnetic field will raise the degeneracy in ℓ of the resonant p-mode oscillations, via perturbations resulting from the Lorentz force. These degeneracy-raising effects will give rise to asymmetric mode-multiplet structures. Both Gough & Thompson (1990), and Dziembowski & Goode (1997) have addressed the implications and potential complications that might result from such phenomena. Here, in an attempt to reveal the presence of an asymmetric frequency structure in the low-degree $\ell = 2$ modes, i.e., to measure the asymmetries

$$[1/2 \cdot (\nu_{n,2,m=-2} + \nu_{n,2,m=+2})] - \nu_{n,2,m=0},$$

we have fitted $\ell = 2/0$ pairs in a series of BiSON power spectra generated from Doppler velocity residuals collected between 1990 May 8 and 1996 Dec 31. In all, we fitted: nine 8-month, six 16-month and a single 32-month frequency spectrum spanning the maximum-to-minimum falling phase of solar cycle 22. Gough & Thompson calculated that asymmetries ranging from ≈ 50 to $150\,\mathrm{nHz}$ might result from a buried magnetic field of the order of $10\,\mathrm{MG}$. Dziembowski & Goode used the BBSO helioseismic data (for $5 \leq \ell \leq 60$) in order to calibrate a model describing the multiplet contamination resulting from the Sun's near-surface magnetic activity. They found that one might expect asymmetries of up to $\sim 200\,\mathrm{nHz}$ ($n = 20$) at the solar activity maximum. They also indicated that the magnitude of the asymmetries should: increase with increasing n (also indicated by Gough & Thompson), and vary substantially with the solar activity cycle.

2. Results and discussion

We used maximum likelihood estimators to perform the fitting (χ^2 2-d.o.f. statistics), minimizing for a model where the frequency of each $\ell = 2$ component was allowed to freely float. The asymmetries were extracted from the frequencies returned by the fitting procedure, and the error bars computed from the formal uncertainties on the fitted component frequencies. Even though a series of simulations we performed indicated that there is possibly insufficient resolution in the 8-month spectra to extract asymmetries of the magnitude indicated by Dziembowski & Goode, and Gough & Thompson, we nevertheless include the 8-month results here for completeness. The global, mean asymmetry (i.e., for $10 \leq n \leq 21$) from the 8-month fits is 53 ± 46 nHz (from a total of 42 independent asymmetry measures), and from the 16-month fits 69 ± 32 nHz (from a total of 24 independent asymmetry measures). We also fitted a 32-month spectrum, and obtained asymmetry measures covering the range $10 \leq n \leq 19$. The mean, 32-month asymmetry is 17 ± 31 nHz (8 independent measures). Given the observed precision of the data, we can find no evidence for any statistically significant increase in the measured asymmetries with frequency (i.e., n).

Clearly, we have insufficient numbers of data to measure – to a satisfactory level of precision – asymmetries of the magnitude suggested by Gough & Thompson, and Dziembowski & Goode. Our fits merely place an upper limit to any mean asymmetry – over the range $10 \leq n \leq 21$ – of ≈ 110 to 200 nHz (3σ). [We are reluctant to ascribe any significance to the $\sim 2\sigma$ result returned by the 16-month fits.] Given that our data set covers the full extent of the falling phase of a solar cycle, it is difficult to predict precisely the size of the global mean we might expect to measure. The calculations of Dziembowski & Goode do suggest, however, that our 3σ exclusion threshold may be of a similar magnitude. Regardless, asymmetries of the size suggested by these authors cannot be excluded by our data.

We also tested the 8 and 16-month data for any solar-cycle dependence in the $\ell = 2$ asymmetries, by computing the asymmetry shift per unit change in the 10.7-cm radio flux at each n. If Dziembowski & Goode are correct, then the calculated coefficients should increase with increasing n, reaching ~ 1.7 nHz per unit RF at $n \sim 20$. Our solar-cycle analysis indicates that – as expected – the precision in the measured asymmetries is insufficient to reveal any functional dependence with n. The 8-month coefficients exclude any solar maximum-to-minimum change in the mean asymmetries at the ~ 270 nHz level (1σ), while those from the 16-month data do so at the ~ 200 nHz level (1σ).

References

Chaplin W. J., Elsworth Y., Isaak G. R., McLeod C. P., Miller B. A. and New R., 1997, *MNRAS*, submitted
Dziembowski W. A. and Goode P.R., 1997, *A&A*, **317**, 919
Gough D. O. and Thompson M. J., 1990, *MNRAS*, **242**, 25

LOW-DEGREE P-MODE SOLAR CYCLE TRENDS FROM BISON DATA

W. J. CHAPLIN, Y. ELSWORTH, G. R. ISAAK, C. P. MCLEOD AND B. A. MILLER
School of Physics & Astronomy, University of Birmingham
Edgbaston, Birmingham B15 2TT UK
E-mail: wjc@star.sr.bham.ac.uk

AND

R. NEW
School of Science & Mathematics, Sheffield Hallam University
Sheffield, S1 1WB

An in-depth discussion of the analysis presented here can be found in an up-coming paper (Chaplin et al., 1997).

1. Trends at frequencies up to $\sim 3900\,\mu$Hz

In order to investigate the solar-cycle dependence of the low-degree p-mode frequencies, we have analysed eighteen 4-month frequency spectra generated from BiSON Doppler velocity residuals collected between 1991 January 01 and 1997 January 05. These data cover the falling phase of solar activity cycle 22, up to the cycle 22/23 boundary. Fig. 1(a) shows frequency shifts, averaged over two orders in n, up to $\sim 3900\,\mu$Hz – as derived from the analysis of the 4-month spectra – normalised to unit change in the 10.7-cm radio flux. The dashed line lying above the BiSON data is a fit to the BBSO 1989-minus-1986 frequency shifts, for $4 \leq \ell \leq 140$ (Libbrecht & Woodard, 1991). The dotted line passing through the data corresponds to the best-scaled fit of the BBSO data to the BiSON data – the best-fit requires the BBSO data to be scaled by 0.71 ± 0.03. This is reasonably consistent – as expected – with the mean, overall ratio of the inverse mode masses of those data used in the BiSON and BBSO analyses (≈ 0.67). Fig. 1(b) shows the BiSON activity-normalised frequency shifts, plotted as a function of inverse mode mass – here, as anticipated, there is a clear correlation between the variables.

2. Trends at frequencies above $\sim 4000\,\mu$Hz

At high frequencies we have fitted the low-ℓ pairs to a single-Lorentzian model in 4 and 8-month averages of short, 7.6-d spectra. The bottom figure below shows the solar-cycle dependence – again normalised to unit change in the 10.7-cm flux – derived

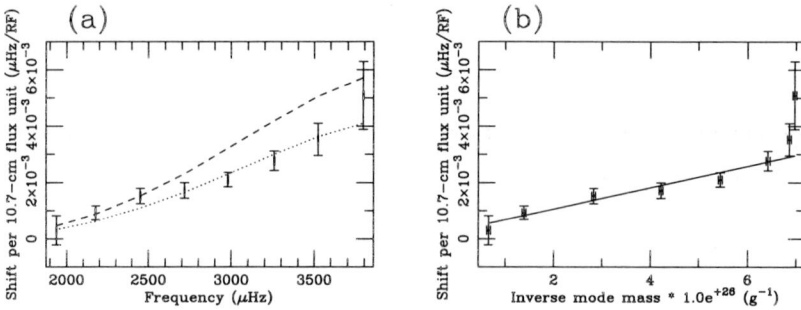

Figure 1. Solar-cycle dependence of fitted BiSON modes at frequencies up to $\approx 3900\,\mu$Hz. In (a) – plotted as a function of frequency. The dashed line lying above the BiSON data is a fit to the BBSO 1989-minus-1986 frequency shifts, for $4 \le \ell \le 140$. The dotted line passing through the data corresponds to the best-scaled fit of the BBSO data to the BiSON data. And in (b) – plotted as a function of inverse mode mass.

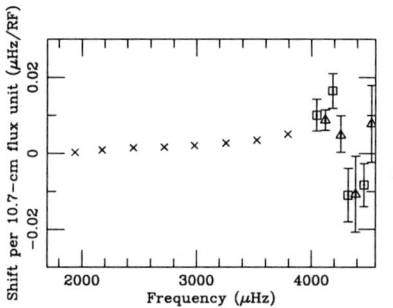

Figure 2. Solar-cycle dependence at high frequency – again normalised to unit change in the 10.7-cm flux – derived from single Lorentzian fits to 4-month averaged $\ell = 2/0$ pairs (triangular symbols with errors) and $\ell = 3/1$ pairs (square symbols with errors). The crosses are the data from Fig. 1(a).

from single Lorentzian fits to 4-month averaged $\ell = 2/0$ pairs (triangular symbols with errors) and $\ell = 3/1$ pairs (square symbols with errors). We also include data from Fig. 1(a) as a lower-frequency reference. If we consider the extracted solar-cycle coefficients, the data are suggestive of a turnover and possible sign reversal (a straight-line fit gives a gradient significant at the 3σ level). Clearly, one must be cautious (e.g., possible systematic complications for the $\ell = 3/1$ fits; see Chaplin et al., 1997).

References

Chaplin W. J., Elsworth Y., Isaak G. R., McLeod C. P., Miller B. A. and New R., 1997, *MNRAS*, submitted
Libbrecht K. G. & Woodard M. F., 1991, *Science*, **253**, 152

COMMENTS ON THE INFLUENCE OF SOLAR ACTIVITY ON P-MODE OSCILLATION SPECTRA

L. GIZON
*W.W. Hansen Experimental Physics Laboratory,
Stanford University, Stanford CA 94305, U.S.A.*

1. Spectral signature of a persistent patch of activity

The systematic p-mode frequency changes which are observed through the solar cycle are believed to be associated with near-surface perturbations confined to the magnetically active latitudes. In this paper, we study the perturbation arising from the presence of a large "active region", corresponding to a localized structural change in a thin region close to the photosphere. We shall ignore the difficult question of the magnitude of the effect, and simply consider some geometric and observational implications for low-degree modes.

The technique (Gizon, 1995) consists of calculating the eigenstates in a frame in which both the active region and the rotation can be assumed to be steady perturbations (see also e.g. Gough and Taylor, 1984). Figure 1 schematically displays the eigenfrequencies of an $l = 2$ multiplet. Rotation was presumed to be known (Kosovichev, Schou, Scherrer, et al., 1997). In a frame co-rotating with the active region, the $(2l + 1)$-fold azimuthal degeneracy is completly lifted. Because nonuniform splitting due to differential rotation is small, and because the spatial extent of the active

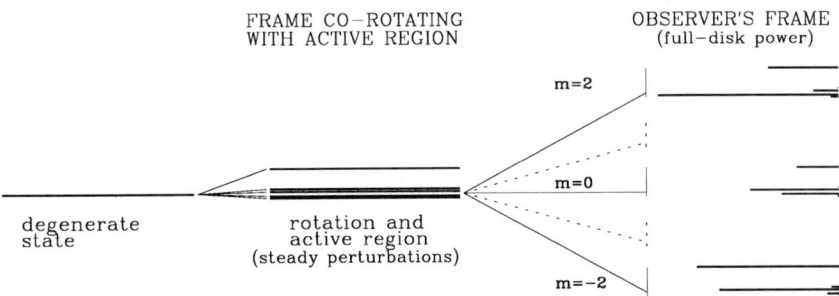

Figure 1. Energy diagram for a quadrupole mode in a frame co-rotating with the active region. Also shown is the full-disk power spectrum seen by an observer on Earth or by SOHO. In this diagram, the *assumed* perturbation from the active centre, lying at latitude 5°, produces a mean frequency shift of the whole multiplet of 80 nHz.

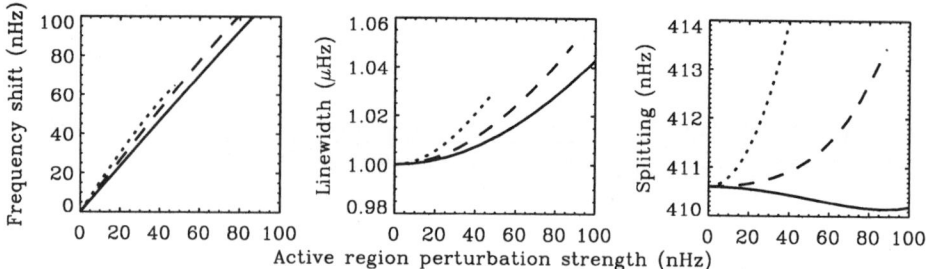

Figure 2. Estimates of the frequency shift, linewidth and rotational splitting of the ($l = 2$, $n = 20$) multiplet, versus active region perturbation strength (expressed in terms of the power-weighted average of the $(l+1)(2l+1)$ individual frequency shifts). Each individual peak composing the multiplet is assumed to have a linewidth of 1 μHz. The solid, dashed and dotted lines refer to the active-region latitudes $5°$, $20°$ and $35°$ respectively.

region is also small, only one of the eigenfrequencies (in co-rotating frame) is significantly affected. The fine separation between the other $2l$ components is essentially due to differential rotation. For an observer orbiting about the Sun's rotation axis at the Earth's orbital angular velocity, the full-disk power spectrum appears to have $(l+1)(2l+1)$ peaks, most of which cannot be resolved (Figure 1).

2. Some observational consequences

We may generate the synthetic profile of an $l = 2$ multiplet by adding the 15 individual profiles, represented by Lorentz curves of equal linewiths. Usually, observers model the expectation of the power spectrum by the sum of three Lorentz functions of unknown amplitudes. Figure 2 shows the errors in the determination of the parameters by use of this approximate model. Estimates of the parameters are obtained by a least-squares technique, where each point is assigned a weight inversely proportional to its value. We see that the centroid frequency and the linewidth are likely to be slightly overestimated. The error in the rotational splitting frequency depends on the latitude of the active region. Of course, the magnitudes of these errors ultimately depend on the physics, and also on the fitting procedure which is used.

Dziembowski and Goode (1997) suggested that seismic information about the solar core should be purged from the Sun's magnetic activity. The latitudinal structure of the near-surface perturbation has to be taken into account in this process. I suggest that longitudinal variations could have some importance as well. A realistic study would have to consider the fact that solar activity is a transient phenomenon.

I thank C.J. Durrant, D.J. Galloway, D.O. Gough and P.R. Wilson for discussions.

References

Dziembowski, W.A. and Goode, P.R.: 1997, *Astron. Astrophys.* **317**, 919.
Gizon, L.: 1995, in Vincent, R.A., et al. (eds.), *Conf. on Solar and Terrestrial Physics STEP'95*, University of Adelaide, Adelaide, pp. 173–176.
Gough, D.O. and Taylor, P.P.: 1984, *Mem. S.A.It.* **55**, 215.
Kosovichev, A.G., Schou, J., Scherrer, P.H., Bogart, R.S., Bush, R., Hoeksema, J.T., et al.: 1997, *Solar Phys.* **170**, 43.

TIME VARIATION OF VELOCITY FLOWS FROM RING DIAGRAMS: A FIRST APPROACH

J. PATRÓN AND I. GONZÁLEZ HERNÁNDEZ
Instituto de Astrofísica de Canarias

AND

D.-Y. CHOU AND THE TON TEAM*
Physics Department, Tsing Hua University

*TON team includes: M.-T. Sun, H.-K. Chang, H.-R. Chen, S.-J. Yeh (Taiwan); A. Jimenez, M. C. Rabello-Soares (Spain); G. Ai, G.-P. Wang (China); H. Zirin, W. Marquette (USA); S. Ehgamberdiev, and S. Khalikov (Uzbekistan).

1. Introduction

The Ring Diagram analysis is a technique designed to infer the presence of horizontal velocity flows under the solar surface by the analysis of the power spectra of solar oscillations in three dimensions: the two components of the horizontal wave number and temporal frequency (Hill, 1988 and Patrón *et al.*, 1995). This procedure is applied as a local analysis technique, performed for a region of the Sun of several heliographic degrees (typically around 15x15 degrees square). As the Sun is rotating, we must track a chosen region as long as it is present at the visible part of the solar disk, and we can continue the tracking after several solar rotations. The comparison of the estimated flow fields obtained at different times can give some ideas about the temporal variations of the flows.

2. Data acquisition and reduction

The TON images (Chou *et al.*, 1995) consist of intensity Ca K II line images with a resolution of about 2 arcsec per pixel. The spatial resolution at the center of the solar disk is of about 1379 Km per pixel. A region of about 15x15 degrees square has been tracked on the whole solar disk image at the rotation rate of the solar equator. This region is centered at the solar equator and Carrington longitude 29.1738 degrees.

The regions have been remapped onto a great-circle space centered at the center of the region. Temporal series of 512 minutes have been constructed for seven dates: 4/5/97, 5/1/97, 5/28/97, 6/27/97, 8/20/97, 8/21/97 and 9/18/97. The time delay between series is about one solar rotation, except between day 5 and 6, which is 24 hours for testing purposes. Finally a tri-dimensional FFT has been performed in the two spatial coordinates and time, with the resulting tri-dimensional power spectra. A horizontal flow field have been fitted to this spectra and finally inverted to get the two components of the flow (East-West and North-South directions, U_x and U_y) as a function of depth (Patrón et al., 1995), as shown in Figure 1. Positive velocities are from South to North and from East to West in the solar disk.

3. Results and conclusions

The temporal resolution of the results, one solar rotation, does not seem to be enough to estimate the temporal evolution of the flows, since the variations do not continue a clear evolution from one rotation to another. In the case of data taken 24 hours apart (8/20/97 and 8/21/97), the flows are more stable, but still big changes occur very close to the surface.

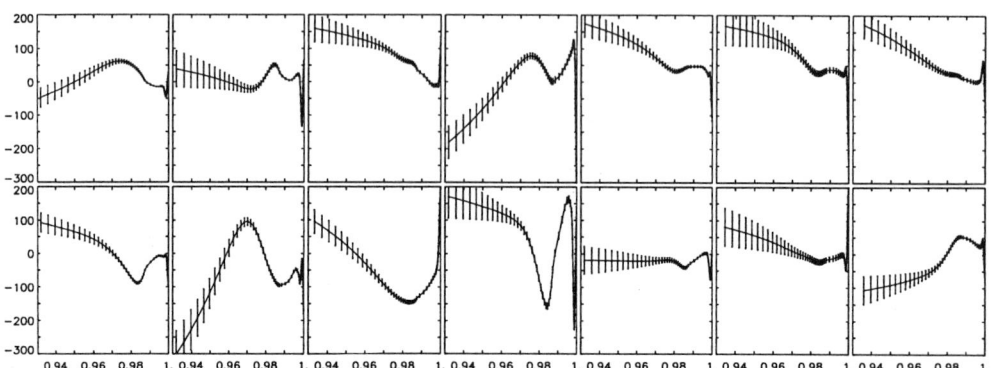

Figure 1. U_x (upper) and U_y (lower) for the 7 time spans (m/s against r/R).

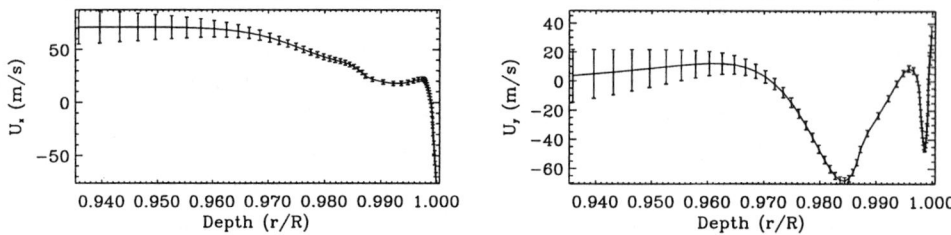

Figure 2. Temporal average of the horizontal flow components.

An average of the seven set of flows have been performed. It is difficult the interpretation of the average in the North-South direction, since we don't have any expectations about steady North-South flows. But in the East-West direction, we have differential rotation with depth. We need to remember that the tracking procedure removes from the data only a constant value corresponding to the estimation of the surface rotation rate, so that, differential rotation still remains in the data. A plot of the average of the components of the flows is shown in Figure 2.

A great part of the computation required for this work was carried out on the Jet Propulsion Laboratory Cray J90 supercomputer.

References

Chou, D., et al, 1995, *Solar Physics*, **160**, 237.
Hill, F. 1988, *Ap. J.*, **333**, 996.
Patrón, J., Hill, F., Rhodes, E.J., Jr., Korzennik, S.G., and Cacciani, A., 1995, *Ap. J.*, **455**, 476.

PUMPING OF ROSSBY WAVES AND VORTICES AT THE BASE OF THE SOLAR CONVECTION ZONE

EVGENIY TIKHOMOLOV
Institute of Solar-Terrestrial Physics SD RAS, Irkutsk, 664033, P.O. Box 4026, Russia

1. Introduction

Rossby vortices excited near the base of the solar convection zone are very appealing objects for interpretation of a number of solar phenomena such as long-lived large-scale magnetic structures, the poleward drift of the axisymmetric components after the polar field reversal, and a peculiar long-term behavior of the nonaxisymmetric components (Tikhomolov and Mordvinov 1996).

Rossby vortices can appear due to break down of the Rossby wave with large wavelength (possibly of the order of the solar perimeter) like on the Earth or due to hydrodynamical instability of the zonal flow developed near the base of the solar convection zone. In any case the important issue that should be addressed is the mechanism capable to force large-scale flows in the region under consideration.

2. The model

It is supposed that mechanism which pumps large-scale flows operates in high latitudes, near 60°. In that region large-scale flows are in quasigeostrophic regime and some simplifications can be involved. To take into consideration the main specific conditions near the base of the solar convection zone a 2-layer model is developed: upper one-third part is convectively unstable, lower two-thirds part are convectively stable. Upper and lower surfaces are nondeformable, impenetrable, stress-free, and have constant temperatures.

The main distinction of the model suggested from the existing models is that the location of the interfacial surface dividing convectively stable and

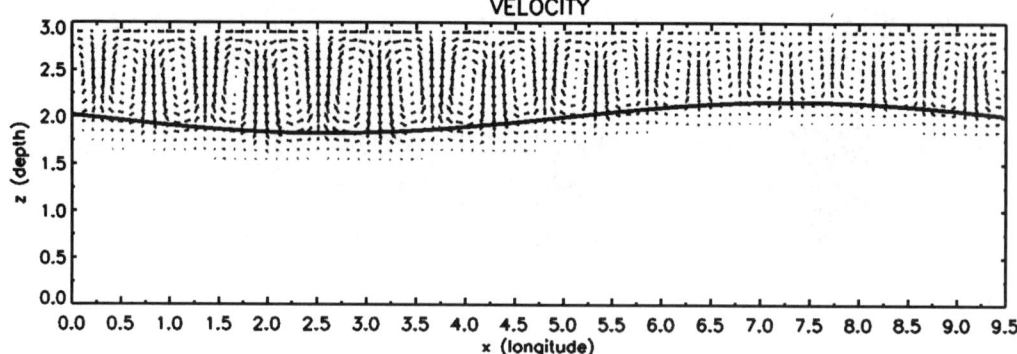

Figure 1. Stationary velocity field established after the initial slight large-scale disturbance of pressure field of stationary convection.

unstable regions slightly varies in depth in accordance with the disturbance of pressure caused by the large-scale flows.

3. Results

2D numerical simulations for the model suggested show that stationary convective pattern is in a rather weak equilibrium. Slight disturbance of pressure field by the Rossby wave or by zonal flow with shear in latitude disrupts this equilibrium and amplitude of the flow increases. Nonlinear effects lead to limitation on the amplitude of the flows and stationary picture with disturbed interface is established. This picture for Rossby wave is shown in Fig. 1. Thick solid line indicate the location of the interface dividing convectively stable and unstable regions.

The physical nature of the forcing of the Rossby wave or zonal flow is the appearance of the large-scale temperature distribution dependent on longitude and latitude. The instability mechanism is akin to the one appearing in a rotating single fluid layer heated from below with a deformable upper stress-free surface (Tikhomolov 1994, 1996). So it can also be referred to as the deformational long-wave instability mechanism.

References

Tikhomolov, E.M. (1994) Sustenance of vortex structures in a rotating fluid layer heated from below, *JETP Letters* **59**, pp. 163–167.

Tikhomolov, E. (1996) Short-scale convection and long-scale deformationally-unstable Rossby wave in a rotating fluid layer heated from below, *Phys. of Fluids* **8(12)**, pp. 3329-3337.

Tikhomolov, E.M. and Mordvinov, V.I. (1996) The peculiar behavior of the large-scale components of the solar magnetic field as a result of Rossby vortex excitation beneath the convection zone, *Astrophys. J.* **472**, pp. 389-398.

OBSERVATION OF LOW-DEGREE MODES FROM SOHO/MDI USING OPTIMAL MASKS

T. TOUTAIN
Nice Observatory
B.P. 4229, F-06304 Nice cedex

A. KOSOVICHEV
HEPL
Stanford University, USA

AND

T. APPOURCHAUX
ESTEC, ESA, The Netherlands

The overlap of peaks of split multiplets in oscillation power spectra and the leakage of other degree modes could significantly affect the measurements of the rotational frequency splitting. We have developed optimal masks which allow to isolate individual components of the multiplets. The method is applied to the Michelson Doppler Imager Low-l data. The results of the mode frequency measurements are reported.

Optimal masks

The MDI instrument provides, among others, a LOI-proxy velocity signal (Scherrer et al.(1995)) consisting of 180 mean velocities, one for each bin of the CCD detector. The signal gathered over 1 year (May 96 to April 97) has a duty cycle better than 99%. Because we are interested only in very low degree modes ($l \leq 3$) we reduce the time series by binning up the 180 pixels to get to the 12 scientific pixels of the LOI (Appourchaux et al., 1997). Instead of applying spherical harmonic masks to the time series we follow Kosovichev (1986) and apply a Singular Value Decomposition technique (equivalent to using the inverse of the leakage matrix) to get optimal masks. Figure 1 shows a $l=1$ m-ν diagram obtained using optimal masks. Crosstalks between modes have disappeared making the fitting of lines easier. Modes are fitted in the Fourier domain, as described by Appourchaux and Gizon(1997). Table I gives the central frequencies we obtain.

References:
Appourchaux T. et al., 1997, Sol. Phys. **170**, 24
Appourchaux, T. and Gizon, L., 1997, these proceedings
Kosovichev, A.G. 1986, Bull. Crimean Astrophys. Obs. **75**, 19
Scherrer,P.H.,et al., 1995, Sol. Phys. **162**, 129

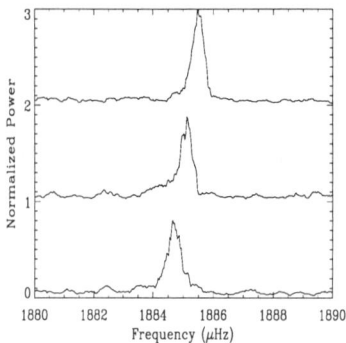

Figure 1. Smoothed m-ν diagram for $l=1$ using optimal masks on the 1-year MDI time series

TABLE 1. Central frequencies (in μHz) with their 1-σ error bars

order	$l=0$	$l=1$	$l=2$	$l=3$
10	1548.31 ± 0.07		1674.51 ± 0.20	1729.06 ± 0.03
11	1686.58 ± 0.04	1749.26 ± 0.06	1810.27 ± 0.04	1865.28 ± 0.03
12	1822.20 ± 0.05	1885.06 ± 0.03	1945.76 ± 0.03	2001.19 ± 0.04
13	1957.40 ± 0.05	2020.76 ± 0.03	2082.08 ± 0.03	2137.75 ± 0.03
14	2093.46 ± 0.05	2156.78 ± 0.04	2217.60 ± 0.03	2273.34 ± 0.03
15	2228.67 ± 0.06	2291.92 ± 0.04	2352.22 ± 0.03	2407.56 ± 0.03
16	2362.74 ± 0.06	2425.46 ± 0.03	2485.80 ± 0.03	2541.53 ± 0.03
17	2496.06 ± 0.06	2559.06 ± 0.04	2619.51 ± 0.03	2676.13 ± 0.03
18	2629.81 ± 0.06	2693.26 ± 0.03	2754.46 ± 0.03	2811.47 ± 0.03
19	2764.10 ± 0.06	2828.06 ± 0.04	2889.58 ± 0.03	2947.00 ± 0.03
20	2898.91 ± 0.03	2963.24 ± 0.03	3024.70 ± 0.03	3082.19 ± 0.03
21	3033.73 ± 0.07	3098.04 ± 0.04	3159.76 ± 0.03	3217.66 ± 0.03
22	3168.57 ± 0.06	3233.02 ± 0.04	3294.93 ± 0.04	3353.41 ± 0.04
23	3303.36 ± 0.07	3368.64 ± 0.05	3430.66 ± 0.05	3489.44 ± 0.06
24	3438.88 ± 0.08	3504.06 ± 0.07	3566.73 ± 0.07	3625.93 ± 0.07
25	3574.40 ± 0.18	3640.25 ± 0.10	3702.90 ± 0.12	3762.67 ± 0.10
26	3710.36 ± 0.22	3776.31 ± 0.20	3839.57 ± 0.20	3899.53 ± 0.15
27	3846.96 ± 0.33	3912.85 ± 0.20	3976.31 ± 0.28	4174.71 ± 0.35

THE INTERNAL ROTATION RATE INFERRED FROM LOWL AND GONG DATA

Using a Forward Method

LI, Y. AND P. R. WILSON
The University of Sydney
School of Mathematics and Statistics,
The University of Sydney, NSW 2006, Australia

Recently the LOWL Group has made available 2-year averages of frequency splitting data obtained mainly for low values of the degree ℓ. Charbonneau *et al.* (1997) find a nearly flat rotational curve in the deep radiative core from these data using a Genetic Forward Modeling method. Subsequently, they test the assumption of latitude independence for $r < 0.5$ by performing a 1.5-D inversion, and find a rotation rate which is 25% greater at the pole than at the equator for $0.1 < r < 0.2$.

Here we use a different forward modelling approach, in which the LOWL data are analysed in conjunction with the GONG data, using a modified version of the method described in Wilson, Burtonclay and Li (1996a).

We selected 998 multiplets from the LOWL data and 957 multiplets from 4-month GONG data (calculated by J. Schou) consisting of the Clebsch-Gordon (C-G) coefficients, $a_i^{n\ell}$, i=1, 3, 5, at frequency range $1500 < \nu < 3500$ nHz. To achieve a self consistent solution from the outer regions through to the radiative core, we combine these two data sets. There are 610 multiplets common to both sets. Thus the combined set includes 1345 multiplets. Differential mean C-G coefficients $\alpha_{2s+1}^{n\ell}$, s=0,1,2 are derived for each corresponding $a_i^{n\ell}$, i=1, 3, 5 using the same method as in Wilson, Burtonclay and Li (1996a).

To avoid the effects of possible systematic difference between the two data sets, we seek a difference d_{2s+1}, which minimizes the sum $S_{2s+1} = \sum_j \{a_{2s+1}^j(GONG) - a_{2s+1}^j(LOWL) - d_{2s+1}\}^2$ where the summation includes only the common multiplets, and j has replaced the multiplet identifier $n\ell$. It is given by $d_{2s+1} = \sum_j (a_{2s+1}^j(GONG) - a_{2s+1}^j(LOWL))/610$. We increase all coefficients in the LOWL data set by this amount, and take the average for the common multiplets.

If the rotation rate $\Omega(r, \lambda)$ at fractional radius r and latitude λ is expressed as, $\Omega(r, \lambda) = \sum_{s=0}^{s_{max}} W_{2s+1}(r)\psi_{2s+1}(\mu)$, where $\mu = \sin\lambda$, and $\psi_1(\mu) = 1$, $\psi_3(\mu) = 1 - 5\mu^2$, and $\psi_5(\mu) = 1 - 14\mu^2 + 21\mu^4$, then the rotation coefficients $W_{2s+1}(r)$ may be found by solving the integral equations

$$\alpha_{2s+1}^j = \int_0^1 W_{2s+1}(r)\frac{1}{j}\{\sum_{j'=1}^j K_{2s+1}^{j'}(r)\}dr \qquad (1)$$

where $K^j_{2s+1}(r)$ is the kernel.

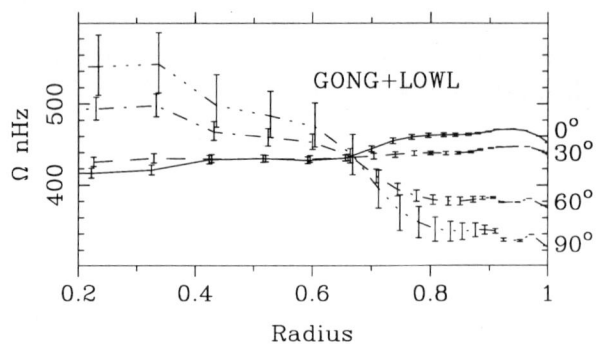

Figure 1. Rotation profiles from combined GONG and LOWL data.

To solve Equation (1), the solar interior is divided into 20 concentric shells. The rotation coefficients at the surface are set equal to values consistent with the plasma rotation rates measured by Snodgrass (1984). Initially, estimates for rotation coefficients at each shell boundaries are given. Between any two shell boundaries, a linear interpolation is carried out to give the rotation coefficients. The initial estimates are then adjusted beginning at the first shell boundary next to the surface and proceeding deeper, till the theoretical differential mean C-G coefficients fit the observations within 1-σ uncertainty range.

The solutions are shown in Figure 1. The error bars were derived by a method described in Wilson, Burtonclay and Li (1997).

The solutions show a flat equatorial angular velocity below $r = 0.66$ and an increased angular velocity at high latitudes, which are in qualitative agreement with Charbonneau et al. (1997).

In order to test the solution further, we consider a solution which is independent of latitude (setting $W_3(r)$ and $W_5(r)$ to zero) for $r \leq 0.66$. The corresponding theoretical differential means fall outside the 1-σ uncertainty intervals for $r < 0.5$ and outside Simultaneous Coverage Probability intervals (Wilson, Burtonclay and Li, 1996b) at a deeper layer.

We conclude that a latitude dependent rotation profile below $r = 0.66$ is a significant feature of this inversion of the combined LOWL-GONG data set.

Acknowledgment
We wish to thank S. Tomczyk and J. Schou for providing LOWL and GONG data.

References
Charbonneau, P., Tomczyk, S., Schou, J., & Thompson, M.J.: 1997, ApJ(submitted).
Snodgrass, H.B. 1984, Sol. Phys. 94, 13.
Wilson, P.R., Burtonclay, D., & Li, Y. 1996a, ApJ. 457, 440.
Wilson, P.R., Burtonclay, D., & Li, Y. 1996b, ApJ. 470, 621.
Wilson, P.R., Burtonclay, D., & Li, Y. 1997, ApJ. (in press).

LOCAL PROPERTIES OF THE SUN'S SEISMIC EVENTS

PHILIP R. GOODE[1] AND LOUIS H. STROUS[2]
Big Bear Solar Observatory[1]
New Jersey Institute of Technology[2]
Newark, New Jersey, USA

AND

THOMAS R. RIMMELE
National Solar Observatory
Sunspot, New Mexico, USA

1. Introduction

We made simultaneous high resolution observations of the solar granular field and the Sun's seismic events. We find that these events occur in the dark intergranular lanes. The events are preceded by the darkening of an already dark lane. On the leading edge of each event, there is a second more abrupt and precipitous darkening lasting a minute or two.

The events are the by-product of the local excitation of the Sun's p-modes. We see the quake energy being directly converted to normal mode energy. The total energy in the effluvia is of the order of magnitude necessary to power the entire p-mode spectrum. The newly created p-modes travel about 30% faster over the nearest bright granules than they do over the local dark lanes.

Until recently, it was widely believed that this deceleration of the upgoing granules induced a steady drumming that fed the resonant acoustic modes. However, Rimmele, et al. (1995) observed that arising in certain places in the dark intergranular lanes there are seismic events which they associated with the excitation of solar oscillations. Nordlund (1997) and co-workers used simulations of convection to argue that these places where events occur are ones in which there is a small, weak local granulation which disappears (this is the phase of gradual darkening of the lane in which we mightn't resolve the small granular features). After the disappearance, the lane catastrophically cools and collapses (this is the second phase for which the observations and simulations give a 1-2 minute timescale).

2. The Data

Our observations were made at the Vacuum Tower Telescope of the National Solar Observatory in Sunspot, New Mexico. The quiet Sun dataset discussed here is from Sept. 5, 1994. The data and the reduction of it are described in detail in Rimmele, et al. (1995), and references therein. One of the basic observational problems here is to distinguish the seismic event power from the dominant resonant mode power. To distinguish the two in our field of view (60"x60" patch of quiet Sun near disk center), Rimmele, et al. measured the velocity as a function of altitude in the photosphere for 65 min by observing the Doppler shift in the 543.4 nm Fe I absorption line. The Doppler shift as a function of depth in the line corresponds to the velocity as a function of altitude in the atmosphere the line spans.

3. The Sun's Seismic Events

We searched our velocity field for phase changes with altitude and found they fit one behavior pattern--uniformly looking like an ascending wave followed by a wave coming back down from above (with a time lag of about 4 to 5 minutes). The signature of these seismic events was detected in the solar photosphere which is not quite isothermal implying that any outgoing wave would be followed by a partially reflected wave. This combination of phase behaviors eased our effort to distinguish power from seismic events from that of normal modes which should show only a small phase change with altitude caused by dissipation, Restaino, et al. (1993)

In order to study the properties of the average event, we superposed slightly more than two thousand of the largest seismic events. This is also a convenient and efficient way to separate significant seismic events from background noise. The superposed events were pinned in time with $T = 0$ being the peak in the product of the square of the acoustic velocity and the vertical phase gradient for each event. After superposing the seismic events, each was oriented such that the intergranular lane was along the x-axis, see the three panels of Figure 1. If the events were purely traveling acoustic waves, the aforementioned product would be proportional to the acoustic or mechanical flux. Regardless, the product is a convenient measure of seismic events.

In Figure 1, we see the time evolution of the "seismic flux" shown as white contours of the averaged events. The events are pinned in time with $T=0$ is the time of peak seismic flux. The seismic flux contours in Figure 1 are normalized to the mechanical flux as defined in Rimmele, et al. The flux is superimposed on the averaged, evolving local granulation which is pinned in time to the peak

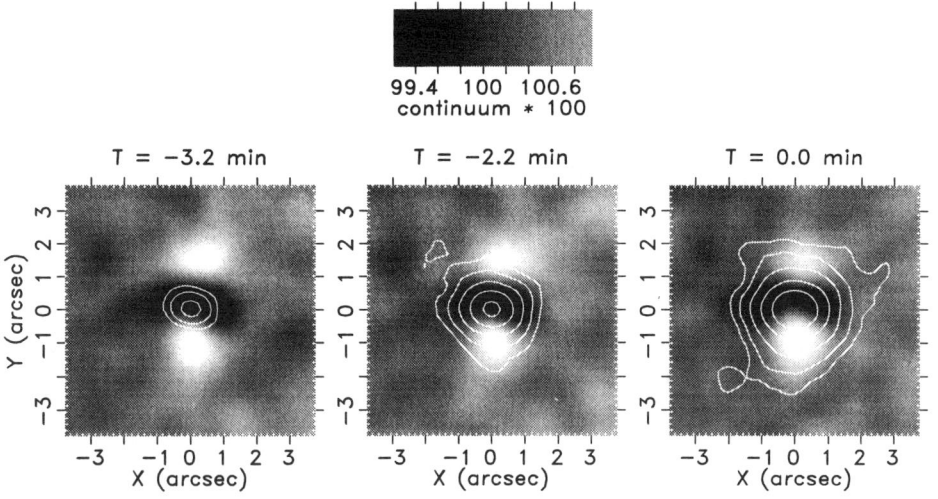

Figure 1. The superposed seismic flux, shown in white contours, is superposed on its local, averaged granulation. The contours are 0.25, 0.5, 1, 2, 4 ×10^7 ergs/cm^2/s. The flux is shown for $T = -3.2, -2.2$ and 0 min. The time steps in the data collection are about 30 s, and one time step before $T = -3.2$ min, the seismic flux is below 0.25 ×10^7 ergs/cm^2/s everywhere in the field of view.

flux. The fact that a dark lane with a bright granule on either side survives the averaging strongly emphasizes this geometry is a common feature of seismic events. The granular contrast is small because granular images are smeared by the averaging. It is also obvious that more than 2" from the center of the field of view, the granular structure is completely washed out by the averaging of many hundreds of events. In the figure, the bright contours represent the outgoing seismic flux. Clearly, the seismic events originate in the lanes, for more detail see Rimmele, et al. (1995).

Further, immediately after the peak in the seismic flux, the lane begins to narrow as though the granules on either side of the lane are being pulled together to fill the void left behind.

From Figure 1, it is also clear that seismic events have a finite duration. Over the three minute span shown, increasing seismic energy can be seen being fed into the aggregated events. After the $T=0$ peak in the figure, the seismic flux gradually subsides. The total duration of the expansive phase of the event is about five minutes because immediately after $T=0$ the flux begins to subside due of reflection from above. The fact that the duration of the expansive phase of the events is closer to five minutes needs to be emphasized for several reasons. The fact that the peak in the observed spectrum of global solar oscillations corresponds to modes with a period of about 5 minutes may well be connected to the comparable temporal duration of the seismic events. That is, because the events are not impulsive, and, in fact endure for a time comparable to the period of the oscillations, resonance may play a role in the excitation of the oscillations.

The five minute timescale of the events is consistent with that calculated in a linear, one dimensional model of seismic events, Goode, Gough and Kosovichev (1992). In their simulations, they showed that the mean velocity and phase properties of data like that of Rimmele, et al. are described well only if the typical event endures for about five minutes at its subsurface point of origin immediately beneath the base of the photosphere. They demonstrated, for instance, if the sub-photospheric model disturbance were more impulsive, the model signal would be too impulsive in the photosphere to describe the data.

4. The Driving of the Oscillations

In Figure 2, we show the superposition of the power at 150 km above the base of the photosphere and the instantaneous phase difference between that altitude and 180 km higher. The specific model altitudes were provided by Keil (1997, private communication). Both quantities in Figure 2 are shown as a function of time and horizontal distance from the event with $T = 0$ being the peak in the seismic flux for the superposed events. We remark that what is generally regarded as being convective power is subsonic and has been filtered out. The $k - \omega$ diagram for our data is shown in Figure 3.

The phase signature characteristic of seismic events is apparent clear out to about 1.4" from the center of the events. Beyond that distance, the phase change with altitude is essentially zero. However, there is excess power from the events going out almost 3". Beyond that distance, no excess power is apparent. The tendency of the power is to decrease as the square of the distance from the events. This tendency is what one might anticipate. The power propagates with a noticeably supersonic speed out to about 1.5". The speed can be estimated from Fig. 2a by measuring the time evolution of the peak in the v^2 power. The large apparent supersonic speed in this inner region probably reflects the finite horizontal size of the events there, as they erupt. Since the lane size is about 1", the horizontal extent of the events is no more than that. So, beyond 1.5" the speed mostly likely represents the true horizontal propagation of the disturbance resulting from the event. In that region, the speed is no more than modestly supersonic. It is likely that most of this power is in f-modes which are *asymptotically* (in terms of horizontal wavelength) surface waves. We first note that the region of the f-mode ridge is apparent in our $k - \omega$ diagram (see Figure 3). The f-modes form the lowest frequency ridge in the p-mode $k - \omega$

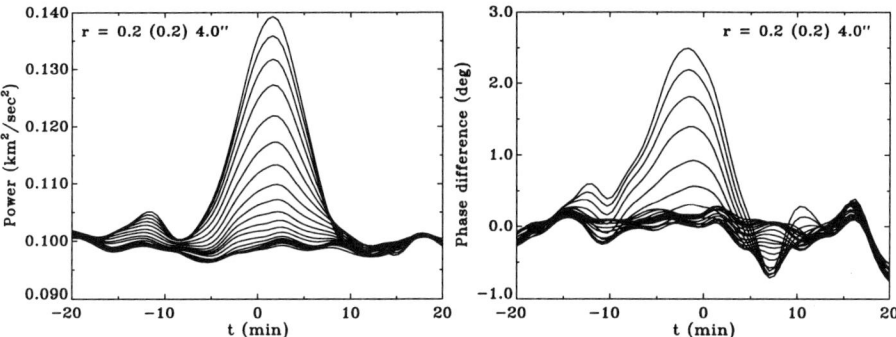

Figure 2. a) The average excess power (v^2) in the neighborhood of more than two thousand seismic events, superposed as in Figure 1, are as shown as a function of distance and time—starting at r=0.2" (the curve with the highest peak power) in steps of 0.2" out to 4.0". Successive distances show successively decreasing peak power out to about 3.0". b) The averaged phase difference between two altitudes in the photosphere (150 km and 330 km above the base of the photosphere) as a function of time and distance in steps of 0.2" as in a). The formal standard error on the averaged phase difference is about 0.1deg at each spatial and temporal point. The largest positive phase change is for r=0.2". A positive phase change corresponds to an upgoing wave. Successive distances show successively decreasing peak positive phase change out to about 1.4". A typical event has a peak phase difference significantly larger than that of the superposed, and therefore smeared result in the figure.

diagram.

Our contention that power has been fed from seismic events into the f-mode part of the spectrum of solar oscillations is greatly strengthened because the acoustic power delivered by the events to beyond 1.4" is: 1) characterized by no vertical phase change which means an essentially infinite vertical phase speed, 2) characterized by a five minute period and a group velocity roughly appropriate for f-modes, and 3) dominated by horizontal wavelengths consistent with those of the f-modes. Thus, power has been fed from seismic events into the Sun's normal modes. However, the normal modes we see are not ones that provide a deep seismic sampling of the Sun's interior. We emphasize that the f-modes are the only ones we could hope to see converted with our technique of following the power from individual sunquakes, since higher order p-modes would skip out of our field of view in a single refraction (see next paragraph).

The events occur just beneath the photosphere, and what we observe is the photospheric effluvia which is converted into atmospheric p-modes. However, much of the energy from these events is directed into the Sun. The process by which this latter energy is converted into p-modes is somewhat different.

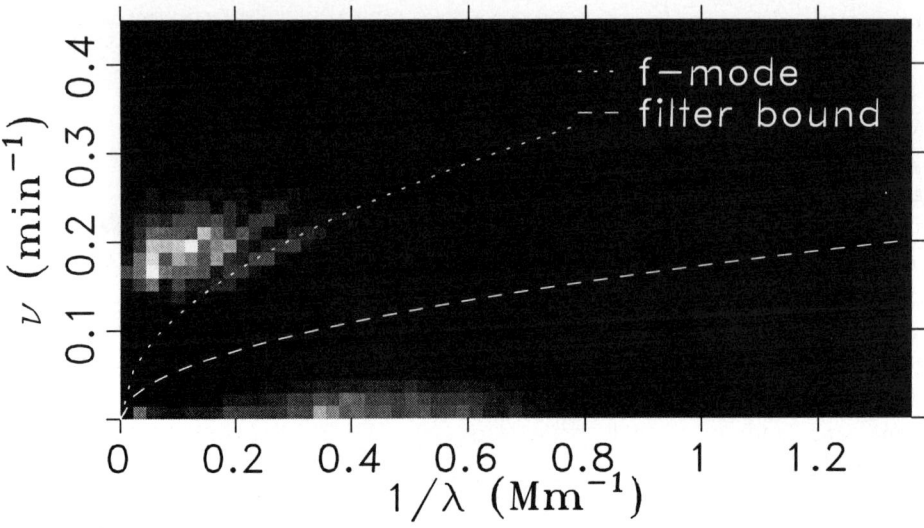

Figure 3. The $k - \omega$ diagram for our data. Model predictions for the f-mode ridge and subsonic filter bound (separating convective power from oscillatory power) are indicated by the bold white curves. Note that the sonic and subsonic powers are well-separated. Note also, that there is significant power in the region of the f-mode ridge.

The inward directed noise is eventually refracted back to the surface where it is partially reflected back into the Sun. Kumar (1993) has shown theoretically that after only a few refractions white noise would be converted into resonant modes. We cannot expect to detect the signature of such skips from our dataset, since the typical distance for a single refraction of, say a five minute period p_1 mode is much greater in extent on the solar surface than our field of view. Thus, we don't (and can't) see power being pumped into all modes, but we see power pumped into a part of the spectrum––the f-mode part for which the skip times are very short. Still, the seismic events can power the entire p-mode spectrum, Rimmele, et al. (1995).

5. Discussion

In the collapse of intergranular lanes that generates the events, one can invoke linear and nonlinear processes. Linear processes would be rarefaction waves generated by the collapse and the subsequent downgoing blob acting like a piston. Nonlinear ones would the implosion of the blob on itself and the infall of material behind the blob. Owing to the non-impulsive nature of the events, we might anticipate that they are dominated by linear effects. This would be consistent with the power in our k-ω diagram being dominated by two distinct regimes––one dominated by the p-modes and the other by the convective power.

The commonly accepted picture of the excitation of solar oscillations is one in which stochastic driving is done by turbulent convection, Goldreich and Kumar (1988), The theory has enjoyed success in explaining the distribution of power within the p-mode spectrum. The theory relies, to some extent, on mixing length formalism in which there is a full symmetry between the role of upgoing and downgoing flows. However, we clearly see from our seismic events there is no such symmetry. Thus, we believe that it would be valuable to account for this asymmetry in a future theoretical effort to quantitatively explain the p-mode spectrum.

Our observations were motivated by the pioneering, large-scale simulations of convection by Nordlund (1985) in which he predicted narrow, supersonic downdrafting plumes. Owing to the many similarities, it would seem our seismic events are associated with the downgoing convective plumes predicted by Nordlund (1985).

This work was supported by AFOSR-92-0094 and NASA SR&T. We gratefully acknowledge many useful conversations with and suggestions from W.A. Dziembowski.

References

Goldreich, P. & Kumar, P. (1988) The Interaction of Acoustic Radiation with Turbulence, *Astrophysical Journal*, **Vol. no. 326**, pp. 462–478
Goode, P.R., Gough, D. & Kosovichev, A. (1992) Localized Excitation of Solar Oscillations, *Astrophysical Journal*, **Vol. no. 387**, pp. 707–711
Keil, S. (1996), *private communication*
Kumar, P. (1993) Solar Oscillations with Frequencies Above the Acoustic Cutoff Frequency, *Astronomical Society of the Pacific*, **Vol. no. 42**, pp. 15-26
Nordlund, Å(1985) Solar Convection, *Solar Physics*, **Vol. no. 100**, pp. 209-235
Nordlund, Å(1997) The Excitation and Damping of p-modes, *these proceedings*
Restaino, S.R., Stebbins, R.T. & Goode, P.R. (1993) Observation of Impulsive Acoustic Events and the Excitation of Solar Oscillations, *Astrophysical Journal*, **Vol. no. 408**, pp. L57-L60
Rimmele, T.R., Goode, P.R., Harold, E. and Stebbins, R.T. (1995) Dark Lanes in Granulation and the Excitation of Solar Oscillations, *Astrophysical Journal*, **Vol. no. 444**, pp. L119-L122
Stebbins, R.T. & Goode, P.R. (1987) Waves in the Solar Photosphere, *Solar Physics*, **Vol. no. 110**, pp. 237-253

OBSERVATION OF SEISMIC EFFECTS OF SOLAR FLARES FROM THE SOHO MICHELSON DOPPLER IMAGER

A.G. KOSOVICHEV
W.W. Hansen Experimental Physics Lab., Stanford Univ., U.S.A.

AND

V.V. ZHARKOVA
Department of Physics and Astronomy, Glasgow University, U.K.

Abstract. Solar flares are the strongest localized seismic disturbances on the solar surface. During the impulsive phase a high-energy electron beam heats the chromosphere, resulting in explosive evaporation of chromospheric plasma at supersonic velocities. This upward motion is balanced by a downward recoil in the lower part of the chromosphere that excites propagating waves in the solar interior. On the solar surface the outgoing circular flare waves resemble ripples from a pebble thrown into a pond. We report on first observations of the seismic effects of a solar flare from the SOHO Michelson Doppler Imager (MDI) and compare the results with a theoretical model. Observation of flare seismic waves provide important information about the flare mechanism and about the subphotospheric structure of active regions.

1. Observations

The flare of 9 July 1996 was the only significant X-ray flare observed during the first year of the SOHO mission. This was a fairly moderate flare classified as X2.6/1B with the corresponding X-ray flux of 0.26 erg cm^{-2} s^{-1}. When the solar activity is near its maximum, observations of flares with several times this energy are not uncommon.

The X-ray flare of 9 July 1996 was detected by BATSE (Burst and Transient Source Experiment) (Schwartz et al. 1993) on board the Compton Gamma Ray Observatory (Fig. 1). The X-ray flux began to increase at 09:07:49 UT and reached a sharp maximum at 09:09:40. The magnetic field measurements from MDI show that the flare energy release was associated with emerging flux of opposite polarity in the active region NOAA 7978 a few hours prior the flare. The flare occurred when MDI was operating in the full-disk mode taking 1024x1024 pixel 60-sec averaged Doppler velocity observations every minute. Analyzing the MDI Dopplergrams, we have detected a strong localized downward mass flow during the X-ray impulse. This flow occupied 2 or 3 pixels of the solar images on the CCD, which corresponds to a linear size of about 3-5 Mm. The dashed curve in Fig. 1 shows the maximum velocity of the downflow as a function of time. The velocity impulse was almost as sharp as one in the X-ray flux. However, the maximum velocity of ~ 1.5 km s^{-1} was observed approximately 1.5 min after the X-ray maximum, at 09:11:00. It is likely that the actual velocity impulse is smoothed in the 60 s-average Dopplergrams, and that in reality it was significantly stronger and sharper. By simply displaying the MDI Doppler-

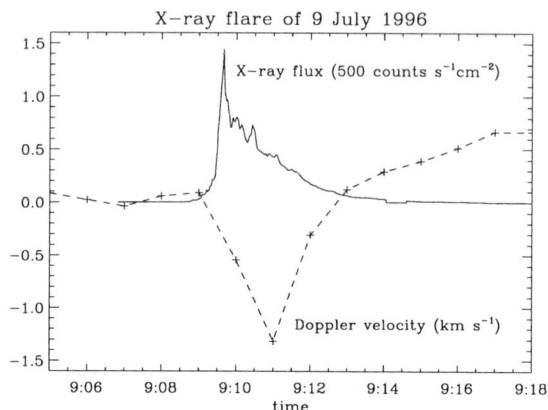

Figure 1. The 1.024-sec averages of the X-ray flux in the energy range 25-100 keV of the solar flare of 96/07/09 from the BATSE flare monitor (solid curve), and the 1-min averages of the Doppler velocity from the MDI (dashed curve).

grams in sequence, we have detected a circular wave packet propagating from the flare. The wave is first detected at about 09:30, approximately 20 min after the flare, at a distance of ≈ 18 Mm from the flare site. It is clearly observed in the sequence of Dopplergrams for about 30 min, until 10:00. After that, the wave amplitude drops rapidly and the wave is lost in the ambient noise.

2. Analysis

To construct seismograms of the solar flare, we have tracked the Dopplergrams of the 62x62-Mm region around the flare to remove the solar rotation, and then remapped the Doppler images into polar coordinates centered at the point of the initial velocity impulse. A difference filter was applied to remove background velocities. After that the data were Fourier-transformed with respect to the azimuthal angle and the Fourier coefficients were plotted as functions of the angular distance from the initial point and time. Figure 2 shows the first three coefficients of the Fourier transform corresponding to the azimuthally averaged signal ($m = 0$), and to the dipole ($m = 1$) and quadrupole ($m = 2$) components. The circular wave ($m = 0$) shows a set of ridges with a positive slope, that begins about 18 Mm from the flare at 09:32, and reaches ≈ 50 Mm at 09:48. The mean velocity of the wave packet was ≈ 45 km s^{-1}. The maximum amplitude of this symmetric wave is approximately 50 m s^{-1}. These ridges are similar to the ridge pattern of the time-distance diagrams obtained by cross-correlating the oscillation signal between different regions on the solar surface. However, to detect the time-distance ridges the cross-correlation function has to be averaged over several hours of data. Thus, flares provide a unique opportunity to do time-distance seismology of

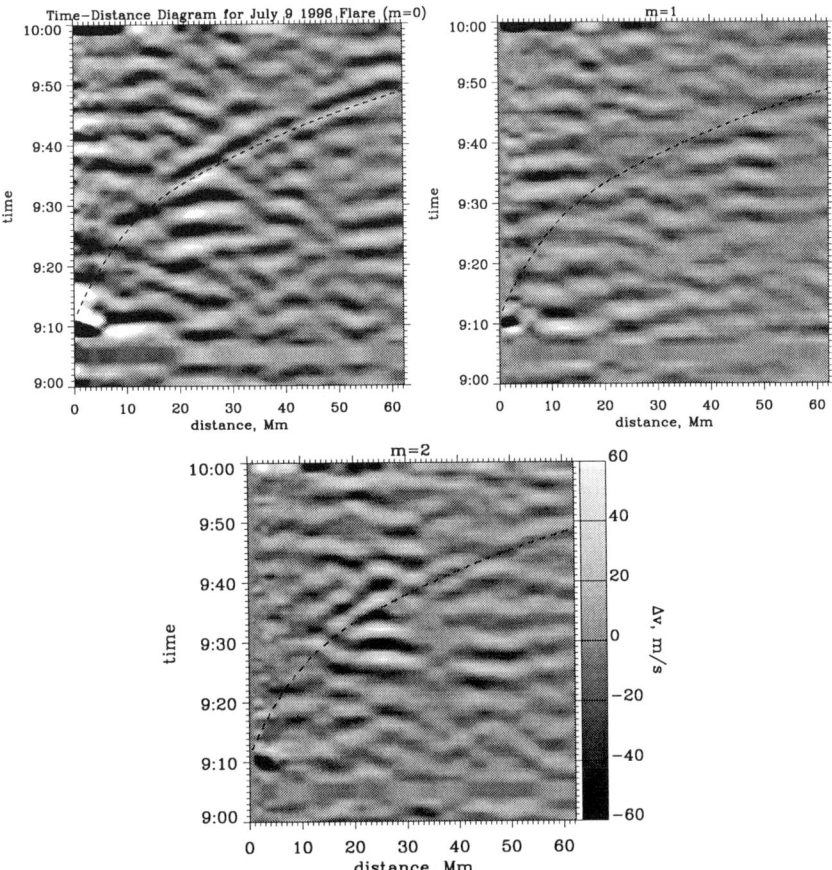

Figure 2. Time-distance diagrams representing the first three azimuthal components of the flare seismogram constructed from 1-min velocity differences. The $m = 0$ component represents symmetrical circular wave. The $m=1$ and 2 plots show the dipole and quadrupole components of the seismic response. The dashed curves show the theoretical time-distance relation for acoustic rays initiated at the flare core at 09:11:00 and propagating through the solar interior.

active regions based on localized impulsive sources. While the dipole component of the flare wave does not have a significant signal, the quadrupole component shows ridges at distances from 15 to 40 Mm from the flare. The deviation from the azimuthal symmetry could result from scattering on large-scale inhomogeneities in the active region.

3. Comparison with a theoretical model

Comparison with a theoretical model of the flare response (Kosovichev and Zharkova, 1995), shown in Fig. 3, leads to interesting conclusions about the mechanisms of the flare impact on the Sun's interior. The thick target model of solar flares (e.g.

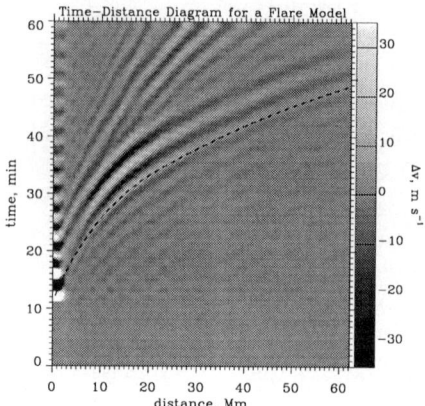

Figure 3. The theoretical flare seismogram computed for a localized momentum impulse of 10^{22} g cm s^{-1} applied to the solar surface at $t = 11$ min. The dashed curve shows the acoustic ray time-distance relation.

Kosovichev, 1986; Zarro et al., 1988, Zharkova & Brown, 1994) predicts that most of the flare momentum is transported to the photosphere in a downward propagating shock wave. This shock wave could explain the observed amplitude of the seismic waves if it carried a momentum of 10^{22} g cm s^{-1}. However, the momentum in the flare core, $\rho S v^2 \tau$, estimated from the MDI data (see Fig.1) for a density, $\rho \sim 10^{-8}$ g cm^{-3}, flare area, $S \sim 10^{17}$ cm^2, flow velocity, $v \sim 10^5$ cm s^{-1}, and impact duration, $\tau \sim 10^2$ s, is only $\sim 10^{21}$ g cm s^{-1} (cf Zarro et al., 1988). This suggests that additional sources of momentum and energy played a significant role in initiating the seismic wave during the impulsive phase of the flare (e.g. heating by high-energy electrons and protons).

We have also found that the observed travel time of the flare wave is ≈ 1 min shorter than the time predicted for a quiet Sun model. This might be explained if the acoustic waves propagated faster in the flare region because of the higher plasma temperature and strong magnetic field, or if the seismic flare wave was initiated before 09:11 when the flare shock reached the photospheric level (e.g. by high-energy particles).

These initial results suggest interesting perspectives for flare seismology and understanding the effects of flare processes in the Sun's interior.

We thank Tom Duvall, Todd Hoeksema, Hugh Hudson and Phil Scherrer for their interest and useful discussions. SOHO is a project of international cooperation between ESA and NASA. This research is supported by the SOI-MDI NASA contract NAG5-3077 at Stanford University.

References

Kosovichev, A.G. 1986, *Bull. Crimean Astrophys. Observatory*, **75**, 6.
Kosovichev, A.G. & Zharkova, V.V. 1995, in: *Helioseismology*, Proc. 4th SOHO Workshop, eds J.T.Hoeksema, V.Domingo, B.Fleck & B.Battrick, ESA SP-376, Noordwijk, 341.
Schwartz, R.A., Fishman, G., Meegan, C., Wilson, R. & Paciesas, W. 1993, *Adv. Space Res*, **13**, 233.
Zarro, D.M., Canfield, R.C., Strong, K.T., Metcalf, T.R. 1988, *ApJ*, **324**, 582.
Zharkova, V.V. & Brown, J.C. 1994, in: *Solar Dynamic Phenomena and Solar Wind Consequencies*, Proc. 3rd SOHO Workshop, ESA SP-373, Noordwijk, 61.

LINE ASYMMETRY AND EXCITATION MECHANISM OF SOLAR OSCILLATIONS

R. NIGAM, A.G. KOSOVICHEV, P.H. SCHERRER AND J. SCHOU

W.W. Hansen Experimental Physics Laboratory,
Stanford University, Stanford CA 94305, U.S.A.

1. Introduction

In his opening address at the conference Dr. Tim Brown posed the line asymmetry problem between velocity and intensity as a puzzle in helioseismology that has been resisting theoretical explanation for many years. It was the observations of Duvall *et al.* (1993) that for the first time indicated that the power spectrum of solar acoustic modes show varying amounts of asymmetry. In particular, the velocity and intensity power spectra revealed an opposite sense of asymmetry. Many doubted the correctness of the experiment and thought it to be a puzzling result (Abrams & Kumar, 1996). Many authors have investigated this problem theoretically and have found that there is an inherent asymmetry whenever there is a localized source exciting the solar oscillations (Gabriel, 1995; Roxburgh & Vorontsov, 1995; Abrams & Kumar, 1996; Nigam *et al.* 1997). This problem has important implications in helioseismology where the eigenfrequencies are generally determined by assuming that the power spectrum was symmetric and can be fitted by a Lorentzian. This leads to systematic errors in the determination of frequencies and, thus, affects the results of inversions (Rhodes *et al.* 1997). In this paper we offer an explanation for the difference in parity of the two asymmetries and estimate the depth and type of the sources that are responsible for exciting the solar p-modes.

2. Observations

Power spectra are computed from the full disk velocity and intensity images obtained from the MDI instrument (Scherrer *et al.* 1995). From these two power spectra shown in Figure 1A we see that the p-mode peaks of the velocity spectrum have negative asymmetry (more power on the low frequency end of the peak) while the peaks of the intensity spectrum have positive asymmetry (more power on the high frequency end of the peak). In the velocity spectrum (Figure 1A(a)), the asymmetry is strongest for low frequency (low radial order) modes and becomes negligible around and above the acoustic cut-off frequency (≈ 5.2 mHz). However, the asymmetry in the intensity oscillations (Figure 1A(b)) increases with frequency for modes below the acoustic cut-

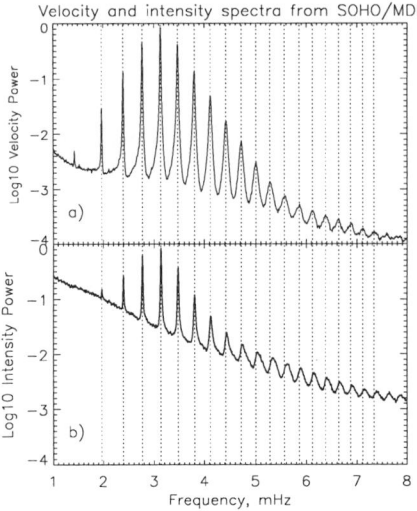 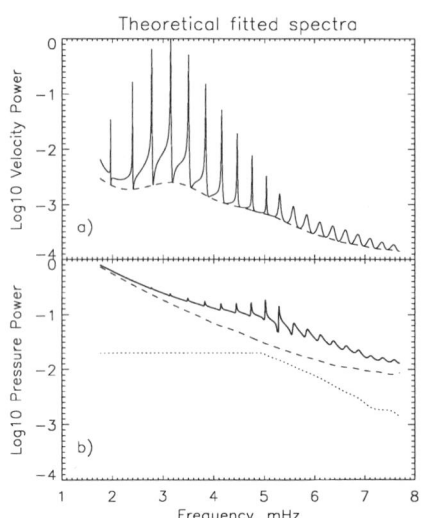

Figure 1. **A.** Normalized power spectra for solar oscillations of angular degree $l = 200$: a) Doppler velocity and b) continuum intensity for the same 3-day period of 21-23 July 1997. The leftmost peak of the velocity spectrum corresponds to the f-mode. The other peaks of both spectra correspond to acoustic (p) modes of radial order from 1 to 21 (from the left to the right). The vertical dotted lines in both panels indicate the locations of the p-mode maxima in the velocity power spectrum to show that a relative shift in frequency occured at and above the acoustic cut-off frequency (≈ 5.2 mHz).
B. Normalized theoretical power spectra for solar p-modes of $l = 200$ produced by a composite source located at a depth, $d = 75$ km beneath the photospheric level and an observing location $r_{\rm obs} = 300$ km above the photosphere, where the observed spectral line is formed: a) velocity spectrum (solid line) with additive uncorrelated noise (dashed line) and b) pressure spectrum (solid line) with additive correlated (dotted line) and uncorrelated noise (dashed line).

off frequency, and then gradually decreases at higher frequencies. We find in the data that the f-mode shows a similar reversal of asymmetry as in the case of p-modes.

3. Numerical model of solar p-mode excitation

Physically, solar acoustic modes are formed by resonances of the acoustic cavity beneath the Sun's surface. The shape and size of the cavity depends on the sound-speed and density stratification and on the mode angular degree. We assume that these waves are generated by turbulence in the convection zone, and apply Lighthill's (1952) method, in which the acoustic sources of various multipole orders are transferred to the right hand side of the wave equation, to calculate the velocity and pressure perturbations. We also assume that the intensity variations recorded by the MDI instrument correspond to Lagrangian pressure (or temperature) perturbations. Then, using a standard decomposition onto spherical harmonics (Gough, 1993) we transform Lighthill's equations into a single second-order wave equation

$$\frac{d^2\Psi}{dr^2} + \left[\frac{\omega^2 - \omega_c^2}{c^2} - \frac{L^2}{r^2}\left(1 - \frac{N^2}{\omega^2}\right)\right]\Psi = S[f, q], \tag{1}$$

where Ψ is proportional to the Lagrangian pressure perturbation δp, r is the radius, ω is the frequency, ω_c is the acoustic cut-off frequency, c is the equilibrium sound speed, N is the equilibrium buoyancy frequency, $L = \sqrt{l(l+1)}$, S is a combination of source terms that include the fluctuating Reynolds stress force (dipole source term), f and the mass source (monopole term), q. The Green's function $G_\Psi(r, r_s)$ of equation (1) using a standard solar model of Christensen-Dalsgaard el al. (1996), is found numerically with the boundary conditions that $G_\Psi = 0$ at $r = 0$ (as the perturbations are negligible much below the lower turning point) and the Sommerfeld radiation condition is applied above the photosphere (this ensures outgoing waves). Damping is added by making the frequency complex.

4. Why is the asymmetry reversed?

The Green's function for the pressure perturbation is calculated from equation (1). The Lagrangian perturbations are then calculated from the Green's function for a single source location and source type by multiplying by a suitable source function and adding solar noise. The noise is assumed to consist of a part $c_p(\nu)$ that is correlated with the source function $s(\nu)$, while $n_p(\nu)$ forms the uncorrelated additive background. One then obtains for the pressure perturbation

$$p(\nu) = s(\nu)[c_p(\nu) + G_p(\nu)] + n_p(\nu) \tag{2}$$

and the velocity perturbation is found in a similar manner, but the correlated part here is much smaller. Equation (2) is a simple model of the observed solar signal, that includes the correlated noise. Here G_p is proportional to G_Ψ. The asymmetry in velocity and intensity power spectra is of opposite sense because a component of the solar noise that is correlated with the source is present. The correlated component of the noise is below a certain threshold to preserve the asymmetry obtained by the above model in the velocity power spectrum but that it is large enough (the correlated component is above a certain threshold) to reverse the asymmetry in the intensity spectrum. The source position r_s from the origin is kept fixed and the power spectrum is computed from equation (2) for the pressure perturbation. It is found that the correlation $c_p(\nu)$ reverses the asymmetry found in $G_p(\nu)$ when computing the power spectrum as seen from Figure 1B(b). The uncorrelated noise plays no role in the reversal of asymmetry. The intensity and velocity fluctuations are computed from the absorption line that is formed at a particular height in the solar atmosphere. The spectral line is effected by the p-mode oscillations and the solar granulation, which modulate and shift the line. It is thus the granulation overshoot that forms the correlated component of the noise as it transports the effect of the source directly onto the line in the atmosphere. It also excites the solar oscillations in the intergranular lanes (Goode et al., these Proceedings). Without adding correlated noise no reversal in asymmetry can be brought about between intensity and velocity.

5. Results and discussions

The velocity and pressure power spectra computed from equations (1) and (2) are shown in Figure 1B. They capture the features of the asymmetry in the observations (Figure 1A). It is important to note that the narrow range of the acoustic source depth (75 ± 50 km) found by comparing the theoretical and observed spectra coincides with the region of superadiabatic convection in the solar model. This region represents the highly unstable upper boundary layer of the convection zone where the convective motions are most violent. At this conference recent observations of Goode together with the simulations of Nordlund suggest that the p-modes are excited in the cooler intergranular lanes. Their data also indicates that the solar noise is correlated to the acoustic source. This might be a plausible explanation for the reversal of asymmetry between velocity and intensity power spectra.

6. Conclusion

In the talk a solution to the puzzle was presented. From our model we see that the intensity and velocity power spectra have opposite sense of asymmetry, because a correlated component of the solar noise is present in the data. A composite (monopole and dipole) source that excites the solar p-modes is found to be located in a thin superadiabatic layer 75 ± 50 km below the photosphere.

We thank Tom Duvall, Douglas Gough, Åke Nordlund, Stuart Jefferies, Sergei Vorontsov and Ron Bracewell for useful discussions. SOHO is a project of international cooperation between ESA and NASA. This research is supported by the SOI-MDI NASA contract NAG5-3077 at Stanford University.

References

Abrams, D. and Kumar, P. 1996, *Astrophys. J.* **272**, 882
Christensen-Dalsgaard, J. and GONG Team 1996, *Science.* **272**, 1286
Duvall, T.L. Jr., Jefferies, S.M., Harvey, J.W., Osaki, Y. and Pomerantz, M.A. 1993, *Astrophys. J.* **410**, 829
Gabriel, M. 1995, *Astr. Astrophys.* **299**, 245
Gough, D.O. 1993, in: Astrophysical Fluid Dynamics, (ed. J.P. Zahn and J. Zinn-Justin, Elsevier, Amsterdam), 399-560
Lighthill, M.J. 1952, *Proc. Roy. Soc. Lon.* **A211**, 564
Nigam, R., Kosovichev, A.G. and Scherrer, P.H. 1997, in Sounding Solar and Stellar Interiors, Proc. IAU Symp. 181, in press
Nigam, R., Kosovichev, A.G., Scherrer, P.H. and Schou, J. 1997, *ApJL.*, submitted
Rhodes, E.J. Jr., Kosovichev, A.G., Schou, J., Scherrer, P.H. and Reiter, J. 1997, *Solar Phys.* **175**, in press
Roxburgh, I.W. and Vorontsov, S.V. 1995, *MNRAS.* **272**, 850
Scherrer, P.H. and MDI Team 1995, *Solar Phys.* **162**, 129

EXCITATION AND DAMPING OF P-MODES

ÅKE NORDLUND
*Theoretical Astrophysics Center, and
Astronomical Observatory / NBIfAFG,
Juliane Maries Vej 30, 2100 Copenhagen Ø, Denmark*

AND

ROBERT F. STEIN
*Dept. of Physics and Astronomy, Michigan State University,
East Lansing, MI 48823, U.S.A.*

1. Introduction

Millions of seismic eigenmodes are excited to low amplitudes in the Sun, with an amplitude spectrum that grows approximately as a power of the frequency until it reaches a maximum at 3 mHz, and then drops off again towards higher frequencies (Libbrecht, 1988; Libbrecht & Woodard, 1991).

We would like to answer the following fundamental questions about these 'solar five-minute oscillations':

1. What excites the oscillations?
2. What damps the oscillations?
3. Why is the dominant period 5 minutes?

These questions are notoriously difficult to attack with analytical methods, because the properties of the oscillations depend critically on the properties of turbulent convection in the near-solar-surface layers. Even before the detailed mode structure was observed (Deubner, 1975), it was suspected that waves would be excited by the turbulent convection (Biermann, 1946; Biermann, 1948; Schwarzschild, 1948; Schatzman, 1949; Unno, 1964; Stein, 1967; Stein, 1968), and oscillations of the solar surface layers had been observed (Leighton et al., 1964; Evans & Michard, 1962). Following the realization that a rich spectrum of global modes exist on the Sun (Ulrich, 1970; Leibacher & Stein, 1971; Deubner, 1975; Ulrich & Rhodes, 1977), a large number of attempts have been made to explain their excitation and damping properties (e.g., Ando & Osaki, 1977; Goldreich & Keeley, 1977; Gough, 1980; Antia et al., 1982; Christensen-Dalsgaard & Frandsen, 1983; Antia et al., 1987; Goldreich & Kumar, 1988; Goldreich & Kumar, 1990; Balmforth, 1992a; Balmforth, 1992c; Goldreich et al., 1994). The outcome of these studies for a long while actually confused the issue, because even the question of the stability of the modes could not be decided with any certainty by analytical means. One of the most impressive and laborious studies is the one by

Balmforth (1992abc), who among many other results showed that a crucial factor for determining the stability of (in particular) the low-frequency modes is the response of the Reynolds stress to the presence of coherent modes. Balmforth showed that a commonly used approximation of these effects in terms of a turbulent viscosity significantly underestimates the stabilizing influence of convection, and that this explained the tendency to find overstability in works that adopted this approximation. He was able to fit the damping of low-frequency modes considerably better than in earlier works (possibly with the exception of Gough (1980)). Nevertheless, he concluded his series of papers with the statement "It would be wrong to rule out thermal overstability as the underlying cause of the five-minute oscillations on the basis of the current understanding".

By now, the consensus is that the modes are indeed stable and stochastically excited, but analytical predictions of their properties still contain major uncertainties. Thus, for example, Goldreich et al. (1994) used a number of (at least five) free parameters to fit the empirically determined stochastic excitation spectrum, openly admitting that some of the parameters had to be assigned unexpected values.

The main problem with analytical estimates of the excitation and damping is indeed that one cannot avoid the use of a number of free parameters, and that the results depend quite sensitively on the values adopted for some of these. Numerical simulations offer a way forward in this situation. The hydrodynamic equations that describe the solar surface layers contain no free parameters *per se*, and provided that sufficiently realistic physics is used, numerical solutions of these equations should provide increasingly accurate representations, as the numerical resolution is increased. The main problems with such an approach are, on the one hand that it is difficult to estimate the actual accuracy of the solutions, and on the other hand that, even with accurate numerical solutions it is non-trivial to extract the desired quantitative and qualitative information from the available data.

In what follows, we provide a brief overview of what numerical simulations can tell us about the excitation and damping of the solar five-minute oscillations. The discussion is based on numerical simulations of the solar surface layers, along the lines of previous work (Nordlund, 1982; Nordlund, 1985; Stein & Nordlund, 1989; Nordlund & Dravins, 1990; Nordlund & Stein, 1990; Rast et al., 1993; Stein & Nordlund, 1997). The numerical models include the relevant physics of the solar surface layers: ionization, radiative energy transfer, etc., and have been checked against a number of observational diagnostics. The results are found to be consistent with statistics of solar granulation (Wöhl & Nordlund, 1985; Nordlund et al., 1997), spectral line widths and shapes (Dravins et al., 1981; Dravins et al., 1986; Dravins & Nordlund, 1990a; Dravins & Nordlund, 1990b; Spruit et al., 1990; Nordlund, 1991; Kiselman & Nordlund, 1995), the depth of the solar convection zone (Nordlund & Stein, 1997). and the p-mode frequency behavior (Rosenthal et al., 1998).

In previous work (Stein & Nordlund, 1991) we have made use of long simulation runs with spontaneously excited modes to show, e.g., that the stochastic excitation due to entropy fluctuations significantly exceeds that due to the turbulent pressure (Reynolds stress) fluctuations. These experiments were 'passive', in the sense that the modes were left to establish the amplitudes that are appropriate for the sparse mode spectra of the relatively small boxes that are used in the numerical experiments.

To study specifically the excitation and damping of the oscillations, we have also performed a number of 'active experiments', where particular modes are initially either excessively excited or excessively damped. By actively exciting particular modes to high amplitude it becomes easier to study the mode damping, and by actively damping the modes it becomes easier to study the mode excitation.

To analyze excitation and damping mechanisms, it is advantageous to separate horizontal mean values and their coherent fluctuations (the modes), and refer to the remaining fluctuations as 'the turbulent convection'—of course with the understanding that correlations among the fluctuating variables also may exhibit a partly coherent behavior, and thus feed back into the behavior of the 'modes'. We summarize such a formalism in the next Section, and demonstrate the separation between 'modes' and 'convection'. In Section 3 we use the formalism to study and discuss the excitation mechanisms, and in Section 4 we treat the damping in a similar way. In the concluding section we summarize what we have thus learnt about the excitation and damping mechanisms.

2. Formalism

The formalism summarized here is an extension of the one used by Stein & Nordlund (1991) and by Rosenthal et al. (1998). It is presented in more detail by Nordlund & Stein (1998).

2.1. 1-D EQUATIONS

For a one-dimensional envelope, the equations for radial near-surface motions may be written (in Lagrangian coordinates):

$$\frac{D}{Dt}(\rho) = -\rho \frac{\partial}{\partial z}(u_z) \tag{1}$$

$$\rho \frac{D}{Dt}(u_z) = -\frac{\partial}{\partial z}(P - \rho\sigma_{zz}) + \rho g \tag{2}$$

$$\rho \frac{D}{Dt}(e + \frac{1}{2}u_z^2) = -\frac{\partial}{\partial z}(u_z P + F) + \rho u_z g, \tag{3}$$

where $u_z P$ is the acoustic flux and F represents other fluxes.

In a true 1-D model, the only other fluxes are due to radiation and heat conduction. The turbulent convective flux that is often included in the energy equation of otherwise one-dimensional models cannot be determined without resorting to a multi-dimensional picture (if only for making order of magnitude or scaling estimates).

2.2. 3-D EQUATIONS

The three-dimensional version of Eqs. (1)–(3) only differ by having additional, horizontal transport terms (but support an incomparably richer spectrum of motions):

$$\frac{D}{Dt}(\rho) = -\rho \frac{\partial}{\partial x_j}(u_j) \tag{4}$$

$$\rho \frac{D}{Dt}(u_i) = -\frac{\partial}{\partial x_i}(P) + \frac{\partial}{\partial x_j}(\rho\sigma_{ij}) + \rho\delta_{iz}g \tag{5}$$

$$\rho \frac{D}{Dt}(e + \frac{1}{2}u^2) = -\frac{\partial}{\partial x_j}(u_j P + F_j) + \rho u_z g \tag{6}$$

There exist several interaction mechanisms between 'convection' and 'modes' (the quotes are there to emphasize that it is non-trivial to actually separate these effects—however in what follows we generally drop the quotes). The convective energy flux carries almost all of the solar luminosity at depths larger than a few hundred kilometers below the optical surface, and the 'turbulent pressure' associated with convection is significant in a shallow layer near the solar surface. Since convection is a non-stationary process, there are significant, incoherent fluctuation in these transport processes. Such fluctuations necessarily give rise to random deviations from the average vertical balance of energy and vertical momentum, and hence, to sources of acoustic 'noise'. Conversely, in the presence of coherent wave-like perturbations around the average state, the convective transport of heat and momentum will be influenced (modulated), in general with some delay relative to the variations in the background state. Such coherent fluctuations of the average transport of energy and momentum in turn feed back onto the wave propagation; i.e., part of the coherent fluctuations of the convective medium are mediated by the convection itself.

2.3. DENSITY-WEIGHTED AVERAGING

In order to turn such qualitative remarks into quantitative expressions, it is helpful to use a consistent set of definitions for averages and fluctuations. Stein & Nordlund (1991) adopted density-weighted averages for per-unit-mass variables (here denoted by lower case), and straight averages for per-volume quantities (here denoted by upper case).

$$\bar{F} = \langle F \rangle_{xy}, \tag{7}$$
$$\bar{f} = \langle \rho f \rangle_{xy} / \langle \rho \rangle_{xy}, \tag{8}$$

where $\langle \rangle$ stands for horizontal averages. With these definitions it follows that

$$\langle F' \rangle \equiv \langle F - \bar{F} \rangle = 0, \tag{9}$$
$$\langle \rho f' \rangle \equiv \langle \rho f - \bar{\rho}\bar{f} \rangle = 0, \tag{10}$$
$$\langle \rho f g \rangle = \bar{\rho}\bar{f}\bar{g} + \langle \rho f' g' \rangle. \tag{11}$$

The 'chain rule' (11) follows from Eqs. (9)–(10). Repeated use of the chain rule results in precise definitions for the total kinetic energy and kinetic energy flux, in terms of contributions from coherent motions and fluctuations:

$$\langle \tfrac{1}{2}\rho u^2 \rangle = \tfrac{1}{2}\bar{\rho}\bar{u}^2 + \tfrac{1}{2}\langle \rho u'^2 \rangle = \bar{\rho}\,\bar{e}_{\text{kin,mode}} + \bar{\rho}\,\bar{e}_{\text{kin,turb}} \tag{12}$$

$$\langle \tfrac{1}{2}\rho u^2 u_z \rangle = \tfrac{1}{2}\bar{\rho}\bar{u}^2\bar{u}_z + \langle \tfrac{1}{2}\rho u'^2 u'_z \rangle + \langle \rho u'^2_z \rangle \bar{u}_z + \langle \tfrac{1}{2}\rho u'^2 \rangle \bar{u}_z$$
$$= \bar{F}_{\text{kin,mode}} + \bar{F}_{\text{kin,turb}} + \bar{P}_{\text{turb}}\bar{u}_z + \bar{\rho}\,\bar{e}_{\text{kin,turb}}\bar{u}_z \tag{13}$$

Equation (12) expresses the total kinetic energy as a sum of the kinetic energy of the mean motion and the kinetic energy of the convection. Equation (13) expresses the total kinetic energy flux as the sum of the kinetic energy flux of the mean, the turbulent kinetic energy flux, the acoustic fluxes associated with the horizontally averaged turbulent pressure and velocity, and the advection of average turbulent kinetic energy.

Carrying on with a similar analysis of the other hydrodynamic equations, using the density-weighted averaging, one may derive a set of exact, horizontally averaged 3-D equations (Nordlund & Stein, 1998).

2.4. SEPARATION BETWEEN OSCILLATIONS AND CONVECTION

For purely radial modes, we consider the fluctuations in time of the barred, horizontally averaged quantities as the variables belonging to the oscillations, and the primed 3-D fluctuations as the ones associated with the convection. Qualitatively similar results are to be expected for low ℓ non-radial modes, since they are nearly radial in the surface layers. In fact, since numerical models of the solar surface region typically have horizontal sizes ~10 Mm, the smallest non-zero ℓ-values that could be studied would have ℓ in excess of 400.

3. Stochastic Excitation

The kinetic energy of the mean obeys

$$\bar{\rho}\frac{D}{Dt}(\frac{1}{2}\bar{u}_z^2) = -\frac{\partial}{\partial z}(\bar{u}_z\bar{P}_{\text{gas}} + \bar{u}_z\bar{P}_{\text{turb}}) + (\bar{P}_{\text{gas}} + \bar{P}_{\text{turb}})\frac{\partial \bar{u}_z}{\partial z} + g\bar{\rho}\bar{u}_z. \tag{14}$$

Apart from boundary effects, the work comes from the term

$$(\bar{P}_{\text{gas}} + \bar{P}_{\text{turb}})\frac{\partial \bar{u}_z}{\partial z} = -(\bar{P}_{\text{gas}} + \bar{P}_{\text{turb}})\frac{D}{Dt}(\ln\bar{\rho}) = -\bar{P}_{\text{tot}}\frac{D}{Dt}(\ln\bar{\rho}). \tag{15}$$

Work arises from correlations in time between fluctuations in the two factors, $\delta\bar{P}_{\text{tot}}(t)$ and $\frac{D}{Dt}(\ln\bar{\rho})$. The fluctuations consist of both coherent and incoherent contributions, where the coherent contributions include the in-phase (pressure relative to density) and out-of-phase response of the convection to the coherent p-mode motions. The in-phase (adiabatic) part of the coherent response perturbs the frequency of the mode, while the out-of-phase (non-adiabatic) part causes damping (or driving) of the mode. The relation between coherent pressure and density perturbations may formally be written

$$\delta\ln\bar{P}_{\text{tot}} = \gamma\,\delta\ln\bar{\rho}, \tag{16}$$

where $\gamma = \gamma(z,\omega)$ is a complex depth- and frequency-dependent factor. The function $\gamma(z,\omega)$ may be measured empirically in the numerical simulations, for example by performing "active" experiments, where a few modes are excited to significant amplitudes. The experiments may be "tuned", by varying the depth of the models, so that a suitable range of eigenfrequencies is covered.

The work done by the convection on the modes is the convolution of the out-of-phase, non-adiabatic, pressure fluctuations with the modal density fluctuations.

$$W = -\int dt \int \frac{dm}{\bar{\rho}}\,\delta\bar{P}_{\text{tot}}\frac{D}{Dt}(\ln\bar{\rho}) = \int dt \int dz\,\delta\bar{P}_{\text{tot}}\frac{\partial\dot{\xi}}{\partial r}, \tag{17}$$

where dm is a Lagrangian mass element, and ξ is a Lagrangian displacement that satisfies

$$\bar{u}_z = \frac{D}{Dt}(\xi) \equiv \dot{\xi}. \tag{18}$$

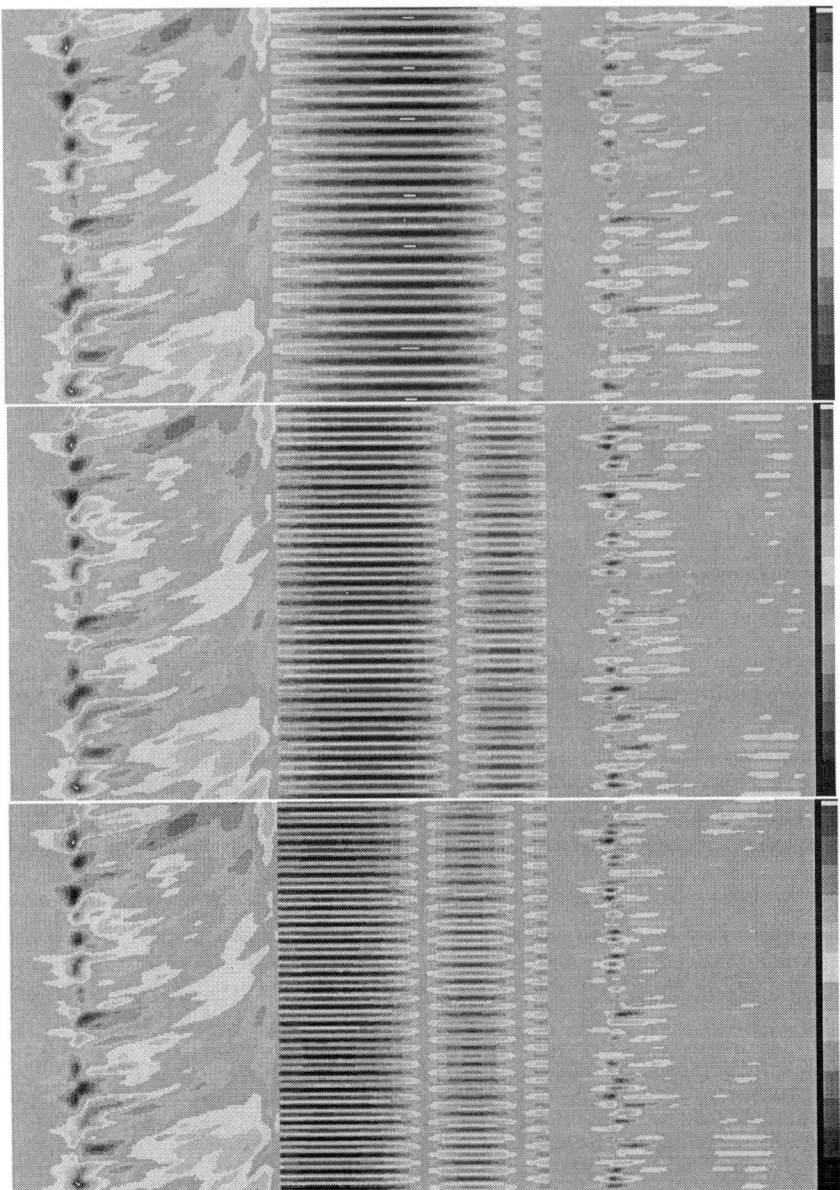

Figure 1. Images showing the integrand in the stochastic work integrals for radial modes at 3, 4, and 5 mHz. The leftmost part of each panel shows the horizontally averaged, non-adiabatic pressure fluctuation from an 80 minutes time sequence of a $6 \times 6 \times 3$ Mm numerical simulation with a spatial resolution of $125 \times 125 \times 82$ (Nordlund & Stein, 1998). The middle part of each panel shows the mode pattern, and the rightmost part shows the product of the mode pattern and the convective fluctuations. (Color versions of this and other figures are available at URL http://www.astro.ku.dk/~aake/talks/Kyoto97.)

Figure 1 shows the the horizontally averaged incoherent (convective) pressure fluctuations, the coherent (modal) density fluctuations and the product of the two, for three particular frequencies. High frequency modes tend to "chop up" the convective fluctuations in time, while low-frequency modes tend to sample more than one convective fluctuation per cycle. In addition, high- and low-frequency modes sample the convective perturbations differently in depth, with high-frequency modes tending to also chop up the convective fluctuations in the depth direction.

For a particular realization of the convection, the net result depends of course on the particular phase with which the modes are overlaid onto the convective fluctuations. Symbolically, we may write this as

$$a_i \to a_i + \delta a_i e^{i(\phi - \phi_0)}, \tag{19}$$

where a_i is the amplitude of a particular mode, and δa_i is the perturbation of that amplitude when the mode is assigned the phase ϕ_0. Averaging over phase one obtains a net contribution to the mode energy

$$|a_i|^2 \to |a_i|^2 + |\delta a_i|^2. \tag{20}$$

A more detailed analysis (Nordlund & Stein, 1998) results in the following expression for the average rate of stochastic excitation per unit surface area:

$$\frac{\Delta \langle E_\omega \rangle_{\text{ens}}}{\Delta t} = \frac{|\int_r dr \, \delta P_\omega \, \partial \xi_\omega / \partial r|^2}{\Delta \nu \int_r dr \, |\xi_\omega|^2 \, \rho \, (\frac{r}{R})^2}, \tag{21}$$

where δP_ω is the Fourier amplitude of the incoherent pressure fluctuations, and $\Delta \nu$ is the frequency resolution with which δP_ω is measured (note that in the limit $\Delta \nu \to 0$ the power per unit frequency interval $|\delta P_\omega|^2 / \Delta \nu$ is independent of $\Delta \nu$ for a random function of time). Because the stochastic fluctuations are concentrated near the surface, the integral samples the mode compressibility $\partial \xi_\omega / \partial r$ in a shallow layer near the surface, where modes below about 3 mHz are evanescent. The result is that there is an overall scaling with frequency that comes from the asymptotic behavior ($\partial \xi_\omega / \partial r \sim \omega^2 \xi$) of evanescent modes (Osaki, 1990; Balmforth, 1992c; Goldreich et al., 1994).

It is noteworthy that the only property of the turbulent convection that enters is the stochastic fluctuation of the total pressure. This is in contrast to the many approximate, analytic expressions for the stochastic excitation that have been derived (Lighthill, 1952; Unno, 1962; Moore & Spiegel, 1964; Unno, 1964; Stein, 1967; Goldreich & Keeley, 1977; Christensen-Dalsgaard & Frandsen, 1983; Goldreich & Kumar, 1988; Goldreich & Kumar, 1990; Balmforth, 1992c; Goldreich et al., 1994). These expressions, one way or another, correspond to the exact expression with its fluctuations of the total pressure, but they have been expanded into a multitude of approximate terms, with possibly obscure cancellation effects. Balmforth (1992c) indeed pointed out that such expressions are non-unique, and that even the distinction between monopole, dipole and quadrupole source terms is ambiguous. Equation (21) short-circuits such ambiguities, by expressing the driving directly in terms of the one relevant quantity: the total pressure fluctuation, $\delta \bar{P}_{\text{tot}} = \delta P_{\text{gas}} + \delta P_{\text{turb}}$. Such an equation is of course only useful if one has access to numerical data that make it possible to evaluate the resulting expression. Fig. 2 shows the result of evaluating Eq. (21) from an 80 minute interval of numerical simulations (Nordlund & Stein, 1998).

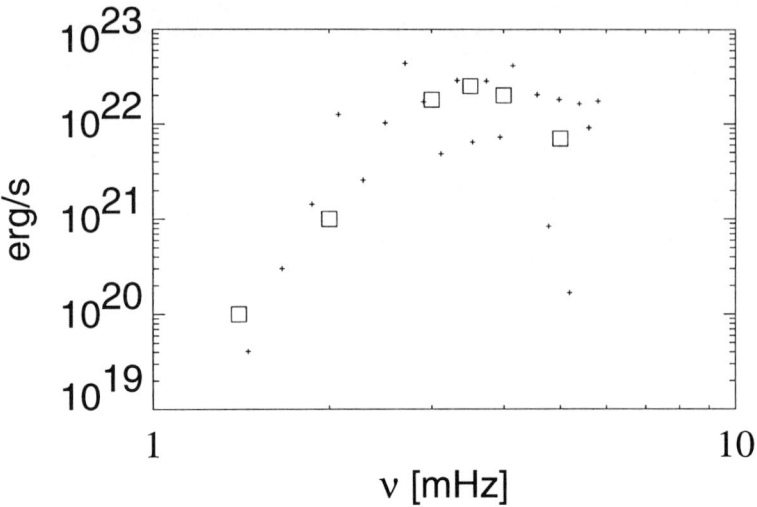

Figure 2. The stochastic energy input, scaled to the total solar area. The small plus symbols are from the numerical experiments, the large squares are determined from solar observations (cf. Fig. 7, Goldreich et al., 1994).

3.1. EXCITATION EVENTS

Using Eq. (21), and the direct visualizations of the work integral in Fig. 1, one may study not only the statistically averaged contributions to the stochastic driving, but also the contributions from particular events. Figure 3 shows the correlation between the instantaneous, horizontally averaged surface intensity and the contributions to the work integral from the surface layers. It is evident that there is a tight correlation, illustrating that it is indeed the fluctuating rate of cooling at the surface that is a main source of the stochastic excitation in the surface layers. Fig. 4 illustrates this further, by showing the contributions to the horizontal average over a short interval of time.

The relatively large dynamic range of intensity fluctuations ($\sim 20\%$ rms) makes it difficult to distinguish individual contributions to the average intensity fluctuation in the fully resolved images, but smearing the images to a resolution roughly corresponding to the size of granules helps. The horizontally averaged intensity fluctuation is most of the time the net effect of the changes in several granules, but occasionally, individual events dominate. The rapid drop in average intensity from the 3rd to the 5th panel in Fig. 4 is due mostly to the rapid cooling of a single granule fragment, resulting in a darkening that gradually disappears, as new granules develop in the area.

It is probably extreme events similar to this one that are the "seismic events" identified by Goode et al. (1992) (cf. Rimmele et al., 1995; Restaino et al., 1993). As shown by Goode et al. (1992), a sufficiently rapid cooling leads to an event characterized by a noticeable phase propagation, consisting of first an upward propagating expansion wave, followed by a ringing "wake". In the context of the current discussion, such events appear as "tips of the icebergs": they are the most extreme events in the

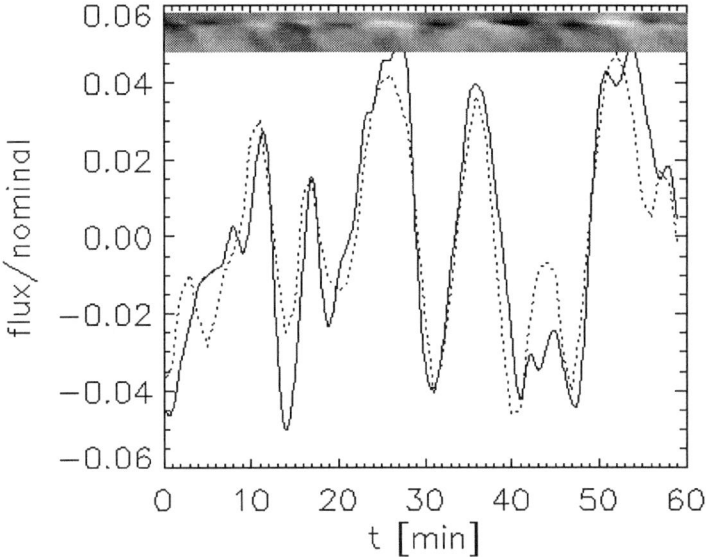

Figure 3. The average surface radiation intensity (dashed) and the non-adiabatic pressure fluctuation (full drawn, and image strip) in the surface layers of a numerical simulation.

ensemble of fluctuations that constitute the acoustic source. A detailed comparison between observations and synthetic spectral line data generated from the numerical simulations is necessary to estimate what fraction of the excitation comes from identifiable seismic events.

4. Damping

Damping is a linear effect, that arises because of a systematic phase difference between the coherent fluctuations of pressure and density. The work done by the coherent motions is again proportional to the work integral, Eq. (17), where $\delta \bar{P}_{\text{tot}}$ is now the coherent, non-adiabatic part of the pressure fluctuations. The resulting expression for the damping rate is

$$\frac{1}{E_\omega} \frac{dE_\omega}{dt} = \frac{2 \int_r dr \ \text{Im}[\delta P_\omega] \ \partial \xi_\omega / \partial r}{\omega \int_r dr \ |\xi_\omega|^2 \ \rho \left(\frac{r}{R}\right)^2}, \quad (22)$$

where $\text{Im}[\delta P_\omega]$ is the imaginary part of the coherent fluctuation of the total pressure. As for the excitation, this integral can in principle be evaluated directly from numerical simulations. It is, however, non-trivial to separate the coherent part of the fluctuation from the stochastic ones in simulations where the damping and excitation are approximately in balance. One obtains a much "cleaner" situation by performing active experiments, where an initial "kick" excites a few oscillatory modes to a relatively high amplitude. By using higher amplitudes than the modes spontaneously attain, one obtains the benefit that the coherent fluctuations dominate over the stochastic ones, and that, consequently, the linear damping dominates over the

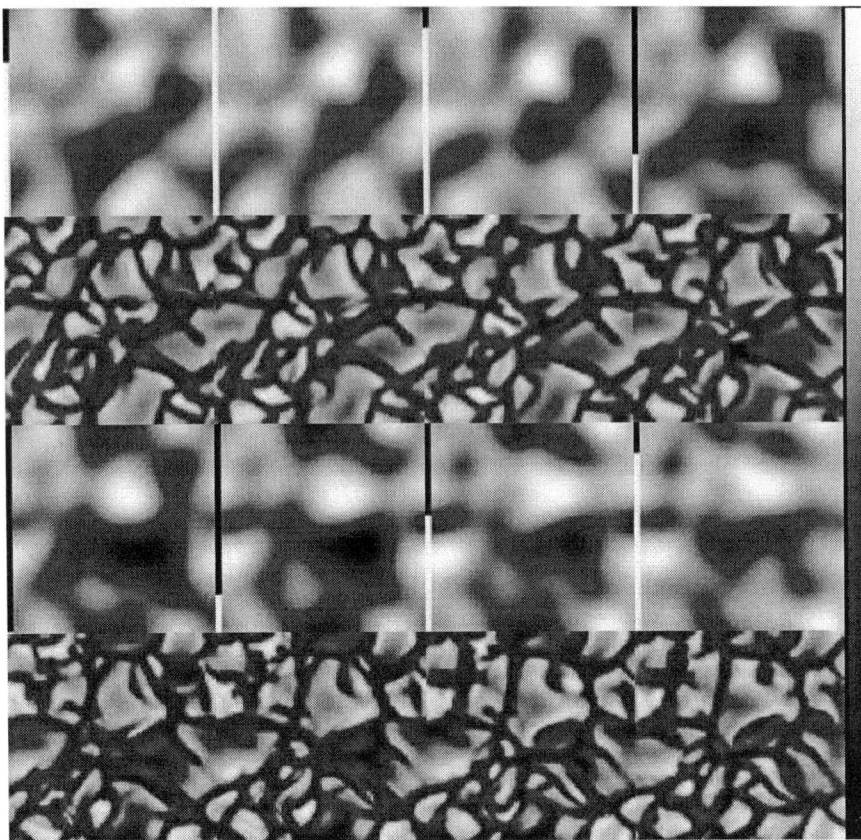

Figure 4. A time sequence of surface radiation intensity images at one minute time intervals, covering eight solar minutes. The upper rows of panels have been convolved with a Gaussian kernel, producing a spatial resolution comparable with the size of granules. The intensity contrast is thus reduced from $\sim 20\%$ to a few percent, making it easier to locate the contributions to changes of the overall average intensity, which is indicated by the thin bars between the individual panels (the range is $\approx \pm$ 4 %).

stochastic excitation. In such situations, Eq. (22) may indeed be used to evaluate the damping rate directly from the numerical data.

With the current set of models, this works well for frequencies of the order of 3 mHz; there, most of the contribution to the integral in Eq. (22) comes from layers close to the surface, and the damping rates are indeed consistent with the observed FWHM line widths ~ 1 μHz near 3 mHz (Libbrecht, 1988; Toutain & Frölich, 1992; Bachmann et al., 1993). For lower frequencies, one encounters two (related) difficulties. First, the lowest eigenfrequencies of the current set of active experiments lies at 2.6 mHz; i.e., it is not possible to measure the damping at lower frequencies from this set of experiments. One might consider performing continuously forced experiments, where an external volume force and appropriate boundary conditions maintain perturbations modeled after known eigenmodes, for those surface parts of the modes that are covered by the numerical simulations. Here one encounters the second difficulty, namely that the contribution to the work integral at low frequencies actually contains significant contributions from deeper layers. This may be appreciated by considering the variation with depth of the factors that enter into the integral in Eq. (22). The factor $\partial \xi/\partial r$ increases rapidly with depth for eigenmodes of low frequency. The overall magnitude of the turbulent pressure also increases with depth, even though the ratio $P_{\text{turb}}/P_{\text{gas}}$ peaks near the surface. Thus, unless the relative fluctuation of the turbulent pressure $\delta P_{\text{turb}}(\omega)/P_{\text{turb}}$ decreases rapidly with depth, one should expect that the work integral picks up significant contributions from deeper layers (cf. Fig. 14, Balmforth, 1992a). In order to account for these contributions, there seems to be no way around extending the models so that one can measure the damping using eigenmodes that are contained within the box.

Given that such measurements of the damping can be made, there is still the task of identifying the physical causes of the phase shift of the coherent fluctuations of the total pressure. The imaginary part of the gas pressure fluctuations is easily understood: it stems from the non-adiabatic terms in the energy equation. The factors that control the phase of the turbulent pressure fluctuations are less evident from first principles. However, using the technique of density-weighted averaging, one may derive a differential equation that describes the time evolution of the turbulent pressure (Nordlund & Stein, 1998):

$$\frac{D}{Dt}(P_{\text{turb}}) = \begin{array}{ll} -\dfrac{\partial \langle \rho u_z'^2 u_z' \rangle}{\partial z} & \text{vertical Reynolds stress transport} \\[1ex] +2\langle u_z'(g\rho - \dfrac{\partial P_{\text{gas}}}{\partial z})\rangle & \text{buoyancy work} \\[1ex] -3P_{\text{turb}}\dfrac{\partial \bar{u}_z}{\partial z} & \text{expansion / compression} \\[1ex] +2\langle \rho u_j' \sigma_{jz} \rangle, & \text{viscous dissipation} \end{array} \quad (23)$$

where the terms have been labeled with their physical interpretation.

It is instructive to visualize the left and right hand side terms in this equation as two-dimensional (depth-time) images (Fig. 5). Examination reveals that the coherent response of the buoyancy work is particularly crucial. In an equilibrium situation, the buoyancy work is positive, and is balanced by the viscous dissipation and redistributed by the transport term. Below the surface, the buoyancy work comes predominantly from the core of the downdrafts, that have a small horizontal area filling factor, and hence large downward velocities and smaller than average gas pressure gradients, so are subject to net downwards forces. The response to a mode perturbation comes mainly from the modulation of the gas pressure gradient: in the upwards acceleration phase that precedes the expansion phase, the increased gas pressure gradient reduces or reverses the buoyancy work. The result is a tendency for the turbulent pressure to be 180 degrees out-of-phase with the density fluctuation.

When combined with the other terms in the equation for the turbulent pressure, the result is that the turbulent pressure fluctuations tend to be in phase with the density fluctuations at high frequencies, because of the compression term, tend to be in anti-phase with $\delta\bar{\rho}$ at low frequencies, and pass between the two extremes through phase shifts that lead to damping of the modes.

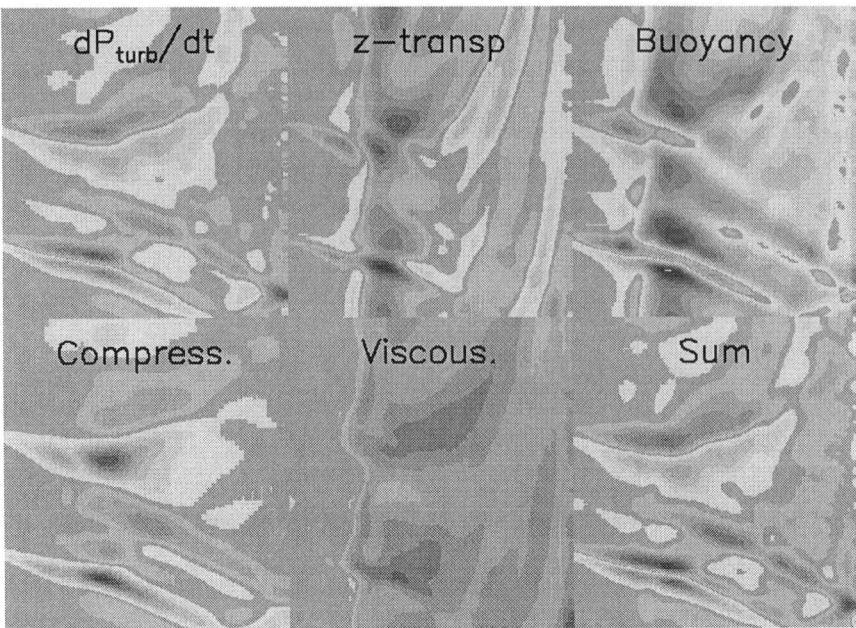

Figure 5. A visualization of the Eq. (23) left hand side, the right hand side terms, and their sum. In each panel, depth increases from left to right, and time increases from bottom to top.

5. Conclusions

The results of numerical simulations are helpful in several ways in relation to the fundamental questions about the solar five-minute oscillations: What excites them, what damps them, and why is their dominating period five minutes?

First of all, the numerical simulations confirm what has been surmised from the very beginning about the origin of the solar five-minute oscillations: the oscillations are stochastically excited by the turbulent convection, and they are damped by the collective response of the convection and radiation, with the main damping agent at low frequencies being the phase-shifted response of the turbulent pressure. This can be concluded both from the fact that oscillations are indeed stochastically excited in the numerical simulations (and are not overstable), to an amplitude that is comparable to the real solar oscillations. It is confirmed by the fact that the excitation power that can be directly measured in the simulations agrees to within the numerical scatter with the empirically determined excitation power, with a frequency dependence that is determined by the same asymptotic scaling relations that have been employed in the analytical estimates of the excitation power. In addition, the numerical simulations allow a detailed study of the physical sources of the excitation, with the possibility to study images of the horizontally averaged excitation (or proxys thereof) as a function of depth and time, or in full detail, as functions of three dimensions and time. The main cause of the excitation turns out to be the stochastic variations of the cooling rate at the surface, and the fluctuations of convective flux and turbulent pressure that are consequences thereof.

The detailed mechanisms behind the linear damping of the modes are more difficult to separate, but the numerical results again confirm the picture that has been previously obtained by analytical methods (Gough, 1980; Balmforth, 1992a): namely that the main agent of damping at low frequencies is the phase-shifted, coherent response of the turbulent pressure to the oscillations. As pointed out by Balmforth, this effect cannot be described as simply a turbulent viscosity. On the contrary, it is strongly influenced by factors such as the coherent response of the buoyancy work to the oscillations, an effect that tends to force the response of the turbulent pressure in anti-phase with the density perturbations. There are additional terms in the differential equation that describes the time evolution of the turbulent pressure, and the combined effect of these terms is a phase lag that is such as to damp the coherent motions. The reason that the damping decreases more slowly with decreasing frequency than expected from naive estimates based on the near-surface amplitude of the evanescent modes is that the work integral picks up contributions from deeper layers at low frequencies. This is in agreement with the results of Gough (1980) and Balmforth (1992a).

The relatively more rapid decrease of the excitation power with decreasing frequency is in also in agreement with previous, analytic work (e.g., Balmforth, 1992c; Goldreich et al., 1994), and is caused by the more pronounced weighting of the excitation work integral towards the surface, relative to the work integral that controls the damping. As a result, the scaling of excitation power with frequency is primarily determined by the frequency behavior of the mode compressibility $\partial \xi_\omega / \partial r$, and peaks near the frequency where the logarithmic derivative $\partial \ln |\partial \xi_\omega / \partial r| / \partial \ln \omega$ is $\sim +2$, thus balancing the ~ -2 slope of the stochastic pressure fluctuations there. For the Sun, the result is that the power peaks at period of about 5 minutes. It is not obvious (at least from a superficial analysis) that the ratio of this frequency to the atmospheric cut-off frequency should be expected to be the same in stars whose surface structure differ substantially from that of the Sun.

Acknowledgments: The work of Å.N was supported in part by the Danish Research Foundation, through its establishment of the Theoretical Astrophysics Center. RFS was supported by NASA grants NAG 1695 and NAG 5-4031 and NSF grant AST 9521785. The calculations were performed at the National Center for Supercomputer Applications, which is supported by the National Science Foundation, at Michigan State University and at UNI•C, Denmark. This valuable support is greatly appreciated.

References

Ando, H., Osaki, Y. 1977, PASJ, 29, 221
Antia, H. M., Chitre, S. M., Gough, D. O. 1987, in J. Christensen-Dalsgaard, S. Frandsen (eds.), Advances in Helio- and Asteroseismology, Vol. 123 of *IAU Symp.*, 371
Antia, H. M., Chitre, S. M., Narasimha, D. 1982, Solar Phys., 77, 303
Bachmann, K. T., Schou, J., Brown, T. M. 1993, ApJ, 412, 870
Balmforth, N. J. 1992a, MNRAS, 255, 603
Balmforth, N. J. 1992b, MNRAS, 255, 632
Balmforth, N. J. 1992c, MNRAS, 255, 639
Biermann, L. 1946, Naturwiss., 33, 118
Biermann, L. 1948, Z. f. Astrophys., 25, 161
Christensen-Dalsgaard, J., Frandsen, S. 1983, Solar Physics, 82, 469
Deubner, F. L. 1975, A&A, 44, 371
Dravins, D., Lindegren, L., Nordlund, Å. 1981, A&A, 96, 345

Dravins, D., Lindegren, L., Nordlund, Å. 1986, A&A, 158, 83
Dravins, D., Nordlund, Å. 1990a, A&A, 228, 184
Dravins, D., Nordlund, Å. 1990b, A&A, 228, 203
Evans, J. W., Michard, R. 1962, A&A, 257, 287
Goldreich, P., Keeley, D. A. 1977, ApJ, 212, 243
Goldreich, P., Kumar, P. 1988, ApJ, 326, 462
Goldreich, P., Kumar, P. 1990, ApJ, 363, 694
Goldreich, P., Murray, N., Kumar, P. 1994, ApJ, 424, 466
Goode, P. R., Gough, D., Kosovichev, A. G. 1992, ApJ, 387, 707
Gough, D. O. 1980, in H. A. Hill, W. Dziembowski (eds.), Nonradial and Nonlinear Stellar Pulsations, Vol. 125 of *Lecture Notes in Physics*, 273
Kiselman, D., Nordlund, Å. 1995, A&A, 302, 578
Leibacher, J., Stein, R. F. 1971, Ap. Letters, 7, 191
Leighton, R. B., Noyes, R. W., Simon, G. 1964, ApJ, 135, 474
Libbrecht, K. G. 1988, ApJ, 334, 510
Libbrecht, K. G., Woodard, M. F. 1991, Science, 253, 152
Lighthill, M. J. 1952, Proc. R. Soc. London, A211, 564
Moore, D. W., Spiegel, E. A. 1964, ApJ, 139, 48
Nordlund, Å. 1982, A&A, 107, 1
Nordlund, Å. 1985, Solar Physics, 100, 209
Nordlund, Å. 1991, in L. Crivellari, I. Hubeny, D. G. Hummer (eds.), Stellar Atmospheres: Beyond Classical Models, Kluwer
Nordlund, Å., Dravins, D. 1990, A&A, 228, 155
Nordlund, Å., Spruit, H. C., Ludwig, H.-G., Trampedach, R. 1997, A&A, 328, 229
Nordlund, Å., Stein, R. F. 1990, Computer Phys. Communications, 59, 119
Nordlund, Å., Stein, R. F. 1997, in F. P. Pijpers, J.Christensen-Dalsgaard, C. S. Rosenthal (eds.), Solar Convection, Oscillations and their Relationship; SCORe'96, Kluwer Academic Press, Dordrecht, (in press)
Nordlund, Å., Stein, R. F. 1998, ApJ, (in preparation)
Osaki, Y. 1990, in Y. Osaki, H. Shibahashi (eds.), Progress of Seismology of the Sun and Stars, Springer-Verlag, Berlin, 145
Rast, M. P., Nordlund, Å., Stein, R. F., Toomre, J. 1993, ApJ, 408, L53
Restaino, S. R., Stebbins, R. T., Goode, P. R. 1993, ApJ, 408, L57
Rimmele, T. R., Goode, P. R., Harold, E., Stebbins, R. T. 1995, ApJ, 444, L119
Rosenthal, C. S., Christensen-Dalsgaard, J., Nordlund, Å., Stein, R. F., Trampedach, R. 1998, A&A, (in preparation)
Schatzman, E. 1949, Ann. d'Astrophys, 12, 203
Schwarzschild, M. 1948, ApJ, 107, 1
Spruit, H. C., Nordlund, Å., Title, A. 1990, ARA&A, 28, 263
Stein, R. F. 1967, Solar Physics, 2, 385
Stein, R. F. 1968, ApJ, 297, 154
Stein, R. F., Nordlund, Å. 1989, ApJ, 342, L95
Stein, R. F., Nordlund, Å. 1991, in D. Gough, J. Toomre (eds.), Challenges to Theories of the Structure of Moderate Mass Stars, Vol. 388 of *Lecture Notes in Physics*, Springer, Heidelberg, p. 195
Stein, R. F., Nordlund, Å. 1997, ApJ, (in press)
Toutain, T., Frölich, C. 1992, ApJ, 412, 870
Ulrich, R. K. 1970, ApJ, 162, 993
Ulrich, R. K., Rhodes, E. J., J. 1977, ApJ, 218, 521
Unno, W. 1962, PASJ, 14, 416
Unno, W. 1964, Trans. Int. Astr. Un., XII(B), 555
Wöhl, H., Nordlund, Å. 1985, Solar Physics, 97, 213

RAMAN SPECTROSCOPY OF SOLAR P-MODES

M.P. RYUTOVA AND T.D. TARBELL
*Stanford-Lockheed Institute for Space Research, LMATC,
3251 Hanover Street, Palo-Alto, CA 94304, USA*

Abstract. According to observational data some regions of quiet sun show an excess, or the "emission" of acoustic waves. We propose a mechanism to explain the excess of the emission of acoustic waves based on nonlinear coupling of magnetic flux tube oscillations and the acoustic waves. We also suggest a new approach in the data analysis, which can be called "Raman spectroscopy of p-modes". This new idea is based on the fact that oscillating flux tubes constructively interfere with the acoustic wave which results in the generation of beat waves with combinational frequencies: the power spectra of scattered waves in addition to main peak will have Stokes and anti-Stokes satellites. Their amplitude and frequency shift reflect the properties and the structure of the observed region.

1. Introduction

The presence of the constant level of the solar p-mode fluctuations is a reflection of the dynamic balance between two processes: generation of p-modes by turbulent motions, and their damping by classical and/or anomalous dissipation mechanisms. Studies of the acoustic wave properties show that the sources and sinks are distributed non-uniformly over the solar surface, with sinks associated predominantly with sunspot and plage regions (Braun 1995; LaBonte & Ryutova 1993). The sources are more likely connected with the turbulent motions outside magnetic regions and, according to Brown's conjecture (1991), may be isolated and well separated from one another in space and time. Localized sources of the acoustic waves (Brown et al. 1992) and excess of the acoustic emission (Bogdan et al. 1993; Braun 1995) were observed in the quiet sun regions with the enhanced network of small scale magnetic elements. We propose a physical mechanism to explain the excess of the emission in the quiet Sun based on the interaction of p-modes with non-steady state motions, and in particular with the oscillating magnetic flux tubes. Note, that only non-steady motions can give rise to the increase of the energy of outgoing waves. We find two major effects that may contribute to this process. One is a *resonance scattering*: the energy of p-modes propagating in the random ensembles of flux tubes damps out due to the

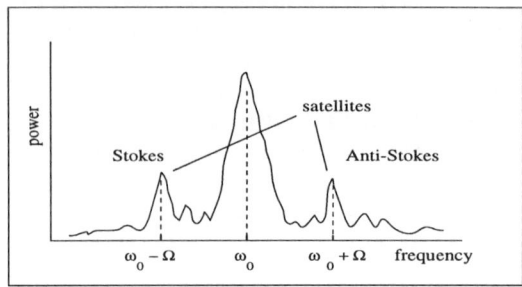

Figure 1. The sketch of power spectrum in Raman scattering.

mechanism similar to Landau damping, which consists in resonance excitation of natural oscillations of magnetic flux tubes (Ryutov & Ryutova 1976; Ryutova & Priest 1993a,b). The transformed energy of p-modes remains for a "long time" in a form of flux tube oscillations. Then, in a time $\tau_{rad} \simeq 1/\omega(kR)^2$, the resonant flux tubes radiate accumulated energy into surrounding plasma (R is a typical flux tube radius). Because τ_{rad} depends on flux tube radius, different flux tubes radiate secondary acoustic waves in different times, which is consistent with the conjecture by Brown. The resonance (Cherenkov) condition is determined by sound speed and parameters of flux tube: $\omega = (\mathbf{k} \cdot \mathbf{n})v_{ph}$, where $\mathbf{n}$ is a unit vector along the tube axis, and v_{ph} is the phase speed of excited tube oscillation, either kink ($m = \pm 1$) or sausage ($m = 0$) mode. Depending on the actual distribution of magnetic elements the power of outgoing waves may be considerably higher than incoming wave power. However, the radiation may occur only if the phase velocity of flux tube oscillations exceeds the sound speed in the ambient plasma: $v_{ph} > c_{se}$. Another effect works in the opposite case when $v_{ph} < c_s$ and is connected with nonlinear coupling of flux tubes oscillations and acoustic waves. This nonlinear coupling leads to appearance of beat waves with combinational frequencies. The power spectrum of outgoing wave will have then along the peak at the main frequency, the Stokes and anti-Stokes satellites that reflect the properties and structure of the observed medium.

2. Raman scattering of p-modes.

Oscillations of flux tubes may be excited by the convective motions or by fast reconnection processes. We consider the situation when oscillations of flux tubes are excited to a non-negligible level and have non-steady character. Then, nonlinear terms in the MHD equations cause a coupling of the incoming acoustic wave, and oscillations of flux tube. Which, in turn, leads to generation of beat waves with combinational frequency, $\omega_{in} \pm \Omega$ and wave number $k_{iz} \pm K_z$ (Figure 1). The frequency shift, Ω does not correspond to any of the eigen-frequencies of the p-modes; this is a perturbation characteristic to each particular region, and may correspond, for example to eigenfrequency of a "scattering center" (see below). The effective distance at which waves interact with given flux tube is of the order of flux tube radius, R. Main component of pressure perturbation in the vicinity of the magnetic flux is axisymmetric: $\delta p \simeq k\xi_{ac}p$; $\delta v \simeq \omega\xi_{ac}kR$; $\delta R/R \simeq \delta p/p$. This is coupled with $m = \pm 1$ flux tube motions. The resulting nonlinear drive exists at the scale $\simeq R$, has $m = \pm 1$ symmetry, and can be expressed as the following "equivalent" displacement $\delta x \simeq \xi_{kink}\, k\xi_{ac} \cdot \delta x$

can be conceived as a "forcing" term for generating the secondary emission. Assuming that $\omega = \omega_{in} \simeq \Omega$, $k_z^{in} \simeq K_z$, one can find an estimate for the power radiated per unit length of flux tube, P_{Raman}. If the density of flux tubes (number of flux tubes in the square of a unit area) is σ, then the power scattered by the volume of the size $L \times L^2$, is $\sigma P_{Raman} L^3$. Incident acoustic energy flow into the same volume is $\xi_{ac}^2 k^2 \rho c_{se}^3 L^2$. The ratio of the two is a measure of the excess of the emitted acoustic power and has a form: $\kappa = \sigma L (kR)^4 \ k \xi_{kink}^2$. Obviously, quantity κ has a meaning of the "optical depth" with respect to the scattering process. For the sausage oscillations, $\kappa = \sigma L (kR)^2 \ k \xi_{sausg}^2$. Using the amplitude of satellites an frequency shift one can infer physical parameters of "scattering centers", and develop the analysis to study the structure of atmosphere through the spectroscopic features of p-modes.

3. Examples and Discussion

We give some examples of measured power spectra of incoming and outgoing waves and their possible interpretation, only to demonstrate the principle of the method. The power spectra was obtained by Doug Braun using the Hankel decomposition method for MDI/SOHO high resolution data set. The target center was a very quiet region. The power spectra of incoming and outgoing waves for four values of spherical degree, ℓ, are given in Table 1. P_{in} and P_{out} are values (in arbitrary units) of net power averaged over a narrow frequency interval (0.23 mHz) near the peaks in outgoing power spectrum. We use first three cases, ℓ=452.6; 699.4 and 740.6, as an il-

TABLE 1. Power spectra and estimates for inferred Alfvén speed.

ℓ	P_{in}	P_{out}	ν (mHz)	$\nu_0 - \nu_\mp$	c_s km s^{-1}	v_A km s^{-1}
452.6	2.14 10^3	1.44 10^4	ν_-=2.67	0.55	31.1	5.30
	3.46 10^4	2.72 10^4	ν_0= 3.22			
	1.31 10^4	3.10 10^4	ν_+=3.77	-0.55		
699.4	4.96 10^3	9.90 10^3	ν_-=2.60	0.60	20.0	3.75
	3.52 10^4	3.51 10^4	ν_0= 3.20			
	2.75 10^4	4.48 10^4	ν_+=3.8	-0.60		
740.6	1.54 10^4	1.69 10^4	ν_-=2.70	0.62	19.6	3.66
	2.32 10^4	5.90 10^4	ν_0= 3.32			
	2.78 10^4	3.25 10^4	ν_+=3.94	-0.62		
905.1	1.35 10^4	5.37 10^3	ν_-=3.06	0.55	17.4	N/A
	2.93 10^4	5.76 10^4	ν_0= 3.61			
	1.31 10^4	3.10 10^4	ν_+=3.77	-0.69		

lustration of possible Raman scattering, and the fourth case, ℓ=905.1 as an example of resonance emission. In the first three cases the satellite peaks in outgoing power spectra exceed their counterparts in the incoming power spectra and are shifted equally from the central peak. The "Stokes" and "anti-Stokes" shifts may be used to infer the characteristic Alfvén speed. Indeed, in the subsurface layers Alfvén velocity is small compared to sound speed, therefore the phase velocities of both, kink and sausage mode of flux tube oscillations is close to the Alfvén speed, i.e. $v_{ph} = \Omega/k \simeq v_A$, where $\Omega = 2\pi|\nu_0 - \nu_\pm|$. Assuming that the wave number is approximately the same as that of incident acoustic wave we can express it through sound speed and write an estimate as: $v_A \simeq c_s \Omega/\omega_{in}$. For given spherical degree ℓ and frequency ν the local sound speed is roughly estimated as $c_s(l,\nu) = 2\pi R_\odot \nu/l$. Results of the inferred Alfvén speed are given in the last columns of the table. For ℓ=905.1 this estimate is not applicable: there are no equidistant satellites. But one can see that the outgoing wave power is considerably higher than the incoming wave power which may be provided by reemission of previously accumulated energy by resonant flux tubes.

The future observations and data analysis should be directed toward detailed study of high resolution magnetograms and wave field associated with compact isolated regions. One should bear in mind some distinct specific features of both possible effects, resonance emission and Raman scattering. For example, in a case of **resonance absorption and radiation of secondary acoustic waves** the necessary condition for the accumulation of acoustic wave power in the oscillating flux tubes is that the radiative damping time must be larger than the inverse Landau damping rate; this implies the condition that the acoustic wavelength must be larger than the average distance between the flux tubes: $\lambda \gg d$. Which means that the regions with high population of small scale structures will contribute to resonance scattering most efficiently. The intensity of main peak in outgoing waves may considerably exceed the main peak of incoming power. In the case of **Raman scattering** the most prominent feature is the appearance of *equidistant* Stokes and anti-Stokes satellites in the power spectrum of outgoing waves. The emitted power is proportional to magnetic filling factor and *size* of the observed area. It is more sensitive to small parameter $(kR)^2$ than the power radiated by resonant flux tubes. On the other hand, in contrast to the resonance case where the condition $(kR)^2 \ll 1$ is required, here the parameter $(kR)^2$ may be finite. Therefore, large regions containing flux tubes comparable with the acoustic wave length, will contribute the Raman scattering most readily. Good candidates here may be the quiet photospheric network near plages and regions containing small pores.

Acknowledgments

We thank Doug Braun for providing the data and for helpful comments. This research is supported by NASA MDI contract NAG5-3077 (PR 9162) at Lockheed.

References

Bogdan,T, Brown, T.M., B.W. Lites & J.H. Thomas, (1993) The Absorption of p-modes by Sunspots, *ApJ*, **Vol. 406**, pp. 723–734

Braun, D.C., (1995) Scattering of p-Modes by Sunspots. I. Observations, *ApJ*, **Vol. 451**, pp. 859-876

Brown, T.M. (1991) The Source of Solar High-Frequency Acoustic Modes: Theoretical Expectations, *ApJ*, **Vol. 371**, pp. 396–401

Brown, T.M., T. Bogdan, B. Lites and J. Thomas (1992) Localized Sources of Propagation Acoustic Waves in the Solar Photosphere, *ApJ*, **Vol. 394**, pp. L65–L68

LaBonte, B. & Ryutova, M. (1993) A Possible Mechanism for Enhanced Absorption of p-Modes in Sunspot and Plage Regions, *ApJ*, **Vol. 419**, pp. 388–397

Ryutov, D.D.& Ryutova, M.P. (1976) Sound oscillations in a plasma with magnetic filaments, *Sov.Phys. JETP*, **Vol. 43**, pp. 491-497

Ryutova, M.P. & Priest, E.R. (1993a;b) The Propagation of Sound Waves in a Randomly Magnetized Medium. *ApJ*, **Vol. 419**, pp. 349-370; pp. 371-381

EXPLORATORY SIMULATION OF SOLAR GRANULES: HOW SHARP IS THE CONVECTION/RADIATION TRANSITION?

KWING L. CHAN
The Hong Kong University of Science and Technology, Hong Kong

AND

Y. C. KIM
Yale University, USA

1. Introduction

Currently, the most successful direct simulation of the solar granules (and the convection/radiation transition layer) is the three-dimensional (3D) model computed by Stein and Nordlund (1989). So far, there is no other similar 3D models available for comparison [however, see Ludwig et al. (1997) for a recent 2D calculation]. We are developing an alternative numerical approach to simulate the 3D radiation hydrodynamics of this layer. In this approach, the Eddington approximation is used to handle the radiation rather than solving the radiative transfer equations along rays, and the ADISM method (Chan and Wolff 1982) which solves the Navier Stokes equations in conservative forms is used to speed up the thermal relaxation of the fluid layer. We are in the process of testing the numerical accuracy of the codes. This paper summarizes the results of a test that illustrate the effects of vertical space resolution on the mean profiles of some important quantities.

2. The Model and Results

The computed domain is a rectangular box $1500km$ wide and $1000km$ deep, with approximately $200km$ above the τ (optical depth) ~ 1 level. We perform calculations with three different meshes: $45 \times 45 \times 83$, $45 \times 45 \times 42$, and $45 \times 45 \times 21$. The vertical resolutions for these meshes are $13.3km$, $26.6km$, and $53.3km$. The computed mean temperature (T) and flux of kinetic energy (F_{ke}) profiles are shown in Figs 1 and 2. The solid, dashed, and dotted (with plus signs to

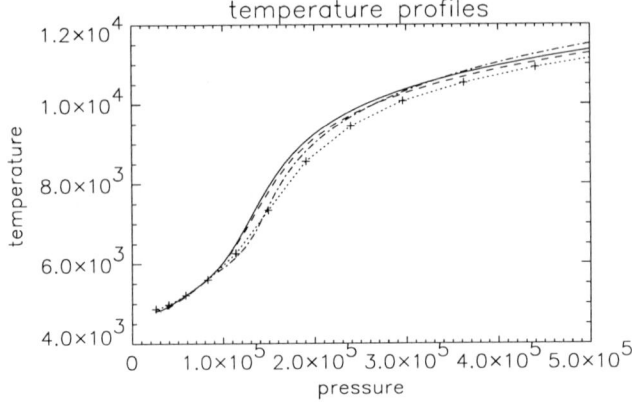

Figure 1. Mean temperature profile for different resolutions

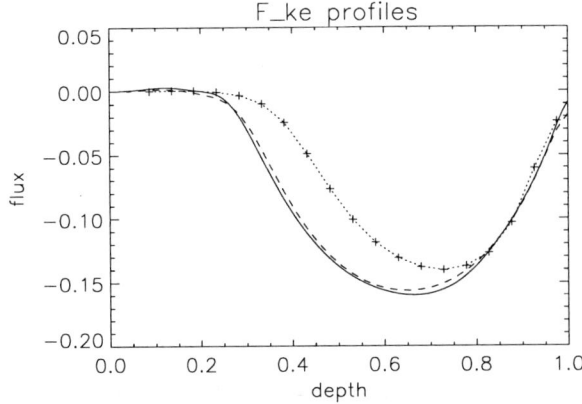

Figure 2. Flux of kinetic energy profiles for different resolutions

show the grid locations) curves are for the high, medium, and low resolution cases, respectively. The dot-dashed curve in Fig 1 shows a profile based on the standard mixing-length theory which is not far off. The high and medium resolution results are close to each other, while the low resolution result differs more significantly, by about $500K$ in the deeper region. In Fig 2 the flux unit is the total energy flux. The low resolution has a relatively more serious effect on F_{ke}.

References

Chan, K. L., and Wolff, C. L. 1982, *J. Comput. Phys.*, **47**, 109
Ludwig, H.-G., Freytag, B., and Steffen, M. 1997, *this volume*
Stein, R. F., and Nordlund, A. 1989, *Astrophys. J.*, **342**, L95

ACOUSTIC IMAGING AND SUBSURFACE ABSORPTION STRUCTUR OF SUNSPOTS

H.-K. CHANG AND D.-Y. CHOU
Department of Physics, National Tsing Hua University,
Hsinchu 30043, Taiwan, ROC

B. LABONTE
Institute for Astronomy, University of Hawaii
Honolulu, HI 96822, USA

AND

THE TON TEAM

Recently, Chang et al. (1997, hereafter Paper I) developed a new method, based on time-distance curves (Duvall et al. 1993), to construct a three-dimensional acoustic intensity image of the solar interior. Here we report results from another data set.

Solar p-mode waves, which are continually generated and dissipated stochastically by the turbulent convection, are scattered and absorbed by local inhomogeneities. If we appropriately add p-mode waves observed at the surface, based on the knowledge of how the waves propagate (time-distance curves), such that acoustic signals emanating from a particular point at a particular time are collected in phase, we can reconstruct the p-mode amplitude at the target point at the target time. The scheme of adding signals from a target point in phase plays the same role as a lens in optics. Thus we call it the "computational acoustic lens" (Paper I). This imaging technique can be used to construct acoustic amplitudes at any point in the solar interior based on the time-distance curve between that depth and the surface, which can be computed from a standard solar model based on the ray theory. The time series of constructed amplitudes provides information of both the intensity and the phase of the wavetrain. The intensity can be used to study p-mode absorption regions, and the phase can be used to probe other properties of local inhomogeneities, such as flow and change in wave speed.

The helioseismic data in this study were taken by the Taiwan Oscillation Network (TON) in the period of July 29 - August 6, 1996. NOAA 7981 is selected as a target region. Detailed data reduction procedure is described in Chou et al. (1997). The acoustic signals constructed with a *normal* time-distance curve are those propagating outward from a target point since the normal time-distance curve describes outgoing waves. With a time-reversed time-distance curve, we can also construct incident acoustic intensity at a target point. The acoustic images constructed with a time-reversed time-distance curve have the same acoustic intensity as those constructed with a normal time-distance curve and show no signature of the sunspot as expected because incident waves in the surrounding area have not been affected by local properties of the target point before they reach it.

To study p-mode absorption in active regions, we subtract the incident acoustic intensity from the outgoing intensity and then divide it by the incident acoustic intensity for each target point to obtain fractions of power of incident waves absorbed in the medium along wave paths. At the surface, the strongest absorption, 29.8%, occurs in the preceding sunspot. We construct absorption images at various depths from the surface to 80 Mm deep at an interval of 10 Mm. A collecting area of the same annular region, 2-14°, is used for all depths. These subsurface absorption images are qualitatively similar to the intensity images constructed with outgoing waves for NOAA 7973 (Paper I). The absorption feature in the sunspot disappears at about 50 Mm below the surface for NOAA 7981 and about 40 Mm below the surface for NOAA 7973. Since the major component of the spatial fluctuation in the quiet Sun is believed to be noise, we can estimate the vertical spatial resolution from the correlation of the quiet Sun feature among different depths. The vertical spatial resolution for both regions is about 20 Mm. Thus we estimate that the depth of the absorption region for sunspots is about 20 - 30 Mm.

In addition to acoustic absorption images, the time series of constructed amplitudes can also provide phase information. The relative phase between two time serieses can be determined from the cross-correlation technique. Here mentioned are two results from the phase study. First, the phase of p-modes advances about one minute for each reflection at the surface. Second, the phase shift between the time serieses constructed by incident waves and those constructed by outgoing waves in the sunspot area is different from the quiet Sun. Phase information has a better vertical spatial resolution than intensity, and may soon prove to be a more powerful tool for probing solar subsurface structures.

HKC, DYC, and the TON project were supported by NSC of ROC under grants NSC-87-2112-M-007-044. BL was supported by NASA grant NAGW-1542 and NASA contract NAS8-37334 of USA.

References

Chang, H.-K., Chou, D.-Y., LaBonte, B., and the TON team: 1997, *Nature* **389**, 825 (Paper I).
Chou, D.-Y., Chang, H.-K., Sun, M.-T., LaBonte, B., Chen, H.-R., Yeh, S.-J., and the TON team: 1997, *ApJ*, submitted.
Duvall, T. L. Jr., Jefferies, S. M., Harvey, J. W., and Pomerantz, M. A.: 1993, *Nature* **362**, 430.

EXCITATION OF OSCILLATIONS - AN UPDATE OF BISON DATA

W. J. CHAPLIN, Y. ELSWORTH, G. R. ISAAK, C. P. MCLEOD AND
B. A. MILLER
School of Physics & Astronomy, University of Birmingham
Edgbaston, Birmingham B15 2TT UK
E-mail: ype@star.sr.bham.ac.uk

AND

R. NEW
School of Science & Mathematics, Sheffield Hallam University
Sheffield, S1 1WB

BiSON is a 6-station, world-wide network of instruments which observe low-degree solar oscillations.

1. Excitation of oscillations - comparison with theory

Theory (Goldreich & Keeley, 1977) predicts that the probability distribution of the mode powers is expected to follow negative exponential statistics under the condition that the interval of time for which the modes are averaged is less than the lifetime of the mode. Simulations (Chaplin et al., 1997) confirm this. The first publication of such an analysis for real data (Elsworth et al., 1995) showed that the oscillations largely follow the predictions with a small excess of the very largest power in the modes. In this paper we extend the previous observations of BiSON data to 80 months of data taken during the declining phase of the solar cycle. The data span the period January 1990 to August 1996. The data have been analysed using Fourier transforms each about 1/2 day in length. The minimum data fill accepted in any one period was 0.7 (Elsworth et al., 1995). Below this fill the data were discarded. The resolution of such short transforms is insufficient to distinguish between the components of the even and odd mode pairs. The mode powers are calculated for mode pairs $\ell=0\&2$ and $\ell=1\&3$.

Comparison with theoretical predictions show that, as seen previously, there is still an excess of very high excitations which can be seen throughout the 80 months.

2. Extending the analysis range

We integrate for less than one lifetime of a mode in order to be able to follow its evolution. At frequencies below about 2.0 mHz the lifetimes of the modes increase (Chaplin et al., 1997) so that the length of the individual sub-set transforms can also increase. An indication of the background noise against which the mode must be detected can be found from the measured power between the modes. The signal to noise derived in this way together with the known linewidths indicates that the individual modes can be followed from 2 mHz to 1.6 mHz using 4-day transforms.

References

Goldreich P., & Keeley D. A., 1977, *ApJ.*, **211**, 934

Chaplin W. J., Elsworth Y., Isaak G. R., McLeod C. P., Miller B. A. and New R., 1997, *MNRAS*, **287**, 51

Elsworth Y., Howe R., Isaak G. R., McLeod C. P., Miller B. A. and New R., 1995, *GONG '94*, 51

Chaplin W. J., Elsworth Y., Isaak G. R., McLeod C. P., Miller B. A. and New R., 1997, *MNRAS*, **288**, 623

ARE SOLAR P MODES CORRELATED ?

T. FOGLIZZO, R.A. GARCÍA AND THE GOLF TEAM[†]
Service d'Astrophysique, DAPNIA/DSM, CE-Saclay
91191 Gif-sur-Yvette, France
[†]GOLF Team: P. Boumier, J. Charra, A.H. Gabriel, G. Grec, J.M. Robillot, T. Roca Cortés, S. Turck-Chièze, R.K. Ulrich

Abstract. We have studied the statistical properties of the energy of individual p modes, extracted from 310 days of GOLF data near the solar minimum. The exponential distribution of the energy of each mode is clearly seen. The modes are found to be uncorrelated with a ±0.6% accuracy, thus supporting the hypothesis of stochastic excitation by the solar convection.

The same analysis performed on the same modes just before the solar maximum, using IPHIR data, rejects the hypothesis of no correlation at a 99.3% confidence level. A simple model suggests that $31.3 \pm 9.4\%$ of the energy of each mode is coherent among the modes studied in IPHIR data, corresponding to a mean correlation of $10.7 \pm 5.9\%$.

1. GOLF and IPHIR data sets

We have considered the set of p modes corresponding to $17 \leq n \leq 25$, $l = 0$ and 1, observed by GOLF between 11th April 1996 and 14th February 1997, filtered through a window of $\Delta\nu = 8\mu Hz$. The two m-components of the mode $l = 1$, however, are not separated. After removal of the data surrounding a gap in the series, the sample is made up of 210 points for each mode.

We have also extracted from IPHIR data the same 11 modes as Baudin et al. (1996) ($l = 0$, $19 \leq n \leq 23$, and $l = 1$, $18 \leq n \leq 23$). The size of the filtering window is reduced to $6\mu Hz$ due to the higher level of noise. The sample is made up of 78 points after removal of the data surrounding two gaps in the series.

2. The distribution of energy of each mode is exponential

Following the picture of p-mode excitation by turbulent motions (Goldreich & Keeley 1977), we have compared the observed energy distribution of each mode with an exponential distribution (Woodard 1984).

A first test measures the variance of the normalized distribution of energy. A second one considers the maximum distance between the observed and theoretical cumulative distributions (Kolmogorov-Smirnov test, hereafter KS). These two quantities are compared with those obtained from artificial exponential series, using a Montecarlo simulation of 10^5 trials.
The output of each test, denoted by $P_{|V|}$ and P_{KS} is therefore a probability that the observed value corresponds to the output of a Montecarlo simulation.
These tests agree very well with exponential distributions, both for GOLF and IPHIR.

3. Correlation coefficient

If excited by the granules, the exponentially distributed modes energies would be independent. The sum Υ_k of their normalized energies should then be distributed as a Gamma distribution Γ_k, which we test with the variance and KS tests. Both tests agree with this "null hypothesis" when applied to GOLF data: $P_{|V|} = 49.9\%$, $P_{KS} = 76.2\%$ for 9 modes $l = 0$, and $P_{|V|} = 95.8\%$, $P_{KS} = 66.5\%$ for 18 modes.
By contrast, the same tests applied to IPHIR data *reject the null hypothesis* with a 99.3% confidence level with the variance test, and a 95.7% confidence level for the KS test ($P_{|V|} = 0.7\%$, $P_{KS} = 4.3\%$).
A refined estimate of the correlation can be obtained by making some assumptions about the origin of the correlation. We build the distribution function Γ_k^λ where a fraction λ of the energy of each mode is due to a common random signal, added to the velocity of each mode, resulting in a correlation $\mathcal{C} = \lambda^2$. This "λ-hypothesis" is also tested with the variance and KS tests. The error bar of the correlation coefficient is defined as the range within which each test remains inside the upper 68.3% region. Both tests, applied to 18 modes of GOLF for various values of λ, give $\mathcal{C} \leq 0.6\%$, which is the sensitivity limit of the test. Applied to IPHIR data, the variance test obtains $\mathcal{C} = 6.1 \pm 3.3\%$, while the KS test obtains $\mathcal{C} = 10.7 \pm 5.9\%$.
No correlation was found between 11 windows of noise centered $20\mu Hz$ to the right of each mode, thus ruling out the possibility of a pointing noise origin.

4. Conclusion

The correlation found in IPHIR data reveals the existence of a source of coherent excitation of p modes, in addition to the well known granules excitation. The difference between IPHIR and GOLF data would indicate a change in the fraction of the energy coming from this additional source, varying from $\lambda = 31.3 \pm 9.4\%$ in IPHIR data to less than 8% in GOLF data. Details about our method can be found in Foglizzo et al. (1997). If related to the change in magnetic activity, a confirmation will be obtained by performing the same analysis on GOLF and VIRGO data when we approach the solar maximum, in a couple of years.

References

Baudin F., Gabriel A., Gibert D., Pallé P.L., Regulo C., 1996, A&A 311, 1024
Foglizzo T., García R.A., Boumier P., Charra J., Gabriel A.H., Grec G., Robillot J.-M., Roca Cortés T., Turck-Chièze S., Ulrich R.K. 1997, A&A (in press)
Goldreich P., Keeley D.A., 1977, ApJ 211, 934
Woodard M., 1984, PhD Thesis, University of California, San Diego

THE INTEGRATED MAGNETIC FIELD OF THE SUN SEEN BY THE GOLF EXPERIMENT

R. A. GARCÍA[1], T. ROCA CORTÉS[2] AND THE GOLF TEAM[†]
[1] SAp, DAPNIA/DSM, CE Saclay. 91191 France
[2] Instituto de Astrofísica de Canarias, E-38205, Tenerife, Spain
[†] GOLF Team: P. Boumier, J. Charra, A.H. Gabriel, G. Grec, J.M. Robillot, S. Turck-Chièze, R.K. Ulrich

Abstract. The secondary objective of the GOLF experiment on-board the SOHO space mission is to measure the line-of-sight component of the disk averaged magnetic field of the Sun. GOLF is an improved disk-integrated sunlight resonant scattering spectrophotometer. Using an extra fixed quarter wave plate placed at the entrance of the instrument, enables a selection of the circularly polarized solar light and therefore, the disk averaged solar line-of-sight component of the magnetic field can be obtained. Unfortunately, due to occasional malfunction of the rotating mechanisms, only a series of 26 continuous days are available. Here, the analysis of this series is presented including the value of the averaged magnetic field.

1. Method

The method of obtaining the magnetic field is based on the measure of the $I \pm V$ Stokes parameters (a full description is in García (1996)). If we assume that the solar magnetic field is weak throughout the atmosphere where the sodium doublet is formed, we can use the so-called weak field approximation. In this case, the displacement due to the solar magnetic field, $\delta\lambda_{B\odot}$ is much smaller than the Doppler width of the original line profile (for a general description see e.g. Stix 1989). In this approximation, we can consider the magnetic field as a small perturbation and make a Taylor expansion of the Stokes parameters in $\delta\lambda_{B\odot}$, up to first order. Adding and substracting the non zero Stokes parameters, i.e. I and V:

$$I \pm V = I(\lambda_o) \pm \delta\lambda_{B\odot} \cdot \left(\frac{dI}{d\lambda}\right)_{\lambda_o} \quad (1)$$

Each GOLF measurement is a combination of this two parameters: I_b^-, I_b^+, I_r^-, I_r^+ which correspond to $I - V$ and $I_b'^-$, $I_b'^+$, $I_r'^-$, $I_r'^+$ to $I + V$; where I' denotes the

σ^- polarized component of the solar light and λ_0 denotes the longitudinal Zeeman wavelength.

Substracting equations 1, and expressing each magnitude in terms of GOLF observables we can obtain an expression that links the longitudinal component of the global magnetic field of the Sun with the GOLF measurements:

$$B_\odot = \frac{\delta B}{2} \cdot \left[\frac{(I_b'^+ - I_b^+ + I_b'^- - I_b^-)}{(I_b'^+ - I_b'^- + I_b^+ - I_b^-)} - \frac{(I_r'^- - I_r^- + I_r'^+ - I_r^+)}{(I_r'^+ - I_r'^- + I_r^+ - I_r^-)} \right] \quad (2)$$

The validity of this method has been tested using a numerical simulation of the GOLF experiment (García, Roca Cortés and Régulo 1997). The simulated magnetic field was recovered with a precision better than 2 parts in 1000.

2. GOLF magnetic field observations

In Fig. 1, the resultant solar magnetic field is plotted as a function of time. A fitting of the form $A \cdot sin(t/P + \phi) + c$ has been calculated giving a main periodicity of $\simeq 13.8$ days with an amplitude $A \simeq 0.281$ and an offset of $c \simeq 0.149\ G$.

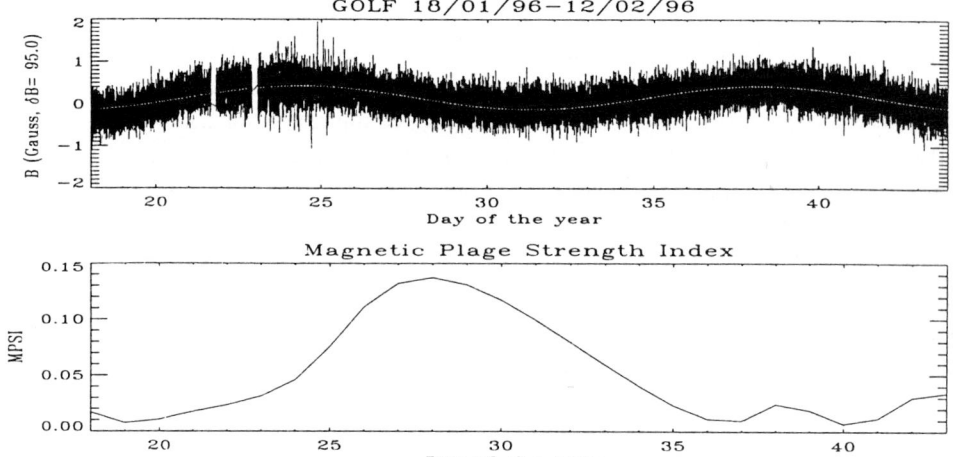

Figure 1. Up: Longitudinal global magnetic field of the Sun computed using equation 2. The gray line shows the best sinusoidal fit. Down: MPSI magnetic index for the same period of time.

This periodicity is related to solar rotation. During this period of time the Mount Wilson MPSI magnetic index (see the lower graph) shows the passage of an active region over the solar disk. The small phase shift (~ 3 days) is well explained by the different weighting functions of the two instruments.

References

-García R.A., 1996, PhD Thesis, Univ. La Laguna, Tenerife, Spain
-García, R.A., Roca Cortés, T. and Régulo C. 1997, *Astron. Astrophys. Supp. in Press*
-Stix, M. 1989, *The Sun. An Introduction*, Springer-Verlag

TEMPORAL BEHAVIOR OF SOLAR P-MODES FROM GONG AND GOLF EXPERIMENTS

V.G. GAVRYUSEV
CAISMI CNR, Largo E.Fermi 5, 50125, Florence, Italy

AND

E.A. GAVRYUSEVA
Osservatorio Astronomico di Capodimonte, via Moiariello 16, I-80131, Naples, Italy
Institute for Nuclear Research of RAN, Moscow, Russia

We used the measurements of solar oscillations taken by GONG and GOLF experiments. The first set of data are the integrated images obtained from the complex GONG observations taken from June 10 of 1995 to January 7 of 1997, 578 days in total, referenced below as **ts0** time series. Radial, dipole and quadrupole modes are well visible in this time series. The second data set is the GOLF time series obtained onboard SOHO mission from April 11, 1996 to June 22, 1997. GOLF observes the "Sun as a star". This time series is similar to **ts0** of GONG but of a better quality (better signal-to-noise ratio; uniform, practically uninterrupted data). Both experiments are significantly overlapped in time. Because of this the direct comparison between them is possible, and the effects visible in both observations support each other.

To study the temporal evolution of the power of a given mode of the oscillations, one has to analyze the sequence of the corresponding power spectra. A 20 day window, which is reasonably close to the typical visibility time of all radial modes of radial order from 13 to 26 (Gavryusev & Gavryuseva, 1996), was chosen. This window runs through the whole data series with a one day step. The standard Fourier transform produces the mean power corresponding to the current window.

The temporal change of the power of each individual mode usually shows a quasi random appearance of the pulses of different duration and amplitude (Baudin 1996; Gavryusev & Gavryuseva 1997). Such behavior is typical for all the modes we have analyzed. Components of the rotational multiplets of a given radial order do not perfectly correlate. Most of them differ significantly in their mean power level over 2.5 years. As a result, the components of the rotational multiplets are strongly asymmetric.

There is a question still to be answered, whether the big pulses appear completely randomly in time, or if there is something coordinating their appearance in a more or less regular manner. One of the current theoretical hypotheses is the following: the big pulses appear just by chance, due to random interference between very frequent excitation events. In such a case one can not expect any stable periodicity in the time sequence of the spectra of the window running throw the long data set. It seems true for some of the modes. But we find a few examples of very regular periodical behavior over 800 days. $\ell = 0$ $n = 15$ and $n = 17$ demonstrate quasi 50-days and 135-days variations respectively. The other modes behave in a different way. But for many of them the autocorrelation analysis shows the presence of significant periodicities.

To reveal something common for all the modes, we calculated the sum of the power of 14 modes of radial order from 13 to 26. The frequencies of the individual peaks in the spectra of each running window generally change with time, and in many cases there is more than one peak around the frequency of an expected mode. Because of this we used the power *integrated* in 7 μHz intervals, centered at the mode frequency, obtained from the mode profile revealed from the complete data series.

A variation of the total p-mode power with the period of solar rotation was clearly detected by Gavryusev & Gavryuseva (1996,1997) in IPHIR measurements (taken in 1988, when solar activity was rising up). During the time of solar minimum, covered by the GONG and GOLF data sets, this periodicity is much less visible in the running spectra, while it can still be detected by a Fourier transform of the series of the power spectra or by their autocorrelation analysis in the both experiments. It seems, that during this time the solar magnetic structure has changed and the new structure is not yet established. We may expect, that later on, when the new magnetic structure will be more regular, the change of p-mode power will become stronger again and better correlated with solar rotation.

It is necessary to investigate this question in detail, but one thing is already obvious: the simple model of completely random frequent excitation fails to explain such long term correlations in time. The excitation processes are stochastic but by no means trivial. And the investigation of the temporal behavior of the p-mode power is of vital necessity if we wish to understand the mechanisms of the excitation of the solar oscillations.

Acknowledgements

This work utilizes data obtained by the Global Oscillation Network Group (GONG) project, managed by the National Solar Observatory, a Division of the National Optical Astronomy Observatories, which is operated by AURA, Inc. under a cooperative agreement with the National Science Foundation.

The data were acquired by instruments operated by the Big Bear Solar Observatory, High Altitude Observatory, Learmonth Solar Observatory, Udaipur Solar Observatory, Instituto de Astrofísico de Canarias, and Cerro Tololo Interamerican Observatory.

References

Baudin F., Gabriel A., Gilbert D., Palle P., Régulo C., 1996, A&A, **311**, 1024
Gavryusev V.G., Gavryuseva E.A., 1996, A&A, **310**, 651
Gavryusev V.G., Gavryuseva E.A., 1996, in Cool Stars, Stellar Systems, and the Sun 95, ASP Conference Series, v. **109**, 129
Gavryusev V.G., Gavryuseva E.A., 1997, Solar Physics, **172**, 27

ON THE FORMATION OF LINE PROFILES OF SOLAR P MODES

I.W. ROXBURGH[1] AND S.V. VORONTSOV[1,2]
[1] *Astronomy Unit, Queen Mary and Westfield College*
Mile End Road, London E1 4NS, UK;
[2] *Institute of Physics of the Earth*
B.Gruzinskaya 10, Moscow 123810, Russia

Abstract. We address the problem of the opposite asymmetry of low-frequency p-mode line profiles observed in intensity and velocity measurements (Duvall et al. 1993). We use a simple model to illustrate that this feature can be explained by including a contribution from the stochastic excitation velocity field to the non-resonant background in the doppler measurements.

Our analysis is based on the simple model of acoustic power spectra described in Roxburgh and Vorontsov (1995). We extend this theoretical description slightly, allowing for an instantaneous point-like source—an excitation which is described by δ-function in both space and time. We make a simple order of magnitude esimate of the source function for monopole and dipole excitations, and also estimate the direct contribution to the doppler velocity signal from the kinetic velocity of the convective eddy (Roxburgh and Vorontsov 1997). We define the total complex amplitude $A(\omega)$ of the doppler velocity measurements as

$$A(\omega) = k_{mon}A_{mon}(\omega) + k_{dip}A_{dip}(\omega) + k_{kin}A_{kin}(\omega), \tag{1}$$

with three weighting coefficients k_{mon}, k_{dip} and k_{kin}. The first two coefficients are the efficiency factors of monopole and dipole excitation; k_{kin} is a factor of "visibility" of the kinematic velocity field. When the kinematic effect is neglected, the line profiles seen in intensity and in doppler measurements are similar. We expect the intensity data to be less sensitive to the kinematic velocity field, and simulate the results of the intensity measurements simply by using the Eq.(1) with k_{kin} set to zero. We also expect the efficiency factors to have nearly the same order of magnitude, if our hypothesis is correct.

Fig.1 shows the power spectrum produced by a composite excitation source. Line profiles depend on the relative magnitude and phase of the even and odd components of the excitation; the observational asymmetry of intensity measurements (smaller amplitude at the lower-frequency wing of the line) can be easily reproduced. The results obtained with the same composite excitation source but with the kinematic effect taken into account are shown in Fig.2. The line asymmetry is changed by the coherent contribution to the acoustic background of the kinematic-velocity signal.

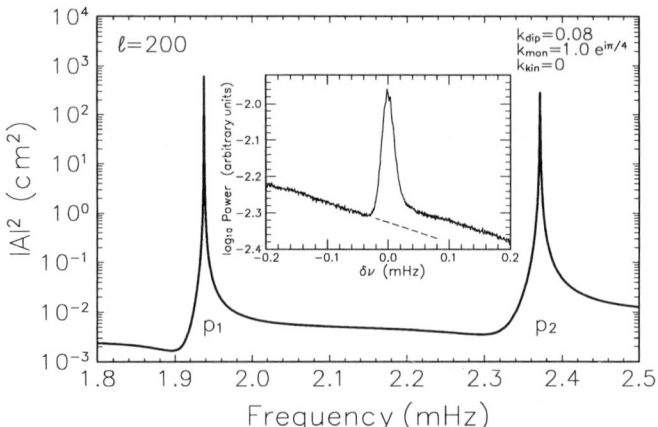

Figure 1. Artificial power specrum for a composite excitation source. Line profile observed in the intensity measurements (Duvall *et al.* 1993) are shown in the insert (courtesy S.M. Jefferies)

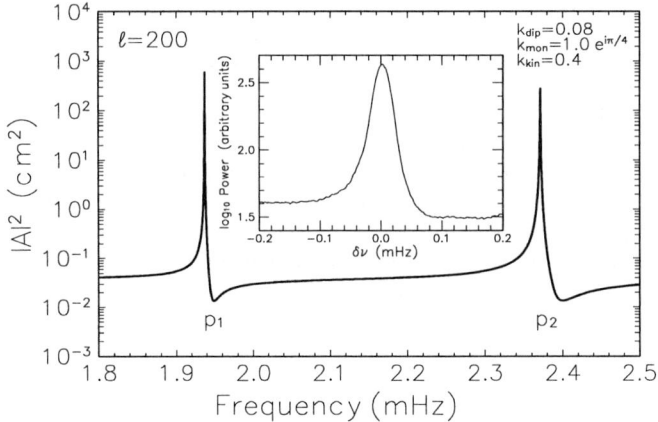

Figure 2. Same as Fig.1, but with additional contribution of the kinematic effect. The insert shows the line profile observed in velocity measurements (Duvall *et al.* 1993, courtesy S.M. Jefferies)

References

Duvall, T. L., Jr., Jefferies, S. M., Harvey, J. W., Osaki, Y. and Pomerantz, M. A. (1993), *ApJ* **410**, p.829

Roxburgh, I. W. and Vorontsov, S. V. (1995), *Mon. Not. R. astr. Soc.* **272**, p.850

Roxburgh, I. W. and Vorontsov, S. V. (1997), *Mon. Not. R. astr. Soc.*, in press

NEW DEVELOPMENTS IN ASTEROSEISMOLOGY

D. W. KURTZ
Department of Astronomy, University of Cape Town,
Rondebosch 7701, South Africa

1. Introduction

The day before the start of IAU Symposium 185 Wojtek Dziembowski and I were sitting in the beautiful 300-year-old garden of Shisendo in Kyoto discussing asteroseismology. Wojtek put his fingertips together to make a little temple of his hands in front of him. Then, slowly moving this temple up and down in front of him, while staring off at the green Japanese mosses in the dappled sun and shade of Shisendo's garden forest he said to me in his studied way, "Don, I am not sure that there is any asteroseismology yet". I thought, "Uh-oh. If others believe that is true, then they may not think I have much to say in my review of new developments in asteroseismology at S185." Yet I do have much to say, so I think I had better define what I mean by asteroseismology.

Sometimes I think that between theoreticians and observers, and between helioseismologists and asteroseismologists, there is potential confusion by what we mean when we say "asteroseismology". It reminds me of the story of a Western drug company which some years ago decided to market their headache remedy, a kind of aspirin, in Japan. To avoid the problem of translating their English into correct Japanese, they decided that there would be no written description on the box containing the pills. Instead, a long, narrow box was designed on which three pictures fit side-by-side. The first picture showed a woman's face with her eyes squeezed shut with the terrible pain of a headache; sparks shot out of her head. In the middle picture the look on her face was neutral as she was seen taking two of the pills with a glass of water. Then, in the last picture, she had a beaming smile; clearly her headache was gone. Well, sales were so low as to be nearly non-existent. The Japanese were *not* interested in this product. And the reason is obvious, of course: The Japanese read from right to left.

Maybe the confusion between helioseismologists and asteroseismologists comes from us "reading" in opposite directions. Helioseismology has incredible depth, with millions of pulsation frequencies and modes, with the physics being pushed to precisions of fractions of a percent. Whereas for stars we have at best a few hundreds of modes (in white dwarfs) and less than a few dozen modes in any main sequence star. That is a pitifully small amount of data to claim we are doing seismology, in comparison with the sun. The paucity of the data is what stimulated Wojtek's comment to me. But with stars we have essentially an unlimited number of subjects on which to work. And the physical conditions are wonderfully varied. There is so much more with which to confront seismology theory than for the single star, the sun.

So here is my definition of asteroseismology: We are doing asteroseismology whenever we can constrain the physics of any star, in any way, by using two or more pulsation frequencies – at a minimum, just two frequencies; not necessarily millions as in the sun. And with that definition, I have a vast amount of exciting new work to discuss in this review of "New Developments in Asteroseismology". For details and depth I urge you to read the asteroseismology papers in these proceedings to see the excitement in the field and the excitement we experienced at the Symposium:

- Helioseismology desperately wants observations of g-modes (just *one* will do to start with) to probe the core of the sun. Read in Kevin Krisciunas's paper about an entire class of g-mode pulsators only slightly hotter than the sun, the γ Dor stars.

- Find out from Conny Aerts how HIPPARCOS discovered over 100 new Slowly Pulsating B (SPB) stars, hotter g-mode pulsators, when before HIPPARCOS there were less than a dozen known. These SPB stars fill a theoretical instability strip that was predicted before the observations!

- Find out from Tim Bedding and from Christoph Keller how the search for solar-type oscillations is going in other G stars.

- Read below, in section 2, my opinion on the possibility of a g-mode in the famous, and now controversial, star 51 Pegasi.

- Discover from Dietrich Baade and Wojtek Dziembowski about the observation and theory of pulsation in the B stars on and above the main sequence.

- See the most exciting new result of this Symposium in the paper by Bob Stobie *et al.*: The discovery of 2 to 3 minute pulsations in a group of sub-dwarf B stars, the romantically-named EC 14026 stars. Very satisfyingly, read about the theoretical prediction of these pulsations in the paper by Gilles Fontaine *et al.*, a prediction made independently of the discovery. These stars are so exciting that I cannot resist discussing them some myself in section 3 below.

- Read about the latest observational results on the rapidly oscillating Ap stars in Jaymie Matthews' review (it is too bad for you, if you missed the actual performance at S185), the latest theoretical work in the paper by Hideyuki Saio and Alfred Gautschy, and some new results from me and my colleagues in section 5 below. Especially, read how HIPPARCOS trigonometric parallaxes and our asteroseismic parallaxes agree. We *are* doing asteroseismology!

- The δ Scuti stars are the main sequence stars with the largest number of pulsation frequencies and modes. Read Michel Breger's observational review and Joyce Guzik's theoretical discussion.

- The white dwarfs are the current asteroseismological superstars. They have hundreds of frequencies and dozens of modes. Asteroseismic masses have a precision of a few hundredths of a solar mass. Seismological distances and the trigonometric parallax of GD358 agree: asteroseismology works. Rotational splitting is observed and studied;

possible magnetic splitting has been seen. There even seems to be the prospect that a massive (M > 1 $M_\odot$) DAV (ZZ Ceti) star may show the effects of differential crystallisation of Oxygen (*i.e.* a change in the internal structure, heat budget, and hence pulsation frequencies caused by Oxygen "snow"). Read all about it in the papers by S185's white dwarf tag-team, Chris Clemens and Steve Kawaler.

- And *don't* miss the poster paper by Brian Warner and Liza van Zyl, who have discovered that GW Lib, a cataclysmic variable with an orbital period of less than 2 hours, has a ZZ Ceti primary star. This pulsating white dwarf is probably about a billion years old and should have a surface temperature of 9000 K. But accretion heating has pumped its temperature up to 12000 K and pushed it into the DAV instability strip. All DAV stars have masses of about 0.6 solar masses, but GW Lib may have a very different mass, and asteroseismology will tell.

That is my conclusion to this review, given right here in the introduction: Asteroseismology is alive, well, and "asteroseismology will tell". If you were not there at Symposium 185, if you missed this IAU GA, I hope you are feeling a tinge of regret. Assuage it by reading the rest of these proceedings.

2. Have g-modes already been discovered in a solar analogue?

In 1995 radial velocity variations were announced by Mayor and Queloz (1995) with an amplitude of about 56 m s^{-1} and a period of 4.23 days in 51 Peg. Marcy *et al.* (1997) independently confirmed these results. The interpretation of these radial velocity variations is that there is a planet of mass at least half that of Jupiter orbiting at a distance of 0.05 au from a near-twin to the sun. It was widely believed prior to this discovery that "gas giant" (really liquid giant) planets could only exist far from their parent star, as they do in our solar system. The discovery of a planet of Jupiter-sized mass with a semi-major axis only 20% that of Mercury is very exciting, and led to much discussion and theoretical modelling of how such an improbable planet could form, evolve and survive.

The radial velocity variations in 51 Peg have an excellent signal-to-noise ratio. There is no dispute about their reality. After their announcement everyone believed in the planet. But then the 27 February 1997 issue of *Nature* hit the news-stands with a flurry of worldwide media coverage. This was because "Dolly", the cloned sheep, was the cover-girl. However, also noted on the cover of that issue, but not so widely reported in the world's newspapers, was a paper by David Gray (Gray 1997) announcing the discovery of line profile variations in 51 Peg with the same 4.23-d period as the radial velocity variations.

The implication of this, and Gray's conclusion, is that there is no planet orbiting 51 Pegasi! The radial velocity shifts in 51 Peg caused by reflex Doppler shifts (as the purported planet orbited it every 4.23 days) should cause no change in the *shape* of the spectral lines, only in their wavelength. The presence of variations in the line profiles can only be explained by stellar pulsation, and a period as long as 4.23 days in a solar-type star demands a single, *very* high-overtone g-mode. A lot of excitement ensued. Some of the people involved with the planet hypothesis and the original discovery of the radial velocity variations were vigorously dismissive of Gray's result, even publicly dismissive of Gray himself. The critics said to Gray "Let us see your data. We do not believe your result." Thus ensued a great scientific battle – one with defenders of an improbable planet battling the proposers of an improbable g-mode. Marcy *et al.* (1997) say, "The only viable interpretation is a companion ..." Gray & Hatzes (1997) say, "... the planet hypothesis is no longer viable ..." This battle is not over. Both sides, and some neutral on-lookers, are gathering more ammunition in further, time-consuming, painstaking observations.

Gray & Hatzes (1997) (from here on G&H) accepted the challenge to publish their data, and I have had a good look at it, so here is my opinion: G&H measured the "bisector" in a spectral line of Fe in 51 Peg at three depths, 85%, 71% and 48% of the continuum intensity. Spectral lines are not symmetric, the bisector in 51 Peg, typical of solar-type stars, has a characteristic "C" shape. If the planet hypothesis is true for 51 Peg, then the expectation is that the radial velocity variations at all three depths will be the same, *i.e.* there will be no change in the bisector shape. G&H claim to have found 4.23-d variability in the bisector shape, which means that the radial velocity variations at the three depths are not identical. A planet cannot cause this.

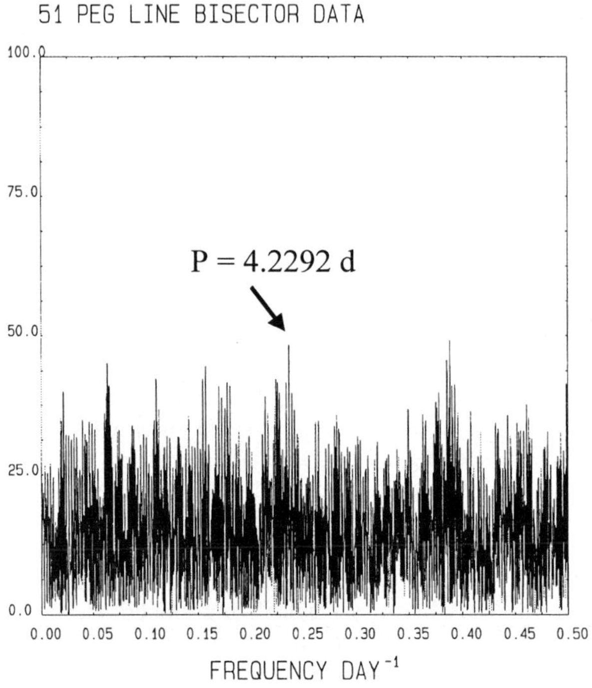

Figure 1. This is an amplitude spectrum of G&H's line bisector data showing that the highest peak is at P = 4.23 days, and that its amplitude barely exceeds the noise. The ordinate scale is in m s^{-1}.

I have Fourier analysed G&H's bisector data. The amplitude spectrum I get is shown in Figure 1. In the absence of any outside knowledge, I would conclude that there is no significant signal whatsoever in this amplitude spectrum. G&H argue that the peak at 4.2292 d is precisely at the period of the radial velocity variations, and that the probability of finding the highest peak in the amplitude spectrum at this particular period is extremely low. They, therefore, conclude that there is real variation in the line bisector, hence the line profile. They point out that they have observed many other solar-type stars, and that the amplitude spectra of the bisector data for those stars show significantly lower noise than that in figure 1.

I concede that there is *some* merit to their argument. They *may* have evidence of profile variability. The reason that I am not convinced is that the amplitude spectrum shown in figure 1 is for data which consists of the *difference* of two signals, both of which have the full radial velocity variation (if there is no profile variation). If *any* systematic effect is present, then a signal in the difference could be generated as an artefact. G&H argue that this is not probable, since the amplitude of the profile variability is nearly as large as the radial velocity variations themselves. The problem, as I see it, is that a small systematic difference might add in complex space to increase the amplitude of a random noise peak at P = 4.23 days to give, as an artefact, the peak in figure 1. The only way to test this is with further data, and such data are being gathered.

The importance of line profile variability cannot be overemphasised. While most of the astronomical community is excited by the idea of an improbable, Jupiter-sized planet at 0.05 au from a solar-type star, we in the asteroseismology community find a g-mode of period 4.23 days even more exciting and improbable. To explain the line profile variability Gray & Hatzes suggest a single, sectoral g-mode of low degree, but very high overtone, $n > 100$. I have heard Paul Butler, a member of Marcy *et al.* (1997), publicly dismiss the pulsation hypothesis as absurd on the basis of the required n, and only one "observed mode". But in the roAp star HD 134214 we see a single p-mode which probably has an overtone of order $n \approx 40$. Given that, I do not "know" that a single g-mode with $n > 100$ is impossible in 51 Peg. The observations will have to decide that. However, it does seem improbable, and it is extremely exciting for asteroseismology, if true.

Contrary to the antagonists' feelings in this controversy, this is a no-lose situation. Either we get an amazing planet and a new view of the formation and dynamics of planetary systems, or we get a high-overtone g-mode and the ability to probe the core of a solar-type star. New physical understanding is the result either way, thanks to the outstanding observations being produced by all groups working on 51 Peg.

3. EC 14026 stars

You haven't yet heard of the EC 14026 stars? Then you are guilty of not reading the important literature! Because there they are: in figure 2, reproduced right from the front cover of the Annual Report of the South African Astronomical Observatory! Just to toot the South African horn a bit louder here, I'd like to point out that in their textbook, *Stellar Interiors*, Hansen and Kawaler (1991) list (in their Table 2.4) thirteen classes of intrinsically variable stars. To that list the γ Dor stars and the EC 14026 stars now need to be added. Of the fifteen classes, two of them – the EC14026 and roAp stars – were discovered at SAAO, and the seminal work of Alan Cousins on γ Dor itself was done at SAAO.

Most of the EC 14026 stars have pulsation periods of only 2 to 3 minutes! You can read the details of the early discoveries in a series of published papers (Kilkenny *et al.* 1997; Koen *et al.* 1997; Stobie *et al.* 1997; O'Donoghue *et al.* 1997), and you can read about the independent theoretical prediction that such pulsations should be driven in Charpinet *et al.* (1997). You can turn right now to the review by Stobie *et al.* of these stars in this volume. Of particular interest, note the lower light curve in figure 2. Here is an EC14026 star in an eclipsing binary with an orbital period of 2.4 hours! In that famous textbook, *Nonradial Oscillations of Stars*, Unno *et al.* (1989) discuss (in their section 6.2) the prospect of determining pulsation degree, ℓ, from the phase shift of a pulsation mode as the pulsator is eclipsed by a companion. At the time of the writing of that textbook, there were no proven cases of such a pulsator in an eclipsing binary. Now there is – the EC 14026 star you see in figure 2. We await the results of the careful modelling of the reflection effect in this system; once that is removed it will be possible to study the phase behaviour of the pulsation through ingress and egress of the primary partial eclipse, and determine the degree of the mode.

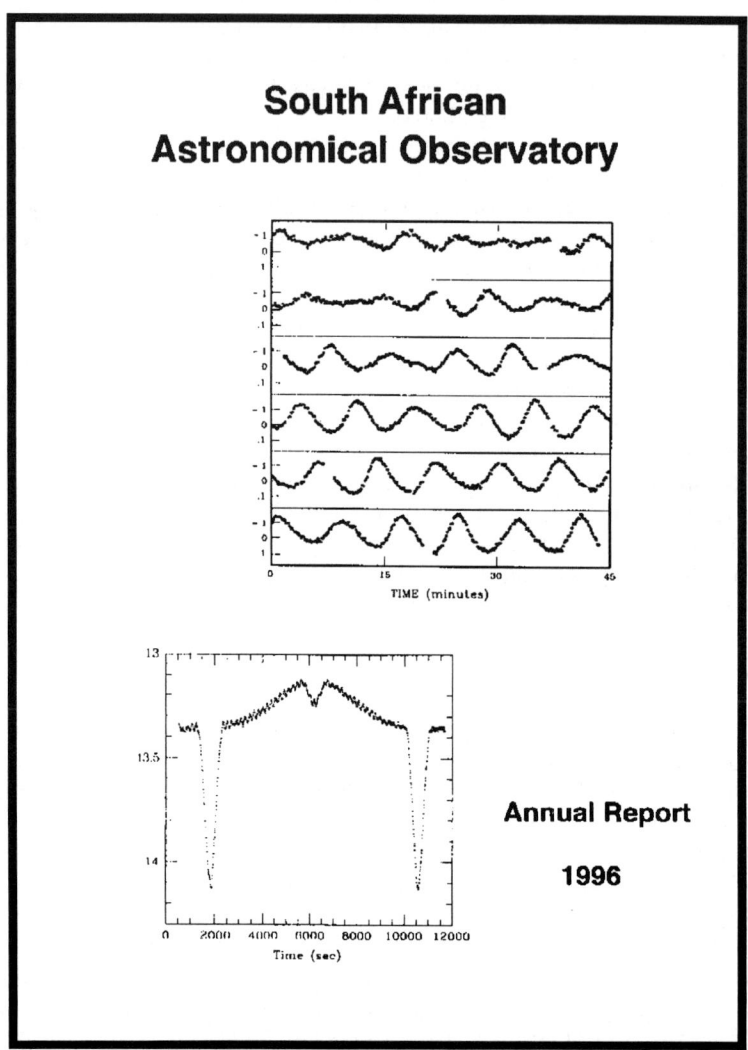

Figure 2. The front cover of the 1996 Annual Report of the SAAO showing a 4.5-hr light curve of one of the longer period, P ≈ 8 min, EC 14026 stars. The peak-to-peak amplitude is about 15%. The bottom amazing light curve shows an EC 14026 star with periods near 140 s and 180 s in an eclipsing binary with an orbital period of 2.4 hr.

I consider the discovery of the EC 14026 stars to be very important. They are sdB stars on the extreme blue end of the Horizontal Branch. The evolutionary pathway to that position in the HR diagram is not completely understood. These stars are nearly pure Helium stars with masses near 0.5 $M_\odot$, with a very thin H atmosphere – only about 2% of the mass of the star. This is insufficient to sustain H burning, so these stars run purely off of He fusion.

The way in which a Red Giant which is H shell burning undergoes a helium flash and then sheds almost all of its H envelope is not fully understood. There are models in which a single star is able to do that; there is another model which requires a binary system to get to the sdB state. The pulsations in the EC14026 stars can elucidate the internal structure of these stars through the techniques of asteroseismology. It may be possible to constrain, or determine, the precise thickness of the H layer this way. Asteroseismology of white dwarfs has successfully done this. A better understanding of the structure of the EC 14026 stars may lead to an understanding of how they have evolved to their current states.

The evolutionary time-scale for He burning has also not been tested observationally. It is possible that in a few years period changes may be measured in the EC14026 stars which are directly the result of internal structure changes caused by the on-going He fusion. Thus the period changes may also constrain the He fusion time-scale.

4. Metallicism and pulsation

Gravitational settling and radiatively driven levitation are important physical effects in many stars. Gravitational settling of helium in the sun is part of the standard solar model (Bahcall & Pinsonneault 1992). One result is that He settling slightly ameliorates the solar neutrino problem by about 7%. Metal settling may help explain the Li and Be abundances seen in the solar atmosphere. Those elements are fused at temperatures of 1 - 2 million K, and the solar model needs very fine tuning to get the convection and overshoot mixing to take the isotopes of Li and Be to just the right depth (Richard et al. 1996).

Theoretical calculations suggest that the pulsation in the EC 14026 stars is driven by the κ-mechanism operating on the iron opacity in a layer in the star in which radiative levitation has substantially increased the iron abundance (Charpinet et al. 1997). Only the combination of diffusion theory and the OPAL opacities leads to an understanding of their pulsation mechanism.

Gravitational settling of helium in Globular Cluster stars releases gravitational potential energy, heats the stellar cores, hence causes the stars to burn more brightly and have shorter lifetimes than in the absence of settling. This gives younger age determinations for the globular clusters and partially eases the discrepancy between their ages and the Hubble Age of the universe (Chaboyer et al. 1996). In even more exotic circumstances gravitational settling of helium droplets (helium rain) powers Jupiter's infra-red emission, and differential crystallisation in massive white dwarfs may result in Oxygen "snow" (see Kawaler, this volume). Gravitational settling also plays a part in the "Li problem" in F stars. Brian Chaboyer's discussion of the Li problem at this Symposium is the clearest, most thorough I have heard.

Of all the stars for which radiative diffusion (both gravitational settling and radiative levitation) is important, the stars which show the greatest effect are the chemically peculiar stars of the upper main sequence. Amongst the A stars these are primarily the non-magnetic Am stars and the magnetic Ap stars. About 30 percent of the stars in the lower instability strip (where it crosses the main sequence among the A and early F stars) are δ Scuti stars which pulsate with amplitudes ranging from a few mmag to nearly a mag. Most of the non-pulsating stars in this part of the instability strip are spectrally peculiar — mostly Am and Ap stars. Only a few Am and Ap stars pulsate, and it is a strength of diffusion theory that it can explain this. In stable A and early F star atmospheres helium settles gravitationally. This shuts off the κ-mechanism operating in the He II ionisation zone which drives δ Scuti pulsation. It was one time thought that a strength of the general exclusion between the pulsators and the Am and Ap stars is that it avoids the necessity of explaining how stars with surface radial velocity variations of km s^{-1} can be stable to turbulence at the level of fractions of a cm s^{-1}, which is typical of calculated diffusion velocities. However, it is now clear that this necessity cannot be avoided.

For a few stars that show abundance anomalies thought to be caused by diffusion processes, and which also pulsate, the implication is that the pulsation is laminar and generates no turbulence. These stars include some evolved Am stars near the cool border of the δ Scuti instability strip; a few marginal Am stars, and one classical Am star — all of which pulsate with amplitudes of a few hundredths of a magnitude, or less — and the rapidly oscillating Ap stars. It also includes the large amplitude peculiar star, HD 40765. In that star Kurtz *et al.* (1995) conclude that 10 km s^{-1} pulsation generates no turbulence at the level of a fraction of a cm s^{-1}.

The study of stellar pulsation (asteroseismology for stars with two or more frequencies) thus probes the atmospheric conditions under which diffusion may occur.

5. Rapidly oscillating Ap stars

The roAp stars are the only main sequence stars which have p-mode pulsations at all similar to those in the sun. In this volume they are reviewed observationally by the electric (literally at S185) lecturer, Jaymie Matthews, and theoretically by Hideyuki Saio & Alfred Gautschy. I present new results from the work of my colleagues and myself here, too.

5.1. ASTEROSEISMIC LUMINOSITIES

The frequency spacings of 12 roAp stars allow theoretical asteroseismic estimates of their luminosities. These stars are spectroscopically so peculiar that these luminosities have been thought to be the best available. However, severe doubt was thrown on this by Dziembowski & Goode (1996) when they calculated that the magnetic perturbation to the pulsation frequencies is so large (of order tens of μHz) that perturbation theory will not work, and the development of the oblique pulsator model (*e. g.* see Shibahashi & Takata 1993; Takata & Shibahashi 1995) was on shaky ground.

With the new HIPPARCOS parallaxes Matthews *et al.* (1997) have been able to test the asteroseismic luminosities. Figure 3 is from their paper, and it shows good agreement: Asteroseismology works. There are two exceptions. One is HD 166473 which, according to the HIPPARCOS parallax, has a luminosity only 40% that of the sun. For a main sequence A star this is obviously incorrect, so we conclude that the HIPPARCOS parallax is wrong. For α Cir the discrepancy is a bigger problem. The asteroseismic parallax is determined from a frequency spacing of 50 μHz (Kurtz *et al.* 1994). The secondary frequencies in this star give a strong indication that 50 μHz is the right value, but they are all of amplitude about 0.2 mmag, so this needs confirmation. In this important case another intensive study of α Cir is called for.

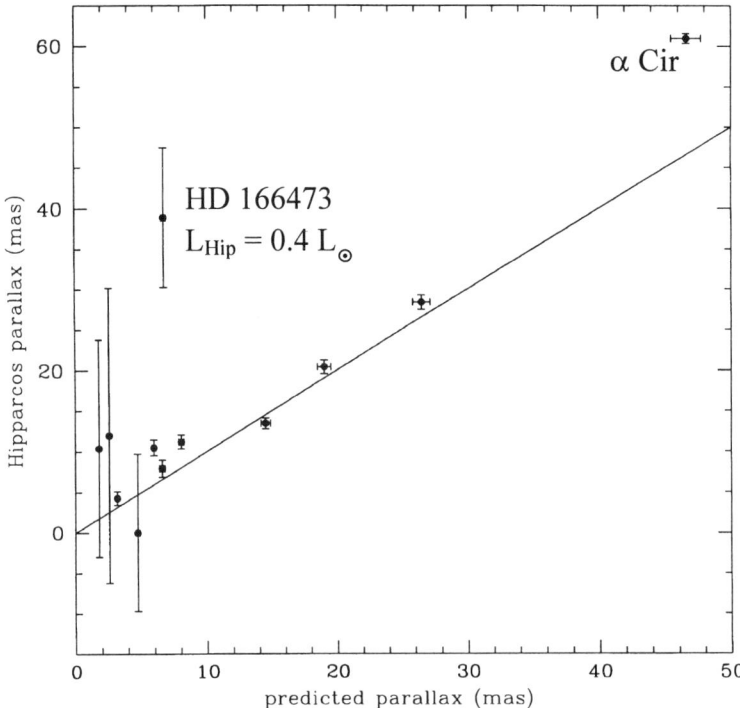

Figure 3. A comparison of the HIPPARCOS and asteroseismic parallaxes. The asteroseismic parallaxes are systematically slight smaller, and two stars (discussed in the text) stand out. Otherwise, the agreement is remarkably good – indicating that the asteroseismic luminosities are correct.

For the remainder it can be seen from Figure 3 that the asteroseismic parallaxes are systematically slightly smaller than the HIPPARCOS parallaxes. That means that the A star models used to predict the luminosity from the frequency separations systematically give too large a luminosity, hence the models are too hot and/or too large in radius, or the magnetic field affects the frequency *spacings*. To get an order of magnitude feel for the discrepancy: If we assume the radii are correct, then the parallax disagreement indicates that the effective temperatures of the roAp stars are about 1000 K cooler than the A star models from which the asteroseismic luminosities were calculated.

I remind you that the atmospheres of the roAp stars are peculiar to pathological. Luminosities are notoriously difficult to determine, and even the effective temperature can lead to decades-long, acrimonious dispute. In the most extreme case, HD 101065, temperature estimates range from less than 6000 K to over 8000 K! This particular star is arguably the most peculiar in the sky. In its visible spectrum the lead role is played by singly ionised Holmium, with strong supporting roles from Dysprosium, Neodymium,

Gadolinium, Samarium, Lanthanum, *etc.* (presuming you know where *"etc."* leads with that series as a starter).

Asteroseismology is providing unique constraints on the structure of the Ap stars.

5.2. THE SIXTH FREQUENCY OF HR 1217

In 1989 a group of 15 astronomers published an analysis of HR 1217 based on a large multi-site observing campaign (Kurtz *et al.* 1989) . The amplitude spectrum of this star has a set of five frequencies which are separated by alternating 33.3 µHz and 34.7 µHz. Shibahashi & Saio (1985) had noted that these separations are most consistent with alternating even and odd degree modes, and that the even degree modes are most likely to be radial. But there were problems. One is that the "even" degree modes show amplitude modulation with the 12.46-day rotation period of the star. Radial modes should show no amplitude modulation with rotation; they look the same from all aspects. The other problem is that there is a sixth mode, which is separated from the fifth by 50 µHz. With the interpretation of the first five frequencies as alternating even and odd ℓ modes, the fundamental frequency spacing implied is 68 µHz. The ratio of 50/68 = ¾, and there is no known way to generate p-modes separated at ¾ Δv_0.

There is another interpretation. That is that the first five frequencies are all dipole modes; then the fifth and sixth frequencies are separated by 1.5 Δv_0 which is theoretically quite acceptable. It does leave the problem that the amplitude modulation is not the same for all the "dipole" modes, the problem that the alternating frequency spacing then has no explanation, and the problem that the mode at Δv_0 above the fifth frequency is missing. But at least there is no need to explain a frequency separation of ¾ Δv_0.

This second explanation is now eliminated. The HIPPARCOS parallax for HR 1217 is $\pi = 20.4 \pm 0.8$ mas; the asteroseismic parallax is 19.1 ± 0.5 mas if $\Delta v_0 = 68$ µHz; it is $\pi = 11.0 \pm 0.5$ µHz if $\Delta v_0 = 34$ µHz. The latter is ruled out. $\Delta v_0 = 68$ µHz and the five frequencies are from alternating even and odd ℓ modes. The separation of the sixth frequency from the fifth is ¾ Δv_0; this is an ongoing mystery that can no longer be ignored in the hopes that it will go away. And what about the amplitude modulation of the even modes which are expected to be radial? That can be explained as the magnetic field modifying the surface appearance of what otherwise is an excited radial mode. Interestingly, in these proceedings that is exactly what Shibahashi & Aerts have suggested is also the explanation for the frequency pattern seen in β Cephei itself.

5.3. FREQUENCY VARIABILITY IN roAp STARS

A group from the University of Cape Town and the South African Astronomical Observatory has been observing roAp stars on a long-term basis for frequency variability for 6 years now. In this on-going project we get one hour of observation of HR 3831 on each possible night over the approximately 8-month season when it is observable. The reason for the emphasis on this star is that it is the best-studied of the roAp stars, it has interesting rotational amplitude and phase variability which we need to remove to study the frequency variability, and the rotational variations are interesting subjects to study in their own right. We also observe for one hour once per week two other roAp stars, HD 134214 and HD 128898 (α Cir). HD 134214 is singly periodic with the shortest known pulsation period for the roAp stars, 5.65 minutes. The very bright star α Cir is nearly circumpolar, and has a single large amplitude pulsation mode with only small amplitudes for other modes and small rotational sidelobes. The frequency variability of HR 3831 is discussed in Kurtz *et al.* (1997); we have not published O–C diagrams for the other two stars recently.

It is clear that there are variations in the pulsation cavities of these stars on time-scales of years. For HR 3831 the variations can be characterised as cyclic with a time-scale of 1.6 years. For α Cir the time-scale is about 6 years, and HD 134214 is harder to characterise. HR 3831 is shown in figure 4; α Cir and HD 134214 in figures 5 and 6. The O–C diagrams show the pulsation phase in radians minus a constant phase. These variations cannot be easily explained as Doppler shifts caused by companions; for HR 3831 many companions would need to be hypothesised. In addition, the Ap stars have a very low incidence of binarity, only about 20% are in short period binary systems. So the frequency variations are intrinsic, and they indicate a cyclic variability in the acoustic cavity – this may be anything which affects the sound speed. One speculation we have made is that this indicates a magnetic cycle. The time scale and amplitude of the frequency variations are similar to those which are seen in the sun over the solar cycle. Magnetic fields in Ap stars are thought to be fossil, however, rather than dynamo generated, so this suggestion has not met with much approval. Whatever the physical mechanism that is at work, we have a new observational phenomenon which will eventually tell us more of the inner workings of the roAp stars via asteroseismology.

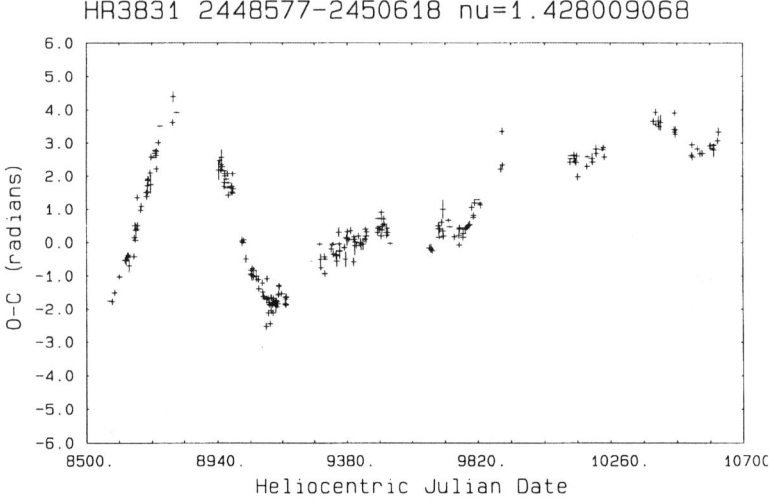

Figure 4. The phase O–C diagram for six years of data obtained in our long-term monitoring program for HR 3831. Each point shows the difference between the least-squares determined phase for each night of observation and the analytical rotational phase variation calculated from a model of the rotational variation. The frequency ranges over about 0.12 μHz.

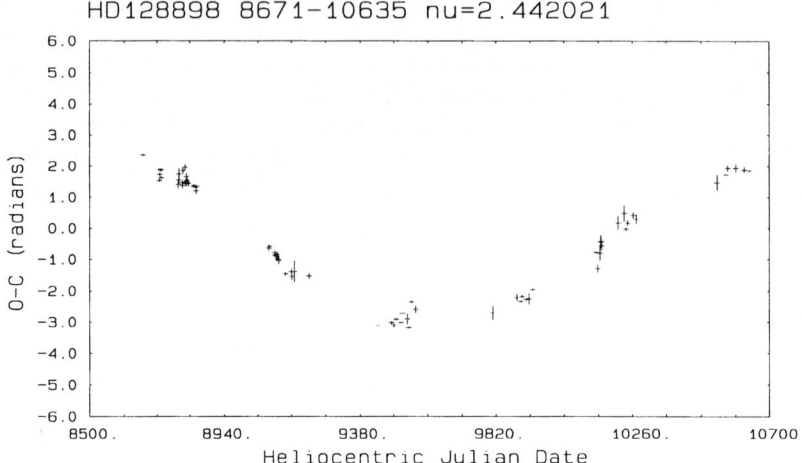

Figure 5. The phase O–C diagram for six years of data obtained in our long-term monitoring program for α Cir.

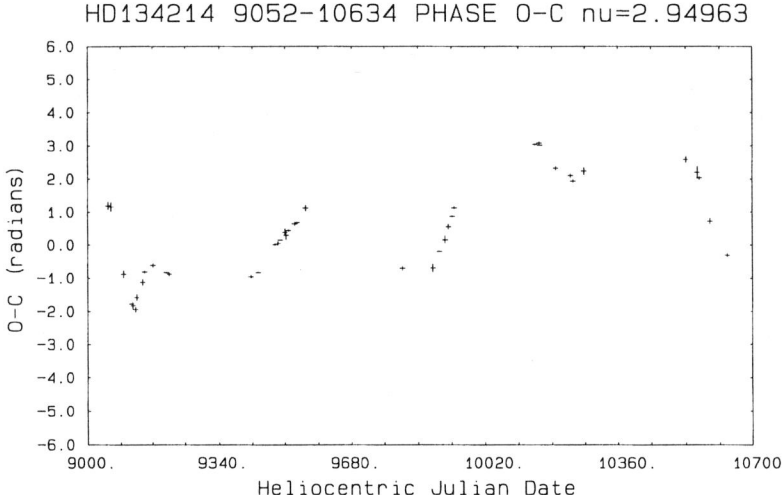

Figure 6. The phase O–C diagram for six years of data obtained in our long-term monitoring program for HD 134214.

5.4. ATMOSPHERIC NODES IN roAp STARS

I was very excited by the announcement by Jaymie Matthews (Matthews *et al.* 1990, 1996) that atmospheric T-τ in roAp star could be determined from the amplitude of their pulsations at different wavelengths. A problem much discussed for the roAp stars is that of the critical frequency: Many of them seem to pulsate with frequencies so high that their vertical wavelength should be of order of the thickness of the surface boundary layer. In that case the mode should become evanescent; without very large driving there is not enough energy for modes with frequencies above the critical frequency to be maintained as standing waves. This problem is discussed in the review of Saio & Gautschy and in the paper by Audard *et al.* in this volume. Matthews found that the T-τ gradient in HR 3831 was much steeper than in a normal A star model; this steepness sharpens the surface boundary and increases the critical frequency. It also results in the derivation of smaller overabundances of the rare earths and lanthanides for a given observed line strength.

Because this looked such a fruitful field, Thebe Medupe and I decided to begin a large project to observe many roAp stars in *UBVRI* colours to apply Matthews' technique of inferring T-τ from limb darkening affects on the pulsation amplitude. Instead of modelling this first on a computer, however, we started with a simple analytic derivation of what we expected, making some reasonable first-order assumptions, such as a black body energy distribution and the Wien approximation. To our surprise we found that limb-darkening is not an important effect (Kurtz & Medupe 1996, Medupe & Kurtz 1997). The pulsation amplitude drops from the blue to the infrared by a factor of two greater than the drop expected for a simple black body. Limb-darkening can only account for about 12.5% of this. Much more sophisticated computer modelling confirms this result.

Medupe and I found that the observed amplitude is expected to be

$$A_{obs} = 1.086 \frac{1}{4}\sqrt{\frac{3}{\pi}\left(\frac{4-\mu_\lambda}{3-\mu_\lambda}\right)} \cos\alpha \frac{hc}{\lambda k T_0} \frac{\Delta T}{T_0}$$

where μ_λ is the wavelength dependent limb-darkening coefficient, T_0 is the temperature at the atmospheric level appropriate to the wavelength observed, and ΔT is the semi-amplitude of the pulsational polar temperature variation. What is clear is that only a strong dependence on ΔT can match the observations. We suggested that we were seeing over a large fraction of the vertical wavelength of a high-overtone p-mode.

Several studies now give strong support to this suggestion. The paper by Baldry *et al.* in these proceedings shows radial velocity measures in which an atmospheric node is evident in α Cir. Pulsation phase versus equivalent width diagrams give further support in studies of α Cir (Baldry *et al.* 1997) and γ Equ (Kanaan & Hatzes 1997). Theoretical models are now also producing pulsation nodes in the observable atmospheres of roAp stars.

We thus have the prospect of a new asteroseismic technique for probing the atmospheres of the most peculiar stars known. Yet again, "asteroseismology will tell".

References

Bahcall, J. N., Pinsonneault, M. H., 1992, Rev. Mod. Phys., 64, 885.
Baldry, I. K., Viskum, M., Bedding, T. R., Kjeldsen, H., Frandsen, S., 1997, MNRAS, in press.
Chaboyer, B. Demarque, P. Kernan, P. J. Krauss, L. M, 1996, *Science*, 271, 957.
Charpinet, S. Fontaine, G. Brassard, P. Chayer, P. Rogers, F. J. Iglesias, C. A. Dorman, B., 1997, ApJ, 483, L123.
Dziembowski W., Goode P.R., 1996, ApJ, 458, 338.
Gray, D. F., 1997, *Nature*, 385, 795.
Gray, D. F., Hatzes, A. P., 1997, ApJ, in press.
Hansen, C. J., Kawaler, S. D., 1991, *Stellar Interiors: Physical Principles, Structure and Evolution*, Springer-Verlag, New York.
Kanaan, A., Hatzes, A. P., 1997, preprint.
Kilkenny, D., Koen, C., O'Donoghue, D., Stobie, R. S., 1997, MNRAS, 285, 640.
Koen, C., Kilkenny, D., O'Donoghue, D., van Wyk, F., Stobie, R. S., 1997, MNRAS, 285, 645.
Kurtz D.W., Garrison, R. F., Koen, C., Hofmann, G. F., Viranna, N. B., 1995, MNRAS, 276, 199.
Kurtz D.W., Matthews J.M., Martinez P., Seeman J., Cropper M., Clemens J.C., Kreidl T.J., Sterken C., Schneider H., Weiss W.W., Kawaler S.D., Kepler S.O., van der Peet A., Sullivan D. J., and Wood H.J., 1989, MNRAS, 240, 881.
Kurtz D.W., Sullivan D. J., Martinez P., Tripe P., 1994, MNRAS, 270, 674.
Kurtz D.W., van Wyk F., Roberts, G., Marang F., Handler, G., Medupe, R., Kilkenny, D., 1997, MNRAS, 287, 69.
Kurtz, D. W., Medupe, R., 1996, Bull, Ast. Soc. India, 24, 291.
Marcy, G. W., Butler, P. R., Williams, E., Bildsten, L., Graham, J. R., Ghez, A. M., Jernigan, J. G., ApJ, 481, 926.
Matthews J.M., Wehlau W.H., Rice, J., Walker G.A.H., 1996, ApJ, 459, 278.
Matthews J.M., Wehlau W.H., Walker G.A.H., 1990, ApJ, 365, L81.
Matthews, J. M., Kurtz, D. W., Martinez, P., 1997, in preparation.
Mayor, M, Queloz, D., 1995, *Nature*, 378, 355.
Medupe, R., Kurtz, D. W., 1997, MNRAS, submitted.
O'Donoghue, D., Lynas-Gray, A. E., Kilkenny, D., Stobie, R. S., Koen, 1997, MNRAS, 285, 657.
Richard, G., Vauclair, S., Charbonnel, C., Dziembowski, W. A., 1996, A&A, 312, 1000.
Shibahashi H., Saio H., 1985, PASJ, 37, 245.
Shibahashi H., Takata M., 1993, PASJ, 45, 617
Stobie, R. S., Kawaler, S. D., Kilkenny, D., O'Donoghue, D., Koen, 1997, MNRAS, 285, 651.
Takata M., Shibahashi H., 1995, PASJ, 47, 219.
Unno W., Osaki Y., Ando H., Saio H., Shibahashi H., 1989, *Non-radial oscillations of stars*, Univ. Tokyo Press.

THEORETICAL ASPECTS OF ASTEROSEISMOLOGY

J. CHRISTENSEN-DALSGAARD
Teoretisk Astrofysik Center, Danmarks Grundforskningsfond, and Institut for Fysik og Astronomi, Aarhus Universitet, DK-8000 Aarhus C, Denmark

1. Introduction

Asteroseismology depends on the ability to determine frequencies for identified modes of stellar oscillation, and to use these frequencies to probe stellar interiors. Thus crucial aspects are the dependence of the frequencies on stellar properties, and the likelihood that modes are excited to observable amplitudes. I provide a brief overview of these issues and some remarks about how the frequencies may be analyzed, as a background for the more detailed presentations elsewhere in the proceedings.

2. The dependence of frequencies on the properties of a star

The frequency dependence can in general be separated into the part arising from the spherically averaged stellar structure and the part resulting from effects of rotation, the latter being treated as a small perturbation. The spherical structure determines frequencies that depend on the degree l and radial order n of the modes, but not on their azimuthal order m; rotation induces a splitting after m.

The dependence of the frequencies of a spherically symmetric star on stellar structure can to a large extent be understood in terms of *characteristic frequencies* of the star. The most basic of these is the *dynamical frequency* $\omega_{\rm dyn} = (GM/R^3)^{1/2}$, where M and R are the mass and radius of the star, and G is the gravitational constant; for stellar models related by homology, the pulsation frequencies scale precisely as $\omega_{\rm dyn}$. Measurement of a single frequency of an identified mode therefore essentially provides information on $\omega_{\rm dyn}$, *i.e.*, on the mean density of the star; such a measurement is interesting in itself, but hardly constitutes seismology.

The detailed properties of the oscillations depend largely on two characteristic frequencies: the Lamb frequency $S_l = [l(l+1)]^{1/2} c/r$, where c is the adiabatic sound speed and r is the distance from the centre; and the buoyancy frequency N, given by

$$N^2 = g \left(\frac{1}{\Gamma_1} \frac{d\ln p}{dr} - \frac{d\ln \rho}{dr} \right) \simeq g^2 \frac{\rho}{p} (\nabla_{\rm ad} - \nabla + \nabla_\mu) \equiv N_0^2 + N_\mu^2 , \qquad (1)$$

where p is pressure, ρ is density, $\Gamma_1 = (\partial \ln p / \partial \ln \rho)_{\rm ad}$ is the adiabatic exponent, and g is gravity. In the last approximation I assumed the ideal gas law for a fully ionized gas. Here $\nabla = \mathrm{d}\ln T/\mathrm{d}\ln p$, ∇_{ad} is its adiabatic value and $\nabla_\mu = \mathrm{d}\ln\mu/\mathrm{d}\ln p$, where μ is the mean molecular weight; also, I introduced

$$N_\mu^2 = -g\frac{\mathrm{d}\ln\mu}{\mathrm{d}r} = -4\pi G\rho\frac{\mathrm{d}\ln\mu}{\mathrm{d}\ln m} \ . \tag{2}$$

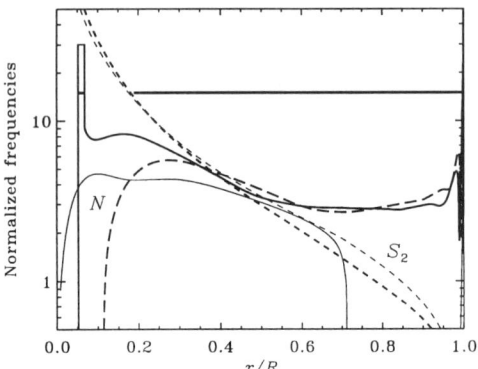

Figure 1. Characteristic frequencies, in units of ω_{dyn}. Thin lines are for a model of the present Sun and heavy lines for $1.8 M_\odot$ models. Long-dashed line: N in the $1.8 M_\odot$ ZAMS model; continuous lines: N in evolved $1.8 M_\odot$ and solar model; short-dashed lines: S_2 in evolved $1.8 M_\odot$ and solar model. The heavy horizontal line shows the frequency of a mode with contributions both from the p-mode and g-mode regions.

The behaviour of S_l and N, in units of ω_{dyn}, is illustrated in Fig. 1 for a model of the present Sun and ZAMS and evolved models of a $1.8 M_\odot$ star. It is evident that S_l scales approximately homologously, as does N_0 for the evolving $1.8 M_\odot$ model. On the other hand, N_μ reflects the gradient in μ in evolved models which has resulted from nuclear burning; this is particularly striking in the $1.8 M_\odot$ star, where the shrinking convective core has left behind a strong gradient in μ, leading to the peak in N.

The local physical nature of a mode is determined by its frequency, relative to S_l and N: where $\omega > S_l$ and $\omega > N$ the mode behaves as an acoustic wave (or a p mode) whereas where $\omega < S_l$ and $\omega < N$ it behaves as a gravity wave (or g mode). The mode is evanescent where the frequency is between S_l and N, growing or decaying exponentially with r. As illustrated in Fig. 1, modes generally have p-mode behaviour at high frequency and g-mode behaviour at low frequency; however, a given mode can have different characters in different parts of the star. The global nature of the mode is then essentially determined by the region where the local energy is largest.

For high-frequency modes in solar-like stars, including the five-minute oscillations of the Sun, $|\omega/N| \gg 1$. Such modes behave essentially as standing sound waves, and the frequency is determined by the sound speed. For low-degree modes (of main relevance for distant stars) the cyclic frequency of a mode of degree l and radial order n satisfies the asymptotic relation $\nu_{nl} \simeq \Delta\nu(n+l/2+\alpha)+\epsilon_{nl}$ (*e.g.* Tassoul 1980). Here $\Delta\nu = (2\int_0^R \mathrm{d}r/c)^{-1}$ is the inverse of the sound travel time across a stellar diameter; this can be estimated as the large frequency separation $\Delta\nu_{nl} \simeq \nu_{nl} - \nu_{nl-1}$. The correction term ϵ_{nl} leads to a small frequency separation $\delta\nu_{nl} = \nu_{nl} - \nu_{n-1\,l+2}$, which is largely determined by the sound-speed gradient in the core of the star. Since for an ideal gas $c^2 \simeq \Gamma_1 k_\mathrm{B} T/\mu m_\mathrm{u}$ (k_B being Boltzmann's constant and m_u the atomic mass unit), $\delta\nu_{nl}$ is sensitive to the composition structure of the star and hence its evolutionary state. As a result of this sensitivity, $\delta\nu_{nl}$ does not scale as ω_{dyn} with stellar mass and radius, while $\Delta\nu_{nl}$ essentially scales homologously.

The departures from the simple asymptotic expression contain potentially important information. In particular, sharp features in the stellar structure may induce oscillations in the frequencies, as functions of mode order. An important example is the localized decrease in Γ_1 associated with the second helium ionization zone; the resulting perturbation in the frequencies has been used to infer the solar helium abundance (*e.g.* Vorontsov *et al.* 1991; Kosovichev *et al.* 1992; Pérez Hernández & Christensen-Dalsgaard 1994), and corresponding information may become available from stellar observations (Lopez *et al.* 1997; Pérez Hernández & Christensen-Dalsgaard 1997). A second example is provided by the oscillations induced by the possible sharp variation in the sound-speed gradient at the edges of convective regions, particularly in the presence of overshoot. Analysis of this feature in the solar case has placed limits on the extent of overshoot beneath the solar convection zone (*e.g.* Basu *et al.* 1994; Monteiro *et al.* 1994; see also Monteiro *et al.*, these proceedings).

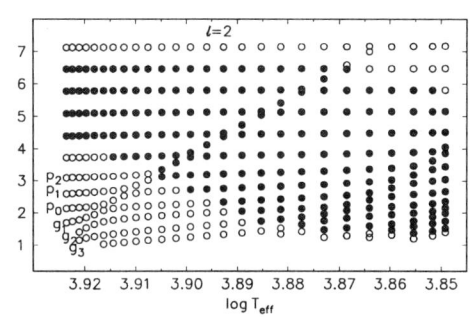

Figure 2. Modes of degree $l = 2$ in an evolution sequence of $1.8 M_\odot$ models, indicated by their effective temperature $T_{\rm eff}$. The frequencies σ are given in units of $\sqrt{3}\omega_{\rm dyn}$. Filled circles indicate unstable modes, open circles stable modes. (From Goupil *et al.* 1996.)

The approximately uniform spacing of the solar-like oscillation spectrum substantially simplifies analysis of of observed modes of oscillations. In other types of pulsating stars, the spectrum of oscillations can be considerably more complex, with a corresponding greater richness in the potentially available information. This is the case for evolved stars with a (possibly earlier) convective core, where the steep maximum in the buoyancy frequency (cf. Fig. 1) allows for g-mode like behaviour at frequencies otherwise characteristic for p modes. As the star evolves, S_l and N_0 scale as $\omega_{\rm dyn}$, *i.e.*, as $R^{-3/2}$. On the other hand, N_μ increases with the increasing composition gradient and central density as the star evolves. It follows that for modes behaving like p modes the frequencies decrease with evolution while g modes trapped in the region of composition gradient have increasing frequencies. This is illustrated in Fig. 2 for a $1.8 M_\odot$ evolution sequence. Since frequencies in units of $\omega_{\rm dyn}$ are shown, the p-mode values are essentially constant, while the g-mode frequencies reflect the increasing N_μ. Where the frequencies of two modes of different nature approach, the modes undergo avoided crossing, where the modes exchange physical nature. Near the avoided crossing the two modes have a mixed character, with substantial amplitude in both the regions of p-mode and g-mode behaviour. Evidently the observational identification of such a pair of modes would provide important information about the properties of the edge of the convective core, possibly including evidence for convective overshoot.

Rotation (or other departures from spherical symmetry) induces a dependence of the frequencies on the azimuthal order m of the mode. For slow rotation, one may neglect quadratic effects in the angular velocity Ω, including the distortion of the equilibrium structure. Then the effect of rotation on the frequencies is solely a lifting of the degeneracy of the frequencies in m. If Ω is assumed to depend only on r

$$\omega_{nlm} = \omega_{nl0} + m\beta_{nl} \int_0^R K_{nl}(r)\Omega(r)\mathrm{d}r \qquad (3)$$

(*e.g.* Gough 1981). Here the kernels $K_{nl}(r)$ have been defined such as to have unit integral. For high-order p modes β_{nl} is close to one, and the rotational splitting $(\omega_{nlm} - \omega_{nl0})/m$ is approximately an average of Ω weighted by the energy density in the mode. For high-order g modes, on the other hand, $\beta_{nl} \simeq 1 - [l(l+1)]^{-1}$. The pattern of approximately uniformly split frequencies indicated by Eq. (3) simplifies the mode identification, although it must be kept in mind that not all the components for a given multiplet nl need be excited, or visible.

With more rapid rotation, second-order effects of rotation must be taken into account. These change the equilibrium structure of the star, leading to frequency changes for all modes; also, they cause an additional contribution to the frequency splitting which, for uniform rotation and $l \gtrsim 2$ can be approximated by

$$\delta\omega_{nlm}^{(2)} \simeq -f_{nl}\frac{\Omega^2 R^3}{GM}\omega_{nl0}P_2\left(\frac{m}{l+1/2}\right), \qquad (4)$$

where $P_2(x) = (3x^2 - 1)/2$ is a Legendre polynomial (*e.g.* Gough & Thompson 1990; see also Dziembowski & Goode 1992). The factor f_{nl} is approximately $1/3$ for high-order or high-degree p modes; thus the magnitude of the quadratic effect in Ω corresponds approximately to the surface ratio between the centrifugal force and gravity. Notice that this relation results in a non-uniform spacing of the split frequencies. For rapidly rotating stars, including many δ Scuti stars, the splitting is comparable with the separation between the multiplet frequencies ω_{nl} of the corresponding non-rotating star, leading to very complex spectra of oscillations.

3. Mode excitation and selection

Stellar modes of oscillation are of little interest unless excited to observable amplitudes. Two distinct types of excitation are relevant. In the Sun, and presumably other solar-like stars, the oscillations are intrinsically stable. Here the modes arise through stochastic driving by the turbulent convection (see also Nordlund & Stein, these proceedings). The resulting amplitudes are determined by the balance between the linear damping rate of the modes (in itself strongly affected by the perturbation in the turbulent flux and pressure) and the energy input from the turbulence, at the frequency of oscillation. This in turn depends on the spectrum of turbulence and the properties of the oscillation eigenfunctions in the region, very near the surface, of nearly sonic convection. In the Sun, these factors combine to provide driving to substantial amplitudes in a fairly broad frequency region, between around 2 and 5 mHz; the resulting amplitudes are largely independent of the degree of the mode, since the scale of the relevant convection is much smaller than the scale of the oscillations.

As a rough guide the energy in a stochastically excited mode corresponds to the energy of a convective eddy with the corresponding time scale (*e.g.* Goldreich & Keeley 1977); this was applied by Christensen-Dalsgaard & Frandsen (1983) to estimates of the expected amplitudes of solar-like oscillations in other stars. From the results Kjeldsen & Bedding (1995) found an approximately linear dependence of the amplitude on the ratio L/M between the luminosity and mass of the star.

Houdek (1996; see also Houdek *et al.* 1997) carried out an extensive survey of the stability of oscillations in main-sequence stars and the expected amplitudes of stochastically excited modes; he used the non-local mixing-length theory of Gough (1976) and Balmforth (1992). The predicted amplitudes were fairly similar to those obtained by Christensen-Dalsgaard & Frandsen (1983).

The second type of excitation occurs where the modes are *overstable*. In this case the oscillation extracts mechanical energy from the flux of energy through the star, operating as a heat engine with matter being heated at compression and cooled at expansion. The contributions to the driving typically arise from specific regions in the star and must compete with damping elsewhere. The general mechanism was first proposed by Eddington (1926) and elaborated to explain the excitation of oscillations in Cepheids by Zhevakin (1953) and Cox & Whitney (1958). It involves a perturbation to opacity such that energy is preferentially trapped at compression. It is typically associated with a local bump in the opacity, caused by the ionization of a specific element. To result in net driving, the bump must be located at the transition between nearly adiabatic and strongly nonadiabatic oscillation (see also the discussion by Cox 1974): this transition occurs at the location m_t in mass such that the thermal time scale $\tau_{\rm th}(m_t)$ of matter outside m_t is equal to the oscillation period, *i.e.*,

$$\omega \tau_{\rm th}(m_t) \simeq \omega L^{-1} \int_{m_t}^{M} c_p T {\rm d}m \sim 1 , \tag{5}$$

where c_p is the specific heat at constant pressure. Thus the condition for instability is that $m_t \simeq m_b$, the mass corresponding to the relevant bump in the opacity.

Instability of Cepheids and δ Scuti stars arises from the opacity feature at the second ionization of helium. It was shown by Cox & Whitney that this accounted for the location of the Cepheid instability strip. The recent revisions of opacity tables (*e.g.* Rogers & Iglesias 1992; Seaton *et al.* 1994) have shown a bump at higher temperature, arising predominantly from bound-bound transitions in iron-group elements, which gives rise to the instability of the β Cephei stars and other B stars (*e.g.* Cox *et al.* 1992; Moskalik & Dziembowski 1992). Interestingly, in the recently discovered hot pulsating subdwarf stars, the EC 14026 stars (Kilkenny *et al.* 1997; see also Stobie *et al.*, these proceedings) radiative levitation appears to concentrate iron in the critical region, enhancing the instability (Charpinet *et al.* 1997; Fontaine, these proceedings).

It follows from Eq. (5) that the location of the transition depends on the frequency of oscillation: with increasing frequency, $\tau_{\rm th}(m_t)$ must decrease, *i.e.*, the transition move closer to the surface. With increasing effective temperature the relevant bumps in the opacity move closer to the surface; thus instability may be expected at higher frequency. Hence, higher-order acoustic modes are excited to the blue side of the instability strip corresponding to the fundamental radial oscillation. Conversely, on the cool side of the instability strip modes of lower frequency, *i.e.*, g modes, may be excited. These predictions are generally confirmed by detailed stability calculations. In particular, the slowly pulsating B stars (*e.g.* Aerts *et al.*, these proceedings) appear to pulsate in g modes, excited by the opacity mechanism operating on the iron bump in the opacity (Dziembowski *et al.* 1993). Similarly, it is tempting to identify the recently discovered γ Doradus stars, at lower $T_{\rm eff}$ than the δ Scuti stars (*e.g.* Balona *et al.* 1994) as caused by excitation of g modes by the second helium ionization bump; here, however, the understanding of the excitation is complicated by effects of convection.

For overstable modes, a separate mechanism is needed to explain the final limiting amplitude and hence the selection of observable modes. The amplitude-limiting mechanism for stars near the main sequence is so far highly uncertain. They generally pulsate in several modes (thus making them more interesting from a seismological point of view), with amplitudes too small to saturate the driving mechanism. It has been proposed that the amplitudes are limited by resonant interactions involving stable modes (*e.g.* Dziembowski & Królikowska 1985; Dziembowski *et al.* 1988), although no detailed predictions of spectra for specific stars have been made so far; even so, this might possibly explain the often complex spectra observed. It is interesting, however, that increasing observational sensitivity appears to reveal steadily more modes, as has been shown dramatically by the observations of FG Virginis (Breger *et al.* 1995, 1997; see also Breger, these proceedings).

4. Analysis of observed frequencies

A prerequisite for using the observed frequencies to obtain a test of the properties of the stellar interior is the identification of the observed modes. For solar-like oscillations this may be possible from the regular nature of the expected spectrum (*cf.* Section 2). Indeed, Christensen-Dalsgaard & Gough (1980) on this basis obtained an identification, confirmed by later observations, of the solar low-degree modes. Furthermore, the data can be analysed in terms of the large and small frequency separations $\Delta\nu_{nl}$ and $\delta\nu_{nl}$ which may be compared with models, even without a precise assignment of n. This provides a measure of the mass and, because of the dependence of $\delta\nu_{nl}$ on the core structure, of the age of the star (*e.g.* Ulrich 1986; Christensen-Dalsgaard 1988). In fact, the values of $\Delta\nu_{nl}$ and $\delta\nu_{nl}$ inferred from observations of η Boo (Kjeldsen *et al.* 1995) were in good agreement with evolution models of the star (Christensen-Dalsgaard *et al.* 1995; Guenther & Demarque 1996). However, it should be noted that the determination of mass and age from the frequency separations presupposes (unrealistically) that all other parameters of the star are known (*e.g.* Gough 1987).

Given the uncertainty in basic stellar parameters, much can be gained by combining the observed frequencies with other, more 'classical' observations of stellar properties. This is particularly true for systems of two or more stars which may be assumed to share certain properties, such as age and chemical composition. Brown *et al.* (1994) developed a method for combining such data for binary systems and demonstrated the potential improvements, resulting from access to frequency data, in the determination of the parameters of the system. A preliminary application of this technique to models of stellar clusters was carried out by Audard *et al.* (1997).

The identification of solar-like modes is simplified by the broad-band nature of their excitation. In δ Scuti stars it appears that the more selective excitation and amplitude limitation mechanisms lead to complex spectra, which are difficult to interpret. Viskum *et al.* (1997) demonstrated how the degrees of the modes could be identified by combining observations in Balmer-line equivalent width and broad-band intensity (see also Breger, these proceedings). As a result, reasonably reliable identifications are available for a substantial number of modes in the very mode-rich star FG Vir. Similar techniques have been applied previously, on the basis of amplitude ratios and phase differences between different types of observations (*e.g.* Watson 1988).

Although a problem for the mode identification, the variety of modes in δ Scuti stars is a great bonus for investigations of the stellar interiors. Here I consider the possibility for carrying out rotational inversion, as presented by Goupil *et al.* (1996). They noted that there was a fairly broad range of unstable modes in the somewhat evolved $1.8 M_\odot$ star which they investigated, including several g modes trapped at

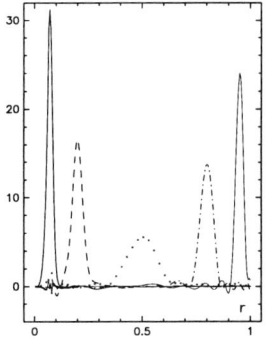 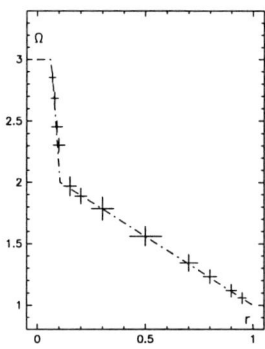

Figure 3. The left panel shows averaging kernels obtained from inversion of modes in an evolved $1.8 M_\odot$ model. The right panel shows the assumed rotation profile (dot-dashed line) and the rotation rates inferred from the inversion. The abscissa is fractional radius. (From Goupil et al. 1996.)

the edge of the core (*cf.* Fig. 2). They argued that modes of degree up to 6 might be visible in observations sensitive to very low-amplitude oscillations and considered the corresponding kernels for the first-order rotational splitting (*cf.* Eq. 3). These were sufficiently varied to allow inversion for the internal rotation rate throughout the star, with fairly high resolution in the core and near the surface. To demonstrate this, Goupil *et al.* carried out a SOLA inversion (*cf.* Thompson, these proceedings), illustrated in Fig. 3, for an artificial rotation profile. The averaging kernels (left panel) provide a measure of the resolution of the inversion; the high resolution in the core reflects the availability of g modes peaked in this region. As shown by the right panel, the inversion does in fact successfully recover the assumed rotation law.

It is fair to say that current observations are not adequate for obtaining the required data, particularly at relatively high degree. However, there is hope that future space experiments, such as COROT (*e.g.* Catala *et al.* 1995) may reach the required sensitivity and frequency resolution. In any case, the analysis demonstrates the potential for investigating stellar interiors when data on g modes are available.

Acknowledgements. I am grateful to Goupil *et al.* for providing Figs 2 and 3, and to M. J. Thompson for useful discussions. This work was supported in part by the Danish National Research Foundation through the establishment of the Theoretical Astrophysics Center.

References

Audard, N., Brown, T. M., Christensen-Dalsgaard, J. & Frandsen, S., 1997. In *Poster Volume; Proc. IAU Symp. 181: Sounding Solar and Stellar Interiors*, eds Schmider, F.X. & Provost, J., Nice Observatory, in press
Balmforth, N. J., 1992. *Mon. Not. R. astr. Soc.*, **255**, 603
Balona, L. A., Krisciunas, K. & Cousins, A. W. J., 1994. *Mon. Not. R. astr. Soc.*, **271**, 905
Basu, S., Antia, H. M. & Narasimha, D., 1994. *Mon. Not. R. astr. Soc.*, **267**, 209
Breger, M., Handler, G., Nather, R. E., *et al.*, 1995. *Astron. Astrophys.*, **297**, 473
Breger, M., Zima, W., Handler, G., *et al.*, 1997. *Delta Scuti Star Newsletter*, No. 11, 21

Brown, T. M., Christensen-Dalsgaard, J., Mihalas, B. & Gilliland, R. L., 1994. *Astrophys. J.*, **427**, 1013
Catala, C., Auvergne, M., Baglin, A., *et al.*, 1995. In *Proc. Fourth SOHO Workshop: Helioseismology*, eds Hoeksema, J. T., *et al.*, ESA SP-376, ESTEC, Noordwijk, p. 549
Charpinet, S., Fontaine, G., Brassard, P., Chayer, P., Rogers, F., Iglesias, C. A & Dorman, B., 1997. *Astrophys. J.*, **483**, L123
Christensen-Dalsgaard, J., 1988. In *Proc. IAU Symp. 123, Advances in helio- and asteroseismology*, eds Christensen-Dalsgaard, J. & Frandsen, S., Reidel, Dordrecht, p. 295
Christensen-Dalsgaard, J. & Frandsen, S., 1983. *Solar Phys.*, **82**, 469
Christensen-Dalsgaard, J. & Gough, D. O., 1980. *Nature*, **288**, 544
Christensen-Dalsgaard, J., Bedding, T. R. & Kjeldsen, H., 1995. *Astrophys. J.*, **443**, L29
Cox, A. N., Morgan, S. M., Rogers, F. J. & Iglesias, C. A., 1992. *Astrophys. J.*, **393**, 272
Cox, J. P., 1974. *Rep. Prog. Phys.*, **37**, 563
Cox, J. P. & Whitney, C., 1958. *Astrophys. J.*, **127**, 561
Dziembowski, W. A. & Goode, P. R., 1992. *Astrophys. J.*, **394**, 670
Dziembowski, W. A., Moskalik, P. & Pamyatnykh, A. A., 1993. *Mon. Not. R. astr. Soc.*, **265**, 588
Dziembowski, W. & Królikowska, M., 1985. *Acta Astron.*, **35**, 5
Dziembowski, W., Królikowska, M. & Kosovichev, A., 1988. *Acta. Astron.*, **38**, 61
Eddington, A. S., 1926. *The Internal Constitution of the Stars*, Cambridge University Press
Goldreich, P. & Keeley, D. A., 1977. *Astrophys. J.*, **212**, 243
Gough, D. O., 1976. In *Problems of stellar convection, IAU Colloq. No. 38, Lecture Notes in Physics, vol. 71*. eds Spiegel, E. & Zahn, J.-P., Springer-Verlag, Berlin, p. 15
Gough, D. O., 1981. *Mon. Not. R. astr. Soc.*, **196**, 731
Gough, D. O., 1987. *Nature*, **326**, 257
Gough, D. O. & Thompson, M. J., 1990. *Mon. Not. R. astr. Soc.*, **242**, 25
Goupil, M.-J., Dziembowski, W. A., Goode, P. R. & Michel, E., 1996. *Astron. Astrophys.*, **305**, 487
Guenther, D. B. & Demarque, P., 1996. *Astrophys. J.*, **456**, 798
Houdek, G., 1996. *Pulsation of solar-type stars, PhD Dissertation*, University of Vienna.
Houdek, G., Gough, D. O., Balmforth, N. J. & Christensen-Dalsgaard, J., 1997. To be submitted to *Astron. Astrophys.*
Kilkenny, D., Koen, C., O'Donoghue, D. & Stobie, R. S., 1997. *Mon. Not. R. astr. Soc.*, **285**, 640
Kjeldsen, H. & Bedding, T. R., 1995. *Astron. Astrophys.*, **293**, 87
Kjeldsen, H., Bedding, T. R., Viskum, M. & Frandsen, S., 1995. *Astron. J.*, **109**, 1313
Kosovichev, A. G., Christensen-Dalsgaard, J., Däppen, W., Dziembowski, W. A., Gough, D. O. & Thompson, M. J., 1992. *Mon. Not. R. astr. Soc.*, **259**, 536
Lopes, I., Turck-Chièze, S., Michel, E. & Goupil, M.-J., 1997. *Astrophys. J.*, **480**, 794
Monteiro, M. J. P. F. G., Christensen-Dalsgaard, J. & Thompson, M. J., 1994. *Astron. Astrophys.*, **283**, 247
Moskalik, P. & Dziembowski, W. A., 1992. *Astron. Astrophys.*, **256**, L5
Pérez Hernández, F. & Christensen-Dalsgaard, J., 1994. *Mon. Not. R. astr. Soc.*, **269**, 475
Pérez Hernández, F. & Christensen-Dalsgaard, J., 1997. *Mon. Not. R. astr. Soc.*, in press
Rogers, F. J. & Iglesias, C. A., 1992. *Astrophys. J. Suppl.*, **79**, 507
Seaton, M. J., Yan, Y., Mihalas, D. & Pradhan, A. K., 1994. *Mon. Not. R. astr. Soc.*, **266**, 805
Tassoul, M., 1980. *Astrophys. J. Suppl.*, **43**, 469
Ulrich, R. K., 1986. *Astrophys. J.*, **306**, L37
Viskum, M., Dall, T. H., Bruntt, H., *et al.*, 1997. In *A half century of stellar pulsation interpretations: A tribute to Arthur N. Cox*, eds Guzik, J. A. & Bradley, P., ASP Conf. Ser., in press
Vorontsov, S. V., Baturin, V. A. & Pamyatnykh, A. A., 1991. *Nature*, **349**, 49
Watson, R. D., 1988. *Astrophys. Space Sci.*, **140**, 255
Zhevakin, S. A., 1953. *Astron. Zh.*, **30**, 161

SEISMOLOGY OF THE ZZ CETI STARS

J. CHRISTOPHER CLEMENS
Palomar Observatory
105-24, California Institute of Technology, Pasadena, CA 91125

Abstract. The pulsations of white dwarf stars are potentially a rich source of information about white dwarf structural properties. Extracting and applying this information to improve our knowledge of white dwarf interiors requires measuring individual eigenperiods in a complex power spectrum, and identifying the character of the eigenmodes they represent. This review will summarize observational progress in these areas for the ZZ Ceti pulsators.

1. Introduction

The ZZ Ceti stars are the coolest and most numerous of the three known classes of pulsating white dwarf stars. They are apparently normal DA (H atmosphere) white dwarfs, with effective temperatures appropriate for the formation of a H partial ionization zone near their surfaces (12,500 K). They exhibit non-radial g-mode pulsations with periods ranging from about $100 - 1000$ s, and amplitudes of $1 - 80$ mmag. The dominant period exhibited by any individual star is anti-correlated with its temperature, i.e., cooler stars pulsate with longer periods (Winget and Fontaine, 1982; Clemens, 1993; Bergeron *et al.*, 1995). The cooler stars also show larger amplitudes and more complex period spectra. The longer periods presumably result from the deeper convection zone, and the correspondingly longer thermal timescale at the region of strongest pulsational driving, although other factors may contribute (Brickhill, 1983; Wu, 1997). The amplitude and complexity trends are not as well-explored theoretically, nor as well-quantified observationally.

The g-mode spectrum of theoretical models is suitably rich to accommodate the observed periods for many different choices of degree l and radial order n. Thus it is difficult to interpret the measured periods (presumed to be eigenperiods) without some independent means of identifying the character of the associated eigenmodes. A variety of methods have been applied to the problem of mode identification. I will discuss some of them in this review, including a promising new effort using Keck II and the Low Resolution Imaging Spectrograph (LRIS; Oke *et al.*, 1995). Before this discussion, I will draw attention to some recent observational results by Kanaan (1996) regarding the boundaries of the instability strip.

2. Driving Mechanisms and the Red Edge

To drive pulsations, heat flowing from the highly conductive white dwarf interior through the radiative and convective zones near the surface must be converted to mechanical work. Most of the published theory has explored the κ mechanism as the valve which modulates heat flow to produce pulsational driving (Dolez and Vauclair, 1981; Winget et al., 1982; Bradley and Winget, 1994). However, Brickhill (1991a), and more recently Wu (1997) have criticized this approach on the grounds that the models assume that the convective flux is not perturbed by the pulsations. Because convection can adjust on a timescale much shorter than the observed pulsation periods, Brickhill (1991) and Wu (1997) argue for the opposite assumption—that convection adjusts instantaneously to the pulsations. Under this assumption, they find pulsational driving via a mechanism Brickhill has named "convective driving". In convective driving, the convection zone itself drives pulsations by storing energy during compression, and releasing it upon expansion.

This is not a theoretical review, so I will not attempt to assess the merits of these models. The ultimate crucible for refining theory lies in observation, and Kanaan (1996) has recently presented observations which are relevant to the issue of pulsational driving and damping. One of the challenges for non-adiabatic pulsation theory is to match the location of the observed edges of the instability strip. Unfortunately (or fortunately, depending on your point of view) the temperature of the blue edge is sensitive to the convective efficiency assumed by the models (Bradley and Winget, 1994). So the precise location may be a good indicator of the appropriate convective efficiency to use in a model, but it is probably not a good way to choose between competing models for the pulsation driving mechanism.

The red edge offers more promise, because the models exhibit very different behavior there. Models relying on κ driving continue pulsating as they cool to quite low temperatures or until some independent mechanism, such as violation of the surface boundary condition (Hansen, Winget and Kawaler, 1985), causes them to cease. Conversely, in the models for convective driving, the convection zone which drives the pulsations soon absorbs all of the flux modulation, and renders the pulsations undetectable at 11,500 K, at least in flux measurements (Brickhill, 1991b; Wu, 1997).

Observationally, the location of the red edge has been difficult to fix precisely, because searches for pulsation via conventional techniques were afflicted by atmospheric variations at low frequencies. Kanaan (1996) has addressed this problem through the use of high-speed CCD photometry, with interesting results. Figure 1 shows Kanaan's measured pulsation limits for stars cooler than the ZZ Cetis, along with the power measured in the known pulsators. The plot shows the gentle onset of pulsations at about 13,500 K, a rise in amplitude at lower temperatures, and then a dramatic cutoff near 11,500 degrees. Below that temperature, no stars are found to pulsate with amplitudes greater than 4 mmag. Some care must be used in interpreting this diagram, since the known large amplitude pulsators vary in the total amount of power they display, and the plot shows the maximum observed for each star. Nonetheless, figure 1 indicates that the red edge is very abrupt, and offers a challenge to theoretical models, which must account both for the existence of pulsations and for their absence at the appropriate temperatures.

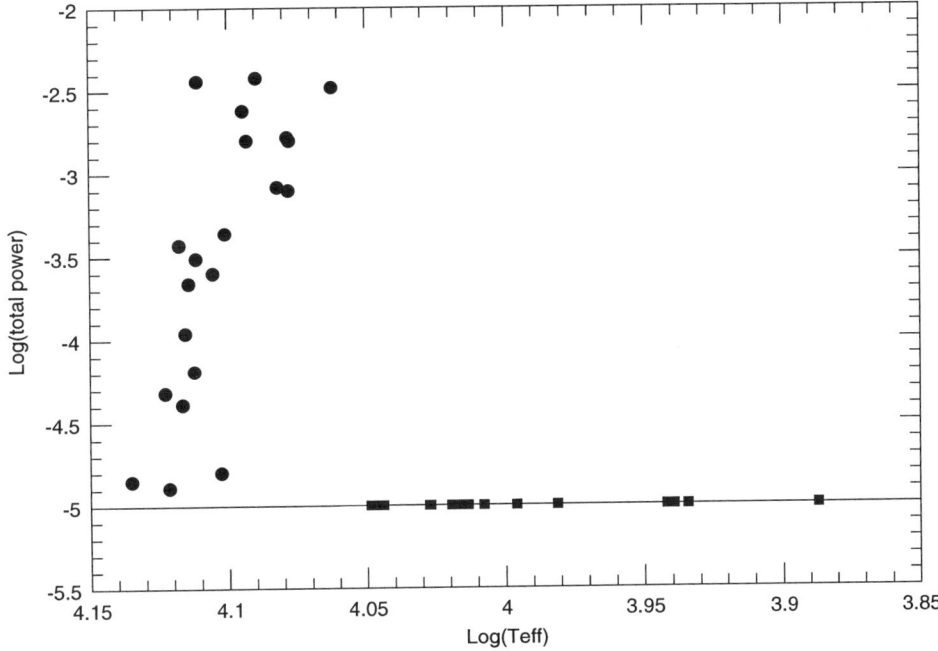

Figure 1. The pulsation power exhibited by ZZ Ceti stars and cooler DAs. Reprinted from Kanaan (1996)

3. Asteroseismology

Fortunately, while helpful, it is not absolutely necessary to know how pulsations are excited to use them for asteroseismological purposes. It is enough to have reliable structural models, whose eigenperiods can be calibrated against the measured periods. Currently, the goals of ZZ Ceti asteroseismology are to measure stellar masses (M_*) and, particularly, the masses of the surface H layers (M_H). Measuring M_H for a significant sample of ZZ Cetis can help resolve an ongoing debate concerning the nature of DA white dwarfs, and the source of the division between DA and non-DA spectral types. Shipman (1997) and Fontaine and Wesemael (1997) give recent reviews of this topic.

Measuring M_H is possible in principle because of the abrupt change in mean molecular weight, and therefore density, at the interface between the surface H layer and the underlying layer (presumed to be He). This interface perturbs the eigenperiods from the value they would have in a homogeneous star (mode trapping; see Winget, 1981). The size and nature of these changes are a diagnostic of M_H. Actually measuring M_H requires comparing the measured mode periods for a star to modes of corresponding n and l in the models. Identifying the n and l of observed modes has been the main challenge, and the main obstacle, to ZZ Ceti seismology.

Clemens (1993) has suggested a scheme for answering limited questions about M_H, without the need for mode identification. In the models for the production of white dwarfs from Asymptotic Giant Branch (AGB) stars, the DAs are those stars

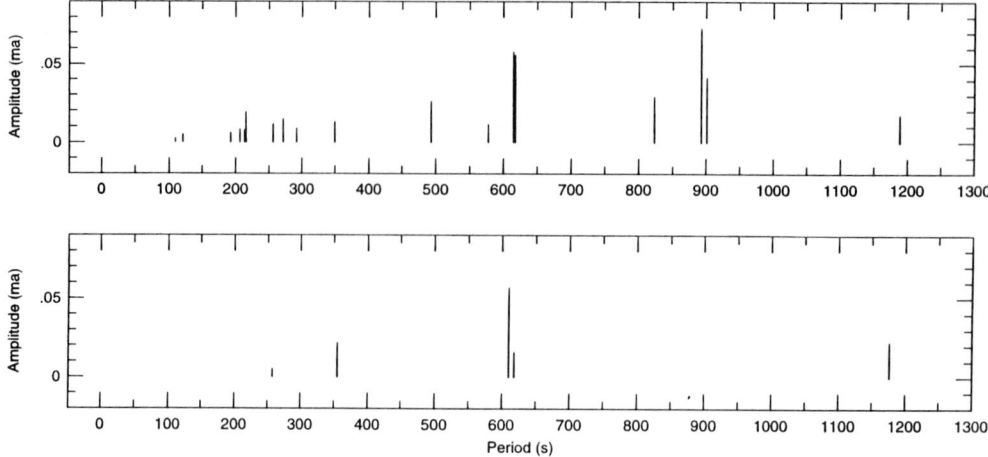

Figure 2. Comparison of the largest modes of ZZ Ceti stars compiled in 1993 (upper panel) with the largest modes of newly discovered ZZ Cetis (lower panel).

that survive their He shell flashes with enough surface H to sustain residual burning until well after they depart the AGB (Iben 1984). We know that DA white dwarfs have a narrow distribution in total mass (Bergeron, Saffer and Liebert 1992), and the Iben models further predict that M_H, ultimately fixed by the physics of nuclear burning, will also be narrowly distributed near $10^{-4} M_\odot$. Since M_* and M_H are the dominant quantities affecting mode periods (Brassard et al., 1992; Bradley, 1996), this theory suggests eigenmodes of a specific n and l in different stars will be narrowly distributed about a single period. Conversely, the competing model composed by Fontaine and Wesemael (1987) to explain observed changes in the ratio of DA to non-DA stars requires a large range of M_H, and a correspondingly wide distribution of periods for the same mode in different stars.

By intercomparing the measured periods of all the ZZ Ceti's, Clemens (1993) found evidence for period groupings, supporting the models of Iben (1984). Clemens suggested that ZZ Cetis discovered in the future could be used to test the reality of these groupings. There have been five ZZ Cetis discovered since that time (Stobie et al., 1995; Giovannini et al., 1997, Vauclair et al., 1997; Jordan et al., 1997). Figure 2 shows the largest modes in each of these stars along with the largest modes of those collected by Clemens in 1993. There remains a clear tendency of modes to group around certain periods. The group near 615 seconds is particularly striking. It is difficult to avoid the conclusion that these stars share a common physical structure, as predicted by the evolutionary models of Iben (1984).

4. Mode Identification

Although figure 2 represents an ensemble argument for similar structure among the ZZ Cetis, it cannot substitute, in the case of any single star, for independent mode identification. Using various schemes, mode identifications have been claimed for a total of five ZZ Ceti stars. In every case but one, the identifications are incomplete, and must be bolstered by additional assumptions to conduct seismological analysis. These analyses typically yield a set of constraints on M_H and M_*.

Two independent groups have identified the 215 s mode in G117-B15A as $l = 1$. Robinson et al. (1995) exploited the differing effects of limb darkening on mode amplitudes at ultraviolet wavelengths. All modes show larger amplitudes in the ultraviolet, where stronger limb darkening diminishes the effects of geometrical cancelation, but modes of $l = 2$ rise faster in amplitude than those of $l = 1$. Robinson et al. found ultraviolet amplitudes consistent with $l = 1$ for the 215 s mode in G117-B15A. Fontaine et al. (1996) measured l for the same star by assuming that the non-linearities in the light curve result from the transfer of linear temperature variations into non-linear flux variations. They found $l = 1$ for all modes. These identifications demand that M_H be greater than $1 \times 10^{-6} M_\odot$. Without assuming some value for n it is impossible to be more specific.

G226-29 represents the best example of an asteroseismological measurement of M_H. The only periods detected in this star form a uniformly spaced triplet at 109.3 s. Fontaine et al. (1992) assumed that this was an $l = 1$ mode rotationally split into 3 components. They found that only for M_H of about $10^{-4} M_\odot$ could any $l = 1$ mode be so short. Kepler et al. (1997) have recently completed the analysis of ultraviolet data from the Hubble Space Telescope and verified that at least two of the triplet members are $l = 1$, so this is now a real measurement of M_H rather than a constraint.

The remaining three stars are GD165, G29-38 and GD154. These have tentative mode identifications based on multiplet splittings, period spacings and trapped-mode spacings, respectively. The values of M_H in units of solar mass are $M_H > 10^7$ for GD165 (Bergeron et al. 1993), $M_H > 4 \times 10^{-7}$ for G29-38 (Bradley and Kleinman, 1997), and $M_H \approx 10^{-10}$ for GD154 (Pfeiffer et al., 1996).

In summary, four out of five of the published identifications give M_H consistent with $\sim 10^{-4} M_\odot$; the fifth does not. The best measurement, that for G226-29, gives exactly the value predicted by models of Iben (1984), and preferred by the ensemble analysis presented by Clemens (1993). Consequently, the results of the few detailed asteroseismological analyses that currently exist are entirely consistent with results Clemens (1993) obtained by intercomparing ZZ Ceti period spectra.

5. The Future

Finally, one more interesting development was presented in the poster session of this symposium by van Kerkwijk et al. Time resolved spectroscopy using LRIS on the Keck II telescope has revealed the radial velocities associated with pulsations in the large amplitude ZZ Ceti star G29-38. Figure 3 shows a time series of the velocity shifts measured from Balmer lines in the spectrum. The lower panel shows the corresponding Fourier transform. Undoubtedly, the measured velocities originate from horizontal motions viewed at the stellar limb; the amplitude of the radial motions in the white dwarf's extreme gravity is immeasurably small. The periods of the radial velocity variations are the same as those measured in flux, with the following notable exceptions: the "combination" peaks appearing at sums and differences of the main power in the flux variations are not present in the velocity measurements.

The detection and measurment of pulsation velocities in ZZ Ceti stars will have numerous consequences we are only beginning to explore. First, they allow a more direct measurement of eigenmode amplitudes, one not afflicted by the non-linearities

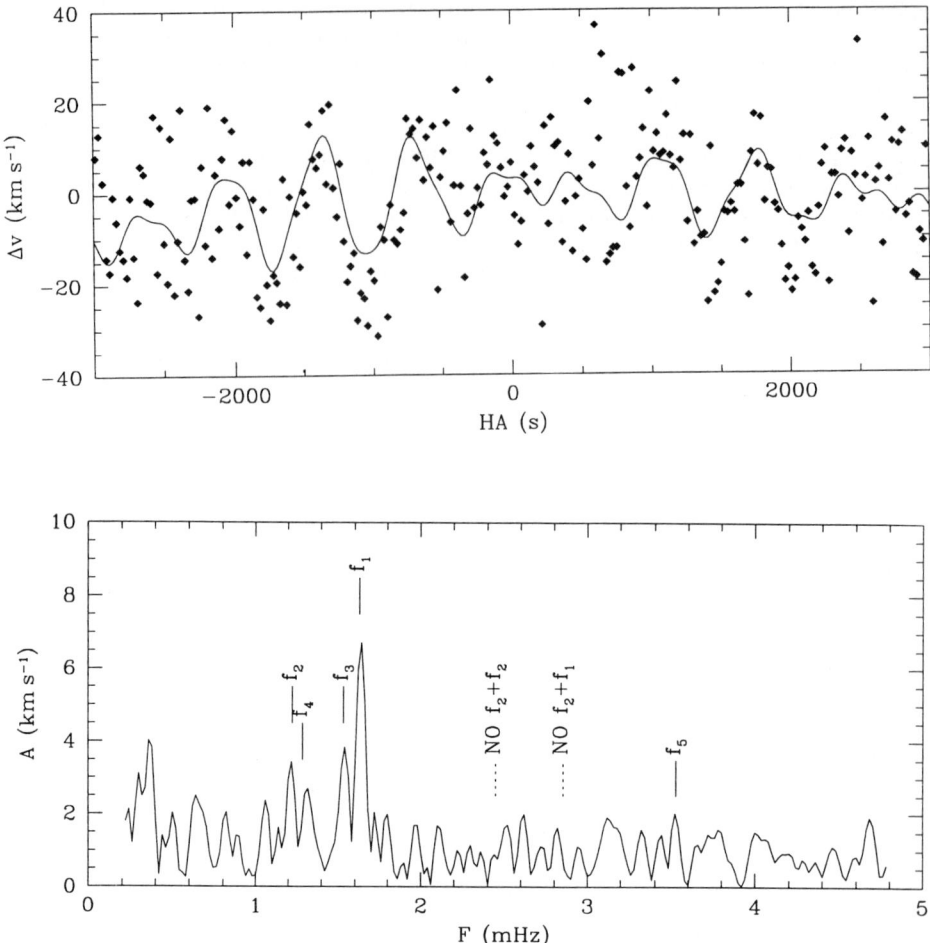

Figure 3. Radial velocity curve for the star G29-38 (upper panel), and its Fourier transform (lower panel). The frequency labels f_1 through f_5 correspond to the five largest frequencies found in the flux modulations. The sum frequencies $f_2 + f_2$ and $f_2 + f_1$ do not appear here, although they are larger than f_5 in the Fourier transform of the flux variations.

displayed in flux. This will allow a more direct comparison to the amplitudes calculated from non-adiabatic pulsation models. Furthermore, they offer, for the first time, a way to select normal modes without the confusion arising from combination frequencies. Finally, they offer a new method of mode identification, because the ratio of velocity to light amplitude varies with l. Continued observations of more ZZ Ceti stars with the Keck telescope promises to provide an entirely new dimension to the study of ZZ Ceti pulsations.

References

Bergeron, P., Fontaine, G., Brassard, P., Lamontagne, R., Wesemael, F., Winget, D. E., Nather, R. E., Bradley, P. A., Claver, C. F., Clemens, J. C., Kleinman, S. J., Provencal, J., McGraw, J. T., Birch, P., Candy, M., Buckley, D. A., Tripe, P., Augusteijn, T., Vauclair, G., Kepler, S. O., and Kanaan, A.: 1993, *Astron. J.* **106**, 1987.

Bergeron, P., Saffer, R. A., and Liebert, J.: 1992, *Astrophys. J.* **394**, 228.

Bergeron, P., Wesemael, F., Lamontagne, R., Fontaine, G., Saffer, R. A., and Allard, N. F.: 1995, *Astrophys. J.* **449**, 258.

Bradley, P. A.: 1996, *Astrophys. J.* **468**, 350.

Bradley, P. and Kleinman, S.: 1997, in J. Isern, M. Hernanz, and E. Garcia-Berro (eds.), *White Dwarfs*, Kluwer Academic Publishers, Dordrecht, p. 445.

Bradley, P. A., and Winget, D. E.: 1994, *Astrophys. J.* **421**, 236.

Brassard, P., Fontaine, G., Wesemael, F., and Tassoul, M.: 1992, *Astrophys. J. Suppl.* **81**, 747.

Brickhill, J. A.: 1983, *Monthly Notices Roy. Astron. Soc.* **204**, 537.

Brickhill, J. A.: 1991a, *Monthly Notices Roy. Astron. Soc.* **251**, 673.

Brickhill, J. A.: 1991b, *Monthly Notices Roy. Astron. Soc.* **252**, 334.

Clemens, J. C.: 1993, *Baltic Astronomy* **2**, 407.

Dolez, N., and Vauclair, G.: 1981, *Astron. Astrophys.* **102**, 375.

Fontaine, G., Brassard, P., Bergeron, P., and Wesemael, F.: 1996, *Astrophys. J.* **399**, L91.

Fontaine, G., Brassard, P., Bergeron, P., and Wesemael, F.: 1996, *Astrophys. J.* **469**, 320.

Fontaine, G., and Wesemael, F.: 1987, in A. G. D. Philip, D. S. Hayes, and J. Liebert (eds.), *The Second Conference on Faint Blue Stars*, IAU Colloq. 95, L. Davis Press, Schenectady, p. 319.

Fontaine, G., and Wesemael, F.: 1997, in J. Isern, M. Hernanz, and E. Garcia-Berro (eds.), *White Dwarfs*, Kluwer Academic Publishers, Dordrecht, p. 165.

Giovannini, O., Kepler, S. O., Kanaan, A., Costa, A. F. M., and Koester, D.: 1997, *Astron. Astrophys.* in press.

Hansen, C. J., Winget, D. E., and Kawaler, S. D.: 1985, *Astrophys. J.* **295**, 547.

Iben, I., Jr.: 1984, *Astrophys. J.*, **277**, 333.

Jordan, S., Koester, D., Vauclair, G., Dolez, N., Heber, U., Hagen, H.-J., Reimers, D., Chevreton, M., and Dreizler, S.: 1997, *Astron. Astrophys.* in press.

Kanaan, A.: 1996, PhD Thesis, University of Texas.

Kepler, S. O., et al.: 1997, *Baltic Astronomy* in press.

Oke, J. B., Cohen, J. G., Carr, M., Cromer, J., Dingizian, A., Harris, F. H., Labrecque, S., Lucinio, R., Schaal, W., Epps, H., and Miller, J.: 1995, *Publ. Astron. Soc. Pacific* **107**, 375.

Pfeiffer, B., Vauclair, G., Dolez, N., Chevreton, M., Fremy, J.-R., Barstow, M., Belmonte, J. A., Kepler, S. O., Kanaan, A., Giovannini, O., Fontaine, G., Bergeron, P., Wesemael, F., Grauer, A. D., Nather, R. E., Winget, D. E., Provencal, J., Clemens, J. C., Bradley, P. A., Dixson, J., Kleinman, S. J., Watson, T. K., Claver, C. F., Matzeh, T., Leibowitz, E. M., and Moskalik, P.: 1996, *Astron. Astrophys.* **314**, 182.

Robinson, E. L., Mailloux, T. M., Zhang, E., Koester, D., Stiening, R. F., Bless, R. C., Percival, J. W., Taylor, M. J., and Van Citters, G. W.: 1995, *Astrophys. J.* **438**, 908.

Shipman, H. L.: 1997, in J. Isern, M. Hernanz, and E. Garcia-Berro (eds.), *White Dwarfs*, Kluwer Academic Publishers, Dordrecht, p. 165.

Stobie, R. S., O'Donoghue, D., Ashley, R., Koen, C., Chen, A., and Kilkenny, D.: 1995 *Monthly Notices Roy. Astron. Soc.* **272**, L21.

Vauclair, G., Dolez, N., Jian Ning, F., and Chevreton, M.: 1997, *Astron. Astrophys.* **322**, 155.

Winget, D. E.: 1981, PhD Thesis, University of Rochester.

Winget, D. E. and Fontaine, G.: 1982, in J. P. Cox and C. J. Hansen (eds.), *Pulsations in Classical and Cataclysmic Variable Stars*, University of Colorado Press, Boulder, p. 46.

Winget, D. E., Van Horn, H. M., Tassoul, M., Hansen, C. J., Fontaine, G., and Carrol, B.

W.: 1982, *Astrophys. J.* **253**, L29.
Wu, Y.: 1997, in J. Isern, M. Hernanz, and E. Garcia-Berro (eds.), *White Dwarfs*, Kluwer Academic Publishers, Dordrecht, p. 467.

ASTEROSEISMOLOGY IN ACTION: PULSATING HOT WHITE DWARFS

STEVEN D. KAWALER
Iowa State University
Department of Physics and Astronomy
Ames, IA 50011 USA

1. Introduction

The development of helioseismology as a forefront area of solar and stellar astronomy has been fascinating to watch. Helioseismology takes full advantage of the benefits of lots of photons and a spatially resolved target. These allow measurement of tiny amplitude photospheric velocity and flux variations, and the precise identification of the oscillation modes corresponding to each observed frequency.

The challenges of stellar (as opposed to solar) seismology include the much smaller fluxes from stars and the lack of spatial resolution. These combine to severely limit the number of modes that are available for detailed study — in most cases, stars show no measurable variations! For stars, oscillations must be of much larger amplitude than seen in the Sun. In addition, the signal from high angular degree modes are wiped out by simple geometric cancellation in the disk-integrated measurements of stars. Thus, while solar seismology involves the study of millions of independent oscillation modes, a "rich" *stellar* pulsator shows a few dozen modes at best.

Asteroseismology can currently be divided into two broad areas – let's call them "virtual" asteroseismology and "active" asteroseismology" Active asteroseismology involves study of known nonradial pulsators – including the pulsating white dwarfs, δ Scuti stars, rapidly oscillating Ap stars, and others that have been discussed in this meeting. In this review, I'd like to illustrate a mature example of active asteroseismology by briefly recounting some of the successes of seismic investigations of pre-white dwarf stars. The ZZ Ceti stars fall outside the scope of this short review, but see the paper by Clemens (1997) in this volume, the review by Winget (1988), and papers by Brassard et al. (1992) and Bradley & Winget (1991) for a flavor of what is happening there.

1.1. WHY CARE ABOUT WHITE DWARFS?

White dwarfs represent the final stage of evolution for most stars, including our own Sun. Locked within their interiors is their prior evolutionary history, and in particular the story of their last nuclear evolutionary stage. By "excavating" the surface layers of a white dwarf, we can determine the circumstances under which its progenitor left the asymptotic giant branch (AGB). The amounts of hydrogen (if any) and helium that remain tell the phase (either between or during thermal pulses) at which the parent AGB star expelled a planetary nebula. In addition, determination of the helium (and hydrogen) layer masses of white dwarf stars clarifies one of the important inputs in white dwarf cooling theory, reducing the uncertainty in the age of the galactic disk as determined through the white dwarf luminosity function.

In addition to the hot white dwarfs, cool pulsating white dwarfs (the ZZ Ceti stars) also place tight constraints on the internal structure of older white dwarf stars. One of the most exciting new areas of work on cool white dwarfs is the prospect of measuring properties of the crystalline interior of a massive ZZ Ceti star, as discussed in Winget et al. (1997). For a recent general review of the properties, evolution, and seismology of white dwarf stars, see the review by Kawaler (1996).

1.2. THE WHOLE EARTH TELESCOPE

Asteroseismological studies of white dwarfs provide constraints ranging from the hot pre-white dwarf PG 1159 stars to the cooler hydrogen–rich ZZ Ceti stars. Exciting results for white dwarfs include measurements of stellar masses, distances, rotation rates (both surface and interior), and internal stratification of composition. Seismological data has demonstrated empirically the action of gravitational settling in white dwarf envelopes, and provided measurements and important upper limits to the cooling rates of white dwarf stars. These are described briefly in the next section. Before discussing these results, some background on the observational aspects of white dwarf seismology are in order.

The principal instrument that is used for seismological observations of white dwarfs is the Whole Earth Telescope (WET). The Whole Earth Telescope is a unique instrument. It was developed by Ed Nather and Don Winget at the University of Texas in 1986, and over the past decade has developed into an active collaboration of over 50 scientists in 14 nations, and continues to produce many scientific results.

The success of the WET is due in large part to the vision of its founders as a collection of scientists working together on mutually interesting science. The common tool needed to make these various scientific inquiries is required to provide uninterrupted time-series photometry that with time-resolution of seconds over time scales of weeks; that is, observations are needed for 24 hours per day over many days. To accomplish this at a reasonable cost requires a network of observatories around the globe staffed by observers with a keen interest in the science – not simply telescope assistants, but active scientists.

Despite being composed of various types of telescopes at observatories of many nations around the globe, the WET is operated as a single instrument and thought of as such by those who participate in it. Target are selected through an informal proposal process. Members of the collaboration suggest targets to a WET committee (known as the "Council of the Wise") who then decide on the priority targets for upcoming observing runs. The PI for each selected target provides a written scientific justification, which is distributed to all collaboration members for use in preparing observing proposals for each of their sites (either locally operated, or relevant national observatories). If time is granted for an observer at a given site, then that site is added to the list of active sites for the run.

During the WET run, data are collected at each site using instruments that are functionally nearly identical: multichannel photoelectric photometers. These allow continuous monitoring of the target star, a comparison star, and a blank sky field. During a WET campaign, data are collected at a given site; following each night, the data are sent via e-mail to a central headquarters (typically in Ames, Iowa). This central HQ is staffed 24 hours/day by several WET astronomers as well as the principal investigator for the primary target. When data arrive at HQ, they are immediately reduced and examined for any problems. After reduction, a given run is incorporated into the growing data set from all sites. If a problem is detected in a given observation, HQ personnel contact the observers (by telephone or e-mail) to discuss the issue and help arrive at a solution. This direct interactivity provides welcome support and encouragement to the observers in remote sites, and ensures high data quality across the network.

Following a run, the data are provided to the PI, who is responsible for final reduction and analysis, and for drafting the paper describing the results. Data rights reside with the PI through the initial publication, which is usually a period of 18 months or so. Authorship of this initial publication includes all participating observers and analysts. Subsequently, the data are made freely available to all who need it, both within and outside the collaboration.

2. Past Results from Whole Earth Telescope Observations

As many of the successes of white dwarf seismology have been studies of the hottest pulsating white dwarfs, I will largely confine this discussion to these stars: the pulsating central stars of planetary nebulae, and the pulsating PG 1159 stars.

2.1. MASSES AND ROTATION RATES

The observed oscillation modes in pulsating white dwarfs are high radial order g-modes of low degree (almost exclusively $l=1$). As a result, the mode periods follow the asymptotic formula

$$P_{nl} = \frac{P_o}{\sqrt{l(l+1)}}(n+\epsilon). \tag{1}$$

This means that the periods of consecutive radial orders are equally spaced in period. This is the principal signature seen in the pulsation spectra of hot white dwarfs. The constant period spacing factor P_o in these stars is primarily determined by the mass of the star (Kawaler & Bradley 1994). Identification of the period spacing has been possible for three of the PG 1159 stars and one planetary nebula central star. The masses range from $0.55 M_\odot$ for the central star of NGC 1501 (Bond et al. 1996) up to $0.66 M_\odot$ for the coolest PG 1159 star PG 0122 (O'Brien et al. 1998).

As with solar p-modes, rotation can be deduced by identifying equal frequency splitting in the pulsation spectrum; a given mode may be split into $2l+1$ components. The frequency splitting is given by the familiar expression

$$\Delta\nu = \Omega(1 - C_{nl}), \tag{2}$$

if the star rotates as a solid body, where C_{nl} is a parameter obtained for a given mode by an integral over the eigenfuctions. For white dwarfs, an additional simplification is that the splitting coefficient is, to a good approximation:

$$C_{nl} \approx \frac{1}{l(l+1)}. \qquad (3)$$

Identification of equally–spaced triplets in the pulsation spectrum therefore allows measurement of the rotation rate of a white dwarf. Also, the presence of triplets rather than quintuplets suggests that most modes are $l = 1$, though several $l = 2$ modes are seen in PG 1159. Rotation periods for white dwarfs range from 5 hours for PG 2131 (Kawaler et al. 1995) to 1.4 days for PG 1159 (Winget et al. 1991), with most being close to 1 day. Note that this mean rotation period of 1 day is not an artifact of diurnal aliases; the Whole Earth Telescope was designed to remove this effect.

2.2. SUBSURFACE STRUCTURE

When the mode spectrum is sufficiently rich, the period spacings can show regular departures from uniformity. This is the result of mode trapping by a subsurface composition discontinuity. Mode trapping allows us to determine the depth of this discontinuity, and therefore provides a key seismological tool for constraining the evolutionary history of white dwarfs.

As of now, one PG 1159 star and one DB pulsator have shown the clearest effects of mode trapping. PG 1159 itself was studied extensively by WET data (Winget et al. 1991). Kawaler & Bradley (1994) showed that the period distribution could be matched by a model with a (helium–rich) surface layer of $3 \times 10^{-3} M_\odot$. Given the surface composition of PG 1159, this thin layer suggests that PG 1159 underwent a final thermal pulse early in its life as a white dwarf, and may have briefly returned to the AGB. Such a "born–again" scenario is plausible given the evolutionary calculations by Iben (1984) and others.

The other star with a measured subsurface composition transition is the pulsating DB white dwarf GD 358. Bradley & Winget (1994) show that the pulsation periods observed by WET (Winget et al. 1994) are consisted with a model that has a pure helium layer with a thickness of $10^{-6} M_\odot$. This is surprisingly thin, given the thickness of the surface layer of PG 1159, which represents its possible progenitor.

2.3. IMPACT ON EVOLUTIONARY MODELS: DIFFUSION

The apparently large difference between the helium layer masses of GD 358 and PG 1159 seems to preclude their having a common origin, with GD 358 being the descendent of PG 1159. In fact, however, the two stars may indeed share common ancestry and a direct evolutionary link. PG 1159 shows a surface that is helium–rich, but has almost as much carbon and oxygen in its outer layers. GD 358 has a nearly pure helium surface.

At the high surface gravities of white dwarfs (with $\log g$ of 7.5 to 8.0) diffusive processes act to drain heavy elements from the surface, thus purifying the outer layers. Though this has been "known" for years, direct evidence has been limited to explaining the trace abundances of heavy elements in white dwarf atmospheres.

Evolutionary calculations of white dwarfs that include diffusive processes have been explored by a number of researchers. However, nearly all are concerned with the abundances of DA (hydrogen–rich) white dwarfs at cooler temperatures, or with the trace metal abundances of hot white dwarf stars. Those that deal with DB white dwarfs had not been fully time–dependent calculations. Dehner (1996) explored the evolution of PG 1159 stars to low temperature including the effects of diffusion, and found that, as expected, the surface helium abundance increases rapidly with time. As the model cools, a pure helium layer collects at the surface, and proceeds downwards. Dehner (1996) and Dehner & Kawaler (1995) found that this pure helium layer reached a thickness of $10^{-6} M_\odot$ when the model reached a temperature of $25,000 K$, which is nearly identical to the conditions deduced by Bradley & Winget (1994) based on WET seismology. Dehner & Kawaler (1996) showed that the pulsation spectrum of their evolutionary model provided a good fit to the observed pulsation periods in GD 358 as well.

Thus seismology of GD 358 demonstrates that the computational aspects of time–dependent diffusion in white dwarfs accurately represent not only the surface chemical evolution of white dwarfs, but also match the interior chemical stratification.

3. Future Directions

Aside from ongoing work in the seismology of pre-white dwarfs of the kind indicated above, there are additional areas of study. Continued measurement of white dwarf masses, rotation rates, and compositional stratification will refine the theory of white dwarf formation and evolution, helping us to lift the veil of ignorance that has clouded the upper-left corner of the H–R diagram.

Beyond this now almost traditional seismological work, there are several new directions that are being explored that use the tools of stellar seismology. This section briefly outlines a few of these new possible directions. Again in this section I concentrate on pre-white dwarf stars. Studies of the cooler DA and DB white dwarfs are equally rich and promising, but lack of space prevents more discussion of these objects.

3.1. CENTRAL STARS OF PLANETARY NEBULAE

The "naked" PG 1159 stars (those without observable nebulae today) were, earlier in their lives, the central stars of planetary nebulae. Indeed, several PNNs show similar spectral features to those seen in the naked PG 1159 stars: CIV and OVI in with emission cores, and so on. Some of these central stars are also nonradial pulsators. Most of them have been discovered as pulsators by Howard Bond and Robin Ciardullo and their collaborators (see Ciardullo & Bond 1996). The pulsation periods in the PNNs are usually much longer than in the naked PG 1159 pulsators. To date, 8 PNNs are known nonradial pulsators. Periods range from about 700 seconds up to several thousand seconds in these stars. The prospect of studying the global properties of PNNs as well as their interior structure through asteroseismology is very exciting. These stars represent the immediate consequences of thermal pulsing on the Asymptotic Giant Branch, as well as the ejection of planetary nebulae. Anything we can learn about their interior structure through asteroseismology will provide very important constraints on this fleeting stage of stellar evolution.

So far, only one of the PNN variables has been fully investigated in an asteroseismological sense. Bond et al. (1996) obtained multisite CCD time-series photometry of the central star of NGC 1501. They were able to identify several modes, and used them to deduce a mean period spacing of 20 seconds. This in turn suggests that the mass of the central star of NGC 1501 is approximately $0.55 M_\odot$, and that it rotates with a period of approximately 1.15 days.

A characteristic of the pulsation spectra of the PNNs is that they are very variable; coherent pulsation modes appear and disappear on short time scales ranging from days to a few years. However, the frequencies of the modes that are manifest at any given time show consistency with a single underlying mode spectrum. Observing these stars over several seasons allows us to collect a "library" of modes, the ensemble being used to deduce the parameters of the spectrum such as the mean period spacing, and rotational splitting values. Similar behavior is seen in some ZZ Ceti stars. Understanding the amplitude instabilities of these stars is difficult, but may be related to the strong winds that are present in PNNs, or to the fact that very high overtone modes are present that are influenced by nonlinear near–surface phenomena.

Another area of mystery about these stars is an apparent correlation between the morphology of the nebula and the pulsation periods of the central star. Pulsating PNNs fall into to groups based on their periods: one group shows periods in excess of 1000 seconds, while the second group show periods that are substantially shorter (averaging 700–800 seconds). Examination of the nebulae themselves shows that the long–period central star are surrounded by fairly symmetric and uniform nebulae. Examples include NGC 246 and NGC 1501. Central stars with short periods are embedded in messy, asymmetric nebulae: examples are the nebula surrounding RX J2117 (Appleton et al. 1993), and NGC 2671. Why should such a correlation exist? Is it merely coincidental, or does the pulsation of the central star influence the nebular morphology? Is it the other way around (does the nebular morphology influence the pulsational instabilities)? Or is there a correlation with stellar mass? Answers to these questions awaits full seismological analysis of a significant fraction of the pulsating central stars.

3.2. ROTATIONAL INVERSIONS

For the richest of the pulsating white dwarfs, PG 1159 and GD 358, several multiplets corresponding to consecutive radial overtones are present in the observed pulsation spectrum. The amount of rotational splitting in general differs from multiplet to multiplet, suggesting that one can estimate the departure of the interior from solid–body rotation. Equation 2 is strictly true only when the rotation rate is constant throughout the interior. If differential rotation exists within a white dwarf, some modes will sample the interior differently than others, producing departures from uniform splittings from mode to mode. Indeed, Winget et al. (1994) suggest that the observed splittings in GD 358 indicate that the outer layers of this star rotate much faster (by a factor of nearly 2) than the inner regions.

Kawaler, et al. (1998) explored the possibilities of using the observed splittings to estimate the internal rotation profile of these two stars. Using inversion techniques and forward calculations, they show that even though the information provided by the pulsations of white dwarf stars is much less extensive than by solar oscillations, some pulsators are rich enough to provide useful constraints on their internal rotation.

The rotational splittings are sensitive to mode trapping: the phase between the rotational splitting and period spacings (as functions of period) allows an observational determination of the sign of any slope in the rotation rate with depth. PG 1159 shows the signature of a rotation curve which is faster at the center than at the surface. Further observations of this star by the Whole Earth Telescope are now available

(Winget, private communication); least–squares analysis of these data, in combination with the first WET data on this star, should be able to test this result.

For the case of the pulsating DB white dwarf GD 358, Kawaler et al. (1998) find that the observed spacings can be reproduced, even only approximately, if the star has a rapidly rotating inner ($r/R < 0.2$) core, a slowly rotating outer layer, and a *counterrotating* midsection; clearly such a rotation curve is unphysical. Thus the earlier suggestion by Winget et al. (1994) is not borne out by the more detailed calculations; Kawaler et al. (1998) conclude that the observed splittings are not consistent with any reasonable internal rotation curve. Still, they conclude that further observations of this star may provide a more consistent description of its internal rotation profile. This work on rotational inversions, though inconclusive, demonstrates the potential of the technique if longer and higher–precision observations can be obtained.

3.3. MEASURING NEUTRINO COOLING VIA SECULAR PERIOD CHANGES

During the pre-white dwarf phase, evolutionary models indicate that neutrino cooling is the dominant source of energy loss for stars near the lower end of the instability strip occupied by the variable pre-white dwarfs. The fraction of the total luminosity contributed by neutrinos increases with increasing mass. Recent asteroseismological analysis of light curves of pulsating pre-white dwarfs near the low temperature bound of the instability strip indicate a trend toward higher masses with decreasing temperature for pulsators within the strip. Thus, O'Brien et al (1997) point out that the cool PG 1159 pulsators should show the largest effects of neutrino cooling. Since the evolutionary changes in these stars will be reflected in the pulsation periods, detection of secular period changes in these stars allows measurement of their cooling rates. Future measurements of secular changes in the period of the dominant modes in these cool, massive pre-white dwarfs can therefore constrain the neutrino emission taking place in their electron degenerate cores, and may provide the first experimental test of thermal neutrino production rates.

4. Conclusions: the Interface Between Pusation and Evolution

With the advantage of having many objects that show nonradial pulsations, white dwarf seismology has advanced to the point where asteroseismological results are now fully integrated into the mainstream of this area of stellar astrophysics. No longer are white dwarf researchers strictly classified as "pulsation types" and "evolution types" and "atmosphere types". Research efforts in white dwarf formation and evolution, as well as in the use of white dwarfs in other astrophysical contexts, routinely include contributions from all of these inputs.

In addition to this mainstreaming of white dwarf seismology, white dwarf pulsators may teach us a lot about the physics of stellar pulsation, and the hydrodynamics of stellar matter. Seismological probes have almost exclusively relied on comparison of observed pulsation frequencies with linear adiabatic pulsation calculations. However, the observed pulsations also show strong nonlinear effects including harmonics, power spectrum peaks at frequencies that are the sums of frequencies of large–amplitude "normal" modes, disappearance and reappearance of normal modes, and other forms of complex amplitude modulation. These nonlinear behaviors contain a wealth of information about the physics of these stars. But because of the many complexities of nonlinear analysis, research is still at an early stage in attempting to mine this information.

It is important to point out that though this brief review has concentrated on WET observations and their interpretations, coordinated network observations are only one approach to the problem. As shown by Clemens (1997), who has observed pulsating white dwarfs at Keck Observatory and with HST, as well as work by Fontaine and coworkers at the CFH telescope, large aperture studies can provide important complementary information. With large telescopes, Clemens (1997) shows that the spectrum of the pulsators responds to the nonradial pulsations of the atmosphere, allowing new measurements of things like the temperature amplitude of the pulsations, and perhaps allowing mode identifications. Fontaine and his group demonstrate that with high S/N, low-amplitude modes can be found that are hidden in network observations with smaller telescopes. These low amplitude modes may help fill in the pulsation spectra, allowing seismological analysis on more complete mode sets.

Finally, significant research is possible with moderate aperture single-site observations. Once a pulsation spectrum has been decoded, follow-up observations at a single site can extend the timing baseline. With such a baseline, phasing of stable observed modes can allow determinations of dP/dt, which in turn places interesting constraints on the evolutionary time scales of the pulsators. Also, for those objects that show modes that come and go on time scales of months, single-site observations can identify these occasional pulsations and help fill in the overall spectrum (see, for example, Kleinman et al. 1994, O'Brien et al. 1996, and Bond et al. 1996).

Acknowledgements

The author thanks the organizers of IAU Symposium 185 and the NASA Astrophysics Theory Program for their support.

References

Appleton, P.N., Kawaler, S.D., & Eitter, J. 1993, AJ, 106, 1973
Bond, H., Kawaler, S., Ciardullo, R. et al. 1996, AJ, 112, 2699
Bradley, P.A. & Winget, D.E., ApJSupp, 75, 463
Brassard, P., Fontaine, G., Wesemael, F., & Hansen, C.J. 1992, ApJSupp, 80, 369
Ciardullo, R. & Bond, H. 1996, AJ, 111, 2332
Clemens, J.C. 1997, these proceedings
Dehner, B.T. 1996, Ph. D. Thesis, Iowa State University
Dehner, B.T. & Kawaler, S.D. 1995, ApJL, 445, L141
Iben, I. Jr. 1984, ApJ, 277, 333
Kawaler, S. 1996, in Stellar Remnants: Saas-Fee Advanced Course 25, ed. G. Meynet & D. Schaerer (Berlin: Springer-Verlag), p. 1-95
Kawaler, S. & Bradley, P. 1994, ApJ, 427, 415
Kawaler, S., Sekii, T., & Gough, D.O. 1998, MNRAS, in press
Kawaler, S., et al. (the WET collaboration) 1995, ApJ, 450, 350
Kleinman, S. et al. (the WET collaboration) 1994, ApJ, 436, 875
O'Brien, M.S., Clemens, J.C., Kawaler, S.D., & Dehner, B.T. 1996, ApJ, 467, 397
O'Brien, M.S. et al. (the WET collaboration) 1998, ApJ, in press
Winget, D.E. et al. (the WET collaboration) 1991, ApJ, 378, 326
Winget, D.E. et al. (the WET collaboration) 1994, ApJ, 430, 839
Winget, D.E., Kepler, S.O., Kanaan, A., Montgomery, M., & Giovannini, O. 1997, ApJL, 487, L191

REPLACING COLOUR BLINDNESS WITH DEPTH PERCEPTION

Rapid spectroscopy and multicolour photometry of roAp stars

JAYMIE MATTHEWS
University of British Columbia
Department of Physics & Astronomy
Vancouver, V6T 1Z4 Canada
matthews@astro.ubc.ca

Warning to the unsuspecting reader: The editors of this volume have (bravely) asked me to "preserve the spirit" of my oral presentation. It is difficult to translate the stirring grandeur of a multi-coloured flashing bowtie into mere words and diagrams. Failing to evoke that grandeur, I have instead settled for occasionally capturing the tackiness of my battery-operated talk in Kyoto.

1. Observing in monochrome

Until recently, most efforts to observe the rapidly oscillating Ap (roAp) stars have concentrated on rapid photometry through a single broad- or intermediate-band filter. This technique can efficiently sample the short pulsation periods of roAp stars (typically 5 – 12 min) even with a telescope of only modest aperture. It is the optimum search strategy for detection of new oscillators.

The single broadband approach is also attractive for global campaigns to resolve the fine structure of roAp eigenspectra. Such a campaign requires many telescopes spaced in longitude (to increase the duty cycle and suppress aliasing in the frequency analysis) monitoring a single bright star for several weeks (to achieve frequency resolution of order 1 μHz). In practice, only 1-m-class or smaller telescopes can be dedicated for so long to a single star (e.g., the WET network) so a broad bandpass is usually necessary to ensure adequate count rates to be sensitive to oscillation amplitudes of a few millimag. These types of monochrome observations have successfully yielded eigenspectra with the sensitivity and resolution essential to asteroseismology (e.g., Kurtz et al. 1989).

However, we've reached a threshold where such data must be supplemented by rapid spectroscopy and photometry at many bandpasses if we are to (a) identify the individual modes in roAp stars, and (b) fully exploit those modes to probe atmospheric structure and/or dynamics with depth.

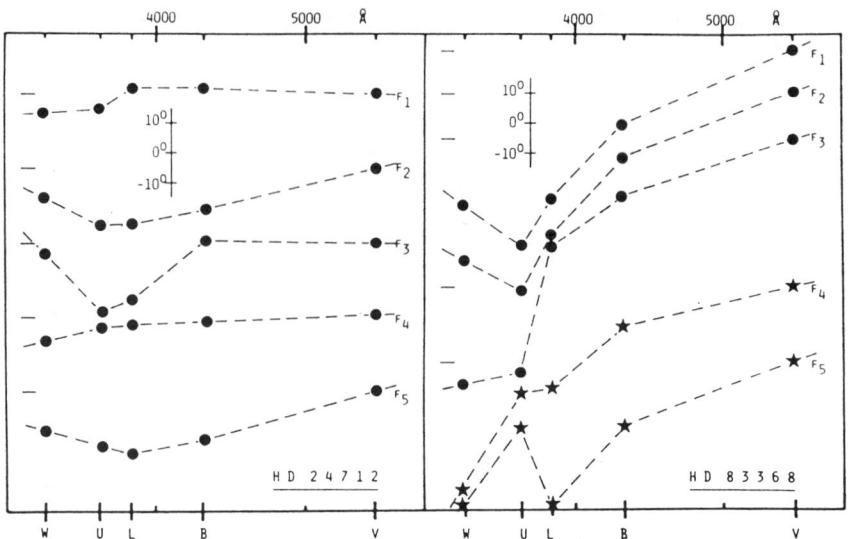

Figure 1. Phase shifts of the oscillations of two roAp stars in five bandpasses (relative to Walraven V), from Weiss (1986).

2. The early days of colour

The first three-colour Hollywood movie produced in the Technicolour process was "Becky Sharp", made in 1935. That's a piece of trivia about colour, but there was nothing trivial about the first rapid *five*-colour (Walraven) photometry of roAp stars by Weiss & Schneider (1984). Their results were the first indication that the pulsation amplitudes of roAp stars drop steeply with increasing wavelength, far more steeply than other known pulsators.

Weiss & Schneider (1984) were also the first to try and use multicolour data as a diagnostic of the physics of roAp stars. Inspired by the success of Balona & Stobie (1979) in identifying modes of δ Scuti stars and Cepheids from phase shifts between their light and colour curves, they tried to apply the same technique to α Cir. The results were ambiguous and inconsistent, even for different runs on the same star. Kurtz & Balona (1984) encountered similar problems in mode identification, based on their Johnson BV data of α Cir.

Schneider (cf. Weiss 1986) also obtained phase-shift information for two other roAp stars, HR 1217 (HD 24712) and HR 3831 (HD 83368), shown in Figure 1. Although the theories of Balona & Stobie (1979) and Stamford & Watson (1981) do not appear to apply to the roAp pulsations, there must be diagnostic information in these phase differences. Currently, Thebe Medupe and Jan Christensen-Dalsgaard are trying to develop a theory to understand the pulsation phase vs. wavelength dependence in roAp stars.

An interesting sidelight to the early applications of multicolour data was Kurtz's (1991, private communication) experiment to use α Cir as its own photometric comparison. Exploiting the steep drop in roAp amplitude with wavelength first recognised by Weiss & Schneider, Kurtz observed α Cir simultaneously in B and I (with a beamsplitter) and obtained reasonable $B-I$ differential photometry even through up to 1.5 mag of extinction (but no bright moonlight). Although the technique has limited applications, it is a reminder that we must always be on the lookout for innovative ways to study stars, such as those described below.

3. Rapid multicolour photometry... but no quick solutions

The steep wavelength dependence of roAp amplitudes was considered a nuisance or a curiousity, until Matthews et al. (1990) suggested it might be a way to probe the atmosphere of one of these chemically peculiar stars. Matthews et al. (1996) used their simultaneous eight-colour optical and infrared rapid photometry of HR 3831 to derive limb darkening coefficients and a $T-\tau$ curve for that star's atmosphere.

Medupe (1997) and Medupe & Kurtz (1997) later demonstrated that limb darkening is not the dominant effect and argued that there is a significant gradient in pulsation amplitude with depth in the upper 2000 km of stars like α Cir. This implies the presence of a radial node very high in the stellar atmosphere. Matthews (1997) has countered that the pronounced abundance stratifications in the upper atmospheres of roAp stars make it dangerous to translate light amplitude vs. wavelength into temperature amplitude vs. depth.

4. Rapid spectroscopy... gradual understanding?

Some of the missing pieces of this puzzle might be found if we tear the puzzling flux distributions of roAp stars into even smaller pieces: going from broad multiband photometry to high-resolution spectroscopy.

The first efforts at rapid spectroscopy of roAp stars were aimed at detecting the radial velocity (RV) variations associated with the light oscillations. There were two general approaches: (1) Fabry-Perot readings of the intensities of the red and blue wings of a single stellar absorption line (e.g., Ando et al. 1988; Belmonte et al. 1989); and (2) cross-correlation of the positions of a large number of lines covering several tens of Ångstroms in high-resolution coudé spectra (e.g., Matthews et al. 1988).

The latter approach – intended to increase the S/N of the measurements by including many lines – yielded the first definitive detection of RV oscillations in an roAp star, HR 1217 (Matthews et al. 1988). The RV semi-amplitude was only ∼200 m/s at a time when simultaneous photometry showed a B semi-amplitude of about 3.4 millimag. This result was in stark contrast to

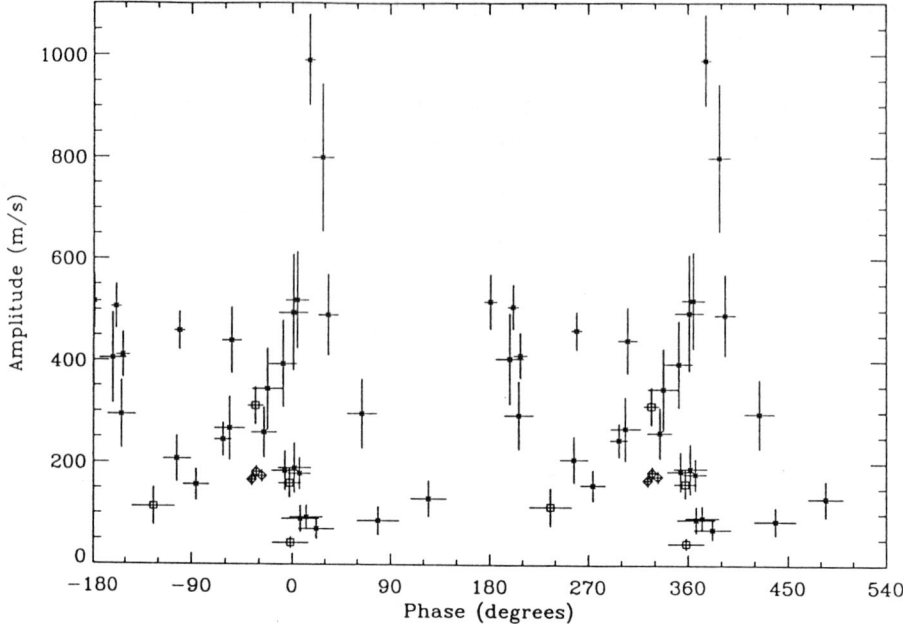

Figure 2. Amplitudes and phases of the principal pulsation mode in α Cir for different wavelength bands, from Baldry et al. (1998).

one Fabry-Perot result, which suffered poorer S/N but seemed to suggest RV amplitudes exceeding 1 km/s in the Ba II λ4930 line in HR 1217 (Ando et al. 1988). Meanwhile, Belmonte et al. (1989), using the λ5317.4 line, claimed a similar RV amplitude for HR 1217 as that reported by Matthews et al., but at a time when the light amplitude was only 1.2 millimag. However, the Matthews et al. estimate of the light-to-RV amplitude ratio proved remarkably consistent with a wide variety of other pulsators – from the Sun to Cepheids – as well as models (Kjeldsen & Bedding 1995).

In retrospect, it is possible that all these apparently contradictary measurements were correct. Recent findings suggest that comparing the multi- and single-line RV measurements was like comparing apples and oranges.

By examining the RV behaviour of many different metal lines in the roAp star α Cir, Baldry et al. (1998) took a bite out of one of those apples (presumably from the Tree of Knowledge) and cast us out of our Garden of mixed fruits (and into my Pit of mixed metaphors!). They found that lines in different wavelength bands showed different RV amplitudes *and phases* (see Figure 2). Independently, Kanaan & Hatzes (1998) found pronounced differences in the RV amplitudes (but *not* phases) of different lines in γ Equ, while Matthews & Scott (1996) noted evidence for line profile variations in certain lines in the same star (Figure 3; see also later in this section).

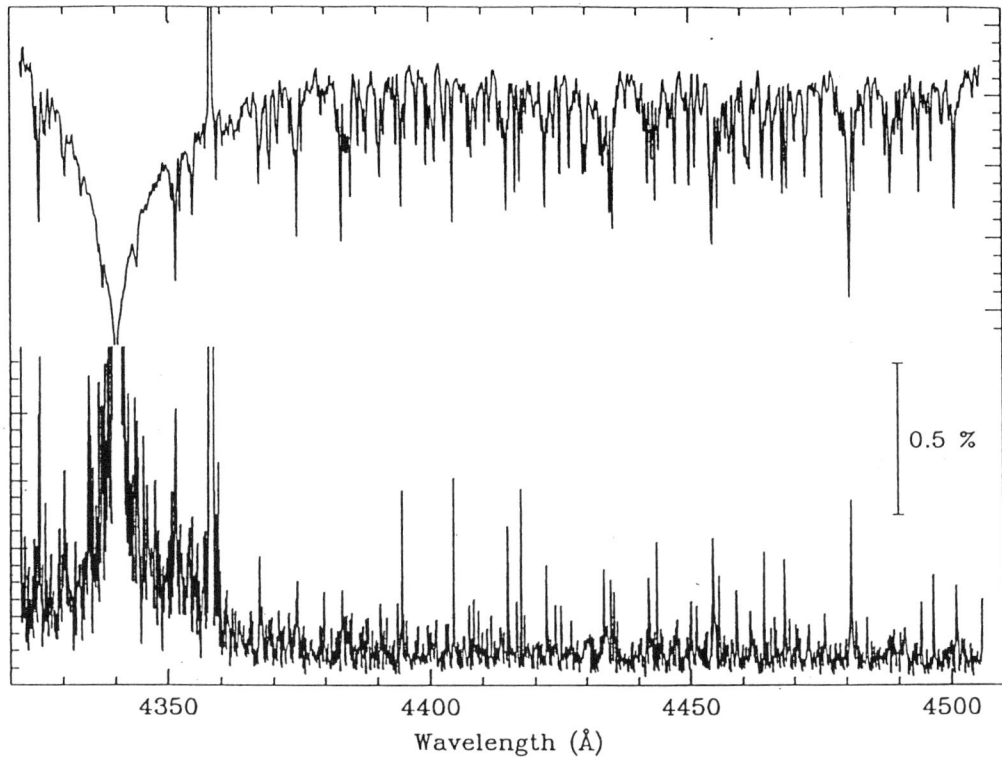

Figure 3. (top) Mean spectrum of γ Equ for 23 August 1988. (bottom) Mean absolute deviation of the individual rapid exposures from the mean. The bar indicates line profile variations with an amplitude of 0.5%. From Scott (1995).

What's going on? Both Baldry et al. (1998) and Kanaan & Hatzes (1998) find a general trend of lower amplitude with larger equivalent width (Figure 4). Is this a clue to the physical mechanism in play here? Baldry et al. interpret the range of amplitudes, and what they believe to be groups of lines varying in antiphase, as further evidence for a radial node high in the atmosphere of α Cir. Kanaan & Hatzes see no significant phase differences among the lines in *gamma* Equ, and believe the varying amplitudes are the result of suppression of oscillations by the magnetic field high in the atmosphere. I believe this to be unlikely, since the magnetic pressure should dominate over gas pressure even much deeper in the atmosphere.

The RV variations of roAp stars are clearly more complicated than we assumed only a decade ago. Individual lines can have amplitudes above 1 km/s while the net variation across a range of spectrum may be less than 10 m/s. That may account for the conflicting results of the cross-correlation

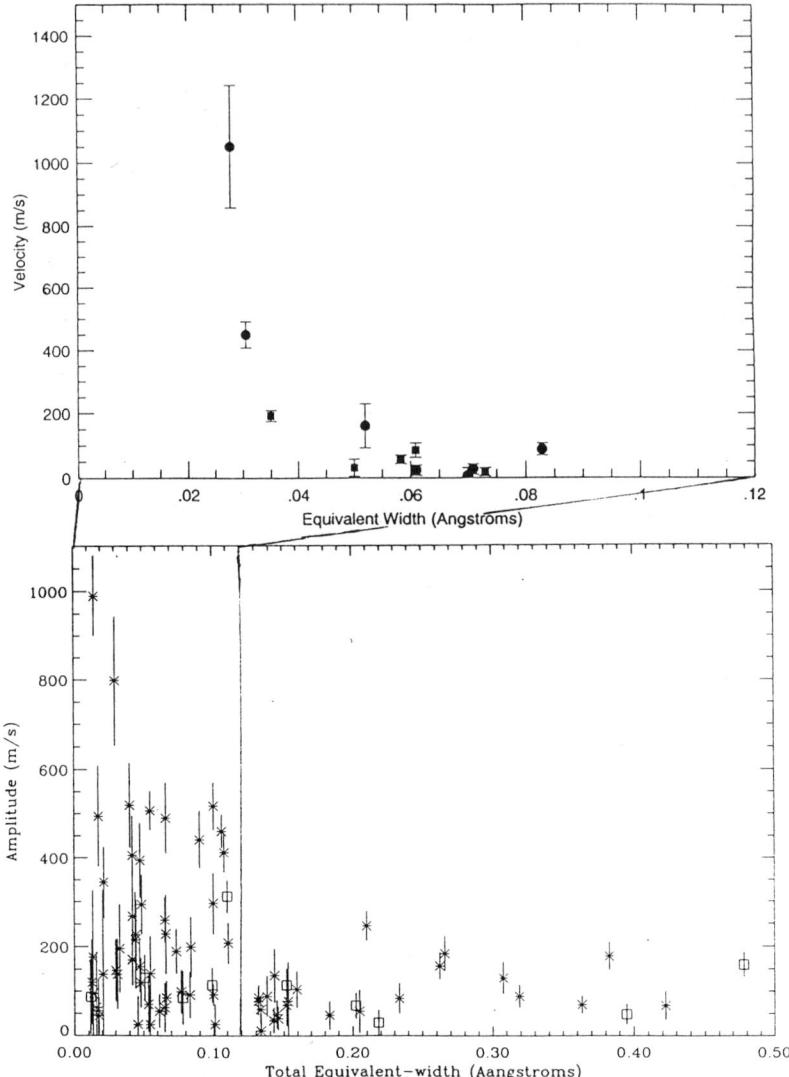

Figure 4. RV amplitude vs. equivalent width: (**top**) γ Equ (Kanaan & Hatzes 1998); (**bottom**) α Cir (Baldry et al. 1998). Note the differences in the horizontal and vertical scales.

and Fabry-Perot measurements of roAp stars in the past. It may be worth re-examining those data in light of the latest developments.

After the failure of mode identification by phase shifts in roAp stars, both Odell & Kreidl (1984) and Baade & Weiss (1987) anticipated the possibility of identifying the nonradial modes of roAp stars by the variations they induce in the spectral line profiles. This technique must overcome two hurdles: (1) the intrinsically low amplitudes of the pulsations, and (2) the inherently narrow lines of Ap stars, which tend to be slow rotators. Undaunted, Schneider &

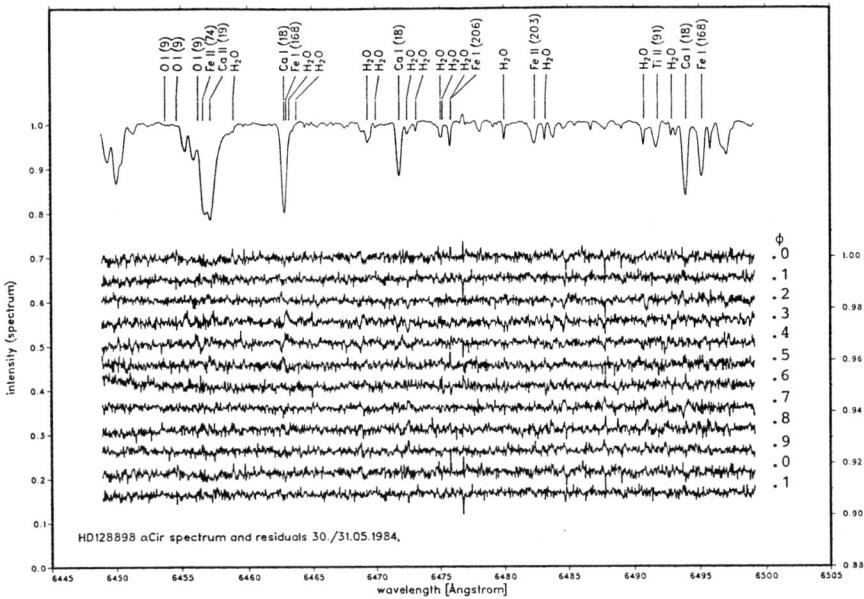

Figure 5. (top) The mean spectrum of α Cir between 6450Å and 6500Å, and (lower) the residuals ordered in pulsational phase. (From Schneider & Weiss 1989).

Weiss (1989) obtained over 300 short-exposure ESO coudé spectra of α Cir which they binned according to the known photometric period and searched for residual variations (see Figure 5) but found none larger than 0.5%.

Matthews & Scott (1996) adopted the same approach with higher-S/N coudé spectra of γ Equ from CFHT obtained in 1988. They see evidence for line profile variations in some lines at the level of nearly 1%. The variations appear consistent with high-overtone ($n \sim 80$) nonradial modes of degree $\ell = 2$ (see Figure 6), but better data are needed. Why only certain lines show profile changes is unclear, but it is likely related to the distribution of elements on the stellar surface and with depth. The amplitude of variability does not appear to correlate with factors like Landé g, excitation energy and $\log(gf)$.

The underlying astrophysics may elude us today, but we are clearly on the threshold of a new era of high-resolution spectral probes of roAp stars.

5. References

Ando, H. et al. (1988) *PASJ*, 40, 249.
Baade D. & Weiss, W.W. (1987) *A&ASuppl*, 67, 147.
Baldry, I.K. et al. (1998) *MNRAS*, in press.
Balona, L.A. & Stobie, R.S. (1979) *MNRAS*, 189, 649.
Belmonte, J.A. et al. (1989) *A&A*, 221, 41

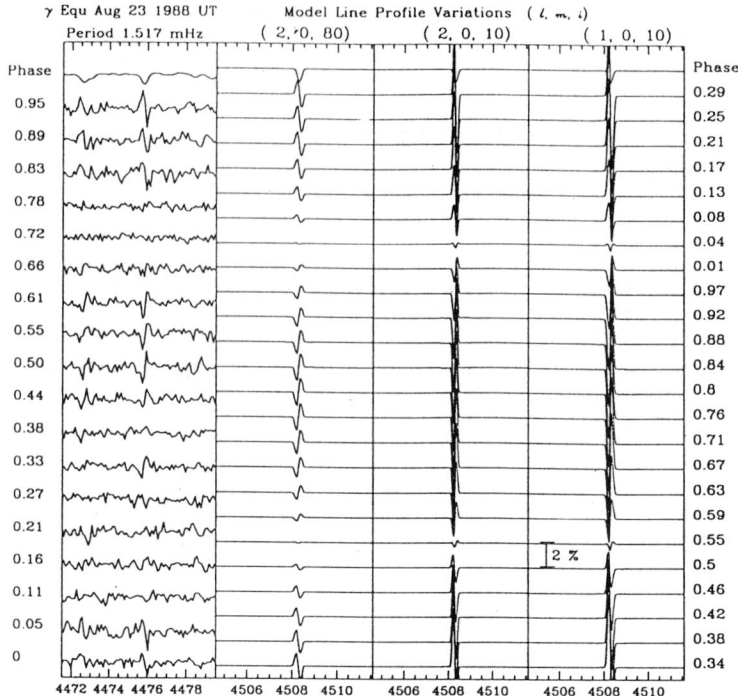

Figure 6. Line profile variations of the Fe I λ4476 line of γ Equ with pulsational phase (left), compared with synthetic residuals for various nonradial modes (right) whose amplitudes are consistent with the measured RV and light amplitudes of the star. The mean observed and model profiles are shown at top. Residuals are offset by 2% of the continuum level.

Kanaan A. & Hatzes, A. (1998) *ApJ*, accepted.
Kjeldsen, H. & Bedding, T.R. (1995) *A&A*, 293, 87.
Kurtz, D.W. & Balona, L.A. (1984) *MNRAS*, 210, 779.
Kurtz, D.W. et al. (1989) *MNRAS*, 240, 881.
Matthews, J.M. (1997) in IAU Symposium 181, pp. 387–394.
Matthews, J.M. & Scott, S. (1996) *ASPC*, 83, 58.
Matthews, J.M. et al. (1988) *ApJ*, 324, 1099.
Matthews, J.M. et al. (1990) *ApJ*, 365, L81.
Matthews, J.M. et al. (1996) *ApJ*, 459, 278.
Medupe, T. (1997) M.Sc. Thesis, University of Cape Town.
Medupe, T. & Kurtz (1997), in IAU Symposium 181, in press.
Odell, A. & Kriedl, T.J. (1984) in 25th Liege Astrophys. Coll., p. 148.
Scott, S. (1995) M.Sc. Thesis, University of British Columbia.
Schneider, H. & Weiss, W.W. (1989) *A&A*, 210, 147.
Stamford, P.A. & Watson, R.D. (1981) *Astrophys. Space Sci.*, 77, 131.
Weiss, W.W. (1986) in IAU Colloquium 90, pp. 219–231.
Weiss, W.W. & Schneider, H. (1984) *A&A*, 224, 101.

RoAp STARS THROUGH THEORISTS' EYES – EXCITATION MECHANISM

ALFRED GAUTSCHY
Astronomisches Institut der Universität Basel, Venusstrasse 7, CH-4102, Binningen, Switzerland

AND

HIDEYUKI SAIO
Astronomical Institute, School of Sciences, Tohoku University, Sendai 980-77, Japan

Abstract. This exposition focuses on the still elusive excitation mechanism of the rapid oscillations in some Ap stars. We visit the domain of overstable magnetic convection which was conjectured as the origin of the observed high radial-order p modes. Mostly we dwell, however, on the classical κ-mechanism. In roAp stars, the zone of partial hydrogen ionization is driving rather effectively. Furthermore, we argue that an atmospheric temperature inversion supports the κ-mechanism driving by preventing oscillation energy from leaking into the atmosphere.

1. Preliminaries

Since the discovery of the short-periodic light variation in a cool Ap star (Kurtz 1978), the number of rapidly oscillating Ap (roAp) stars has grown to about 30. A number of excellent recent reviews addressing various properties of these stars is already available. The most comprehensive article discussing observational as well as theoretical issues of the class of roAp stars was written by Kurtz (1990). Matthews (1991) contributed an observation oriented review emphasizing the aspect of asteroseismology. The most recent observational developments were compiled in the paper of Martinez and Kurtz (1995). Shibahashi's (1991) exposition highlighted theoretical problems remaining with the roAp stars, in particular he contemplated the effects of the magnetic field on the oscillation spectrum. Recently, the perturbation approaches adopting higher-order magnetic multipoles to approximate the stellar field reached new complexities to phenomenologically match the observed oscillation spectra of roAp variables (Shibahashi 1993, Takata & Shibahashi 1995). We aspire here merely to complement the before-mentioned expositions and focus therefore on the excitation mechanism of the rapid oscillations. Part of the article reviews the literature, but some new computations and results are also presented.

We begin by summarizing those observed properties which might prove important when considering the excitation mechanism responsible for the roAp stars. They are main-sequence or post – main-sequence stars with $1.5 - 2M_\odot$. On the HR diagram, all roAp stars seem to reside within the δ Sct region of the instability strip; i.e. rapid oscillations seem to be excited only in cool Ap stars (6800 K $< \log T_{\text{eff}} <$ 8400 K). In other words, a clear-cut instability domain seems to exist for the roAp stars. The periods of light variability range from 5 to 15 min (corresponding to frequencies between 1 and 3 mHz); this is much shorter than the periods of δ Sct star oscillations which lie between 30 min and 6 hr. The oscillation periods, together with the integrated-light variability, of roAp stars indicate high-order nonradial p modes of low degree ($\ell = 1$ and 2) are excited. Despite the overlap of the roAp-stars instability region with the δ Sct one, the roAp stars show *no* δ Sct – type oscillations. In most cases, only few modes in a small frequency range are simultaneously excited. (A notable exception is HD 60435, in which many modes exist – ranging from 0.7 mHz to 1.5 mHz. The lifetimes of some of its modes are shorter than the 7.7 day rotation period [Matthews, Kurtz, Wehlau 1987].) The pulsation amplitudes are modulated with the period of rotation. Kurtz (1982) found the amplitude modulation to be understandable as axisymmetric nonradial oscillations with $\ell = 1$ (and 2) whose pulsation axis coincides with the magnetic axis of the star (oblique pulsator model). The strong (dominantly dipolar) magnetic fields encountered in Ap stars are of the order of a kilo-Gauss.

2. Contemplated excitation mechanisms

The roAp star excitation mechanisms proposed so far can be divided into two groups: First, resonant driving by overstable magnetic convection and second, classical κ-mechanisms.

2.1. OVERSTABLE MAGNETIC CONVECTION MECHANISM

The idea of overstable magnetic convection as a driving agent for roAp pulsations was proposed by Shibahashi (1983). Its physical meaning was elaborated further by Cox (1984). In a zone with a super-adiabatic temperature gradient, a perturbed blob of matter is pictured to ascend or descend and to develop into 'classical' convection in the absence of a magnetic field (or rotation). If a strong enough magnetic field prevails, however, convective motion is stabilized in the sense that the motion of the blob of matter turns oscillatory with a frequency ω_0 given by

$$\omega_0^2 = N^2(k_h^2/k^2) + (\boldsymbol{B} \cdot \boldsymbol{k})^2/(4\pi\rho). \tag{1}$$

The square of the Brunt-Väisälä frequency is denoted by N^2 (which is negative when the temperature gradient is superadiabatic), $\boldsymbol{k}$ is the vectorial wavenumber whereas k_h denotes its horizontal component, and $\boldsymbol{B}$ stands for the magnetic induction. The oscillation period – $2\pi/\omega_0$ – is of the order of the thickness of the super-adiabatic zone divided by the Alfvén speed. The heat exchange between the blob and surrounding matter makes the amplitude of the oscillation grow; i.e. an overstable oscillation develops. This behavior is frequently referred to as *overstable convection*. Under nearly adiabatic conditions, the amplitude-growth is proportional to

$$\exp\left[\frac{1}{2}\nu_T k_h^2 \frac{g}{H_P} \frac{\chi_T}{\chi_\rho} \frac{(\nabla_T - \nabla_{\text{ad}})}{\omega_0^2} t\right], \tag{2}$$

where ν_T is the thermal diffusivity, g is gravity, H_P the pressure scale-height, $\chi_T = (\partial \ln P/\partial \ln T)_\rho$, $\chi_\rho = (\partial \ln P/\partial \ln \rho)_T$, $\nabla_T = d\ln T/d\ln P$, and $\nabla_{\text{ad}} = (d\ln T/d\ln P)_{\text{ad}}$. Both Shibahashi (1983) and Cox (1984) argued that the periods of overstable convection in a thin super-adiabatic zone in the outer envelope of an Ap star are of the same order as the observed periods of roAp stars. The growth time was estimated to be a few weeks to months (Cox 1984).

For overstable convection to serve as an excitation mechanism for roAp oscillations, we must assume that the resonance between the overstable convective motion and the global high-order p modes works well enough to raise the amplitudes of the p modes to an observable level. A quantitative global analysis to confirm or to deny this conjecture is still lacking. If the resonance hypothesis works indeed, the mechanism implies that only Ap stars with strong magnetic fields show rapid oscillations. If theorists are unable to advance sufficiently rapidly to elaborate on this hypothesis, observers might step in with accurate magnetic-field measurements for an appropriate sample of Ap stars. Shibahashi (1983) and Cox (1984) hypothesized furthermore about why only *cool* Ap stars show rapid oscillations. Global oscillations can be resonantly excited only for sufficiently thick superadiabatic zones wherein oscillatory convection is also energetic enough to force global modes to observable amplitudes. For the A-type main-sequence stars it was conjectured that this happens where the roAp stars are located.

Whether global modes of low ℓ are excited by the above mechanism is far from clear. According to eq. (2), the growth rate is higher for overstable convection with larger horizontal wavenumber. Therefore, we would expect global modes with short horizontal wavelengths to experience strong resonance effects and to be most easily excited. However, only low-ℓ modes ($\ell = 1$ and 2) are detected in roAp stars; this seems to contradict the theoretical prediction, even if the effect of low visibility of high-ℓ modes is accounted for (see Dziembowski & Goode 1996). On the theoretical side, a *global analysis* is overdue and unavoidable in order to strengthen the case for overstable convection as a driving agent; this is a very difficult, if not impossible task.

2.2. PARTIAL IONIZATION AND THE κ MECHANISM

The κ-mechanism takes effect when the stellar radiative energy flow is blocked in the compressed phase and released in the expanded phase of an oscillation. Roughly speaking, this occurs – under weak nonadiabatic conditions – if

$$\frac{d}{dr}\left(\kappa_T + \frac{\kappa_\rho}{\Gamma_3 - 1}\right) > 0, \tag{3}$$

where the opacity derivatives κ_T and κ_ρ are defined as $\kappa_T \equiv (\partial \ln \kappa/\partial \ln T)_\rho$ and $\kappa_\rho \equiv (\partial \ln \kappa/\partial \ln \rho)_T$, and $\Gamma_3 - 1 \equiv (d\ln T/d\ln \rho)_{\text{ad}}$ (see e.g. Unno et al. 1989). In the stellar envelope, κ_ρ is nearly constant and of the order of unity while the variation of κ_T plays the essential rôle in the κ-mechanism. The value of κ_T increases outward close to an opacity peak; this is where the κ-mechanism is most effective (outward refers to the direction of increasing radius in the star). Opacity peaks are usually

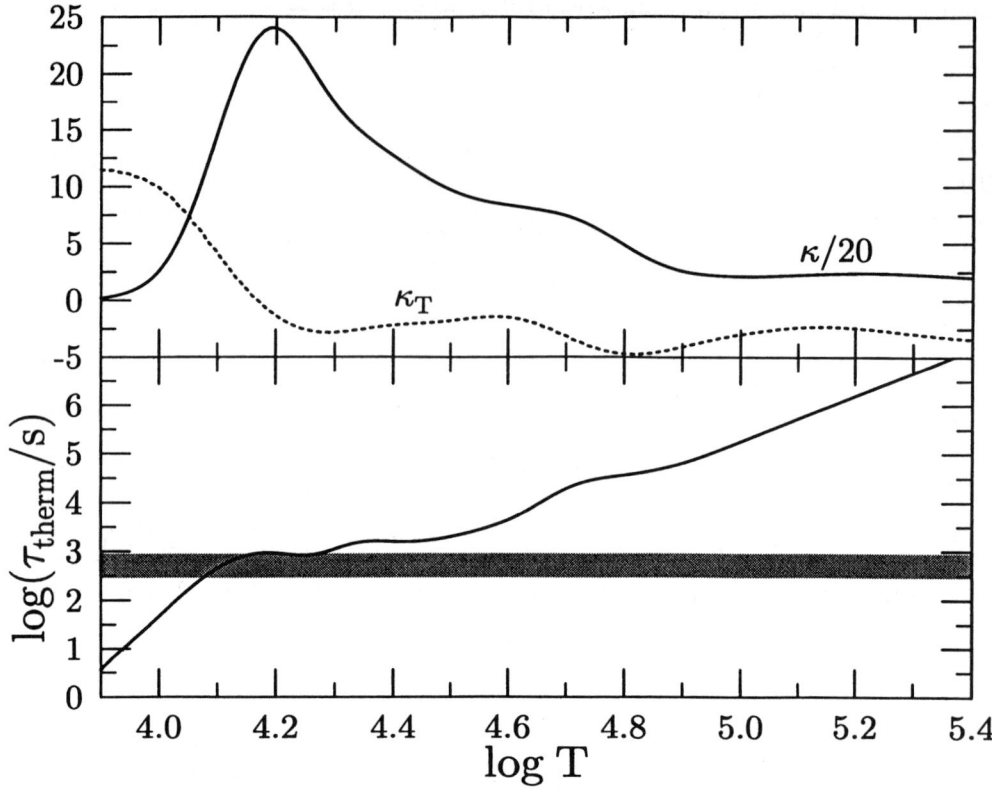

Figure 1. Run of the opacity κ, κ_T, and the thermal timescale in a $1.5 M_\odot$ ZAMS model with respect to temperature. The grey bar indicates the observed period range of roAp stars.

associated with ionization zones where in the inner part of the peak Γ_3 tends to decrease outward. This behavior of the adiabatic exponent helps to further satisfy inequality (3).

Driving through the κ-mechanism in an ionization zone is most effective when the thermal timescale τ_{therm} of the stellar envelope above that ionization zone is comparable with the pulsation period (Cox 1974). The thermal timescale of a $1.5 M_\odot$ ZAMS model is shown in Fig. 1 as a function of temperature. The τ_{therm} was approximated as $(M_* - M_r) C_p T / L_r$. As seen in Fig. 1, the thermal timescale at the HeII ionization zone of an A-type main-sequence star is of order 10^4s which corresponds to the periods of δ Sct – type variability. The κ-mechanism at the HeII ionization zone is indeed the dominant excitation agent of the δ Sct pulsations. On the other hand, the thermal timescale at the H (and HeI) ionization zone lies between a few hundred and about 10^3 seconds which is comparable with oscillation periods of the roAp stars (which is marked with the grey band in Fig. 1). This indicates that the κ-mechanism at the H-ionization zone can potentially excite the short-period modes seen in the roAp stars.

At least two problems have to be overcome before the κ-mechanism can be re-

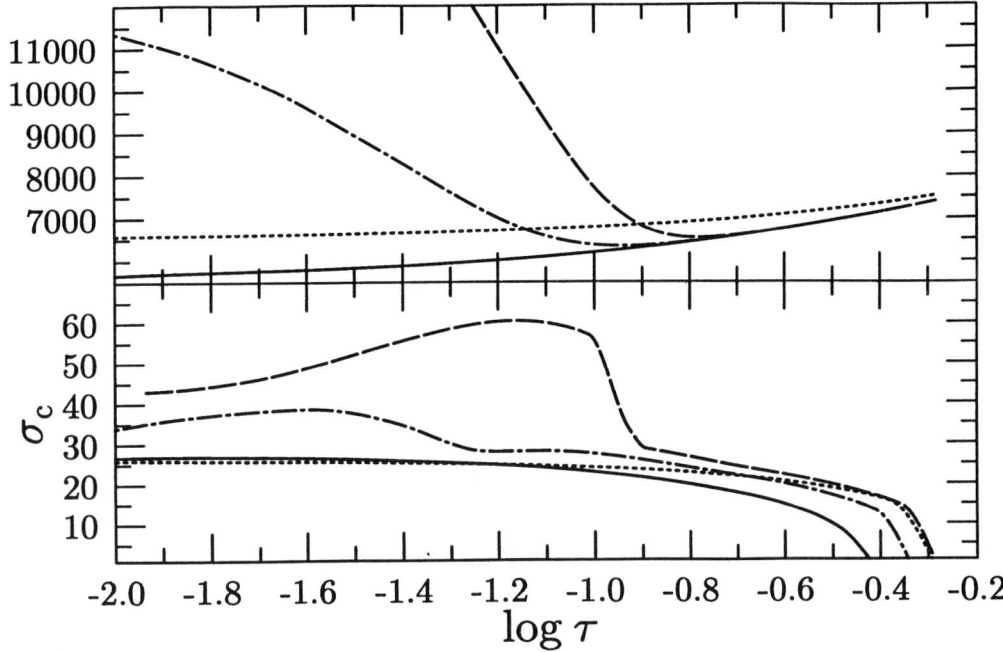

Figure 2. Assumed temperature distributions (upper panel) and corresponding spatial variation of the critical frequency. The horizontal axis represents the logarithm of the optical depth τ. In this normalization, $\sigma_c \simeq 30$, for example, corresponds to 2 mHz.

garded as *the* excitation mechanism for roAp-star variability. One problem is that the δ Sct – type pulsations must be suppressed in roAp stars. Usually, helium is assumed to be drained due to gravitational settling from regions where otherwise partial HeII ionization contributes to pulsational driving. This reasoning works also to argue for the absence of δ Sct – type oscillations in Am stars whose peculiar spectra are explained by elemental diffusion. It is not certain, however, whether gravitational settling is efficient enough to drain He as deep as $T \sim 5 \times 10^4$K, in particular in non-static envelopes. To be candid, it is even controversial whether He is *depleted* or *enhanced* by diffusion. Dolez, Gough, & Vauclair (1988) argued that He should be increased at the magnetic poles if a stellar wind is included in the full picture. The other problem to be resolved has to do with the critical frequency. In an Eddington-type atmosphere the critical frequency is lower than the typical roAp-star oscillation frequencies (Shibahashi & Saio 1985). Despite some uncertainty in the definition of the critical frequency it is clear that the higher the oscillation frequency lies above the critical frequency, the higher the mode-energy leakage at the surface.

Dziembowski & Goode (1996) examined the pulsational stability for $1.8M_\odot$ and $2.0M_\odot$ models which have almost the same effective temperature of about 8.4×10^3K. They noted that in the $2.0M_\odot$ model the driving at the H ionization zone almost excites high order p modes (i.e., the driving is exceeded only slightly by the radiative damping), and that in the $1.8M_\odot$ model the maximum driving occurs for frequencies above the critical frequency. Their result suggests that if some physical effect can be thought of to increase the critical frequency in roAp stars, rapid oscillations are possibly excited by the κ-mechanism in the partial hydrogen-ionization zone. Such an effect must be attributable to Ap-star peculiarities because no rapid oscillation is detected in δ Sct or Am stars. One obvious property of Ap stars is the strong magnetic field. The direct effect of a magnetic field is, however, to damp oscillations (Roberts & Soward 1983; Dziembowski & Goode 1996). Alfvénic waves are generated in a thin surface layer by coupling p-mode oscillations with the magnetic field. As they propagate inward these Alfvénic waves are expected to be dissipated because their wavelength becomes very small.

3. Temperature inversion

A temperature inversion, as it prevails in the solar chromosphere, is not unconceivable in Ap stars due to the presence of strong magnetic fields and a superficial convective region due to the partial H, HeI ionization zone. We note that a temperature inversion in the atmosphere can increase the critical frequency significantly. The adiabatic critical frequency ω_c for high order p modes may be written as

$$\omega_c^2 \simeq \frac{GM}{R^3} \frac{\Gamma_1}{2} \left(\frac{d\ln\rho}{d\ln r} - \frac{V}{2} \right) = \frac{GM}{R^3} \frac{\Gamma_1 V}{2} \left(\frac{1}{\chi_\rho} - \frac{\chi_T}{\chi_\rho} \nabla_T - \frac{1}{2} \right) \qquad (4)$$

(cf. Shibahashi & Saio 1985), where $V = -d\ln P/d\ln r$. As seen in eq. (4), a negative ∇_T increases the critical frequency for adiabatic p modes. (Since high-order p modes as seen in roAp stars should be quite nonadiabatic, the exact critical frequency is, however, somewhat uncertain.)

Figure 2 shows the spatial variations of σ_c (which corresponds to ω_c normalized by $\sqrt{3GM/R^3}$) for some examples of the $T-\tau$ relation with and without temperature inversion. Clearly a temperature inversion in the atmosphere can sufficiently increase the critical frequency to suppress atmospheric energy leakage of roAp-type short-period oscillations. Actually, we have confirmed that in a chemically homogeneous model with a temperature inversion, roAp type oscillations as well as δ Sct – type oscillations are overstable. The temperature inversions used for these computations are rather strong and would possibly leave observable chromospheric evidence which was not seen as reported by Shore et al. (1987). We notice, however, that if the atmospheric temperature inversion is produced only by a *strong* magnetic field in A-type stars, we do expect no temperature inversion and hence no short period oscillations in Am or δ Sct stars. Therefore, a temperature inversion is, after all, a rather elegant concept to increase the cut-off frequency, so that we currently follow up this avenue also with weaker ones.

Assuming slight gravitational settling of helium in the outermost part of the envelope we found δ Sct – type pulsations to be indeed suppressed. One example is shown in Fig. 3, where the imaginary part (appropriately scaled: $\Sigma_I = \text{sign}(\sigma_I) \cdot \log(1 + |\sigma_I|/10^{-4})$) of the eigenfrequency is plotted as a function of period for a population I model with helium depleted by 0.1 (in mass fraction) in the regions with $\log T < 4.7$. The lack of excited long-period oscillations occurs even for a slight decrease of helium because in such an inhomogeneous envelope the hydrogen ionization zone acts as a damping zone. This result suggests that helium need not to be drained completely from the HeII ionization zone to suppress the δ Sct – like oscillations.

To obtain a definite answer on the excitation mechanism of roAp oscillations, the effect of magnetic fields must be included in the stability analysis. The magnetic pres-

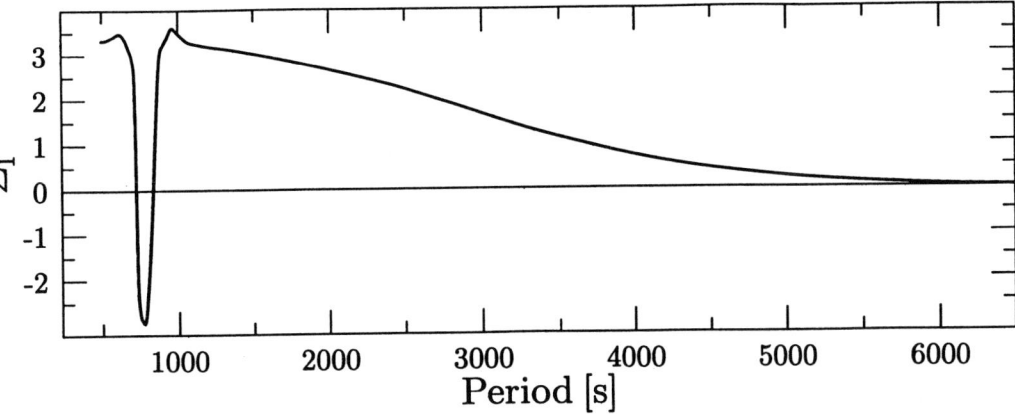

Figure 3. Appropriately scaled imaginary parts Σ_I of the eigenfrequency versus period for a model with a inhomogeneous envelope. A negative Σ_I indicates an overstable the mode.

sure of a kilo-Gauss magnetic field is much higher than the gas pressure only in the outermost zone. Therefore, the magnetic effect on the oscillation frequency is expected to be very small, but the effect on the stability might be considerable. Solving linear pulsation equations including magnetic effects in the adiabatic approximation, Dziembowski & Goode (1996) showed that the damping effect of magnetic fields can reach the magnitude of radiative damping. A self-consistent, fully nonadiabatic analysis is, however, needed. This is very demanding because the magnetic field requires that the latitudinal dependence of an eigenfunction for a given m can no longer be represented by a single associated Legendre function $P_\ell^m(\cos\theta)$. The importance of each ℓ component changes as the strength of magnetic field changes. An additional complexity enters through avoided crossings, i.e. mode interactions, which occur between modes having different latitudinal dependencies.

4. Conclusion

If Ap stars can develop a temperature inversion in their atmospheres, and if a slight helium settling obtains in their envelopes, the problem of exciting rapid oscillations in some of them could be solved. Before a definite answer is obtained in this direction, however, a self-consistent, fully nonadiabatic analysis including the effect of magnetic fields is necessary.

Acknowledgments: A.G. was financially supported by the Swiss National Science Foundation through a PROFIL2 fellowship. H.S. acknowledges financial support from the Swiss National Science Foundation and the Japan Society for the Promotion of Science, which enabled him to stay and work on this project at the Astronomisches Institut der Universität Basel.

References

Cox, J.P. 1974, Rep. Prog. Phys., 37, 563
Cox, J.P. 1984, ApJ, 280, 220
Dolez, N., Gough, D.O., Vauclair, S. 1988, in 'Advances in Helio- and Asteroseismology', IAU Symp. No. 123, eds. J. Christensen-Dalsgaard and S. Frandsen, (Reidel, Dordrecht), p. 291
Dziembowski, W.A., Goode, P.R. 1996, ApJ, 458, 338
Kurtz, D.W. 1978, IBVS, No 1436
Kurtz, D.W. 1982, MNRAS, 200, 807
Kurtz, D.W. 1990, ARA&A, 28, 607
Martinez, P., Kurtz, D.W. 1995, in 'Astrophysical Applications of Stellar Pulsation', eds. R.S., Stobie and P.A. Whitelock, ASP Conference Series, Vol. 83, 58
Matthews, J.M. 1991, PASP, 103, 5
Matthews, J.M., Kurtz, D.W., Wehlau, W. 1987, ApJ, 313, 782
Roberts, P.H., Soward, A.M. 1983, MNRAS, 205, 1171
Shibahashi, H. 1983, ApJ, 125, L5
Shibahashi, H. 1991, in 'Challenges to theories of the structure of moderate-mass stars', eds. Gough, D.O. & Toomre, J., (Springer, Berlin) p. 393
Shibahashi, H. 1993, in 'GONG'94: Helio- and Asteroseismology', eds. R.K. Ulrich, E.J. Rhodes Jr. and W. Däppen, ASP Conference Series, Vol. 76, p. 618
Shibahashi, H., Saio, H. 1985, PASJ, 37, 245
Shore, S.N., Brown, D.N., Sonneborn, G., Gibson, D.M. 1987, A&A, 182, 285
Takata, M., Shibahashi, H. 1995, PASJ, 47, 219
Unno, W., Osaki, Y., Ando, H., Saio, H., Shibahashi, H. 1989, 'Nonradial Oscillations of Stars' (University of Tokyo press)

A SEARCH FOR SOLAR-LIKE OSCILLATIONS IN α Cen A

T.R. BEDDING
School of Physics, University of Sydney 2006, Australia

H. KJELDSEN
Teoretisk Astrofysik Center, Danmarks Grundforskningsfond, Aarhus University, DK-8000 Aarhus C, Denmark

AND

S. FRANDSEN AND T.H. DALL
Institut for Fysik og Astronomi, Aarhus Universitet, DK-8000 Aarhus C, Denmark

Abstract. We have been using a new method to search for solar-like oscillations that involves measuring temperature changes via their effect on the equivalent widths of the Balmer hydrogen lines. We observed α Cen A over six nights in 1995 with the 3.9-metre Anglo-Australian Telescope and the European Southern Observatory's 3.6-metre telescope in Chile. Oscillations were not detected, with an upper limit only slightly higher than the expected signal.

1. Introduction

Many attempts have been made to detect stellar analogues of the solar five-minute oscillations. As with helioseismology, it is hoped that the measurement of oscillation frequencies in other stars will place important constraints on stellar model parameters and provide a strong test of evolutionary theory. However, despite several claims in the literature, it is fair to say that there has been no unambiguous detection of solar-like oscillations in any star except the Sun (see reviews by Brown & Gilliland 1994; Kjeldsen & Bedding 1995; Bedding & Kjeldsen 1998).

We have been using a new method to search for solar-like oscillations that involves measuring temperature changes via their effect on the equivalent widths of the Balmer hydrogen lines. We found strong evidence for solar-like oscillations in the G subgiant η Boo (Kjeldsen et al. 1995; Bedding & Kjeldsen 1995), with frequency splittings that were later found to agree with theoretical models (Christensen-Dalsgaard et al. 1995a,b; Guenther & Demarque 1996). Since then, the improved luminosity estimate for η Boo from Hipparcos measurements has given even better agreement (Bedding et al. 1998). However, a search for velocity oscillations in η Boo by Brown et al. (1997) failed to detect a signal, setting limits at a level below the value expected on the basis of the Kjeldsen et al. result. More recently, Brown et al. (private communication) have obtained a larger set of observations which they are currently processing.

2. Observations of α Cen A

We chose η Boo as the first target for the equivalent-width method because this star was expected to have an oscillation amplitude about five times greater than the Sun. This turned out to be the case (assuming the detection is real). We then turned to α Cen A, a more challenging target because of its smaller expected oscillation amplitude (comparable to solar; Bedding et al. 1996). Being a near twin of the Sun and extremely nearby, this star is an obvious target for detecting oscillations (e.g., Brown et al. 1994). Previous attempts to detect oscillations using Doppler methods were reviewed by Kjeldsen & Bedding (1995) and include two claimed detections at amplitudes 4-6 times greater than solar (Gelly et al. 1986; Pottasch et al. 1992) and two negative results at amplitudes about 2-3 times solar (Brown & Gilliland 1990; Edmonds & Cram 1995).

We observed α Cen A over six nights in April 1995 from two sites:

- at Siding Spring Observatory in Australia, HK and TRB used the 3.9-metre Anglo-Australian Telescope with a coudé echelle spectrograph (UCLES). We recorded three orders centred at Hα and three orders at Hβ. The weather was about 85% clear.
- at La Silla in Chile, SF and THD used the European Southern Observatory's 3.6-metre telescope with a Cassegrain echelle spectrograph (CASPEC). We recorded three orders centred at Hα. The weather was 100% clear.

3. Results and simulations

Data processing of the 20,000 spectra was carried out by HK using the method outlined in Bedding & Kjeldsen (1998). The power spectrum of the resulting time series of equivalent-width measurements is shown in Fig. 1. No obvious excess of power is seen – note that earlier reports of a positive detection were premature (Kjeldsen et al. 1996; Frandsen 1997). The average noise level in the amplitude spectrum (square root of power) is 4.7 ppm, which is somewhat higher than expected purely from photon noise. One extra noise source arises from wavelength-dependent fluctuations in the continuum, which appear to arise from a colour term in the scintillation (Jakeman et al. 1976; Dravins et al. 1997).

The strongest oscillation modes in α Cen A, as measured in Hα equivalent width, are expected to have amplitudes of about 8 ppm, while the solar peak amplitude is about 6 ppm (Bedding et al. 1996).

To set an upper limit on oscillation amplitudes from our observations, we have generated simulated time series consisting of artificial signal plus noise. Each simulated series had exactly the same sampling function and allocation of statistical weights as the real data. The injected signal contained sinusoids at the frequencies calculated by Edmonds et al. (1992), modulated by a broad solar-like envelope centred at 2.3 mHz (which is the expected frequency of maximum mode power – Kjeldsen & Bedding 1995). In each simulation, the phases of the oscillation modes were chosen at random and the amplitudes were randomized about their average values. All these character-

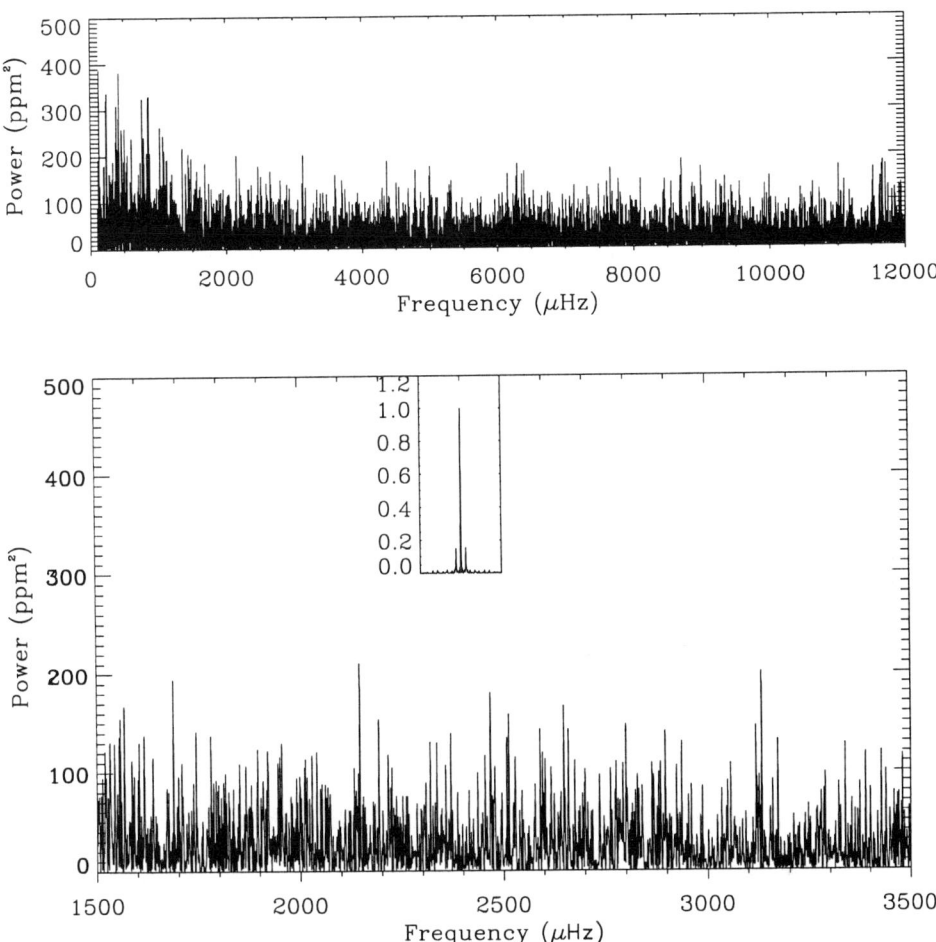

Figure 1. Power spectrum of equivalent-width observations of α Cen A. The lower figure is a close-up of the region where signal would be expected, and the inset shows the power spectrum of the window function.

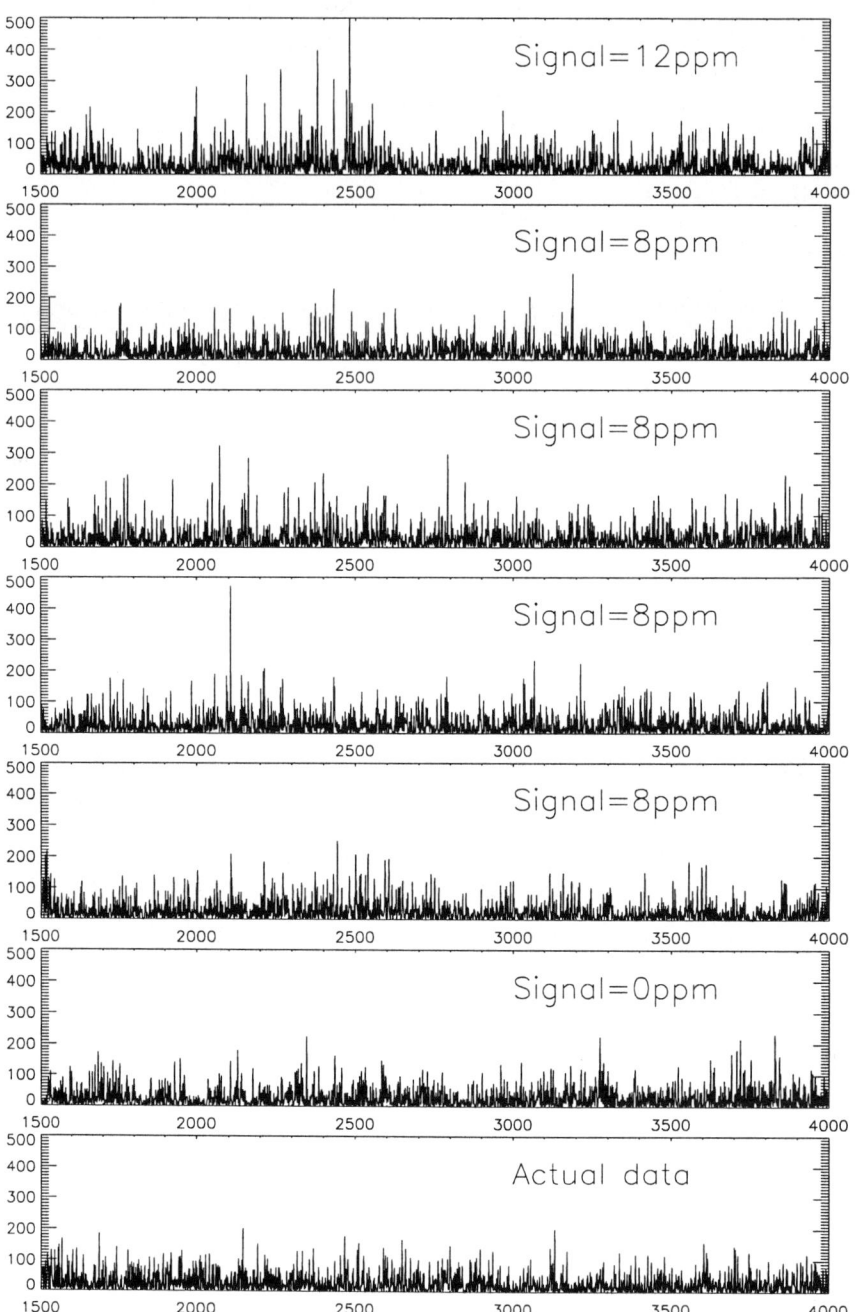

Figure 2. Simulated power spectra, using the same sampling times and data weights as the actual observations.

istics were chosen to imitate as closely as possible the stochastic nature of oscillations in the Sun. Before calculating the power spectrum, we added normally-distributed noise to the time series, so as to produce a noise level in the amplitude spectrum of 4.7 ppm (consistent with the actual data).

Some results are shown in Fig. 2. The top panel shows that a signal with an amplitude of 12 ppm would be easily detectable in our data. It is interesting to note that some of the signal peaks in this simulated power spectrum have been strengthened significantly by constructive interference with noise peaks. For example, a signal peak of 12 ppm which happens to be in phase with a 2-σ noise peak (2×4.7 ppm) will produce a peak in power of 450 ppm^2. This illustrates the point made by Kjeldsen & Bedding (1995; Appendix A.2): the effects of noise must be taken into account when estimating the amplitude of a signal.

The next four panels in Fig. 2 show simulations in which the strongest modes had amplitudes of 8 ppm, the value expected for α Cen A. In some cases, excess power is seen and there is perhaps some hint of regularly spaced peaks. However, the actual data (bottom panel) are clearly consistent with an 8 ppm signal and we conclude that the observations did not have sufficient sensitivity to detect a signal of this strength. We can probably set an upper limit of about 10 ppm.

In summary, our results rule out oscillations at a level slightly less than twice solar, making α Cen A the most stable known extra-solar star. The corresponding upper limit in velocity is about 50 cm/s.

ACKNOWLEDGEMENTS

The observations would have been impossible without the excellent support we received from staff at both observatories. We are especially grateful to Roy Antaw, Bob Dean, Sean Ryan, John Stevenson and Gordon Shafer at the AAO and to Luca Pasquini, Peter Sinclaire and Nicolas Haddad at ESO. We also thank both committees (ATAC and OPC) for allocating telescope time and the AAO Director for granting the sixth AAT night. This work was supported financially by the Australian Research Council and by the Danish National Research Foundation through its establishment of the Theoretical Astrophysics Center.

References

Bedding, T. R., Kjeldsen, H., 1995. In: Stobie, R. S., Whitelock, P. A. (eds.), *IAU Colloquium 155: Astrophysical Applications of Stellar Pulsation*, A.S.P. Conf. Ser., Vol. 83, Utah: Brigham Young, p. 109

Bedding, T. R., Kjeldsen, H., 1998. In: Donahue, R. A., Bookbinder, J. A. (eds.), *Proc. Tenth Cambridge Workshop on Cool Stars, Stellar Systems and the Sun*, A.S.P. Conf. Ser., , San Francisco: ASP (in press)

Bedding, T. R., Kjeldsen, H., Reetz, J., Barbuy, B., 1996, MNRAS 280, 1155

Bedding, T. R., Kjeldsen, H., Christensen-Dalsgaard, J., 1998. In: Donahue, R. A., Bookbinder, J. A. (eds.), *Poster Proc. Tenth Cambridge Workshop on Cool Stars, Stellar Systems and the Sun*, San Francisco: ASP (in press)

Brown, T. M., Gilliland, R. L., 1990, ApJ 350, 839

Brown, T. M., Gilliland, R. L., 1994, ARA&A 33, 37

Brown, T. M., Christensen-Dalsgaard, J., Weibel-Mihalas, B., Gilliland, R. L., 1994, ApJ 427, 1013

Brown, T. M., Kennelly, E. J., Korzennik, S. G., et al., 1997, ApJ 475, 322
Christensen-Dalsgaard, J., Bedding, T. R., Houdek, G., et al., 1995a. In: Stobie, R. S., Whitelock, P. A. (eds.), *IAU Colloquium 155: Astrophysical Applications of Stellar Pulsation,* A.S.P. Conf. Ser., Vol. 83, Utah: Brigham Young, p. 447
Christensen-Dalsgaard, J., Bedding, T. R., Kjeldsen, H., 1995b, ApJ 443, L29
Dravins, D., Lindegren, L., Mezey, E., Young, A. T., 1997, PASP 109, 725
Edmonds, P. D., Cram, L. E., 1995, MNRAS 276, 1295
Edmonds, P. D., Cram, L. E., Demarque, P., Guenther, D. B., Pinsonneault, M., 1992, ApJ 394, 313
Frandsen, S., 1997. In: Provost, J., Schmider, F.-X. (eds.), *Proc. IAU Symp. 181, Sounding Solar and Stellar Interiors,* Dordrecht: Kluwer (in press)
Gelly, G., Grec, G., Fossat, E., 1986, A&A 164, 383
Guenther, D. B., Demarque, P., 1996, ApJ 456, 798
Jakeman, E., Pike, E. R., Pusey, P. N., 1976, Nat 263, 215
Kjeldsen, H., Bedding, T. R., 1995, A&A 293, 87
Kjeldsen, H., Bedding, T. R., Viskum, M., Frandsen, S., 1995, AJ 109, 1313
Kjeldsen, H., Frandsen, S., Bedding, T. R., Dall, T. H., Christensen-Dalsgaard, J., 1996, SONGNews 1 (http://www.noao.edu/noao/song)
Pottasch, E. M., Butcher, H. R., van Hoesel, F. H. J., 1992, A&A 264, 138

RESULTS FROM THE HIPPARCOS MISSION ON STELLAR SEISMOLOGY

L. EYER AND M. GRENON
Geneva Observatory, CH-1290 Sauverny, Switzerland

1. Introduction

Photometric results from the HIPPARCOS mission are of interest to stellar seismology. HIPPARCOS provides a general description of the HR diagram in terms of stability, microvariability and variability. The satellite has detected systematically variables with amplitudes exceeding a threshold, strongly magnitude dependent but as small as a few mmag for bright stars. When the signal is periodic, periods and amplitudes are determined. Because of the peculiar time sampling of HIPPARCOS only first modes and associated amplitudes are generally obtained. This information is particularly relevant to seismology in the case of pulsating stars as δ Scuti stars, slowly pulsating B stars and β Cephei stars.

2. The Main Mission HIPPARCOS

The main mission produced more than 13 million photometric measurements in the Hp wide band for 118 204 stars. That is an average of 110 measurements per star over a time span of 3.3 years. The observed stars were issued from a selection, which consisted of a survey of around 53 000 stars complete to the magnitude $Hp < 7.9 + 1.1\sin(b)$ for stars earlier than G5 and $Hp < 7.5 + 1.1\sin(b)$ for stars later than G5, where b is the galactic latitude, and a set of 65 000 fainter stars selected for their astrophysical or astrometrical interest.

2.1. THE TIME SAMPLING AND THE PHOTOMETRIC PRECISION

The two ingredients, time sampling and precision, are critical in order to know what kind of variability can be detected. With HIPPARCOS, these properties are rather heterogeneous. The basic time sampling can be seen through the example in Fig. 1. The smallest intervals occur in the successive sequence: 20-108-20-etc... minutes. The compact groups of such transits are often separated by around one month. The detection limit for period is mainly 40 minutes, however lower periods can be detected in principle (cf. Eyer and Bartholdi 1998). The photometric precision depends strongly on the magnitude. The data are heteroscedastic (individual transits have different precisions). We give in Table 1 the *mean* behaviour of the precision versus the magnitude Hp (cf. Eyer and Grenon 1997).

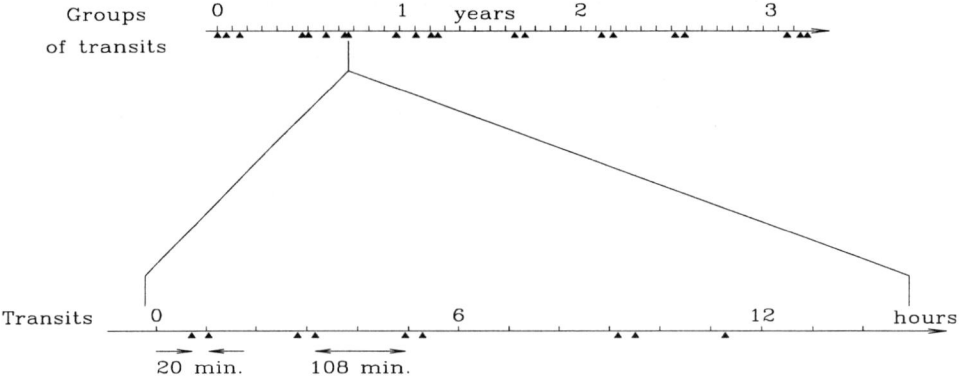

Figure 1. A typical time sampling (example HIP 059862).

TABLE 1. Precision on Hp magnitude of individual transits as function of Hp

Hp	3	5	7	8	9	10	11	12
Precision (mmag)	3	5	7	12	17	25	36	55

2.2. THE RESULTS

A general analysis, done in a collaboration between the Geneva Observatory and the Royal Greenwich Observatory, resulted in the publication of two Catalogues about variables stars and three associated Atlases (cf. van Leeuwen et al. 1997 and Grenon et al. 1997). The catalogue of Periodic Variables contains information about period, epoch, extrema and their associated precisions for 2712 stars. The Catalogue of Unsolved variables gives information about the amplitude of variability for 5442 stars. No periodic solutions were found for these stars. 4086 variables were new discoveries and 1955 stars received names by Samus (1998).

3. Focus on seismology

3.1. NEW VARIABLES

The HIPPARCOS all-sky monitoring was free from biases in favour of particular spectral types, or time sampling as from the Earth; it allowed a systematic detection of variables and hence many new discoveries.

In Table 2 we display, for few variable types of seismological interest, the total number of stars known before HIPPARCOS, and the number discovered by HIPPARCOS with an amplitude and a period.

A study of the β Cephei stars and SPBs was done by Waelkens et al. (1997), who multiplied nearly by 10 the number of known SPBs. A thorough discussion of these stars is presented by Aerts et al. (these proceedings).

TABLE 2. Low amplitude pulsators observed by HIPPARCOS

Variable type	Known (suspected)	New
β Cep	59 (80)	4
SPB	11	72 − 103
δ Scuti	271 (19)	35
γ Dor	11 (17)	∼ 20
roAp	29	0

For cooler stars, especially γ Dor stars and δ Scuti stars, a similar study will be done in order to characterize their pulsation modes and evolutionary status. A rough estimate of the number of new suspected γ Dor stars is around 20 according to Eyer (1998).

In Fig. 2, we put together all HIPPARCOS stars with $\sigma_{Hp} < 0.2$ mag (dots), and data from literature (open circles or crosses; squares and triangles for γ Dor stars and suspected γ Dor stars, circles for δ Scuti stars, and crosses for roAp). Full circles are the data from HIPPARCOS for different period intervals (different grey levels). There is a clump around $M_v \simeq 2.8$ and $B - V \simeq 0.3$ of unevolved main sequence stars with higher periods intermediate between those of δ Scuti stars and γ Dor stars (0.25 < P < 0.5). These new variables are at the red edge of the instability strip and of the Böhm-Vitense gap. On main sequence, the blue edge of the instability strip does not extend bluewards of (B-V) = 0.14.

Rapidly oscillating Ap stars are too rapidly pulsating and have too small amplitudes to be detected by HIPPARCOS. In addition, their periods are not necessarily stable over 3 years.

3.2. DETECTION OF MULTIPLE PERIODS

From simulations (using sinusoidal signals and the CLEAN method), we learn that several periods may be often recovered from the data. However, the possibility to find several periods depends not only on the relative intensities of the amplitudes and of the noise level but also on the value and separation of the periods (for a given star). Furthermore, it depends on the peculiar sampling of each star. Therefore it is, according to our experience, risky to deduce multiperiodic behaviour from the HIPPARCOS data alone, because stars like δ Scuti stars are not always stable over a period of 3 years (example GX Peg), and that the spectrum can lead to wrong conclusions because of the aliasing problem. Period confirmation and a-posteriori studies can be, however, fruitful (e.g. the two main periods in 1 Mon can be recovered). This can lead to the conclusion of the *long term stability* of the pulsation characteristics.

4. Short time scale variations and further studies

Using an estimator of the scatter of the pairwise differences in different but short time intervals (cf. Eyer and Genton 1998), we detected very short-term variables (or

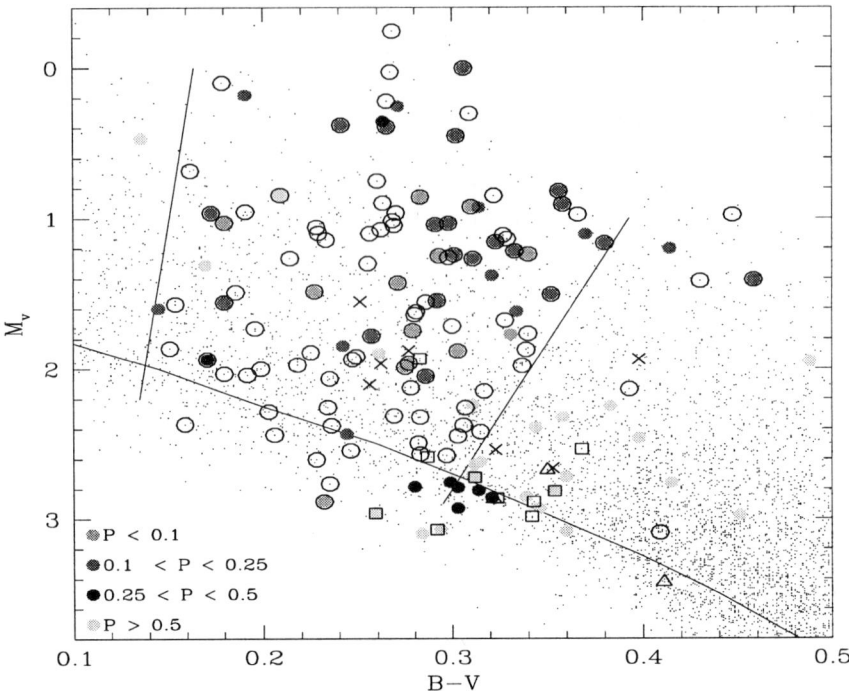

Figure 2. HR-diagram for stars near the instability strip. Filled symbols : HIPPARCOS variables.

light curves with locally large derivatives). We found three variables, HIP 071119, HIP 044025, which are δ Scuti stars and HIP 029055 which is the large amplitude and known RR Lyrae VW Dor; these stars are not in the catalogue of periodic variables. Thus, the catalogue of unsolved variables will reveal many more interesting cases as expected.

References

Eyer, L., and Grenon, M. (1997) Photometric Variability in the HR-diagram, *ESA SP-420*
Eyer, L., and Genton, M. (1998) in preparation
Eyer, L., and Bartholdi, M. (1998), A&A, in preparation
Eyer, L. (1998) Hipparcos et les étoiles variables, PhD Thesis, Observatoire de Genève
Garcia, J.R. et al. (1995) *A&AS*, 109, 201
Grenon, M. et al. (1997) *ESA SP-1200*, Volume 12
Krisciunas, K. and Handler, G. (1995) *IBVS* 4195
van Leeuwen, F. et al. (1997) *ESA SP-1200*, Volume 11
Samus, N. et al. (1998) *IBVS*, in preparation
Sterken, C. and Jerzykiewicz, M. (1993) *Space Science Reviews*, 62, 95
Waelkens, C. et al. (1997), A&A submitted

SLOWLY PULSATING B STARS : NEW INSIGHTS FROM HIPPARCOS

C. AERTS, P. DE CAT AND C. WAELKENS
Instituut voor Sterrenkunde, Katholieke Universiteit Leuven, Celestijnenlaan 200 B, B-3001 Leuven, Belgium

Abstract. The photometric experiment on board of Hipparcos has discovered, among other types of variables, a large amount of new Slowly Pulsating B Stars. We have selected the fourteen brightest stars of this sample, together with five previously known Slowly Pulsating B Stars, for long-term spectroscopic and photometric monitoring. The selected stars have spectral types ranging from B 2 up to B 9 and are thus nicely spread across the instability strip. We here present the results of a preliminary analysis of our data and point out that our sample is unique in the sense that it allows us to perform seismology of massive early-type stars.

1. Selection of the program stars

Of the 267 new B-type variables discovered by Hipparcos, some 100 turn out to be Slowly Pulsating B Stars (hereafter called SPBs). Waelkens et al. (1997) have classified all these new variables and determined the position of the SPBs in the HR diagram. They find that the new SPBs fully cover the instability strip calculated by Pamyatnykh (1997).

SPBs are the most interesting massive stars for seismological purposes, since they pulsate in many high-order g-modes which penetrate deep into the stellar interior and for which asymptotic pulsation theory applies. In this respect, they can be viewed as intermediate-mass main-sequence analogues of the white dwarfs, for which seismological studies have been very successful in the recent past. Besides this, all the newly discovered members of the group deserve to be studied for the simple reason that we have for the first time a large, unbiased sample of SPBs regarding spectral type and periodicity.

From an observational point of view, a large disadvantage of SPBs is that long-term monitoring is necessary in order to obtain meaningful results. The beat-periods in these stars are of the order of months/years and have to be covered to disentangle the complete frequency spectrum and to perform mode identifcations. An advantage, however, is that many of the newly discovered SPBs are sufficiently bright to study

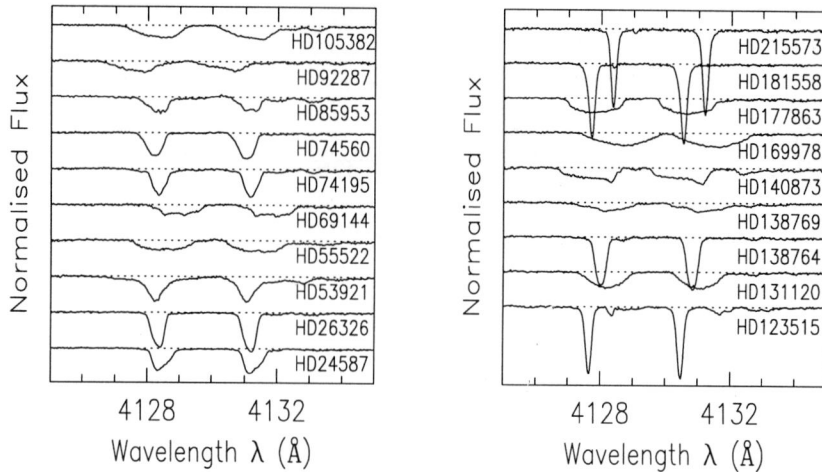

Figure 1. One arbitrarily chosen spectrum of the Si II doublet centered around 4130Å for each of the 19 target SPBs. The rotational velocities derived from our data range from 10 to 100 km/s

their variations by means of line profile studies. Moreover, accurate photometry can be performed with small telescopes.

We have selected the 14 brightest southern Hipparcos SPBs, together with 5 bright ones previously found by Waelkens (1991), for long-term photometric and spectroscopic monitoring. These stars have spectral types ranging from B 2 up to B 9 and are thus well spread across the instability domain to have a general overview of the pulsations in the complete temperature range of SPBs. The photometric data are gathered with the Swiss Telescope of the Geneva Observatory and the high-resolution line profiles are taken with the CAT telescope of the European Southern Observatory, both situated at La Silla, Chile. We started our monitoring in 1996 and have been allotted telescope time with both instruments for this year and in 1998.

2. Preliminary analysis of our data after 6 months of monitoring

We did not obtain photometric data of the SPBs in the course of 1996 and therefore focus on the spectroscopy here. Line-profile variations were observed in March, April and July 1996, each time during 7 nights. We have obtained between 10 and 70 high-resolution, high S/N spectra of the Si II triplet centered around 4130Å for 16 of the targets (for 3 stars, we only obtained a few spectra during the first year). Our spectroscopic data gathered in the course of 1996 have revealed the existence of line-profile variability with the expected time scales in the 5 previously known SPBs and in all 14 candidate SPBs that were discovered by Hipparcos, confirming the pulsational nature of the latter.

A typical spectrum for each of the targets is shown in Fig. 1. HD 74195, HD 74560, HD 123515, HD 177863, and HD 181558 were discovered by means of ground-based photometry by Waelkens (1991), while the others are found with Hipparcos. HD 105382 and HD 138769 are listed in the BSC as respectively Be and shell star, but were clas-

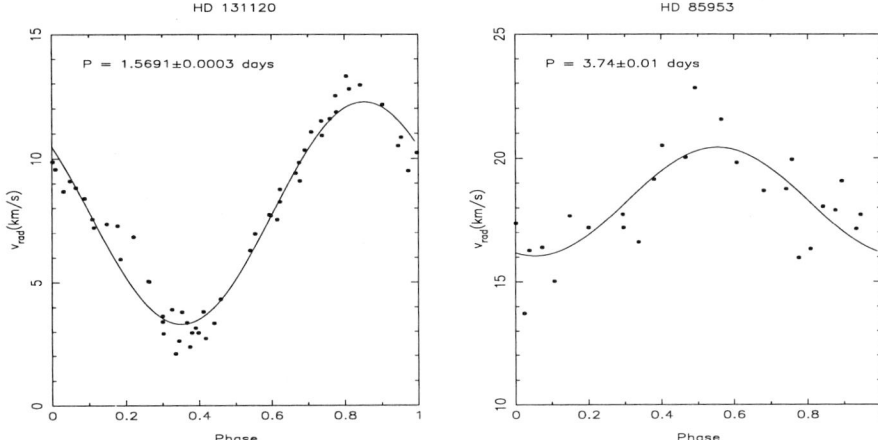

Figure 2. Radial velocities derived from our Si II spectra for the two SPBs that are situated in the common part of the β Cep and SPB instability strip (see Waelkens et al. 1997). The dots are the observations while the full line is a fit for the period indicated in the graph. In both cases, this period agrees with the one found from the Hipparcos photometry

sified as SPB by Waelkens et al. (1997). We observed Hα of the two stars and found variable emission in the case of HD 105382, but absorption for HD 138769. Their Si II line-profile variations are similar to those of the other SPBs. We included them in our sample because it is important to find a link between variable Be stars and confirmed pulsators.

Interestingly, many of the Hipparcos stars that were observed tend to be broader-lined than the previously known SPBs, and these stars present more subfeatures in their line profiles. This probably means that rotation is accompanied by higher-degree modes, which are hardly detected in the photometry. We derive rotational velocities between 10 and 100 km/s from our data, but these may very well be overestimates since we are not yet able to estimate the total pulsational broadening. For the fastest rotators this nevertheless means that the rotation periods are of the same order of magnitude as the pulsation periods, implying that the theoretical framework to study these pulsations must be based on a rotating star, i.e. cannot rely on velocity expressions in terms of one spherical harmonic.

Another result is that the equivalent width variations are large (about 15%) and point towards important temperature variations during the pulsation cycle. Theoretical models that take into account temperature variations in line-profile variations are up to now based on adiabatic pulsation theory. Our data shows that there is a great need to generalise these models to non-adiabatic pulsations. We are currently working on such a generalisation.

As an example, we show in Fig. 2 a phase diagram of the radial-velocity variations for the two SPBs discovered by Hipparcos that are situated in the common part of the instability domain of the β Cep stars and the SPBs (see Waelkens et al. 1997). The main frequency found in our data is the same as the one found in the Hipparcos photometry and confirms their SPB nature. HD 131120 clearly has one dominant mode, while HD 85953 probably is multiperiodic. It is not clear to us yet whether or not these stars also exhibit β Cep-like pulsations (i.e. p modes).

A frequency analysis of the radial-velocity variations shows that for ten of the fourteen Hipparcos SPBs the dominant frequency coincides with the frequency derived from the Hipparcos photometry or multiples of this frequency. The latter case appears when the target turns out to be a binary. HD 123515 was already known to be a binary. Our spectra show that four of the SPBs found by Hipparcos and one previously known SPB also belong to multiple systems and have orbital periods of the order of days. All the binaries with short orbital periods turn out to have large rotational velocities and complicated line-profile variations. This suggests that the binarity results in a particular spectrum of excited modes. For the determination of the orbital parameters of some of the binary SPBs, we refer to Aerts et al. (1997).

3. Future prospects

At present, we are still continuing the gathering of photometric and spectroscopic data. We expect to have a total time base of at least two years for the spectroscopy and of one year for the photometry. In total, we will have covered 21 weeks with photometry and some 10 weeks with spectra. This should allow us to derive the periods and the character of at least the dominant modes. We plan to identify the modes with both the moment method (Aerts 1996), which is an accurate identification technique based on line-profile variations in the case of slow to moderate rotators, and with the method of photometric amplitudes (Heynderickx et al. 1994). Other follow-up observations can then be planned according to the first results.

Our final aim is to disentangle the frequency spectrum of the SPBs and to identify the modes. If we succeed in doing so, important issues such as the determination of accurate masses, the extent of convective overshoot, and the internal rotation law can be determined with high precision. Hipparcos provides us with an opportunity to select a sample of intermediate-mass early-B type stars for which we can obtain such results. In this respect, the discovery of the large amount of SPBs by Hipparcos is an important step towards the understanding of the internal structure of massive stars.

Acknowledgements

CA acknowledges a postdoctoral fellowship of the Fund for Scientific Research of Flanders. A special thank is addressed to Michel Grenon and to Laurent Eyer for involving us in the classification of the variable B-type stars found with Hipparcos and for fruitful discussions. We are grateful to the Observatoire de Genève and to the European Southern Observatory for the generous awarding of telescope time to make this long-term project possible.

References

Aerts, C., 1996, A&A 314, 115
Aerts, C., De Mey, K., De Cat, P., Waelkens, C., 1997, In *A Half Century of Stellar Pulsation Interpretations*, PASP Conference Series, in press
Heynderickx, D., Waelkens, C., Smeyers, P., 1994, A&AS 105, 447
Pamyatnykh, A.A., 1997, In *A Half Century of Stellar Pulsation Interpretations*, PASP Conference Series, in press
Waelkens, C., 1991, A&A 246, 468
Waelkens, C., Aerts, C., Kestens, E., Grenon, M., Eyer, L., 1997, A&A, submitted

THE ACOUSTIC CUT-OFF FREQUENCY OF ROAP STARS

N. AUDARD

Institute of Astronomy, Cambridge, England

F. KUPKA AND W. W. WEISS

Institute for Astronomy, Vienna, Austria

AND

P. MOREL AND J. PROVOST

Observatoire de la Côte d'Azur, Nice, France

For 5 out of 28 known rapidly oscillating magnetic chemically peculiar (roAp) stars, the largest observed frequency seems to exceed the theoretical acoustic cut-off frequency, which is determined by the outermost stellar regions. We show that a better modelling of the atmosphere reconciles the theory with the observations for at least the roAp star α Cir.

We have compared models and adiabatic frequencies for pulsating Ap stars with $T(\tau)$ laws based on Hopf's purely radiative relation and on Kurucz model atmospheres. For α Cir we find models with Kurucz atmospheres which have indeed a cut-off frequency beyond the largest observed frequency. For HD 24712 only models which are hotter by about 100 K and less luminous by nearly 10% than what is actually the most probable value would have an acoustic cut-off frequency large enough.

One may thus speculate that the old controversy about a mismatch between observed largest frequencies and theoretical cut-off frequencies of roAp star models is resolved.

More details can be found in Audard et al. (1997)

1. Stellar models and cut-off frequencies

We have computed representative models for CP2 stars with the CESAM code (Morel 1997). Effects from a magnetic field are neglected in our stellar models.

As the displacement for low-degree modes is essentially vertical, we shall consider only radial modes and, because we investigate modes of high radial orders, we adopt the Cowling approximation to calculate the the cut-off frequency. We have computed the acoustic cut-off frequency according to the 3 following formulations: Vorontsov & Zarkhov (1989), Gough (1986), and in the approximation of an isothermal atmosphere.

2. Results

The main properties of the roAp star HD 24712, based, among others, on a HIPPARCOS parallax, can be reproduced with models of $1.63\,M_\odot$, $Z = 0.02$ and an age of about 900 Myr. We have also computed Hopf and Kurucz models with appropriate age for HD 128898 (α Cir). For the first time, a Kurucz model atmosphere was calculated with an opacity distribution function specific to the composition of α Cir (Piskunov & Kupka 1997). Stellar models with $1.93\,M_\odot$, $Z = 0.03$ and an age of 400 Myr, fit the observed values. For α Cir (HD 128898), models well within the error box can be found with a cut-off frequency which matches the observations. A similar situation exists for HD 134214 which, unfortunately, has a considerably larger error box. However, for HD 24712 only models in the lower left corner of the error box fit.

Due to uncertainties in the observations and the theory, models of different mass and age can fit the same star in the H-R diagram within the error box. Evolutionary tracks and lines of constant cut-off frequency for Kurucz models are plotted in Fig. 1.

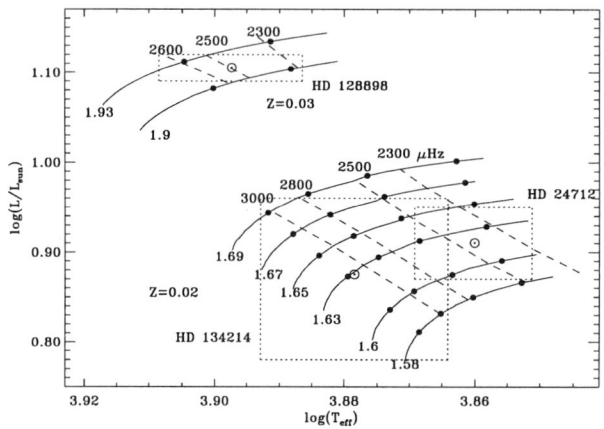

Figure 1. HR diagramme for stars with $1.58\,M_\odot$ to $1.69\,M_\odot$ ($Z = 0.02$), and with 1.90 and $1.93\,M_\odot$ ($Z = 0.03$). Dots indicate ages of 400, 600, 800 and 1000 Myr for $Z = 0.02$, and 400 and 600 Myr for $Z = 0.03$. The roAp stars are indicated by circles and error boxes. Dashed lines of constant cut-off frequency in μHz are drawn for Kurucz models.

References

Audard, N., Kupka. F., Morel.P., Provost. J., Weiss. W.W. (1997) Acoustic cut-off frequency of roAp stars, *A&A*, accepted
Gough, D.O. 1986, in *Hydrodynamic and magnetohydrodynamic problems in the Sun and stars*, Ed. Y. Osaki, University of Tokyo press, p. 117
Morel, P. 1997, CESAM: a code for stellar evolution calculations, *A&AS*, in press
Piskunov, N.E., Kupka, F. 1997, *A&A*, in preparation
Shibahashi, H. 1991, in *Challenges to theories of the structure of moderate-mass stars*, Eds. D. Gough & J. Toomre, Lecture Notes in Physics No. 388, Springer-Verlag, p. 393
Unno, W., Osaki, Y., Ando, H., Saio, H., Shibahashi, H. 1989, *Nonradial Oscillations of Stars*, University of Tokyo press, 2nd edition
Vorontsov, S.V., Zharkov, V.N., 1989, *Astro. Sp. Phy. Rev.*, **vol, 7**, part 1

ASTEROSEISMOLOGY WITH THE SPACE MISSION COROT

A.BAGLIN AND THE COROT TEAM
*DESPA, Observatoire de Paris, URA CNRS 264,
5 place J. Janssen, Meudon CEDEX, France.*

The major participants to the COROT Seismology Team are C. Catala, M. Auvergne (Instrument scientist), A. Vuillemin, T. Appourchaux, E. Michel, M.J. Goupil, P. Boumier, T. Hua, B. Foing, W. W. Weiss, B. Mosser, F. Bonneau and A. Magnan (Project managers), R. Bellenger, M. Decaudin, G. Epstein, P. Levacher...

1. Introduction

Seismology from the ground has already provided many important results, as discussed for instance in Frandsen (1997) and in this symposium. The limitations of this approach are essentially the day-night alternance and the meteorological instabilities for photometry, the lack of photons and the broadening of spectral lines due to rotation for spectroscopy. Seismology from space is less limited, as photometry at a very high accuracy on long uninterrupted observing runs is possible.

Several projects, i.e. (Hudson et al., 1986), (Praderie et al., 1988), (Badialdi et al. , 1996), (Baglin and Auvergne, 1997), have been proposed since the first workshop dedicated to the subject (Mangeney and Praderie, 1984). EVRIS was the first one to be selected (Baglin, 1993), but unfortunately it has been lost in the crash of the MARS 96 spacecraft.

For the moment the only project which is decided and funded is COROT (for Convection and ROTation), in the framework of the "Petites Missions" program of CNES, the French Space Agency. The launch is scheduled in 2002. Several European countries contribute: Spain, Austria and SSD/ESTEC/ESA.

COROT has two independent scientific objectives, both requiring very long uninterrupted observations of the same stars with a very high photometric accuracy. Stellar seismology is the original and primary objective, and defines the characteristics of the instrument. The secondary objective is the search for transits of extraterrestrial planets over the disk of their parent star.

2. Basic scientific specifications for a photometer dedicated to asteroseismology in space

The basic principle relies on the ability to detect a periodic oscillation in a noisy signal and to measure its principal parameters (frequency, amplitude, lifetime) with a certain accuracy.

2.1. DETECTION THRESHOLD

The photometer "counts" events (in fact electrons) directly related to the photon flux emitted by the object, polluted by the different sources of noise, including the unavoidable photon statistics.

To make things simple, let us assume that the noise is white, with a variance σ. The periodic signal has an amplitude a, a life-time τ and is observed during a period T; the mean value of the number of counts per second is N_e. Defining $t = inf(T, \tau)$, the signal to noise ratio $\frac{S}{N}$, in the power spectrum is expressed as

$$\frac{S}{N} = \frac{N_e^2 t a^2}{4\sigma^2} \qquad (1)$$

The challenge is to maximize $\frac{S}{N}$, which in any case is limited by the unavoidable photon noise. So, it is specified that the instrument be photon noise limited, in a given frequency interval, expressed as $\sigma^2 \leq \alpha \sigma_p^2$, where α is larger than, but very close to 1; $\alpha \sim 1.1$ is considered as the best realistic limit. The variance of the photon noise is $\sigma_p^2 = N_e$.

We also know very little about τ. For stochastically excited oscillations, τ is known only for the Sun, but should remain of the order of several days. The duration of the observations will always be longer than that, so that $t = \tau$, i.e. t is determined only by the star's behavior.

Then, in any seismologic photometer, we are left with the relation:

$$\frac{S}{N} = \frac{N_e \tau a^2}{4\alpha} \qquad (2)$$

which relates a and N_e, at a given $\frac{S}{N}$, where

$$N_e = sd \int F(\lambda) \eta(\lambda) q(\lambda) \frac{d\lambda}{hc}, \qquad (3)$$

s is the surface of the photon collector in cm^2;
d is the duty cycle, supposed to be constant in time;
$F(\lambda)$ is the photon flux received at the earth, outside the atmosphere, from a star of apparent bolometric magnitude m_b, as a function of the wavelength λ. It scales as $10^{-0.4(m_b-6)}$ and depends on the atmospheric parameters of the target, essentially its effective temperature;
$\eta(\lambda)$ is the transmission factor of the optical system;
$q(\lambda)$ is the total efficiency of the detector as a function of wavelength.
η and q have to be optimised, but do not vary much ($\eta q \sim 0.45$).

Thus, the minimum detectable amplitude a of an oscillation is related to its life-time, to the size of the collector and the magnitude of the target, through the relation

$$\left(\frac{a}{7\ 10^{-7}}\right)^2 = C \left(\frac{5}{\tau}\right) \left(\frac{0.9}{d}\right) \left(\frac{572}{s}\right) 10^{0.4(m_b-6)}, \qquad (4)$$

where τ is expressed in days, and s in cm^2; $\eta q = 0.45$, and C is a constant slowly varying with the effective temperature of the target (C=1 for $T_{eff} = 6000\ K$).

For a detection at a confidence level of 99% in a white noise, one needs $\frac{S}{N} \sim 4$. As was shown e.g. in Baglin (1989), it is already possible to reach the level of the solar oscillations on a few very bright stars with an entrance pupil smaller than 10 cm diameter. If the size of the telescope increases, the number of objects accessible at the same confidence level increases, and for the same objects the detection threshold is lowered and the confidence level of a given mode is increased. The solar oscillations ($a \sim 2.5\ ppm$) will be detected with a 27 cm collector at a confidence level of 99% for a 6th magnitude star. We will see later that this is the goal of COROT.

2.2. FREQUENCY MEASUREMENTS

In Fourier space, the frequency resolution $\delta\nu$ is the inverse of the total duration of the observation T. However, the accuracy of the measurement of the frequency of a mode depends on the time properties of its excitation, as shown by Libbrecht (1992), and this accuracy approaches $\delta\nu$ only if the signal to noise ratio is large. The need for precise frequency determinations to interpret the data requires long duration observations and high signal to noise ratios. Approximately 150 days and $\frac{S}{N} \sim 4$ are needed to reach an accuracy of $0.1 \mu Hz$.

As the detection threshold is approached with an observing run covering the lifetime of the mode, increasing the length of the run improves essentially the accuracy of the frequency determination. If, as seen in the Sun, a given mode is not always present, it will also increase the probability to detect more modes.

3. The COROT program of asteroseismology

3.1. EXPLORATORY PROGRAM

As COROT is the first mission dedicated to asteroseismology, a preliminary program called "exploratory" will determine the domain of stellar parameters for which oscillations are detectable, and the relation between the amplitudes of the solar-like oscillators and their characteristics. Though several theoretical predictions of the amplitudes of oscillations excited stochastically by convective motions exist (Houdek, 1994; Goldreich et al., 1994), a large uncertainty remains, in particular on the treatment of the superadiabatic external layers and on the scaling of the turbulent velocity field.

To do so one has to observe a sample of objects with a variety of stellar parameters, i.e. mass, age, chemical composition, state of rotation...but with moderate signal to noise ratio and frequency resolution: $\delta\nu \sim 0.5\mu Hz$ is sufficient for that purpose, corresponding to observing runs of 10 to 20 days. Stars down to the 9th magnitude are appropriate targets (see Section 4.4), and 5 to 10 will be observable at the same time (see Section 4.5).

Several tens of stars will have to be followed, corresponding at least to 2 to 4 months of observation.

3.2. CENTRAL PROGRAM

More ambitious and more time consuming than the exploratory one, the central program corresponds to the second step in the development of space asteroseismology. It aims at observing very precisely a small set of objects, selected for their diagnostic power. The choice of the targets will be determined by the results of the exploratory phase.

Based on the solar case, we fix the accuracy of the frequency measurement at $\delta\nu \sim 0.1 \mu Hz$ to have access to mode profiles and rotational splitting and to measure precisely the distribution of the mode frequencies. For a 6th magnitude star (see $ 2.1), the detection threshold is below 1 ppm. For A and F stars close to the main sequence, it will then be possible to measure the size of the convective cores (Michel et al., 1994; Roxburgh et al., 1996), the size of the outer convective zones and their helium content (Gough and Kosovichev, 1993) or the rotation profile of δ Scuti stars (Goupil et al., 1996).

A least 5 runs are planned, during which one bright star (the main target) and several fainter ones in the surrounding field of view will be followed.

3.3. STELLAR VARIABILITY SURVEY

The secondary objective of COROT is the search for extraterrestrial planets. To do so, COROT observes during the central program of seismology in a neighboring region of the sky. It will follow of the order of ten thousand objects simultaneously down to magnitude 16, with an exposure time of 15 minutes, producing light curves of more than 50 000 stars, with an accuracy of a few 10^{-4}. The continuous time coverage, as required for seismology, will give access to the precise time behavior from 1 hour to 100 days.

In addition to the search for transits of exoplanets, this wealth of information will be used to study variable phenomena, erratic or periodic, on these time scales. At this level of accuracy, rotation modulation, activity and spots will be detectable, and a statistical study of variability in the HR diagram will be possible. This serendipity program will complement the HIPPARCOS variability results, which had a poorer time coverage, a smaller photometric accuracy, and a limiting magnitude around 12, but a larger sample of objects (Eyer and Grenon, 1997).

4. The instrument

The design of the instrument has to satisfy the four major constraints on the detection threshold, the noise level, the total duration of an observing run and the duty cycle. The constraints on the length of the observing runs and on the duty cycle depend essentially on the satellite and particularly on the orbit.

4.1. THE FRAMEWORK OF THE "PETITES MISSIONS" PROGRAM

The budget of a space program is a determining factor in defining a seismology mission, as it fixes the order of magnitude of the collector. The CNES "Petite Mission" program corresponds to a total cost of 200 MFF, but accepts some foreign complementary participation. It corresponds to a telescope of approximately 25 to 30 cm diameter. The PROTEUS platform, which has to be used for that mission can attain only low earth orbits. Fortunately, with a polar inertial orbit, observing runs up to 150 days are possible, on selected regions of the sky.

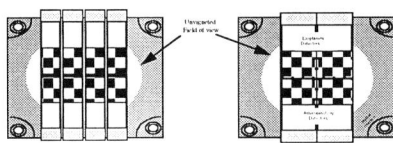

Figure 1. Two possible pavements of the focal plane with frame transfer CCDs; half of the chips are devoted to asteroseimology, the others to the exoplanet program.

4.2. SOURCES OF NOISE

As already mentionned, the condition is that globally, in the frequency domain of scientific interest, the photon noise dominates. This frequency domain, in which oscillations are preferably looked for, has been fixed in the range of $[\,0.1, 10\,]\ mHz$.

There are three main sources of external noise: the detection system, the pointing system and the environment. The major outside perturbators are the stray light from the earth, the high energy particles from the South Atlantic Anomaly and temperature variations forced by eclipses. They have two very different types of temporal behaviors: some are white (i.e. noise related to the detectors), but most of them are modulated at the orbital frequency (approximately $150 \mu Hz$). As we are looking for frequencies, constraints are different when dealing with a white or periodic signal: white noise is directly comparable to the photon noise, and has to remain 10 times smaller. A periodic perturbation will create a signature in the frequency domain which could be taken as a stellar frequency. To avoid such a confusion the perturbation amplitude has to remain below the fixed detection threshold.

As the modulation at the orbital frequency is unavoidable, it is necessary to exclude a frequency interval around the orbital frequency from the scientific domain. In this interval, it will be almost impossible to extract valuable seismologic information, and efforts will be made to keep it as narrow as possible. A suitable correction procedure will help reducing the effect.

4.3. THE DETECTORS

The major constraints on the detectors are the need:

- for an imaging system with a wide field, to be able to observe a reasonnable fraction of the sky and many objects at the same time;
- to maximise the quantum efficiency;
- to reduce all intrinsic noises such as readout noise and dark current;
- to minimize the time between exposures and the readout time;
- to minimize the noise produced by unaccurate pointing;
- to minimize the sensitivity to cosmic rays.

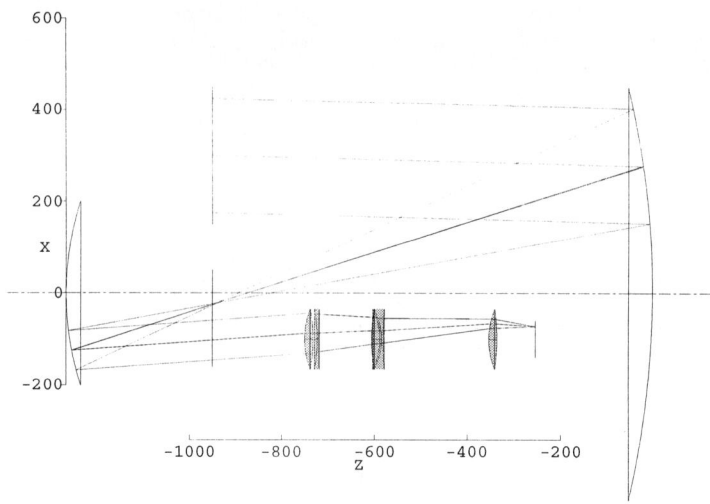

Figure 2. The optical scheme of the afocal telescope, with two parabolic mirrors and a dioptric objective.

These conditions lead us to choose frame transfer MPP (multi pinned phase) and back illuminated chips.

4.4. THE TELESCOPE

The design of the telescope is strongly constrained by the size of the entrance pupil (chosen here to be 27 cm) and by the need to minimize the amount of stray light from the earth entering the telescope. Varying periodically during the orbit, the flux of stray light at an altitude of $\sim$ 800 km is of the order of $10^{17} photons\ cm^{-2}\ s^{-1}$. It has to be strongly reduced by a powerful baffling system. For that reason the preliminary concept (Baglin and Auvergne, 1997) of a classical off-axis, three mirror system has been changed. The best protection is reached with an off-axis afocal parabolic system (Figure 2).

The field of view will be of the order of 4 square degrees. It will contain a bright star (6th magnitude) and several fainter ones down to the 9th magnitude.

Pointing has to be treated with extreme care. The line of sight, during an observing run, has to be stabilised to better than 0.5 arcsecond, to keep the photometric fluctuations due to nonuniformity of the quantum efficiency of the detector below one tenth of the photon noise. This pointing stability will be mainly provided by the PROTEUS platform; if necessary, it will be complemented by an internal fine pointing system.

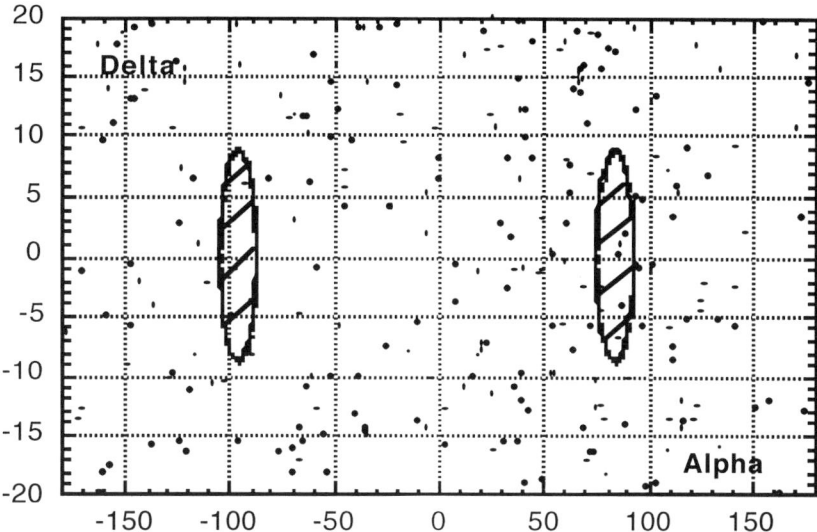

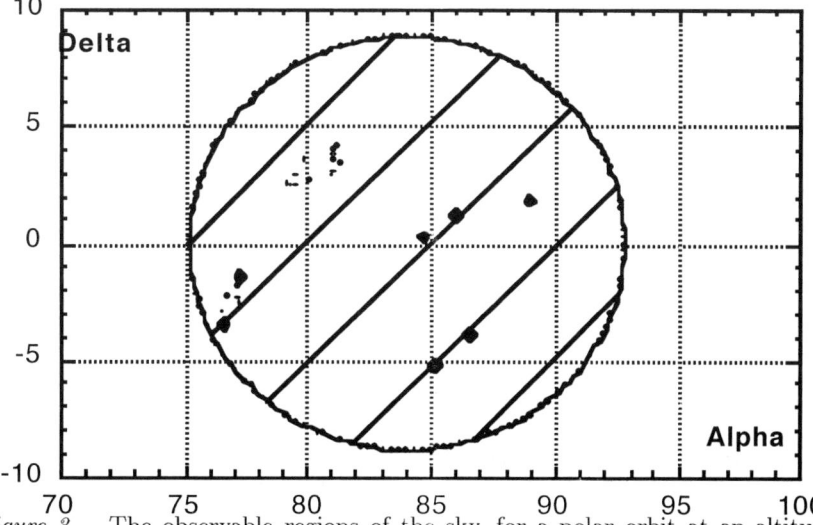

Figure 3. The observable regions of the sky, for a polar orbit at an altitude of 900 km, and a longitude of 0 degres. On the upper panel, dots represent all the candidates for the main targets of the central program. The lower panel represents an enlargment of the observable zone. Dots indicate the best region for the central program, corresponding to a field containing a main target and several fainter stars of interest (see $ 3.2)

5. Mission profile

The only low earth orbits which allow the observation of the same star for several months are polar inertial orbits, the direction of observation being almost perpendicular to the orbital plane, i.e. close to the equatorial plane. To avoid eclipses and straylight from the earth, these directions are limited to small circles, as seen in Figure 3.

For a given position of the orbital plane, fixed by the date of the launch, two conjugated zones are acceptable, 180 degrees apart.

The proposed mission profile starts with 2 to 4 months devoted to the exploratory phase. Then the sequence of 5 to 6 successive long runs of 150 days start, centered on a bright seismology target, during which the exoplanet program also works. At the end of a sequence, the satellite will point in the opposite direction, to start a new long run on a target belonging to the conjugated zone.

The exoplanet program requires observations at intermediate galactic latitude (20 to 30 degrees) to optimise the target density on the sky.

Though all those requirements look quite restrictive, we have shown that they allow the achievement of both the exploratory and the central program.

Complementary ground based observations will help to select the final targets.

References

Badialdi, M., Catala, C., Favata, F., Fridlund, M., Frandsen, S., Gough, D.O., Hoyng, P., Pace, O., Roca-Cortes, T., Roxburgh, I.W., Sterken, C. and Volonté, S., 1996, STARS, Seismic Telescope for Astrophysical Research from Space, Report on the phase A study, ESA report SCI(96), 4.
Baglin, A. (1989) *Solar Physics*, **133**, 155.
Baglin, A., Weiss, W., Bisnovatyi-Kogan, G.S., (1993), in *Inside the Stars* IAU Colloquium **137**, A. Baglin and W.W. Weiss eds., A.S.P. Conf. Ser. **40**,758.
Baglin, A., Auvergne, M. (1996) *in IAU Symposium 181*, in press.
Eyer, L., Grenon, M. (1997), *Hipparcos symposium* Venice, ESA publ. SP-402, p. 467.
Frandsen, S. (1997) *in IAU Symposium 181*, in press.
Goldreich,P., Murray, N., Kumar, P. (1994) *Ap. J.*, **424**, 466.
Goupil,M.J., Dziembovsky, W.A., Goode, P.R., Michel, E. (1996)*A. & A.*, **305**, 498.
Gough, D.O., Kosovichev, A.G. (1993) *Asp. Conf. Ser.*, **42**, 351.
Houdek, G. (1995) *in 4th SOHO Workshop: Helioseismology*, ESA Publ. SP-376.
Hudson, H.S., Brown, T.M., Christensen-Dalsgaard, J., Cox, A.N., Demarque, P., Harvey, J.W., Mc Graw, J.T., Noyes, R.W., 1986, A concept study for an Asteroseimology Explorer, Proposal submitted to NASA.
Libbrecht, K.G., (1992) *Ap. J.*, **387**, 712.
Michel. E., Goupil, M.J., Cassisi, S., Baglin, A., Auvergne M., Buey, T. (1995) *in 4th SOHO Workshop: Helioseismology*, ESA ubl. SP-376, p. 543.
Mangeney, A., Praderie, F., (1984) Proceedings of the workshop on Space Projects in Stellar Activity and Variability, Observatoire de Paris.
Praderie, F., Mangeney, A., Lemaire, P., Puget, P., Bisnovatyi-Kogan, G. (1988), in *Advances in Helio- and Asteroseismology*, J. Christensen-Dalsgaard and S. Franden eds., p.549.
Roxburgh, I.W., (1996) *STARS*, ESA D/SCI(96)4.

BISECTOR VELOCITIES OF Hα IN THE roAp STAR α CIR

Probing the Principal Pulsation Mode

I. K. BALDRY AND T. R. BEDDING
Chatterton Astronomy Dept, School of Physics,
University of Sydney, NSW 2006, Australia.

AND

M. VISKUM, H. KJELDSEN* AND S. FRANDSEN
Institute of Physics and Astronomy, University of Aarhus,
DK-8000 Aarhus C, Denmark.
*Theoretical Astrophysics Center, University of Aarhus.

α Circini is the brightest of the known rapidly oscillating Ap (roAp) stars. Previous observations of this star in photometry (Kurtz et al. 1994) have shown that it has one dominant pulsation mode, which is a pure oblique dipole mode ($\ell=1$) with a frequency of 2442 μHz ($P = 6.825$ min). Kurtz et al. (1994) measured the amplitude of the principal mode to be 2.55 mmag (Strömgren v).

Baldry et al. (1997) showed that the velocity amplitude and phase of the principal pulsation mode in α Cir varied significantly from line to line. However, it was difficult to interpret the data because of blending effects. In this paper, we look at the Hα line in more detail using the same set of observations taken during two weeks in May 1996. These observations include 6366 intermediate-resolution spectra taken from the 1.88-m telescope at Mt. Stromlo, Australia and the Danish 1.54-m at La Silla, Chile.

The Hα line in each of the spectra was divided vertically into 22 contiguous sections. For each section, a time-series of bisector velocity measurements (4900 from Mt. Stromlo, 1466 from La Silla) was produced. Each time-series was high-pass filtered and then cleaned for bad data points. Next, a weighted least-squares fitting routine was applied to each time series. In order to analyse the principal pulsation mode in α Cir, we have measured the amplitude and phase of each time-series at 2442.03 μHz and estimated the rms-noise level by averaging over surrounding frequencies (1100–2300 μHz and 2600–4400 μHz).

The velocity amplitude and phase of the principal mode at different heights in the Hα line is shown in Figure 1. From height 0.4 to 0.8 in the line (where 0.0 is zero intensity and 1.0 is the continuum level), the amplitude decreases from 300 ms^{-1} to zero and then increases again, with a change in phase of 150°. We believe the velocity node (at ∼0.65) and phase jump between height 0.4 and 0.8 is caused by Hα line formation effects, above 0.8 we do not trust the results due to significant line blending. There is good agreement between the two independent data sets from La Silla and Mt. Stromlo which we have analysed separately for this paper.

These results suggest there is a radial node in the atmosphere of α Cir because the bisector velocity reflects the velocity at different heights in the atmosphere depending

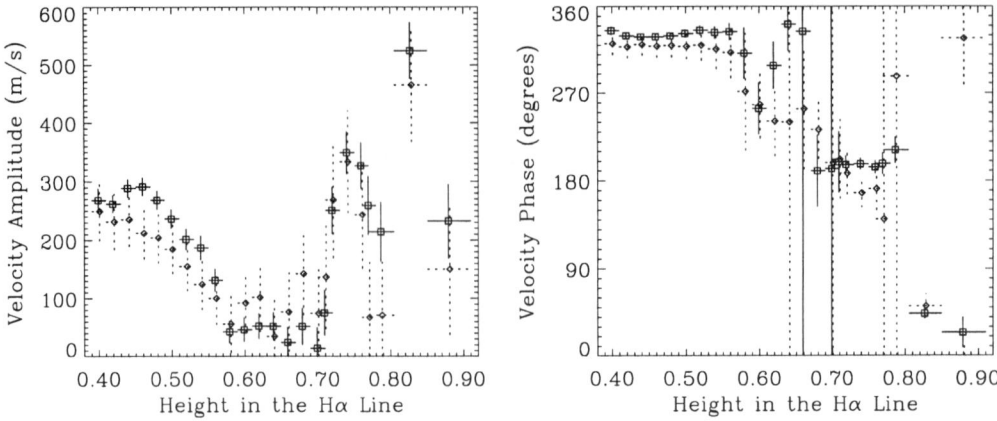

Figure 1. Amplitudes and phases of the principal pulsation mode for the bisector velocity at different heights in the Hα line. The solid lines represent the Mt. Stromlo data and the dotted lines represent the La Silla data. For each measurement, the vertical line is an error-bar while the horizontal line shows the extent of the section in the Hα line.

on the height in the Hα line. There are other possible explanations such as a weak emission component or changes in the continuum which affect the bisector due to the asymmetry of the line. We think that a radial node is the most plausible explanation because of the supporting evidence of the amplitude and phase variations of metal lines (Baldry et al. 1997) and the steep decline of photometric amplitude with increasing wavelength (Medupe & Kurtz 1997). This would imply a surprisingly small radial node separation in the atmosphere of roAp stars (Matthews 1997). Theoretical work needs to be done to determine how this would be possible.

It is interesting to note that similar behaviour has been seen in the Sun. Deubner et al. (1996) have observed a phase discontinuity in the solar oscillations using spectroscopy. The discontinuity occurs in (Velocity − Line Intensity) phase spectra at a frequency of 7000 μHz.

Further results and analysis on the spectroscopy of α Cir will be presented in Baldry et al. (1998).

References

Baldry I.K., Bedding T.R., Viskum M., Kjeldsen H. & Frandsen S. (1997), Spectroscopy of the roAp star α Cir. I. Velocities of Hα and metal lines. *MNRAS*, in press.
Baldry I.K., Viskum M., Bedding T.R., Kjeldsen H. & Frandsen S. (1998), Spectroscopy of the roAp star α Cir. II. .... *in preparation.*
Deubner F.-L., Waldshik Th. & Steffens S. (1996), Dynamics of the solar atmosphere. VI. Resonant oscillations of an atmospheric cavity: observations. *A&A*, Vol. 307, pp. 936–946.
Kurtz D.W., Sullivan D.J., Martinez P. & Tripe P. (1994), The pure dipole pulsation mode of the rapidly oscillating Ap star α Cir: .... *MNRAS*, Vol. 270, pp. 674–686.
Matthews J.M. (1997), Probing the interiors of roAp stars: stellar tubas that sound like piccolos. *Proc. IAU Symp. 181, Sounding Solar and Stellar Interiors*, in press.
Medupe R. & Kurtz D.W. (1997), Determining temperature amplitude as a function of depth in the atmospheres of roAp stars. *MNRAS*, submitted.

THE SINGLE LIFE
OF RAPIDLY OSCILLATING AP STARS

S. HUBRIG
University of Potsdam
Am Neuen Palais 10, D-14469 Potsdam, Germany

N. KHARCHENKO
Main Astronomical Observatory
Golosiiv, 252650 Kiev - 22, Ukraine

AND

G. MATHYS
European Southern Observatory
Casilla 19001, Santiago 19, Chile

1. Introduction

Rapidly oscillating Ap (roAp) stars are cool magnetic Ap SrCrEu stars which pulsate in high-overtone ($n \gg l$), low-degree ($\ell \leq 3$) p-modes, with periods from 6 to 15 minutes and typical amplitudes of a few millimagnitudes. 28 such stars are currently known. The roAp phenomenon is confined to a well-defined region of the Strömgren photometry parameter space (Martinez 1993). However, this region also contains other Ap stars, in which no pulsation could be detected, despite sometimes thorough searches. These apparently constant Ap stars (non-oscillating Ap stars, or noAp stars) appear remarkably similar to the roAp stars in many respects (e.g. colour indices, abundances, magnetic fields). Here we present recently found indications for differences between both groups.

2. Kinematical Properties

Recently, we have studied the kinematical properties of roAp and noAp stars by using new HIPPARCOS proper motions and parallaxes.

First, we have compared the positions in the H-R diagram of a sample of roAp stars and of a sample of noAp stars. Out of 28 roAp stars known, 14 were observed by HIPPARCOS. Our comparison sample of noAp stars contains 34 stars. We have selected all the Ap stars

- in which oscillations have been sought and have not been detected;
- whose photometric indices in the Strömgren system lie within the limits of occurrence of the roAp phenomenon (as defined by Martinez 1993);

– for which HIPPARCOS data exist.

We have found that roAp stars, as a group, are (-0.48 ± 0.36) mag above the ZAMS, while the noAp stars are (-1.30 ± 0.64) mag above the ZAMS. Hence, in agreement with previous results (Mathys et al. 1996), roAp stars are found to be less luminous than their non-oscillating counterparts. This is also consistent with the recent result obtained by North et al. (1997). The difference in the absolute magnitudes between roAp and noAp stars found in this study is 0.82 magnitude. On the other hand, from the comparison of the kinematical characteristics (elements of the galactic orbits, total space velocities, and dispersion of the space velocities) calculated from HIPPARCOS data, we conclude that both groups are nearly identical, with some hints of older kinematics for roAp stars. This leads us to assume that roAp and noAp stars are of the same or only slightly different age, but that the roAp stars are less luminous, less massive, and less evolved (consistent with the results obtained by North et al. 1997). The difference between the masses of roAp stars and noAp stars is particularly important for the understanding of the origin of their oscillations. Plausibly, convection starts becoming efficient for the roAp stars. More generally, the difference of internal structure associated with the mass difference can probably explain why oscillations are observed only in the roAp stars.

3. roAp stars versus noAp stars: binarity

Until now, no roAp star is known to be a spectroscopic binary (SB). We obtained at least two measurements of the radial velocity of 14 of the 28 roAp stars, but we found no evidence for variation in any of these stars. The strong contrast with the rate of occurrence of binarity of 46% reported for cool Ap stars by Gerbaldi et al. (1985) is an indication that there is a real deficieny of binaries among roAp stars.

Relevant radial velocity data are scarce for noAp stars, due to the combination of their relative faintness (many have magnitudes between 8 and 10) and of the southern declination of most of them. However, all the noAp stars for which enough information is available (that is, 11 stars whose radial velocity has been repeatedly measured) either are SBs or show hints of binarity.

The interpretation of a difference in duplicity between roAp and noAp stars, and its meaning for the understanding of the origin of the pulsations in roAp stars are far from obvious. Even if the different internal structure of roAp stars is the reason for their pulsation, it is difficult to understand why no roAp star would be found to be a binary. That the latter is indeed true remains to be more definitely established, through additional observations. Such observations will help to establish which conditions must prevail for the appearance of rapid oscillations in cool Ap stars.

References

Gerbaldi, M., Floquet, M. and Hauck, B.: 1985, *Astron. Astrophys.* **146**, 341
Martinez, P.: 1993, Ph. D. Thesis, University of Cape Town
Mathys, G., Kharchenko, N. and Hubrig, S.: 1996, *Astron. Astrophys.* **311**, 90
North, P., Jaschek, C., Hauck, B., Figueras, F., Torra, J.-C., Mermilliod, J.-C. and Künzli, M.: 1997, ESA SP-402 (in press)

MAGNETIC PROPERTIES OF RAPIDLY OSCILLATING AP STARS

G. MATHYS
European Southern Observatory
Casilla 19001, Santiago 19, Chile

AND

S. HUBRIG
University of Potsdam
Am Neuen Palais 10, D-14469 Potsdam, Germany

1. Introduction

Rapidly oscillating Ap stars generally pulsate in multiple modes, characterized by different frequencies. The amplitudes of these modes may furthermore be modulated with the rotation frequency of the star. For the two roAp stars whose magnetic fields have been sufficiently studied, the maximum pulsation amplitude coincides in phase with one of the extrema of the mean longitudinal magnetic field. Two interpretations of this property have been proposed: the *oblique pulsator model*, according to which the pulsation modes are aligned with the *magnetic* axis of the star, and the *spotted pulsator model*, which assumes that the pulsation modes are symmetric about the *rotation* axis of the star, and that pulsation amplitude modulation is due to the inhomogeneity of the stellar surface (which, itself, is related to the magnetic field geometry). At present, no definite choice between these two models can be made, though the oblique pulsator model is often preferred.

Another unsolved question is that of the excitation mechanism of the pulsation. Plausible candidates seem to be the κ-mechanism *restricted to the magnetic poles* in the He II or Si IV ionization zones, or magnetic overstability.

In short, the presence of a magnetic field is *required* by *all* the mechanisms driving the pulsation that have been proposed as well as by *both* models suggested to explain the observed modulation of the amplitude of the rapid oscillations. This implies that the proper interpretation of the seismological behaviour of the roAp stars requires the knowledge of their magnetic fields. The latter is still very incomplete: here we summarize its current status.

2. Magnetic Field Observations in roAp Stars

Unless otherwise noted, the data summarized here are presented in greater detail in Mathys et al. (1997) and Mathys and Hubrig (1997), and references therein. The reader is referred to those papers for additional information.

Among the 28 roAp stars known:
- 14 definitely or very probably have a magnetic field: HD 6532, HD 19918, HD 24712, HD 80316, HD 83368, HD 101065 (Wolff and Hagen 1976), HD 119027, HD 128898, HD 134214, HD 137949, HD 150562, HD 166473, HD 176232, and HD 201601;
- of these 14, only 2 (HD 24712 and HD 83368) have been studied throughout their rotation cycle;
- magnetic fields have been sought in 4 additional stars, but only upper limits could be obtained: HD 193756, HD 203932, HD 217522, and HD 218495;
- for 10 stars, there have been no magnetic field determinations so far. These stars are HD 9289, HD 12932, HD 42659, HD 60435, HD 84041, HD 86181, HD 161459, HD 185256, HD 190290, and HD 196470.

3. Conclusion

In spite of the importance of magnetic fields for the proper understanding of the pulsational properties of the roAp stars, these fields have only been scarcely studied until now. An additional observational effort is necessary. Accordingly, we have undertaken a program to study more systematically the magnetic fields of the roAp stars.

References

Mathys, G. and Hubrig, S.: 1997, *Astron. Astrophys. Suppl. Series* **124**, 475
Mathys, G., Hubrig, S., Landstreet, J.D., Lanz, T. and Manfroid, J.: 1997, *Astron. Astrophys. Suppl. Series* **123**, 353
Wolff, S.C. and Hagen, W.: 1976, *Publ. Astron. Soc. Pacific* 88, 119

SEISMIC DETECTION OF BOUNDARIES OF STELLAR CONVECTIVE REGIONS

MÁRIO J.P.F.G. MONTEIRO
DMA-FCUP and CAUP, *Universidade do Porto, Portugal*

JØRGEN CHRISTENSEN-DALSGAARD
TAC and Inst. Fysik og Astronomi, *Aarhus Universitet, Denmark*

AND

MICHAEL J. THOMPSON
Astronomy Unit, QMW, *University of London, England*

The edge of a convective region inside a star gives rise to a characteristic periodic signal in the frequencies of its global p-modes (e.g. [1], [4]), such that the frequencies ω are then essentially a smooth function of the mode order n plus a periodic component $\delta\omega_p = A(\omega)\cos[2(\omega\bar{\tau}_d+\phi_0)]$. Here the amplitude is $A(\omega)=\sqrt{A_1^2/\omega^4+A_2^2/\omega^2}$, with A_1 and A_2 being values that depend weakly on frequency ω: A_1 is always present in general, but A_2 will be non-zero only if there is overshoot; $\bar{\tau}_d$ is essentially the acoustical depth τ (i.e. the sound travel time) of the edge of the convection zone measured from the surface of the star; and ϕ_0 is a constant related to the phase of the eigenfunctions. To facilitate the comparison between different stars, we consider the amplitude evaluated at a fiducial frequency by defining $A_d \equiv A(\tilde{\omega})$. For the Sun, we chose as the reference frequency $\tilde{\omega}=2500\,\mu$Hz. If we take this value and scale it for other stars (using just a standard homology scaling for frequencies), we find $\tilde{\omega}/2\pi = 2500\,\mu\text{Hz}\times(M/M_\odot)^{1/2}(R/R_\odot)^{-3/2}$.

Concentrating on the case without convective overshoot, we have

$$A_d = \frac{A_1}{\tilde{\omega}^2} \propto \frac{g}{\tilde{\omega}^2 \tau_t}\frac{d\nabla_r}{d\tau}, \qquad (1)$$

where τ_t is the total acoustical radius of the star, and gravitational acceleration g and radiative temperature gradient ∇_r are evaluated at the convective boundary. This gives the dependence of the amplitude A_d on the position of the boundary. To study the variation of A_d with stellar mass we need to know how the different functions of the stellar structure in equation (1) depend on the mass. Assuming that opacity $\kappa \propto \rho^\alpha T^{-\beta}$, an homology scaling gives $d\nabla_r/d\tau \propto LR^{-3\alpha+\beta-3/2}M^{-5/2-\beta+\alpha}$. Adopting $\alpha=1$ and $\beta=7/2$, we find that

$$A_d = A_{d,\odot}\,(L/L_\odot)\,(R/R_\odot)^{-1}\,(M/M_\odot)^{-5}, \qquad (2)$$

where we have used as reference the value $A_{d,\odot}$ found for the Sun.

The feasibility of detecting the signal is measured by the dimensionless ratio $A_{\rm d}/\tilde\omega = 3.5\times 10^{-5}(L/L_\odot)(R/R_\odot)^{5/2}(M_\odot/M)^{-11/2}$. Thus the signal we seek in the frequencies is typically a few parts in 10^5.

For stars of mass similar to the Sun ($0.8 \lesssim M/M_\odot \lesssim 1.2$) we have roughly that $R \propto M^{1.5}$ and $L \propto M^5$ (Figs 22.2, 22.3 of [2]); thus the amplitude of the signal and its ratio to the reference frequency scale as

$$A_{\rm d} = A_{{\rm d},\odot}\,(M/M_\odot)^{-1.5} \quad \text{and} \quad A_{\rm d}/\tilde\omega = 3.5\times 10^{-5}\,(M/M_\odot)^{-5/4}\;. \qquad (3)$$

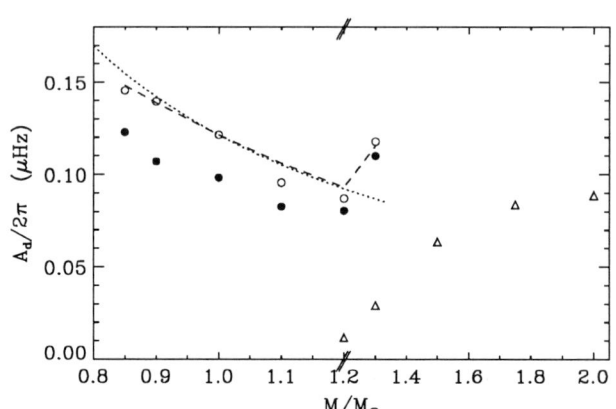

Fig. 1: Expected dependence on stellar mass of the amplitude of the periodic signal in the frequencies arising from the edge of a convective envelope (open circles) or from the edge of a convective core (triangles), as predicted by equation (1). Also shown are values from expressions (2) (dashed line) and (3) (dotted line) for the expected amplitude. The measured values for the signal in the frequencies (with degree ≤ 4) are represented by the filled circles.

From the p-mode frequencies of ZAMS models for different stars we have determined the amplitude of the signal arising from edges of convective regions using the least-squares method described by [4]. The modes used have degrees l between 0 and 4. These modes might in principle be observed for distant stars. The numerical results of the fits are shown in Figure 1. It is clear that the signal that one expects to detect from the edge of convective envelopes in stars of mass up to $1.3 M_\odot$ can indeed be measured in frequency data that we can hope to obtain for such stars. We have also calculated the amplitude of the signal that would be produced by the convective core of a star with mass $\gtrsim 1.5 M_\odot$; as shown by the figure this should also be large enough that it ought to be detectable.

Finally, we make two remarks. First we note that the predicted amplitudes in velocity ($v_{\rm osc}$) for solar-like oscillations in other stars (e.g. [3]), for this mass range, are expected to increase with mass according to $v_{\rm osc} \propto L/M \propto M^4$. Such a behaviour would compensate the decrease with mass found here for the detectability of the signal. Secondly, the composition inhomogeneity arising from evolution implies a strong increase in the detectability of the core for older higher-mass stars. This is expected to be important for studying high-mass stars with a large convective core and a negligible envelope; in particular, overshoot from convective cores may modify the signal in measurable ways.

[1] Christensen-Dalsgaard J., Monteiro M.J.P.F.G., Thompson M.J. (1995), MNRAS 276, 283
[2] Kippenhahn R., Weigert A. (1990), *Stellar Structure and Evolution*, Springer–Verlag
[3] Kjeldsen H., Bedding T.R. (1995), A&A 293, 87
[4] Monteiro M.J.P.F.G., Christensen-Dalsgaard J., Thompson M.J. (1994), A&A 283, 247

ON THE SEISMIC SIGNATURE OF THE HEII IONIZATION ZONE IN STELLAR ENVELOPES

MÁRIO J.P.F.G. MONTEIRO
DMA-FCUP and CAUP, Universidade do Porto, Portugal

AND

MICHAEL J. THOMPSON
Astronomy Unit, QMW, University of London, England

Sharp variations of the structure of the star create a characteristic signal in its frequencies of oscillation (e.g. [3]). The zone of the second ionization of helium is such a localized feature of the structure whose properties depend mainly on the abundance of helium and the equation of state. Considering that such a signal should easily be detectable provided the frequencies are measured to rather better than 1μHz accuracy (the COROT project should measure oscillation frequencies with an accuracy of 0.1μHz), we present here a tool to study this aspect of stellar structure.

The second ionization zone of helium causes a distinct 'bump' $\delta\Gamma_1$ in the adiabatic exponent Γ_1. This gives rise to a perturbation $\delta\omega$ of the frequencies (relative to a fictitious star without such a bump),

$$\delta\omega \sim -\frac{3}{4\tau_t} \left(\frac{\delta\Gamma_1}{\Gamma_1}\right)_{\bar{\tau}_d} \frac{\sin^2(\omega\beta)}{(\omega\beta)} \cos[2(\omega\bar{\tau}_d + \phi_0)] . \qquad (1)$$

Here $\bar{\tau}_d$ is the acoustic depth below the surface at which the bump is located; ϕ_0 is a constant determined by the phase of the eigenfunctions at the surface of the star; $(\delta\Gamma_1/\Gamma_1)_{\bar{\tau}_d}$ is the relative magnitude of the bump in Γ_1; β measures approximately the half (acoustic) thickness of the bump (see Fig. 1a); and τ_t is the total acoustic radius of the star (viz the sound travel time from centre to surface).

In this study, we fit expression (1) to frequencies of stellar models (Table 1), using a non-linear least-squares method (from [6]). Four parameters are used,: $\bar{\tau}_d$, ϕ_0, β and $a_0 \equiv (-3/4\tau_t)(\delta\Gamma_1/\Gamma_1)_{\bar{\tau}_d}$. Only modes with degrees $l \leq 5$ are used in the fit, since we are interested in what might be learned about the distant stars. In order to be able to interpret the results of the fit more easily, we introduce $\delta_{obs} = -4\tau_t a_0/3$. This corresponds to the value of $(\delta\Gamma_1/\Gamma_1)$ at $\bar{\tau}_d$. The relation of δ_{obs}, β and $\bar{\tau}_d$ to the helium ionization bump is shown schematically in Fig. 1a. The expected values of $\bar{\tau}_d$, β and δ_{obs} are: $\bar{\tau}_d \approx 500$–600s; $\beta \approx 100$s; and $\delta_{obs} \approx 0.05$.

The measured values of $\bar{\tau}_d$ are consistent with the expected values. The parameters that are directly relevant to the equation of state and the helium abundance are a_0 (or equivalently δ_{obs}) and β. The correlation of the latter with the equation of state is clear for both stellar masses (see Fig. 1b).

Table 1. ZAMS stellar models of $1M_\odot$ with varying helium abundances (Y), equations of state (EOS), opacities and formulations for convective energy transport. Abbreviations are: SEOS = simple Saha EOS for H and He ionization with ad hoc pressure ionization at high pressure; CEFF = the EFF EOS ([5]) with Coulomb correction term (e.g. [2]); SOP = simple power law opacities; CT76 = opacity tables from [4]; MLT = standard mixing length theory; CM = convection formulation of [1]. Model Z_0 has SEOS with HeII ionization potential set to zero.

The parameters obtained by fitting equation (1) to model frequencies are also given: the acoustic depth $\bar{\tau}_d$ and half-width β are in seconds; a_0 is in μHz; ϕ_0 and δ_{obs} are dimensionless.

Mod.	Y	EOS	Opacity	Conv.	$\bar{\tau}_d$	ϕ_0	$a_0/2\pi$	β	δ_{obs}
Z_0	0.2379	no HeII	SOP	MLT					
Z_1	0.2378	SEOS	SOP	MLT	529	3.5	1.64	108.5	−0.041
Z_2	0.2378	SEOS	SOP	CM	532	3.5	1.63	107.1	−0.041
Z_3	0.2421	CEFF	SOP	MLT	527	3.2	1.82	94.9	−0.045
Z_4	0.2356	CEFF	CT76	MLT	515	3.5	1.85	95.1	−0.046
Z_5	0.2356	CEFF	CT76	CM	518	3.4	1.83	93.0	−0.046

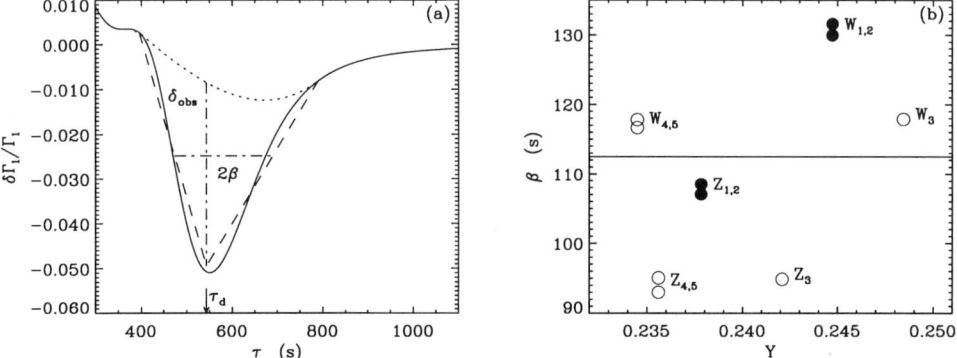

Fig. 1: (a) Schematic of the properties of the ionization zone measurable in the signal $\delta\omega$; the half-thickness β of the bump, the amplitude δ_{obs}, and the location τ_d. Model differences Z_1-Z_0 are shown (solid). (b) Inferred values of half-thickness β versus helium abundance. Models W_{1-5} have same EOS, opacity and convection treatment as Z_{1-5} but have mass $1.1M_\odot$. Filled circles are for the SEOS; open circles are for the CEFF EOS (see Table 1).

We have shown how the HeII ionization region can be studied using the characteristic signal associated with that layer. The signal can be measured using low degree modes (the ones we hope to observe in other stars). We are gratified that we obtain from the signal values for the parameters with the expected magnitudes (for both $1M_\odot$ and $1.1M_\odot$ stellar models). This shows that the signal is indeed coming from the HeII ionization zone, and that our fitting procedure works.

The width β discriminates well the equation of state (see Fig. 1b). Another of our goals has been to measure the helium abundance Y. The area of the bump, as determined by $\beta\delta_{obs}$, is a good measure of Y, but further tests are needed to establish how well this can be determined from frequencies.

[1] Canuto V.M., Mazzitelli I. (1991), ApJ 370, 295
[2] Christensen-Dalsgaard J., Däppen W. (1992), A&AR 4, 267
[3] Christensen-Dalsgaard J., Monteiro M.J.P.F.G., Thompson M.J. (1995), MNRAS 276, 283
[4] Cox A.N., Tabor J.E. (1976), ApJS 342, 1187
[5] Eggleton P.P., Faulkner J., Flannery B.P. (1973), A&A 23, 325
[6] Monteiro M.J.P.F.G., Christensen-Dalsgaard J., Thompson M.J. (1994), A&A 283, 247

THE PROCYON CAMPAIGN:

Observations from Kitt Peak

C. A. PILACHOWSKI, S. BARDEN, F. HILL, J. W. HARVEY,
C. U. KELLER AND M. S. GIAMPAPA
National Optical Astronomy Observatories
PO Box 26732
Tucson, Arizona, 85726-6732
U.S.A.

Time series spectra of the F5IV star Procyon (α CMi) were obtained at the Kitt Peak National Observatory during a 35-night observing run in January-February 1997. The observations were obtained as part of an international collaboration to detect and study acoustic p-mode oscillations in solar-type stars. Spectra covered the wavelength range from 4000 to 5300 Å, with a resolving power of approximately 3500 (1.3 Å resolution). The sampling rate was one observation per minute, and the typical S/N ratio per pixel after averaging along columns is in excess of 1000. We obtained 12,888 spectra. A sample spectrum is shown in Figure 1

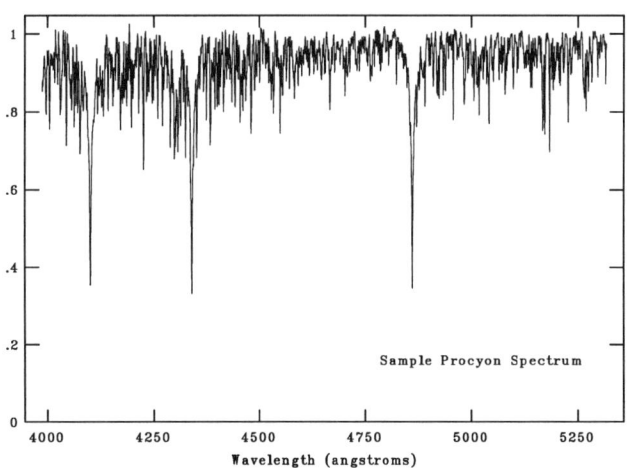

Figure 1. Sample Spectrum of Procyon

In Figure 2, we show the time distribution of the data during the 35-night run. Each point is plotted to show the signal level (in ADU in a single line of the image)

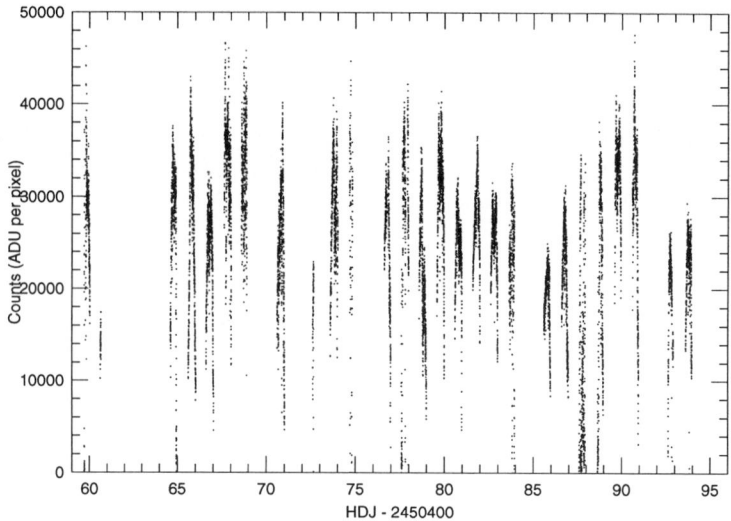

Figure 2. Observations of Procyon

per pixel in the continuum. During the 35-night run, we obtained 24 lengthy time series (longer than 5 hours), and a few shorter series.

The spectra are being analyzed to identify any periodic signals due to acoustic oscillations in Procyon. In addition to measuring the equivalent widths of the three Balmer lines covered by our spectra (H-beta, H-gamma, and H-delta) we will also examine the spectra for variations in the average metal line strength.

Reductions of the data are complete, and the equivalent widths of the Balmer lines have been measured. Preliminary analysis of the full time series is underway. We will also average the power spectra determined from each individual night's data. The frequencies of the expected p-mode oscillations have been predicted by Chaboyer et al. (1997) from a new grid of stellar evolution models for Procyon based on the revised astrometric mass of Girard et al. (1996). These models suggest that p-mode oscillations of Procyon should occur at frequencies near 1100 μHz, with frequency splittings of 54 μHz.

References:

B. Chaboyer, P. Demarque, and D. B. Guenther, presented at "A Half Century of Stellar Pulsation Interpretations: A Tribute to Arthur N. Cox," A.S.P. Conf. Series (in press)

T. Girard, H. Wu, J. T. Lee, S. E. Dyson, E. P. Horch, W. F. van Altena, C. Ftaclas, R. L. Gilliland, K. G. Schaefer, and H. E. Bond 1996, BAAS, 28, 919

DISCOVERY OF NON-RADIAL PULSATIONS IN THE WHITE DWARF PRIMARY OF A CATACLYSMIC VARIABLE STAR

BRIAN WARNER AND LIZA VAN ZYL
Department of Astronomy
University of Cape Town, Rondebosch 7700. South Africa

1. Introduction

Non-radial pulsations in isolated white dwarfs have been known for 25 years and it has been shown that the hydrogen-rich (DA) white dwarfs have a high probability of pulsating if they lie in the instability strip with effective temperature between 11500 and 13200 K - the ZZ Ceti stars (e.g. Kepler and Nelan 1993). Analysing techniques developed for such stars allow derivation of masses, luminosities, rotation periods, hydrogen surface layer masses and other properties (e.g. Kepler and Bradley 1995). A number of binary systems are known in which the primary is a white dwarf; dominant in this class are the cataclysmic variable (CV) stars. Until now no CV primary has been found to have non-radial pulsations.

GW Lib had an outburst of large amplitude in 1983 and was thought to be a nova. Spectra at minimum light (mag 18.5) show it to be in fact a dwarf nova (Duerbeck and Seitter 1987). No other outbursts have been observed, so it is deduced to be a CV of very low rate of mass transfer and long outburst interval (i.e., an extreme SU UMa star), similar to WZ Sge or AL Com. Our observations of GW Lib in March and April 1997, using the University of Cape Town CCD photometer on the 40-in reflector at the Sutherland site of the South African Astronomical Observatory, with integrations (in white light) down to 6 secs, show the presence of multiperiodic oscillations. In the March observations coherent periods at 378.72 and 648.07 secs were present throughout the week. In April these were present at greatly reduced amplitude and other oscillation modes with periods near 390, 400, 680, and also probably at 890, 1020 and 1080 secs had become visible. During the latter week there was considerable variation of amplitude of the various modes.

These periods and amplitude variations are characteristic of the larger amplitude ZZ Ceti stars like G29-38 and G38-29 (Kleinman 1995). In GW Lib the typical amplitude is 3 %, but about half of the light in the visible region comes from accretion-related luminosity; the true amplitude is therefore about twice what is observed. A periodic signal near 119 min, with a harmonic at half this period, was also occasionally seen, which almost certainly is modulation at the orbital period, and is compatible with the expectation that the outburst characteristics imply SU UMa behaviour (e.g. Warner 1995). The presence of a rich spectrum of non-radial oscillation modes in GW Lib should eventually lead to the first really accurate determination of the physical

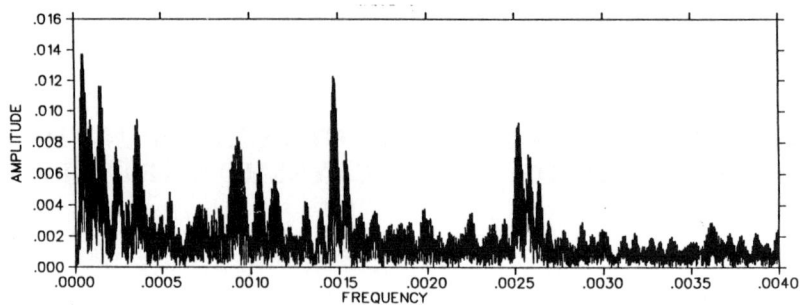

Figure 1. Amplitude spectrum of 4 nights of observations of GW Lib acquired in April 1997

properties of a CV primary - which will be valuable as a standard against which to test the more conventional spectroscopic and photometric methods of determining masses of the primaries. The primaries in CVs have a wide range of surface temperatures (see Table 2.8 of Warner 1995), which, as almost all CV primaries are old enough to have cooled to below 9000 K, probably shows that it is the average accretion rate that determines the surface temperature (see p. 478 of Warner 1995). To be in the instability strip (at least, as found in single stars) a CV primary must have just the low rate of mass transfer that ensures that it will be a very long interval dwarf nova. Observations of the changes in oscillation frequency and amplitude after the next superoutburst will inform studies of non-radial oscillations and also models of structural changes that occur during accretion events.

There is another long-interval SU UMa star which may have non-radial oscillations; this is SW UMa, which at quiescence has a light curve similar to that of GW Lib and in which the dominant period is 950 secs (Shafter *et al.* 1986). WZ Sge has been thoroughly observed and exhibits modulations near 29 secs which are probably related to the rotation of the primary. Its T_{eff} of 14 900K (Sion *et al.* 1995) places it outside the (isolated) white dwarf instability strip. If SW UMa does turn out to have non-radial oscillations it would be of great interest to determine its T_{eff}, and that of GW Lib, and compare them with WZ Sge.

References

Clemens, J.C., 1993. *Baltic Astr.* **2**, 407.
Duerbeck, H.W. and Seitter, W.C., 1987. *Astrophys. Sp. Sci.* **131**, 467.
Kepler, S.O. and Bradley, P.A., 1995. *Baltic Astr.* **4**, 166.
Kleinman, S.J., 1995. *Baltic Astr.* **4**, 270.
Shafter, A.W., Szkody, P. & Thorstensen, J., *Astrophys. J.* **308**, 765.
Sion, E.M. *et al.*, 1995. *Astrophys. J.* **439**, 957.
Warner, B., 1995. *Cataclysmic Variable Stars*, Cambridge University Press, Cambridge.

ASTEROSEISMOLOGY OF δ SCUTI STARS: OBSERVATIONS

MICHEL BREGER
Institut für Astronomie
Türkenschanzstr. 17, A-1180 Wien, Austria

Abstract.
We review recent observational developments which provide important asteroseismological tools. Extensive multisite campaigns of individual δ Scuti stars show that 24 or more pulsation modes with $\ell = 0$ to 2 can be detected photometrically. Spectroscopically, also about 30 modes have been detected, and these can be identified with modes of ℓ values up to 20. Since each technique favors the detection of specific types of modes, hundreds or thousands of modes must be excited in δ Scuti stars.

We examine the quantities which can be matched between observations and theoretical models specifically computed for each star. Recent progress in tbe mode identification of multiple pulsation modes is illustrated by presenting an application of the phase shift method for the two stars, FG Vir and 4 CVn, recently measured by the Delta Scuti Network.

1. Asteroseismology and δ Scuti stars

The most promising candidates for asteroseismology on and near the main sequence are the short-period δ Scuti pulsators, which are stars with spectral types A and F and luminosity classes III–V. The key to asteroseismology is the comparison between observations and realistic stellar models. In particular, we match

- the values of the frequencies of the multiple modes together with the independently determined (n, ℓ, m) quantum numbers, i. e. the identification of the pulsation modes,
- the width and position of the limited frequency range in which the pulsation modes are excited,
- the observed regularities of frequency spacing (e. g. adjacent radial orders, rotational splitting),
- the growth and decay rates through amplitude variations.

An example of applying asteroseismological tools is the predicted existence of a single, observable gravity mode for slightly evolved main sequence stars, which can be used to determine the star's interior rotation rate and to measure the amount of convenctive overshooting (Dziembowski & Pamyatnykh 1991). The star CD -24 7599 is an excellent candidate to search for this mode.

Several observational and theoretical difficulties for the seismological study of δ Scuti stars need to be overcome:

- The typical δ Scuti star varies with small amplitudes of $\sim$ 10 mmag. Consequently, very precise data are required to identify the individual pulsation modes.
- The excited pulsation modes are p-modes of low radial overtone, for which the asymptotic theory of frequency spacing does not apply.
- The majority of the multiperiodic δ Scuti stars are fast rotators with $v \sin i \geq$ 100 km s^{-1}: rapid rotation destroys the equidistant frequency spacing of the rotationally split modes that is expected for slow rotation.
- Possible differential rotation may result in a further complication of the frequency spectrum, but offers an additional motivation to apply asteroseismology to these stars.
- The multiperiodicity makes it necessary to conduct extensive multisite campaigns in order to reduce aliasing.
- Finally, most well-observed δ Scuti stars are evolved objects. The theoretical frequency spectra of such evolved objects are very dense. If correct, this increases the number of pulsation modes that need to be identified observationally as well, requiring even larger observing campaigns. (However, the extensive 1996/7 unpublished measurements by the Delta Scuti Network of the evolved star 4 CVn have not detected this predicted dense spectrum.)

2. New observations: oceans of nonradial modes

The observational strategy adopted by most researchers can be summarized as follows: observationally detect the largest possible number of excited pulsation modes and identify these modes with predictions from models computed specifically for the star. During the last decade, large photometric campaigns have been very successful in detecting multiple modes and determining the frequencies. Large networks with telescopes spaced around the world are necessary in order to reduce the aliasing caused by regular day-time observing gaps and to increase the amount of information which can be collected during an observing season.

The major networks are:

- The Delta Scuti Network, which during the last few years cooperated with the Whole Earth Telescope (DSN/WET, see Breger et al. 1995, Handler 1997a). The network concentrates on long studies of a few selected stars, e.g. 300+ hours and 20+ detected frequencies per star. Furthermore, long-term photometric stability as well as a relative precision of 3 millimag or better per single observation is achieved by the partners. Since its inception, the network has carried out 17 multisite campaigns, which includes some shorter campaigns as well.
- STEPHI (e. g. Michel et al. 1992) and STACC Networks (Frandsen et al. 1996). Both networks specialize in studying δ Scuti stars in clusters, e. g. the variables in the Praesepe cluster. Since these δ Scuti stars in a specific cluster usually are at the same distance and share the same chemical composition, the results for different stars can be directly compared. STACC is a relatively new network using CCD detectors with high duty cycles.

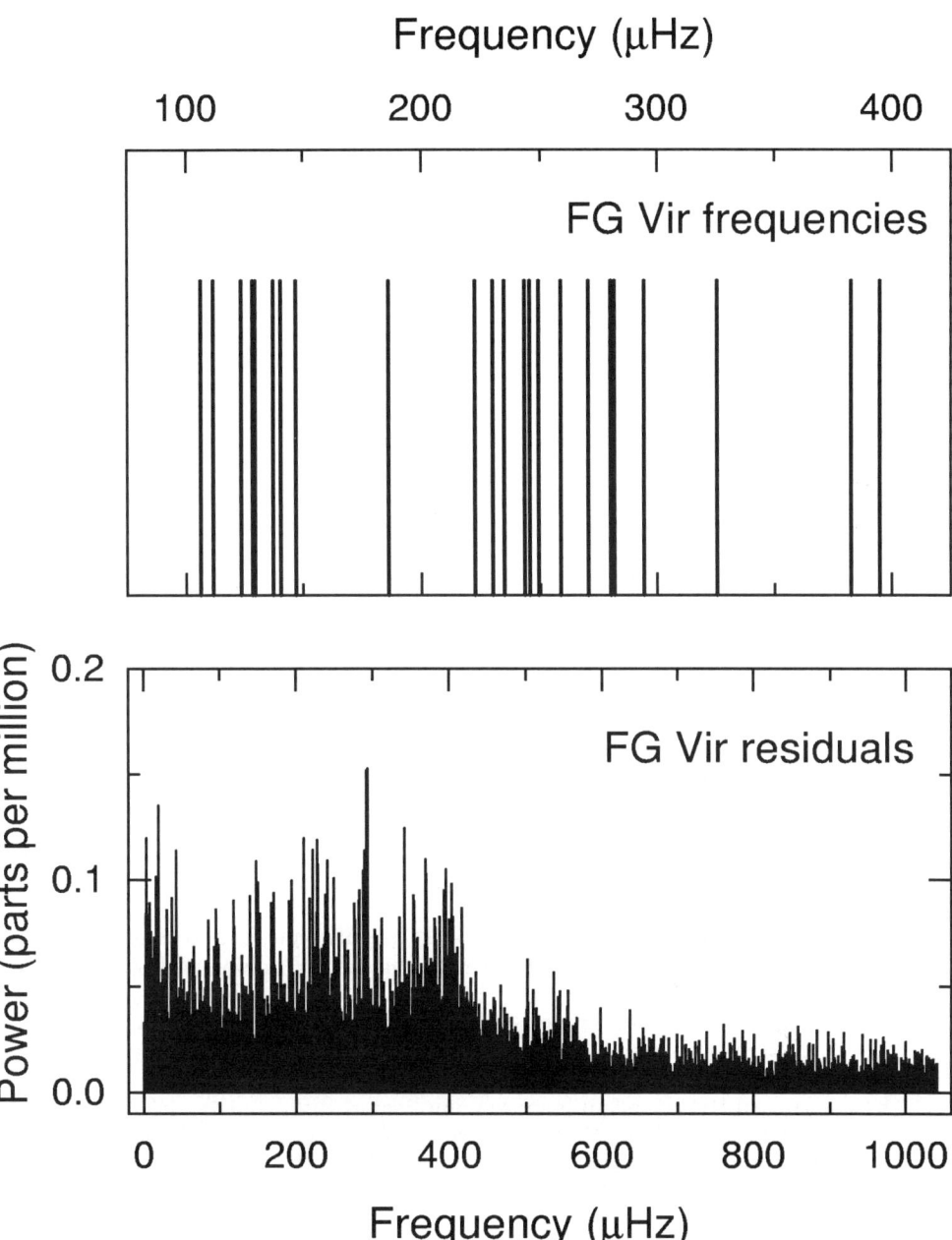

Figure 1. Pulsational behavior of FG Vir (1992-5). Top: The 24 frequencies with power signal/noise≥12. Amplitudes range from 0.0004 to 0.021 and are not shown to scale in the figure. Bottom: Power spectrum of the residuals after prewhitening the 24 frequencies. The pattern indicates that additional pulsation modes are excited in the same frequency region.

— The Merate group (e. g. Mantegazza, Poretti & Bossi 1996) specializes in both photometric and spectroscopic campaigns of selected stars.

The large photometric campaigns have shown that in δ Scuti stars a very large number of modes are simultaneously excited. The excitation often occurs in a very narrow frequency region. One of the δ Scuti stars with the largest number of known modes is FG Vir, for which 24 low-degree pulsation frequencies have been detected so far (Breger et al. 1997b). Fig. 1 shows the 24 frequencies of FG Vir as well as the power spectrum of the residuals. We note the sharp drop of power near 400 μHz, suggesting the presence of additional excited pulsation modes shortward of 400 μHz.

The photometric techniques favor the detection of low-degree modes. While these low-degree modes can also be discovered spectroscopically, (e.g. Mathias & Aerts 1996), spectroscopy has become very important for detecting sectoral high-degree modes (e. g. Kennelly & Walker 1996). Presently unpublished spectroscopic measurements of the star τ Peg by Kennelly, Brown and colleagues show a similar rich mode spectrum with degrees up to $\ell = 20$ in a narrow frequency range. Such behavior is also confirmed by the recent results by Mantegazza (1997) for the fast rotating star V837 Cen. For this star, two prograde modes are deduced from line profiles (with m = -10 and -14 or -15, respectively), which are not the modes seen photometrically. These high-m modes do not have the same frequency in the co-rotating frame as the low-ℓ modes. V837 Cen differs in this respect from τ Peg. The implication of such a narrow frequency region in terms of asteroseismology still needs to be explored.

While the majority of the observations required for asteroseismology are obtained by the major networks, important results are also published by astronomers outside these networks. Examples are the resolution of the complex behavior of θ Tuc (Paparó et al. 1996) and the eight modes determined and modelled for the star δ Scuti itself by Templeton et al. (1997).

In the next sections, we will concentrate on two stars with the largest number of photometrically determined modes: FG Vir and 4 CVn.

3. Identification of pulsation modes

Most successful applications of asteroseismology require that, in addition to the detection of a large number of pulsation frequencies, the actual modes of these frequencies need to be identified. This applies especially to the ℓ and m values. For δ Scuti stars, the identification of nonradial modes has been one of the most persistent problems. One of the reasons is that not all the modes in an expected pattern are excited to an observable level, e.g. of the five possible $\ell = 2$ modes, only two or three might be seen. In this regard the situation is more complicated than that for PG 1159-035, for example (Winget et al. 1991). The solution lies in detecting more modes (by larger and better observing campaigns) and by applying specific mode-identification tools. The presently used methods of mode identification are:

— measure phase differences between the light curves at different wavelengths (e.g. Garrido et al. 1990),
— compare equivalent width variations of the Balmer lines with photometric or radial velocity amplitudes (Bedding et al. 1996),
— frequency values: patterns, frequency differences, Q values.

For photometric campaigns carried out in at least two colors, the phase difference method is a powerful tool for mode identification. The phase shifts expected for different ℓ values are a function of the values of the pulsation constants Q of the individual

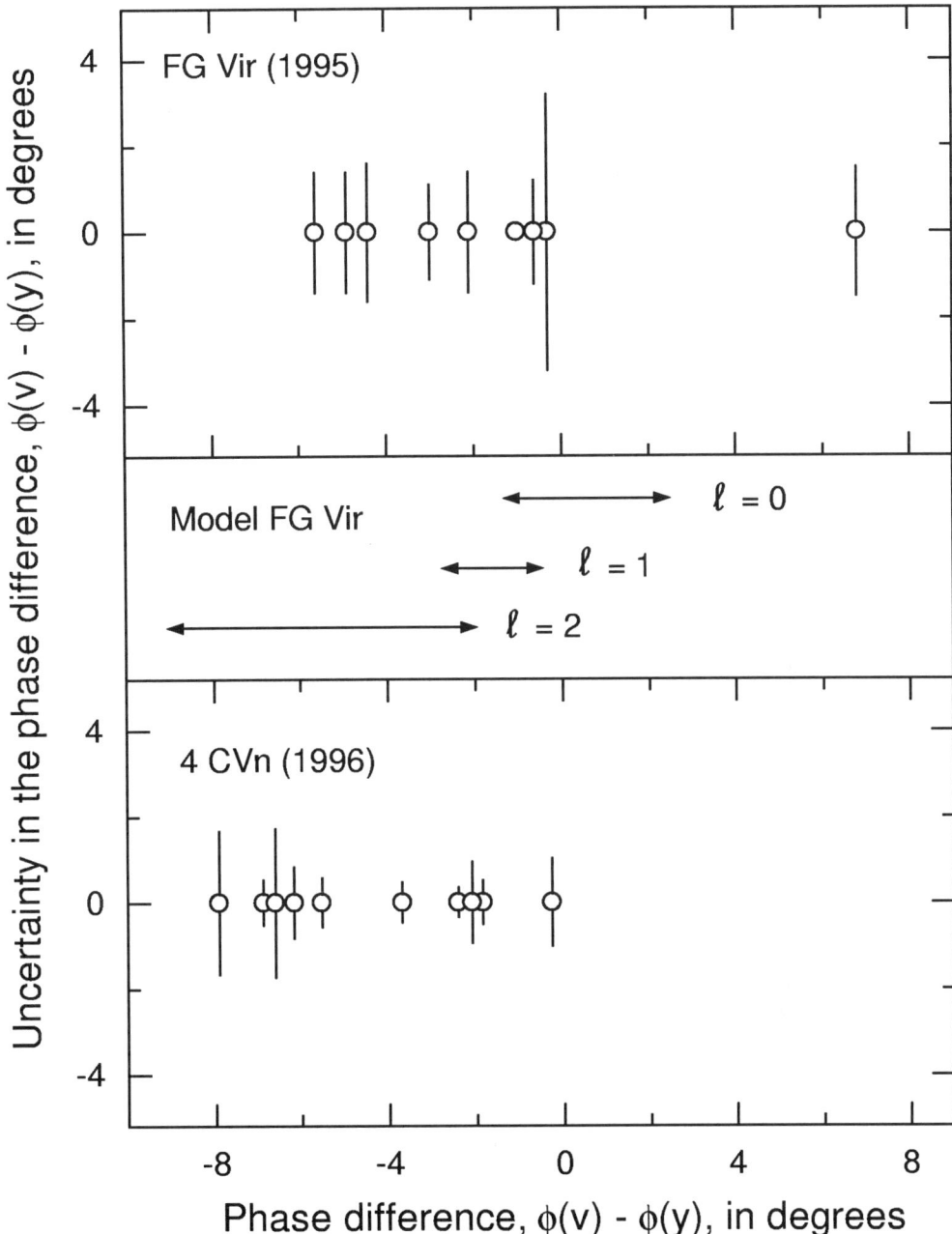

Figure 2. Illustration of how photometric phase differences can be used to determine ℓ values. The results for FG Vir and 4 CVn are preliminary analyses of Delta Scuti Network campaigns. The middle panel shows phase differences for different ℓ values covering all the different frequencies excited in FG Vir (models computed for FG Vir by R. Garrido). For individual modes, the ℓ determination becomes considerably less ambiguous.

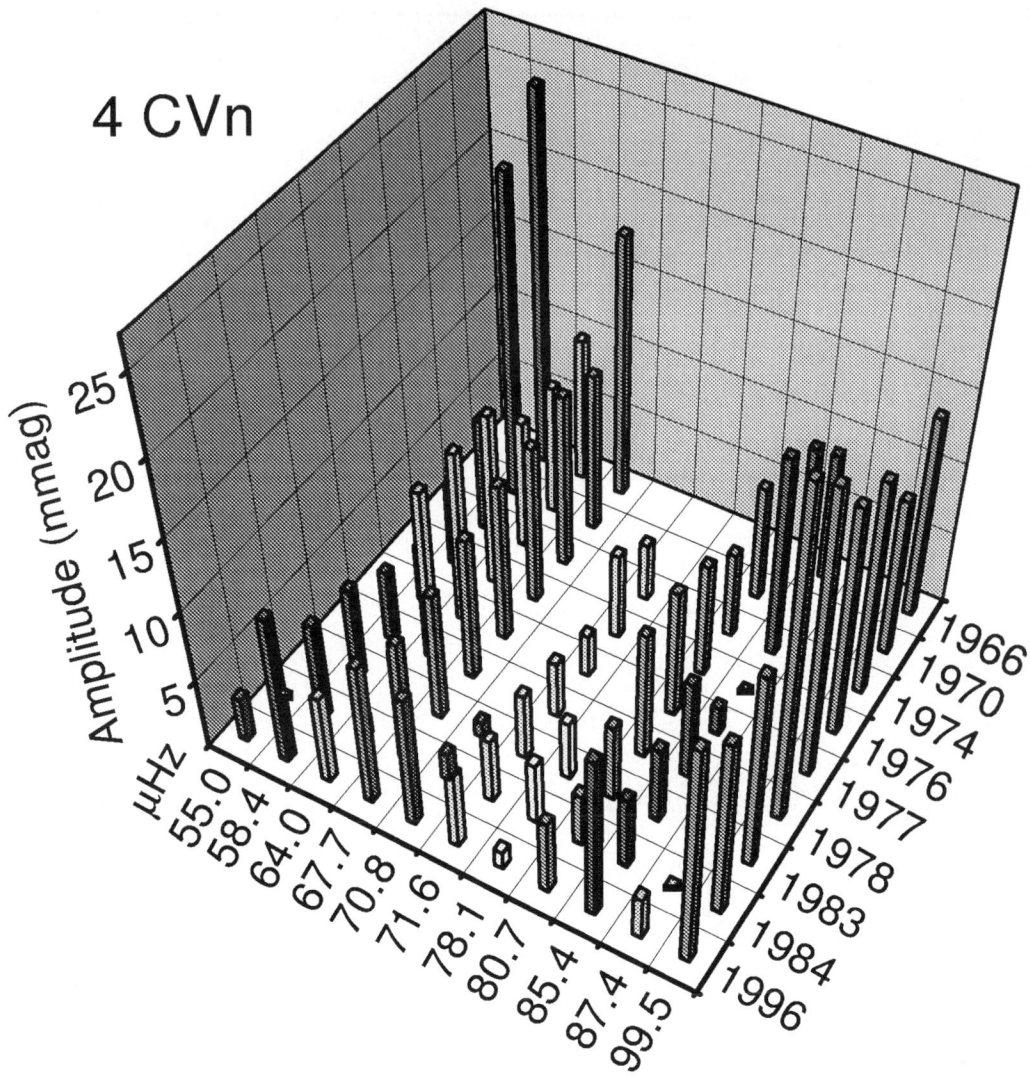

Figure 3. Amplitude variations of the major pulsation modes of 4 CVn from 1966 to 1996. The lower axis shows the frequency of the pulsation mode plotted. Note that most amplitude variations have time scales of a decade or longer.

modes as well as the stellar parameters. The models (e.g. Garrido et al. 1990) predict both phase shifts and amplitude ratios. The work by Watson (1988) indicates that the use of the amplitude ratios still needs to be examined in more detail, since a number of stars show amplitude ratios not predicted by the models. The phase shifts, on the other hand, are less sensitive on the details of modelling and can be used. In fact, some initial successes are already obtained:

- The phase shifts observed for the star 4 CVn by the Delta Scuti Network (unpublished data) satisfy a consistency check: in the phase shift diagram (Fig. 2), it is possible to pinpoint four adjacent pulsation modes in a central group with phase shifts near −2 to −3 degrees and the same ℓ value of 1. The frequency differences between these modes are similar, viz. 14.1, 13.7 and 13.2 μHz. They correspond to the expected values of four adjacent radial quantum numbers, as long as these modes have the same ℓ and m values.
- For the star FG Vir, the mode identifications obtained by us with the phase difference method are in excellent agreement with those obtained by Viskum (1997) from the equivalent width method.

The methods of pulsation mode identification are based on the interaction between observations and theory. Despite the very encouraging initial successes, at this stage two warnings appear prudent: (i) the uncertainties of the observational results need to be kept in mind during the interpretations (and reduced in size in the future), and (ii) the theoretical models need to be further refined and individually calculated for every star.

4. The evolved δ Scuti star 4 CVn

During 1996 the Delta Scuti Network carried out an extensive photometric campaign of the δ Scuti variable 4 CVn (F3 III-IV) and a new neighboring γ Doradus variable, HD 108100 at six observatories (Breger et al. 1997a). In order to increase the number of detected frequencies and to study the short-term amplitude variability, an additional campaign of 4 CVn was carried out during 1997. While the 1997 data are not yet reduced, preliminary results for 1996 are available: 335 hours of high-quality photometry obtained during 55 nights led to the detection of 25 statistically significant frequencies of pulsation. The first results can be summarized as follows:

- 8 of the 25 detected pulsation frequencies are not independent frequencies, but combinations, $f_i \pm f_j$. This number of combination frequencies is unusually large for a δ Scuti variable.
- The observed pulsation frequencies divide into three groups: in the region between 50 and 120 μHz for the excited modes, the region below 40 μHz for combination frequencies $f_i - f_j$ and the region above 120 μHz for the combination frequencies $f_i + f_j$.
- 4 CVn shows a large amount of amplitude variability, while the frequencies are relatively stable (Breger 1990). It is interesting that the time scales of amplitude variability seen in the evolved star 4 CVn are considerably longer than in the unevolved star CD -24 7599 (Handler et al. 1997b). The amplitude variability of the 11 pulsation modes with highest amplitudes are shown in Fig. 3.
- Pulsation models calculated by us with the Dziembowski code predict a dense forest of hundreds or thousands of excited frequency peaks in the power spectrum of this evolved star caused by the many g-modes from the stellar core. This would even hold for low-degree modes. This large number of excited frequencies is not seen by the observations, at least with amplitudes above the observational limit of about 0.4 millimag. In fact, the frequency spectrum resembles those of unevolved δ Scuti stars, once the frequency shift caused by the lower density is accounted for. So far, the pulsation theory used in the models cannot predict amplitudes and their variation in time. Nevertheless, the present results for 4 CVn raise the

tantalizing question whether there exists for evolved δ Scuti stars a presently unknown mechanism which damps pulsation in the stellar core.

During the last few years considerable progress has been made in observational asteroseismology of δ Scuti stars. With presently available equipment, the photometric observational limits can be improved relatively easily:

- The use of CCD detectors can increase the duty cycle while preserving the important low-frequency stability.
- Automatic photoelectric telescopes can considerably increase the amount of photometric data. Measurements in at least two colors can allow mode identifications. As an example, for the 1997 campaign of 4 CVn the Vienna APT has been successfully used.
- The present limit of 10^{-4} in photometric amplitude can be improved with longer small-telescope campaigns of 1000+ hours. This should raise the number of detected frequencies per star to more than 50 and allow the detection of complete multiplets of rotational splitting.
- The discovery of close frequencies in some δ Scuti stars suggests that the frequency resolution of the campaigns needs to be improved by longer coverage during the observing season. Together with other measurements from other years, a frequency resolution of ~ 0.05 μHz should be possible.

Acknowledgements

This work has been partially supported by the Austrian Fonds zur Förderung der wissenschaftlichen Forschung, project number S7304.

References

Bedding, T. R., Kjeldsen H., Reetz, J., Barbuy B., 1996, MNRAS 280, 1155
Breger, M., 1990, A&A 240, 308
Breger, M., Handler, G., Nather, R. E., et al., 1995, A&A 297, 473
Breger, M., Handler, G., Garrido, R., et al., 1997a, A&A 324, 566
Breger, M., Zima, W., Handler, G., et al. 1997b, A&A, in press
Dziembowski, W. A., Pamyatnykh, A. A. 1991, A&A 248, L11
Frandsen, S., Balona, L. A., Viskum, M., et al., 1996, A&A 308, 132
Garrido, R., Garcia-Lobo, E., Rodriguez, E., 1990, A&A 234, 262
Handler, G., Pikall, H., O'Donoghue, D., et al., 1997a, MNRAS 286, 303
Handler, G., Pamyatnykh, A. A., Zima, W., et al., 1997b, MNRAS, in press
Kennelly, E. J., Walker, G. A. H., 1996, PASP 108, 327
Mantegazza, L. 1997, A&A 323, 844
Mantegazza, L., Poretti, E., Bossi, M., 1996, A&A 308, 847
Mathias, P., Aerts, C., 1996, A&A 312, 905
Michel, E., Belmonte, J. A., Alvarez, M., et al., 1992, A&A 255, 139
Paparó, M., Sterken, C., Spoon, H. W. W., Birch, P. V., 1996, A&A 315, 400
Templeton, M., McNamara, B. J., Guzik, J. A., 1997, AJ, in press
Viskum, M. 1997, private communication
Watson, R. D. 1988, ApSS 140, 255
Winget, D. E., Nather, R. E., Clemens, J. C., et al., 1991, ApJ 378, 326

DELTA SCUTI STARS: THEORY

J. A. GUZIK
Los Alamos National Laboratory
XTA, MS B220, Los Alamos, NM 87545
E-mail: joy@lanl.gov

1. Introduction

The purpose of asteroseismology is not only to derive the internal structure of individual stars from their observed oscillation frequencies, but also to test and extend our understanding of the physics of matter under the extremes of temperature, density, and pressure found in stellar interiors. In this review, I hope to point out what we can learn about the Sun by studying δ Scuti stars, as well as what we can learn about stars more massive or evolved than the Sun. I will discuss some of the difficulties in theoretical approaches to asteroseismology for δ Scuti stars, using FG Vir, δ Scuti, and CD $-24°$ 7599 as examples.

2. Why Study δ Scuti Stars?

The study of δ Scuti stars can provide a bridge in our knowledge between the Sun and more massive or evolved stars. We have enough confidence in our modeling and input physics from the reasonably good agreement between solar models and helioseismic inversions that we can now apply these modeling tools to stars somewhat more massive than the Sun. δ Scuti stars are main sequence or slightly post-main sequence (luminosity class III-V) stars of 1.4 - 3 $M_\odot$. δ Scuti stars are burning hydrogen either in a convective core, or in a shell outside a H-depleted core, predominantly by CNO-cycle processing, rather than the pp-chains of the Sun. The Sun has a convective envelope extending through the outer 30% of its radius, and no convective core. Main sequence δ Scuti stars have convective cores that comprise $\sim$20% of the inner radius, and very thin convection zones in the H and He ionization regions near the surface. The Sun's pulsations are stochastically excited by convection, and have radial velocity amplitudes of meters/sec, whereas δ Scuti pulsations are driven by the κ effect (H and He ionization valving), and have amplitudes up to tens of km/sec. The Sun's surface rotation velocity is $\sim$2 km/sec, and its internal rotation profile has been determined rather well from helioseismic inversions. The measured $v \sin i$ values of δ Scuti stars range from 10 to over 200 km/sec (Solano and Fernley 1997). It may be possible to measure portions of the internal rotation profile using oscillation frequencies (Goupil et al. 1996).

3. Constraints for Seismology

We have measured most properties of the Sun much more accurately than we can for other stars. We know its mass, luminosity, age, and surface abundance ratio Z/X quite accurately. For δ Scuti stars, we can determine luminosities if we know their distance (the Hipparcos parallax measurements will help for some δ Scuti stars). We can determine the effective temperature, surface gravity, and composition from photometric colors and spectra. For δ Scuti stars in clusters, we can better estimate the age, luminosity, and abundances by deriving these properties for the cluster as a whole (see, e.g., Hernandez et al. 1998 (Praesepe); Audard et al. 1996 (Hyades)).

Since the photospheres of both the Sun and δ Scuti stars are too cool to exhibit He lines in the visible, we cannot measure the photospheric He abundance directly. For the Sun, we now have strong evidence that diffusive settling has decreased the abundance of He and possibly of heavier elements in the envelope convection zone. The photospheric abundances of δ Scuti stars show wide variety and anomalies (Russell 1995). We have yet to determine how these abundances may have been affected by diffusive settling, radiative levitation, or mass loss.

Since we can resolve the disk of the Sun, we have been able to observe millions of p-modes of degrees ℓ=0-1000. For δ Scuti stars, since we cannot resolve the disk, we can only detect by photometry the low-degree modes (ℓ=0-2, and perhaps 3) with intensities or radial velocities that do not average out over the disk. By examining variations in spectral line profiles of rapidly rotating δ Scuti stars, a number of researchers (e.g. Kennelly and Walker 1996) have been able to detect some higher-degree nonradial modes. We can expect to observe dozens of predicted unstable radial and nonradial modes, pure p-modes, modes with a mixed p- and g-mode character, and possibly pure g-modes. The modes with g-type nodes are potentially very useful in probing the convective core size and measuring the distance of convective overshooting (Dziembowski and Pamyatnykh 1991; Audard et al. 1995).

Mode identification for solar p-modes is straightforward, since we can detect all of the expected modes, and measure the frequencies very accurately. But mode identification is problematic for δ Scuti stars. Not all of the predicted modes are observed, and it is difficult to find a pattern that gives clues to the mode identification. Progress has been made recently in mode identification for a few δ Scuti stars (e.g., 4 CVn and FG Vir) using spectroscopic methods (Breger et al., these proceedings; Viskum et al. 1998), but the identification of some modes remains ambiguous.

4. Approaching Asteroseismology

An example of a star that is promising for asteroseismology is the δ Scuti star FG Vir (Breger et al. 1995, 1997). FG Vir is observed to have T_{eff}=7500 ± 150 K and $\log g$~4.0. The estimated rotation velocity is ~42 km/sec (Mantegazza et al. 1994). Breger et al. (1997) report 21 observed frequencies for FG Vir, including some that likely have mixed p- and g-type character. The luminosity derived from the Hipparcos parallax, in addition to spectroscopic mode identification, indicate that the sixth highest amplitude mode at 140.7 μHz is probably the radial fundamental mode. Several other modes have been identified by spectroscopic methods.

It is difficult to determine the intrinsic metallicity of FG Vir. The measured photospheric [Fe/H] is 0.65 dex, much higher than solar, whereas [C/H] = −0.63, much lower than solar. Such anomalies are expected due to diffusive settling and radiative levitation of elements (Smith 1996). The abundances of other elements less affected by settling or levitation, such as [Ca/H] and [Ti/H], are respectively 0.24 and 0.19 dex, which indicates that FG Vir may have an intrinsic Z of about 0.03.

Our goal is to find a model that matches all of these observational constraints for FG Vir. However, we also need to consider the uniqueness of the model. There are many areas of parameter space to explore, such as the helium (Y) or heavy element abundance (Z), element mixture, and internal rotation profile, which may make it difficult to find a unique model. We may be able to use asteroseismology to determine the Y abundance (Monteiro et al., these proceedings), the Z abundance (Guzik et al. 1998), the internal rotation profile (Goupil et al. 1996), or indicate the need for improved opacities (Brown et al. 1994), or convective overshoot (Monteiro et al., these proceedings; Dziembowski and Pamyatnykh 1991; Audard and Roxburgh 1997). Applying our asteroseismological tools to many δ Scuti stars will be necessary to distinguish between solutions that may produce the same effect. For example, can we discriminate between the effects of increased Z or convective core overshoot, both of which enlarge the convective core?

Here we examine the effect of metallicity on the frequencies of candidate FG Vir models. Guzik et al. (1998) identified two models in their evolution and pulsation calculations of mass 1.82 and 1.95 $M_\odot$, with Z=0.02 and 0.03, respectively, that match many of the observational constraints of FG Vir (T_{eff}, $\log g$, radial fundamental mode at 140.7 μHz). These models do not include convective overshoot. Table 1 compares the model properties. Without additional pulsation frequencies, these models would be indistinguishable from each other observationally, except perhaps for their photospheric abundances (which may not be representative of the interior abundance due to diffusion/levitation). However, the interior structures of these models, particularly the convective core size and degree of core H-depletion, are significantly different.

TABLE 1. FG Vir Model Properties

Model Property	1.82 $M_\odot$	1.95 $M_\odot$
Z	0.02	0.03
R ($R_\odot$)	2.26	2.31
M/R^3	0.1575	0.1580
T_{eff}(K)	7368	7412
$\log g$	3.99	4.00
$\log L/L_\odot$	1.13	1.16
$X_{convcore}$	0.257	0.355
$R_{convcore}$ ($R_\odot$)	0.175	0.220
$M_{convcore}$ ($M_\odot$)	0.155	0.181
Age (Gyr)	0.879	0.731

Figure 1 shows the calculated radial (ℓ=0) and nonradial (ℓ=1 and 2) modes for the two models plotted against each other, assuming for now no rotational splitting (m=0). If all of the calculated frequencies were identical, the points would lie on a straight line. One can see that the predicted radial mode frequencies are nearly identical, which is not surprising since these models have nearly the same mean density. For the nonradial modes, there are a few interesting differences. For ℓ=1, there is

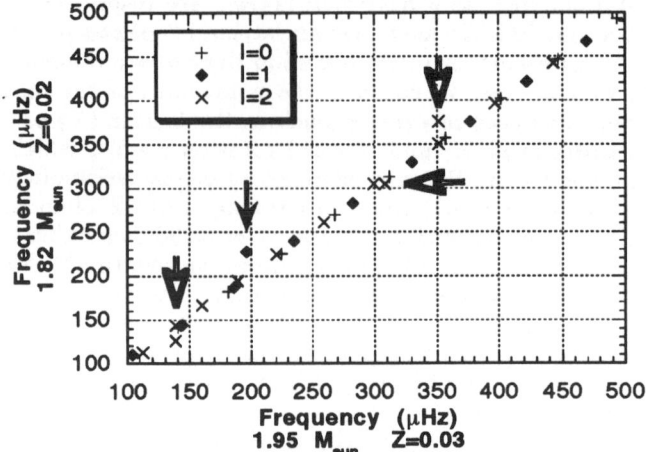

Figure 1. Calculated frequencies of the two candidate FG Vir models from Table 1 plotted against each other. A few nonradial modes are shifted (solid arrow) in frequency or added (open arrows) due to the differences in internal structure between models.

a 30 μHz shift between models in the mode near 200 μHz, as a consequence of the difference in convective core size. For $\ell=2$, the more evolved 1.82-$M_\odot$ model has two closely-spaced frequencies near 140 μHz, and again at 350 μHz, whereas the 1.95-$M_\odot$ model has only one mode near these frequencies. The 1.95-$M_\odot$ model has two closely-spaced modes near 300 μHz, whereas only one mode exists near this frequency for the 1.82-$M_\odot$ model. It is important to note that the modes with nearly identical frequencies in each model may have one more or less radial node, or a p-type instead of a g-type node, and therefore significantly different eigenfunctions.

How well do these models match the frequencies reported for FG Vir? Taking into account plausible first-order rotational splitting for the nonradial modes, and adopting the additional spectroscopic mode identifications of Breger *et al.* (these proceedings) and Viskum *et al.* (1998), neither model simultaneously matches all 21 frequencies (see Guzik *et al.* 1998). The fit would be improved if we could find a model in which a mode is added (compared to the 1.95-$M_\odot$ model) or shifted (compared to the 1.82-$M_\odot$ model) near 130 μHz, and also in which the $\ell=2$ mode near 260 μHz is shifted to $\sim$ 250 μHz. These clues can be used in the next iteration to evolve models that more closely match the observed frequencies.

5. Prospects for Asteroseismology of More-Evolved δ Scuti Stars

It may prove difficult to do asteroseismology on the more evolved δ Scuti stars that are burning hydrogen in a shell outside the H-depleted core, due to the very dense spectrum of predicted modes (mainly high-order g-modes) that have not yet been detected, and the difficulty in mode identification given these predictions. As an example, we consider δ Scuti itself. Templeton *et al.* (1997) reported six frequencies in the range 54-99 μHz for this star. Templeton *et al.* constructed evolution and pulsation models with masses 2.1 and 2.4 $M_\odot$, and Z=0.02 and Z=0.06, respectively, that match the observed spectral type, luminosity (given the Hipparcos parallax), and identified radial fundamental mode frequency (57.731 µHz) of δ Scuti. For the 2.4-M$_\odot$, Z=0.06 model with the larger mode spacing, there are 3 unstable radial modes, 13 unstable ℓ=1 modes, and 19 unstable ℓ=2 modes. Accounting for rotational splitting of the nonradial modes into 2ℓ+1 frequencies, this model has 137 possible modes! A range of ℓ=2 modes, and often at least one ℓ=1 mode can match the five remaining observed frequencies. It is possible that other predicted modes have a very low amplitude and will eventually be detected, but we may need to determine instead why only a few modes (and which modes) are excited to observable amplitudes. Dziembowski and Krolikowska (1990) propose that the ℓ=1 modes may be preferentially selected due to partial mode trapping. For δ Scuti, this would narrow the choice of frequency matches to only one or two ℓ=1 modes per observed frequency.

As mentioned by Breger (these proceedings), there may be observational indications for the evolved δ Scuti star 4 CVn that this dense spectrum of predicted g-modes does not occur at all, as though the g-type nodes have been somehow suppressed in the core. Breger finds that the pattern of observed modes in 4 CVn is consistent with a sequence of radial modes and rotationally-split nonradial p-modes. If such patterns are found in the frequencies of other evolved δ Scuti stars, explaining these observations will pose an interesting challenge to the theoreticians!

6. The Importance of Rotation

To match observed frequencies in detail, and also to derive the internal rotation profile for a δ Scuti star, it is critical to take into account the rotational splitting of modes. To date, rotational effects have been taken into account mainly using perturbation theory approaches. The angular eigenfunctions of nonradial modes are described by spherical harmonics (Y_ℓ^m), in which ℓ is the number of node lines on the surface, and m the number of node lines through the rotation axis. To first order, rotation breaks the m frequency degeneracy, by accounting for the Coriolis force, which is proportional to $\Omega \times v$. When Ω is a function of radius only (i.e. the rotation rate has no latitudinal dependence), the nonradial modes are split into 2ℓ+1 equally-spaced prograde and retrograde modes. First-order corrections neglect the distortion of the star from a sphere, so the radial mode frequencies are not affected.

For moderately-rotating δ Scuti stars, second-order corrections become significant. Second-order corrections account for the centrifugal force ($\Omega \times \Omega \times r$), and include distortion of the star from a sphere. These corrections shift radial modes as well as nonradial, and destroy the symmetrical m-splitting of nonradial modes. Second-order corrections are derived under various approximations by Saio (1981; solid body rotation, include overall shift in multiplet frequencies), Gough and Thompson (1990; include magnetic fields, and overall shift in average multiplet frequency due to distortion, but neglect latitudinal dependence of rotation), and Dziembowski and Goode (1992; include radial and latitudinal dependence of Ω, but neglect overall shift in multiplet frequencies). Accidental degeneracies between modes of different degree ℓ also affect mode frequencies, and must be taken into account (Soufi et al. 1998).

Figure 2 illustrates the first- and second-order rotational splitting corrections for a model of CD −24° 7599 (Bradley and Guzik 1997), with assumed uniform rotation velocity of 75 km/sec (Handler et al. 1997). The solid horizontal lines indicate the 13 observed frequencies of Handler et al. Second-order corrections are estimated using the formula of Saio (1981), strictly valid for n=3 polytropes and uniform rotation. As can be seen, the first-order corrections split the modes evenly, and are ~10 µHz

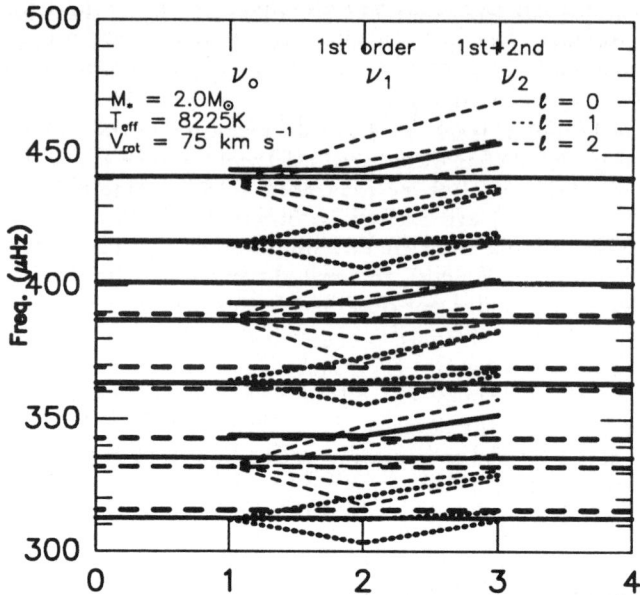

Figure 2. A schematic illustration of the importance of rotational splitting on the frequencies of CD −24° 7599. ν_0 are frequencies without rotational splitting, ν_1 includes first order rotation corrections, and ν_2 includes both first and second-order corrections.

per Δm. The second-order corrections shift all of the multiplets to higher frequency, and break the even spacing within a multiplet. The radial modes are shifted by ∼10 μHz. Considering the predicted spectrum of frequencies, with modes of different ℓ values now lying very close to each other, it will be important to include the effects of mode coupling. It will also be very difficult to disentangle mode identifications without additional information.

Third-order corrections are also significant for even moderately rotating δ Scuti stars. Soufi et al. (1998) are working out a third order perturbation theory formalism. Ultimately perturbation theory may be inadequate for the most rapidly rotating stars (v ∼ 200 km/sec) and we may need to calculate the evolution and pulsations of two- or three-dimensional differentially rotating distorted models. There has been progress in this direction, e.g. the two-dimensional evolution/hydrodynamics modeling of Deupree (1995, 1997), and the three-dimensional pulsation modeling of Clement (1993).

7. What Can We Learn about the Sun from δ Scuti Stars?

Since δ Scuti stars cover a wide range of parameter space in mass, composition, temperature, rotation rate, etc., there are a number of areas where improvements in the physics of modeling δ Scuti stars will feed back into solar modeling. First, we may be able to verify or improve microphysics (opacities, equation of state, diffusion treatments, nuclear reaction rates). Brown et al. (1994), applying inversion techniques to hypothetical frequency spectra of sun-like stars, find that oscillation frequencies can be used to distinguish among models using different opacities or equations of state.

The photospheric composition and pulsation driving of δ Scuti stars are enigmatic. Diffusion time-scales in the stellar envelope are much shorter for the δ Scuti stars than for the Sun due to their very small envelope convection zones. The amount of helium diffused from the pulsation driving layers directly affects the pulsational stability (Cox et al. 1979), amplitudes (Guzik 1993), and light curve shapes (Guzik 1992). Helium settling and selective levitation of some elements may be mitigated by mass loss; a not-unreasoanble mass loss rate of 10^{-12} $M_\odot$/yr would be required to "keep up" with diffusion (Guzik and Cox 1991). If we can use pulsation properties to learn more about abundances in the pulsation driving region of δ Scuti stars, we may be able to better test or calibrate calculations and theories of diffusion, mixing, or mass loss proposed for the Sun to explain the solar Li abundance, or the discrepancy between inferred and calculated sound speed at the convection zone base for standard solar models including helium and element diffusion.

Surface boundary conditions are another area in which we may gain insight from δ Scuti stars. Many of the remaining discrepancies between observed and calculated solar oscillation frequencies are due to problems with solar models near the superadiabatic gradient at the top of the solar convection zone ($\sim$8,000 to 10,000 K). Increased low-temperature opacities, nonequilibrium radiation diffusion, nonadiabaticity, turbulent pressure, and more sophisticated convection treatments have all been proposed to alleviate this discrepancy (see Guzik et al. 1996). Li and Stix (1994) discuss moving the outer boundary of δ Scuti models to a smaller optical depth to reproduce the instability strip blue edge. Audard et al. (1998) discuss treatments of model atmospheres for δ Scuti and Ap stars that improve agreement with the acoustic cutoff frequency. Perhaps investigating problems such as these will provide support for one or more proposed improvements in solar atmosphere modeling.

There are a number of questions raised by helioseismic measurements of the solar internal rotation. How did the Sun's current rotation profile evolve? Do other stars show latitudinally-dependent rotation rates that persist throughout the envelope? Are other stars more rapidly rotating in their cores than at their photospheres? Hopefully, systematics of rotational splitting due to differential rotation will answer these questions for other stars, and help us determine whether the Sun is typical or anomalous.

Finally, there is hope to study g-mode driving, amplitudes, detectability, and mode selection, to learn more about what we can expect to observe for g-modes in the Sun.

8. Discussion

M. Paparo: I would like to call your attention to θ Tuc, for which ten frequencies were obtained which have extremely regular spacing (Paparo et al. 1996). This gives some guidelines for mode identification. Mode identification based on colour data is coming out soon (Sterken 1997). The modeling of θ Tuc is probably easier than the complex case of FG Vir.

J. Guzik: Yes, I am aware of this star, and my graduate student M. Templeton is trying to model it.

Acknowledgements

The author is especially grateful to P. Bradley, A. Cox, M. Templeton, M. Breger, W. Dziembowski, D. Kurtz, and J. Christensen-Dalsgaard for useful discussions and assistance. This work was funded by a NASA Astrophysics Theory grant.

References

Audard, N., Provost, J. and Christensen-Dalsgaard, J. 1995, *Astron. Astrophys.* **297**, 427.
Audard, N., et al.: 1996, *Bull. Astron. Soc. India* **24**, 305.
Audard, N. and Roxburgh, I.W.: 1997, in F. Schmider and J. Provost (eds.) *IAU Colloquium 181: Sounding Solar and Stellar Interiors*, in press.
Audard, N., Kupka, F., Morel, P. and Weiss, W.W. Atmosphere Models of A-F Stars. Effects on the Acoustic Cut-off Frequency, 1998, *Astron. Astrophys.*, submitted.
Bradley, P.A. and Guzik, J.A.: 1997, in F. Schmider and J. Provost (eds.) *IAU Colloquium 181: Sounding Solar and Stellar Interiors*, in press.
Breger, M. et al.: 1995, *Astron. Astrophys.* **297**, 473.
Breger, M. et al.: 1997, *Astron. Astrophys.*, in press.
Brown, T.M., Christensen-Dalsgaard, J., Weibel-Mihalas, B., and Gilliland, R.: 1994, *Astrophys. J.* **427**, 1013.
Clement, M.J.: 1993, *Astrophys. J.* **406**, 651.
Cox, A.N., King, D.S., and Hodson, S.W.: 1979, *Astrophys. J.* **231**, 798.
Cox, J.P.: 1984, *Publ. Astron. Soc. Pacific* **96**, 577.
Deupree, R.G.: 1995, *Astrophys. J.* **439**, 357.
Deupree, R.G.: 1997, Stellar Evolution with Arbitrary Rotation Laws III. Convective Core Overshoot and Angular Momentum Distribution, *Astrophys. J.*, in press.
Dziembowski, W.A. and Krolikowska, M.: 1990, *Acta Astronomica* **40**, 19.
Dziembowski, W.A. and Pamyatnykh, A.A.: 1991, *Astron. Astrophys.* **248**, L11.
Dziembowski, W.A. and Goode, P.R.: 1992, *Astrophys. J.* **394**, 670.
Gough, D.O. and Thompson, M.J.: 1990, *MNRAS* **242**, 25.
Goupil, M.J., Dziembowski, W.A., Goode, P.R., and Michel, E.: 1996, *Astron. Astrophys.* **305**, 487.
Guzik, J.A. and Cox, A.N.: 1991, *Delta Scuti Star Newsletter*, ed. M. Breger, **3**, 6.
Guzik, J.A.: 1992, *Delta Scuti Star Newsletter*, ed. M. Breger, **5**, 8.
Guzik, J.A.: 1993, in J. Nemec and J. Matthews (eds.) *New Perspectives on Stellar Pulsation and Pulsating Variable Stars*, Cambridge U. Press, Cambridge, p. 243.
Guzik, J.A., Cox, A.N., and Swenson, F.J.: 1996, *Bull. Astron. Soc. India* **24**, 161.
Guzik, J.A., Bradley, P.A., and Templeton, M.R. et al.: 1998, in P. Bradley and J. Guzik (eds.) *A Half Century of Stellar Pulsation Interpretations*, in press.
Hernandez, M.M. et al.: 1998, in P. Bradley and J. Guzik (eds.) *A Half Century of Stellar Pulsation Interpretations*, in press.
Handler, G. et al.: 1997, *Astron. Astrophys.* **286**, 303.
Kennelly, E.J. and Walker, G.A.H.: 1996, *Publ. Astron. Soc. Pacific* **108**, 327.
Li, Y. and Stix, M.: 1994, *Astron. Astrophys.* **286**, 815.
Mantegazza, L., Poretti, E., and Bossi, M.: 1994, *Astron. Astrophys.* **287**, 95.
Paparo, M., Sterken, C., Spoon, H., and Birch, P.V.: 1997, *Astron. Astrophys.* **315**, 400.
Russell, S.C.: 1995, *Astrophys. J.* **451**, 747.
Saio, H.: 1981, *Astrophys. J.* **244**, 299.
Smith, K.C.: 1996, *Astrophys. Space Sci.* **237**, 77.
Solano, E. and Fernley, J.: 1997, *Astron. Astrophys. Suppl. Ser.* **122**, 131.
Soufi, F., Goupil, M.J., and Dziembowski, W.A.: 1998, *Astron. Astrophys.*, in preparation.
Sterken, C.: 1997, *Astron. Astrophys.* **325**, 563.
Templeton, M.R. et al.: 1997, *Astron. J.* **114**, 1592.
Viskum, M. et al.: 1998, in P. Bradley and J. Guzik (eds.) *A Half Century of Stellar Pulsation Interpretations*, in press.

THE DISCOVERY OF NON-RADIAL GRAVITY-MODE PULSATIONS IN γ DORADUS-TYPE STARS

K. KRISCIUNAS
Department of Astronomy
University of Washington
Box 351580
Seattle, Washington 98195-1580 USA

Abstract. Over two dozen early F-type variable stars have been identified which constitute a new class of pulsating stars. These stars typically have periods between 0.5 and 3 days with V-band variability of several hundredths of a magnitude. Given the time scale of the variability, the pulsations would have to be non-radial gravity-mode pulsations. The pulsational nature of some of these stars has been proven by means of coordinated multi-longitude photometric campaigns, radial velocity (RV) variations and line profile (LP) variations, indicating low degree spherical harmonics ($\ell = 3$ or less). Evidence is that these stars are younger than 300 Myr; one would surmise that a rapid onset of convection in their outer photospheric layers puts an end to the pulsations. Further observation and modeling of these stars is important for our understanding of stellar evolution, the search for g-modes in the Sun, and is even relevant to the interpretation of radial velocity variations of solar-type stars (e.g. 51 Peg) in the search for extrasolar planets.

1. Introduction

Since the discovery by Cousins & Warren (1963) that the early F-type dwarf star γ Doradus is variable, about two dozen variable stars of similar spectral type and luminosity class have been identified which vary up to 0.1 mag on time scales much slower (e.g. 0.5 to 3 days) than the fundamental radial pulsational period (e.g. 1 to 3 hours) for stars of this density . Krisciunas & Handler (1995) give a list of 17 *bona fide* γ Dor stars and candidates. A color-magnitude diagram of these and other stars is shown in Fig. 1. Examples of light curves of two of the best studied examples are found in Figs. 2 and 3. Updated information on these stars can be found at this website:

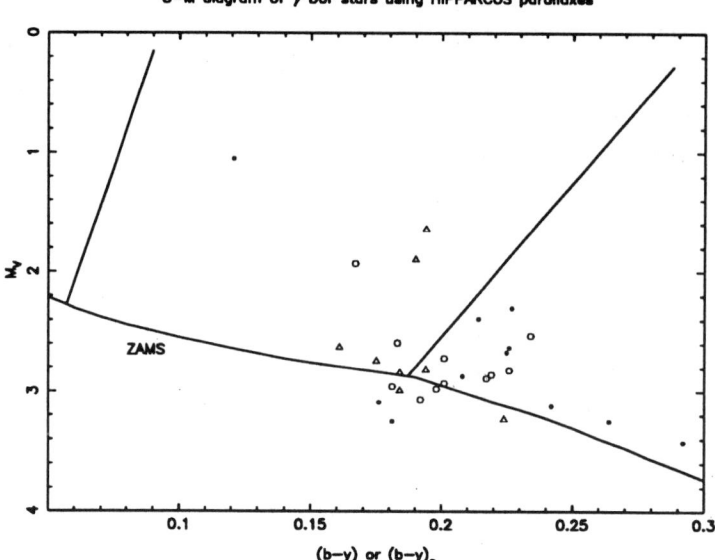

Figure 1. Color-magnitude diagram of *bona fide* γ Dor stars (circles) and candidates (dots). Stars in the open cluster NGC 2516 are represented by triangles. The position of the zero age main sequence and the borders of the δ Scuti instability strip are also indicated. All but 3 of the field stars have absolute magnitudes derived from HIPPARCOS parallaxes.

http://www.ast.univie.ac.at/~gerald/gdorlist.html

As one can see in Fig. 1, the γ Dor-type stars with HIPPARCOS parallaxes are found *on* the main sequence and overlap the cool edge of the δ Scuti instability strip, an extension to fainter absolute magnitudes of the classical Cepheid instability strip. It would of course be of interest to delineate as accurately as possible the boundaries of this region of the HR diagram.

2. What is causing the variability?

Details of the light curve of a variable star (i.e. period, amplitude, shape, color variations) usually provide sufficient clues to determine the cause of the light variations. Spectroscopic information (i.e. radial velocities – RVs – and line profiles – LPs) can be used to prove one's case. As Sherlock Holmes (1890) says: "When you have eliminated the impossible, whatever remains, *however improbable*, must be the truth." This of course assumes that one can compile a *complete* list of suspects, or, in our case, make a complete list of causes of an observed phenomenon.

Eclipsing binaries show very regular, repeatable light curves, with one period and one or two minima per cycle (the primary one often being a magnitude deep). γ Dor stars typically show multiple periods and a full range of a few hundredths to 0.1 mag in the Johnson V-band. We can easily state that γ Dor stars are not eclipsing binaries.

If we take the standard period-mean density equation for pulsating stars and

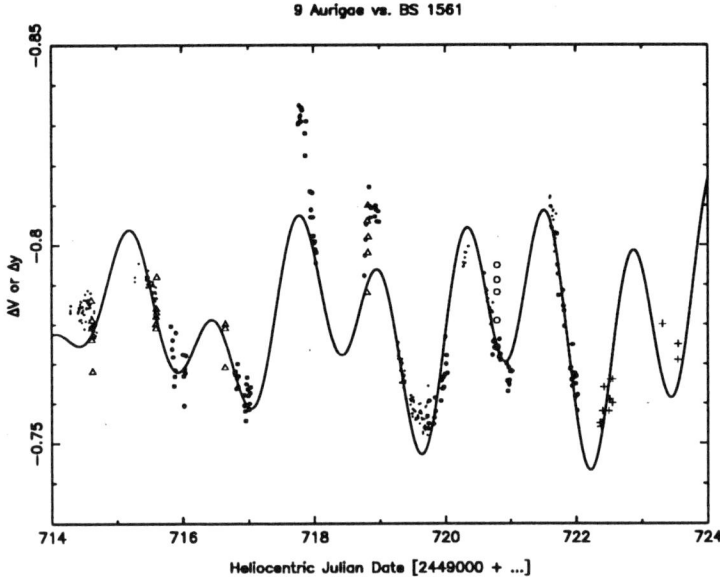

Figure 2. Light curve of 9 Aurigae vs. BS 1561 during part of the 1994/5 campaign. We fit the three frequencies found by Zerbi et al. (1997), namely 0.7948, 0.7679, and 0.3429 d^{-1}. Small dots represent data by Garrido, Rodríguez, and Zerbi at Sierra Nevada Observatory, Spain. Larger dots represent data by Krisciunas, Roberts, Crowe, and Pobocik at Mauna Kea, Hawaii. Other data are by Luedeke in Albuquerque, New Mexico (triangles), Guinan and McCook from Mt. Hopkins, Arizona (open circles), and Sperauskas in Lithuania (+'s). Note that even from day to day the amplitudes are irregularly variable.

express the fundmental radial pulsation period (Π) in terms of the radius and mass of the star in solar units, $\Pi = Q\, R_\star^{3/2}\, M_\star^{-1/2}$. γ Dor stars are typically of spectral type F0 V and have $R_\star \approx 1.73 R_\odot$ (Poretti et al. 1997) and $M_\star \approx 1.6 M_\odot$ (Popper 1980). $Q \approx 0.033$ days for the fundamental radial mode of δ Scuti stars (Fitch 1981, Table 2A), which are close cousins of the γ Dor stars. Thus, a γ Dor star would have a fundamental radial pulsation period of 1.4 hours, with a range of perhaps a factor of 2. Radial overtones and non-radial pressure-mode pulsations would have periods shorter than this.

Given the time scales of photometric variability, γ Dor stars are not pulsating in the fundamental radial mode, radial overtones, or by means of non-radial p-modes. While some are in binary or multiple systems, none is known to have a close interacting companion. One star *formerly* on the list has been shown to be an ellipsoidal primary of a binary system. The northern prototypical γ Dor star, 9 Aurigae, was once thought to be a single-line spectroscopic binary with a period of 391.7 days, but that interpretation of what are actually 2.89-day RV variations of variable amplitude is definitely incorrect.

After finding some "slow" variables with the colors of early F stars in the cluster NGC 2516, Antonello & Mantegazza (1986) suggested that early F-type stars could exhibit spots or non-radial gravity-mode oscillations. Mantegazza *et al.* (1993) and

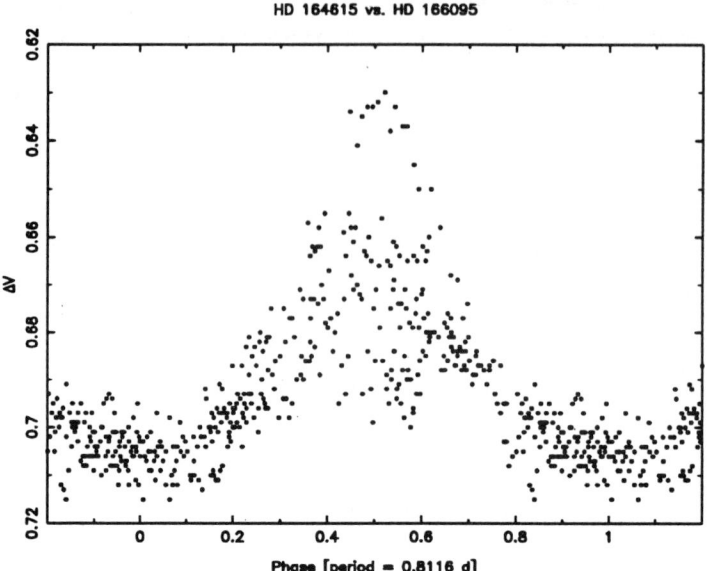

Figure 3. Light curve of HD 164615 vs. HD 166095 during the 1995 campaign (Zerbi et al. 1997). These are the data from Sierra Nevada Observatory only. HD 164615 shows only one period, but the amplitude of the photometric variations is clearly variable.

Krisciunas (1994; following discussions with Luis Balona and Jaymie Matthews) independently suggested that stars like γ Dor constitute a new class of variables. The former clearly favor rotational modulation of spots (which could be dark or bright spots) as the explanation, while I advocated that these were stars exhibiting non-radial g-modes.

Now, it is generally believed that stars with spectral types earlier than F7 do not show evidence of star spots (Giampapa & Rosner 1984). However, Güdel, Schmitt & Benz (1995) report the surprising result that the F0V star 47 Cas exhibits evidence for strong coronal activity. *If* a γ Dor star were exhibiting rotational spot modulation, three testable results can be checked: (1) Is the principal period compatible with the size of the star and the projected equatorial rotational speed? (2) The RV minimum should be 90 degrees out of phase with the luminosity maximum. Finally, (3) spot models only work for stars with one photometric period, or perhaps two *closely spaced* periods.

In the case of 9 Aur under the assumption of a spot model, the projected equatorial rotational speed and the principal photometric period imply that the star is within 16 degrees of being viewed pole-on (Krisciunas et al. 1995a). This means that there would be very little horizon beyond which any hypothetical spots could disappear. Though one can invoke a spot model with a variable projected area of spots as the star rotates, it would be a rather contrived spot model that could account for variability up to ≈ 0.1 mag with two or more *bona fide* periods. (9 Aur has shown evidence of 5 periods of variable amplitude.)

In the cases of 9 Aur and γ Dor, which show a RV range of about 4 km sec^{-1}, the most negative RV more nearly *coincides* with the phase of minimum light of one of the demonstrable periods, rather than being 90 degrees out of phase as predicted by the spot model (Krisciunas et al. 1995a, Balona et al. 1996).

Aerts & Krisciunas (1996) modeled 9 Aur as a non-radial gravity-mode pulsator with $\ell = 3$, $|m| = 1$. This was on the basis of CORAVEL autocorrelation diagrams obtained by Roger Griffin. The photometric variations are primarily driven at $f_1 = 0.795$ d^{-1} while the RV variations and LP variations are driven at the second most significant photometric frequency of $f_2 = 0.346$ d^{-1}. We found that, "the amplitude of the radial part of the pulsation for f_1 is a factor of 4 larger than the one for f_2, while its angular dependence is the same. Since the photometric variability is determined most of all by the radial part of the pulsation, it is quite understandable that the photometric variability is dominated by the mode with frequency f_1."

Balona et al. (1994) tried to model γ Dor itself by means of a differentially rotating spotted star model, but concluded that non-radial g-modes were a better explanation. Since the confirmation of a third frequency for this star and the analysis of LP and RV variations (Balona et al. 1996), the spot model is rejected and the g-mode model is confirmed.

A third γ Dor star which has been confirmed to be a non-radial g-mode pulsator is HD 164615 (Zerbi et al. 1997; Hatzes, in preparation). It exhibits photometric variations most likely at only a single frequency of 1.2321 d^{-1}, a frequency which has been stable for over 10 years. Like other γ Dor stars, it has a variable photometric amplitude. Its LP variations (see Fig. 4) can be modeled by an $\ell = 2$ sectoral mode non-radial pulsator with a mean pulsation amplitude of ≈ 7 km sec^{-1}. The pulsation amplitude seems to be a function of phase. $\ell = 3$ and 4 can also fit the profiles, but the amplitude of the pulsations is lower.

Mantegazza et al. (1994) provide some evidence that the multi-periodic γ Dor star HD 224638 also shows LP variations.

3. The γ Dor phenomenon is related to age

Eggen (1995) and Krisciunas et al. (1995b; following some discussion with Balona) independently suggested that γ Dor stars are all relatively young. Many of the field stars have space velocities like young disk stars. γ Dor is embedded in a β Pictoris-like disk or envelope. Eight candidates are found in the cluster NGC 2516 (age 120 ± 20 Myr), one in the Pleiades (age ≈ 80 Myr), and one in M 34 (Krisciunas & Crowe 1997), whose age is ≈ 250 Myr. Krisciunas et al. (1995b) found no γ Dor stars in the Hyades, whose age from HIPPARCOS data is 625 ± 50 Myr (Perryman et al. 1997). Eggen suggests that the γ Dor stars all lie in the "Böhm-Vitense decrement", a gap in the main sequence of many star clusters at $T_{eff} \approx 7700$ K. γ Dor stars have such temperatures and presumably have not yet experienced a rapid onset of convection in their photospheres. Once convection sets in, the pulsations presumably cease. Modeling of γ Dor stars, however, has not yet been accomplished to any satisfactory degree. Gautschy & Löffler (1996) were unable to find overstable low-degree oscillation modes for purely radially symmetric stars without any interaction with convection, rotation, or magnetic fields.

If we are correct that γ Dor stars are all younger than ≈ 300 Myr – their main sequence lifetimes are ≈ 3 Gyr – then we now have a means of obtaining an upper bound to the ages of some stars which are not in clusters. Also, this provides a means of determining an upper limit to the ages of some white dwarfs, such as the

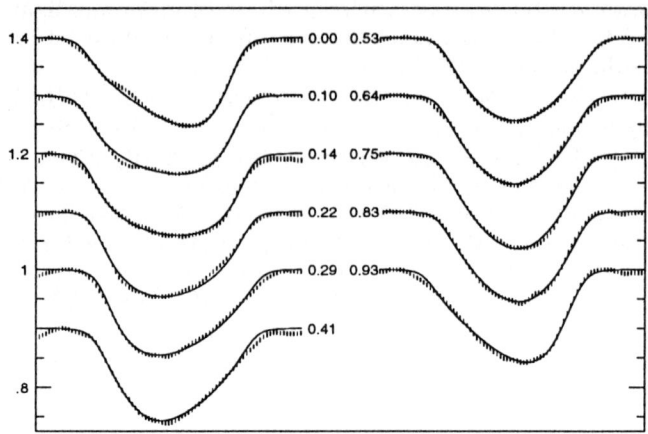

Figure 4. Profiles of the Fe II 531.8 nm line in HD 164615, measured by Hatzes. These are single spectra with phases based on an epoch of JD 2449521.710 and a period of 0.8116 d (Zerbi et al. 1997). The solid lines are fits based on an $\ell = 2$ sectoral mode model with pulsation amplitude 7 km sec^{-1}.

E-component of 9 Aur (Krisciunas et al. 1993), a star whose companionship to its supposed primary should be confirmed.

4. Discussion

Many types of variable stars are characterized by specific types of observational evidence (i.e. light curve amplitude and shape, period(s), correlations of photometric color and brightness, RV variations, LP variations, and signatures at non-optical wavelengths such as infrared or ultraviolet excesses). We of course want to understand the *physical* mechanism for a star's behavior.

The sum total of observational evidence on γ Dor stars indicates that the most likely explanation of their behavior is non-radial gravity-mode oscillations. This conclusion is based on their typically multiple periods, the time scale and amplitude of such variations, and the correlation (including phase) of those variations with RV and LP variations. Early F-stars with one period (even if of variable amplitude) could still be modeled by means of a (bright) spot model. However, it would be a cruel trick of Nature, in my opinion, if some stars in the γ Dor region of the HR Diagram were variable due to g-modes, while others were variable due to spots.

One of the most exciting discoveries of the past two years has been evidence for extrasolar planets (Boss 1996). All we know *for certain* is that there are solar-type stars with RV variations like those expected *if* there are Jupiter-mass planets orbiting them. Some of these variations are on time scales comparable to that of γ Dor stars, and there is *marginal* evidence that 51 Peg exhibits LP variations — something that should *not* be observed if the RV variations of the star are due to an orbiting planet (Gray 1997; but see also Hatzes et al. 1997).

Consider for a moment the RV variations of 9 Aurigae, which in 1993/4 followed its 2.9 d period and which had an amplitude of 2.00 km sec^{-1}. The corresponding photometric amplitude was 12.2 mmag. 51 Peg has a RV amplitude of 59 m sec^{-1} (Mayor & Queloz 1995). *If* 51 Peg were exhibiting non-radial pulsations like a scaled down version of 9 Aurigae, the corresponding photometric amplitude would be less than 4 *ten*-thousandths of a magnitude, clearly beyond our detection capabilities, the claims of Henry et al. (1997) notwithstanding. Thus, if we were to demonstrate that a 51 Peg-type star were pulsating, it would have to be on the basis of RV and LP variations.

Now the question arises, and Tim Bedding brought it up after one of the talks at this Symposium: Can the presence of a close planet tidally induce non-radial *g*-modes in solar-type stars? Terquem et al. (1998) have looked into this question and found that a companion orbiting 51 Peg with a period of 4.23 d induces a radial velocity at the stellar surface, the maximum of which is between 10^{-2} and 6 m sec^{-1} for a companion mass between 10^{-3} and 1.0 $M_\odot$. Such minuscule induced RV variations are not observable for Jupiter-like planets. Because of the extreme predictability of the RV variations in 51 Peg-type stars and the ragged RV variations for non-radial pulsators, the planetary would hypothesis would be the more likely of the two for 51 Peg-type stars. However, because Nature is more devious than we are, we must obtain high resolution spectra of 51 Peg-type stars to see if any do show LP variations. If they do, such variations could be due to the long-sought *g*-modes in stars as cool as the Sun.

References

Aerts, C., and Krisciunas, K. (1996), Mode Identification of the Slowly Pulsating F0V Star V398 Aurigae (9 Aur), *Monthly Notices Royal Astron. Soc.* **278**, pp. 877-882

Antonello, E., and Mantegazza, L. (1986), Variable Stars in the Southern Open Cluster NGC 2516, *Astron. Astrophys.* **164**, pp. 40-44

Balona, L. A., Krisciunas, K., and Cousins, A. W. J. (1994), γ Doradus: Evidence for a New Class of Pulsating Star, *Monthly Notices Royal Astron. Soc.* **270**, pp. 905-913

Balona, L. A., Böhm, T., Foing, B. H., et al. (1996), Line Profile Variations in γ Doradus, *Monthly Notices Royal Astron. Soc.* **281**, pp. 1315-1325

Boss, A. P. (1996), Extrasolar Planets, *Physics Today* **49**, No. 9, pp. 32-38

Cousins, A. W. J., and Warren, P. R. (1963), Variable Stars Observed during the Cape Bright Star Programmes, *Monthly Notes Astron. Soc. S. Afr.* **22**, pp. 65-74

Eggen, O. J. (1995), Young Pulsating Stars in the Böhm-Vitense Decrement, *Information Bulletin on Variable Stars*, No. 4210

Fitch, W. S. (1981), $\ell = 0, 1, 2$, and 3 Pulsation Constants for Evolutionary Models of δ Scuti Stars, *Astrophys. J.* **249**, pp. 218-227

Gautschy, A., and Löffler, W. 1996, Night Thoughts on Slowly Variable F-Type Stars, *Delta Scuti Star Newsletter*, No. 10, pp. 13-24

Giampapa M. S., and Rosner R. (1984), The Appearance of Magnetic Flux on the Surfaces of the Early Main-Sequence F Stars, *Astrophys. J. Letters* **286**, pp. L19-L22

Gray, D. F. (1997), Absence of a Planetary Signature in the Spectra of the Star 51 Pegasi, *Nature* **385**, pp. 795-796

Güdel M., Schmitt J. H. M. M., and Benz A. O. (1995), A Bright X-ray and Radio Corona on the F0V Star 47 Cas?, *Astron. Astrophys.* **293**, L49-L52

Hatzes, A. P., Cochran, W. D., and Johns-Krull, M. (1997), Testing the Planet Hypothesis:

a Search for Variability in the Spectral Line Shapes of 51 Pegasi, *Astrophys. J.* **478**, pp. 374-380

Henry, G. W., Baliunas, S. L., Donahue, R. A., Soon, W. H., and Saar, S. H. (1997), Properties of Sun-Like Stars with Planets: 51 Pegasi, 47 Ursae Majoris, 70 Virginis, and HD 114762, *Astrophys. J.* **474**, pp. 503-510

Holmes, S. (1890), quoted in A. C. Doyle, *The Sign of Four*, London: Spencer Blackett, ch. 6 (first published in *Lippincott's Magazine*, February 1890)

Krisciunas, K., Aspin, C., Geballe, T. R., et al. (1993), The 9 Aurigae System, *Monthly Notices Royal Astron. Soc.* **263**, pp. 781-788

Krisciunas, K. (1994), The Sign of Four: a New Class of Cool Non-Radially Pulsating Stars?, *Comments on Astrophys.* **17**, pp. 213-226

Krisciunas, K., and Handler, G. (1995), A List of Variable Stars Similar to γ Doradus, *Information Bulletin on Variable Stars*, No. 4195

Krisciunas, K., Griffin, R. F., Guinan, E. F., Luedeke, K. D., and McCook, G. P. (1995a), 9 Aurigae: Strong Evidence for Non-Radial Pulsations, *Monthly Notices Royal Astron. Soc.* **273**, pp. 662-674

Krisciunas, K., Crowe, R. A., Luedeke, K. D., and Roberts, M. (1995b), A Search for γ Doradus-Type Variable Stars in the Hyades, *Monthly Notices Royal Astron. Soc.* **277**, pp. 1404-1410

Krisciunas, K., and Crowe, R. A. (1997), A Search for γ Doradus-Type Variables in the Open Cluster M 34, *Information Bulletin on Variable Stars*, No. 4430

Mantegazza, L., Poretti, E., Antonello, E., and Zerbi, F. (1993), Photospheric Activity among Early F-type Stars, in *Inside the Stars*, ASP Conf. Series, **40**, pp. 651-653

Mantegazza, L., Poretti, E., and Zerbi, F. M. (1994), Unusual Variability among Early F-type Stars: HD 224638 and HD 224945, *Montly Notices Royal Astron. Soc.* **270**, pp. 439-448

Mayor, M., and Queloz, D. (1995), A Jupiter-Mass Companion to a Solar-Type Star, *Nature* **378**, pp. 355-359

Perryman, M. A. C., et al. (1997), The Hyades: Distance, Structure, Dynamics, and Age, to be published in *Astron. Astrophys*, astro-ph/9707253

Popper, D. M. (1980), Stellar Masses, *Annual Review Astron. Astrophys.* **18**, pp. 115-164

Poretti, E., Koen, C., Martinez, P., et al. (1997), Discovery and Analysis of γ Doradus Type Pulsations in the F0 IV Star HR 2740 $\equiv$ QW Pup, *Monthly Notices Royal Astron. Soc.*, in press

Terquem, C., Papaloizou, J. C. B., Nelson, R. P., and Lin, D. N. C. [1998], Oscillations in Solar-Type Stars Tidally Induced by Orbiting Planets, in Proceedings of the IXèmes Rencoutres de Blois, *Planetary Systems: the Long View*, Blois, June 1997, Editions Frontières, in press

Zerbi, F. M., Rodríguez, E., Garrido, R., et al. (1997), The γ Dor Variable HD 164615: Results from a Multi-Site Photometric Campaign, *Monthly Notices Royal Astron. Soc.*, in press

PULSATIONS OF OB-STARS: NEW OBSERVATIONS

DIETRICH BAADE
European Southern Observatory
Karl-Schwarzschild-Str. 2
D-85748 Garching, Germany
Internet: dbaade@eso.org

Abstract. Improved observing and data analysis strategies have initiated a considerable expansion of the empirical knowledge about the pulsations of OB stars. Possible correlations between physical parameters and associated pulsation characteristics are becoming more clearly perceivable. This starts to include the asteroseismologically fundamental areas of g-modes and rapid rotation. The β Cephei instability strip continues to be the only locus where radial pulsations occur (but apparently not in all stars located in that strip). Except for spectral types B8/B9 near the main sequence, where pulsations are hardly detected even at low amplitudes, any major group of stars in the Galaxy that are obviously not candidate pulsators still remains to be identified. However, the incidence and amplitudes of OB star pulsations decrease steeply with metallicity. The behaviour of high-luminosity stars is less often dominated by very few modes. In broad-lined stars the moving-bump phenomenon is more common than low-order line-profile variability. But its relation to nonradial pulsation is not clear. The beating of low-ℓ nonradial pulsation modes that have identical angular mode indices may be the clockwork of the outbursts of at least some Be stars. The physics of this episodic mass loss process remains to be identified.

1. Introduction

Ever since their discovery, pulsating OB stars have suffered from the lack of theoretical permissibility. The eventual confirmation through improved opacity calculations of the κ mechanism as a viable driving mechanism (cf. Dziembowski, these proceedings) has given the field a considerable impetus. It has led to the very successful exploitation during the past couple of years of (i) multi-target databases resulting from spacecraft observations (Hipparcos) or from the monitoring of clusters with CCD's, (ii) very long time series of high-resolution spectra, (iii) improved techniques for their analysis, and (iv) a larger wavelength coverage of the observations.

The results open new perspectives for the study of the actual physical processes. Some groups of stars, such as supergiants and Be stars, which originally were defined on the basis of persistent spectroscopic differences and inferred physical properties, can also by means of their pulsation properties be distinguished. This is intriguing because it gives rise to the question of a possible causal relationship between pulsational and other properties and its directionality.

After short reviews of the properties of all known groups of pulsating OB stars (Sects. 2-5), an attempt is made to summarize the results in terms of physical parameters and to identify areas where theoretical support could be useful (Sect. 6).

2. β Cephei Stars

For β Cephei stars, a convenient defining criterion remains the shortness of their periods which range from 2 to 6 hours. Only in β Cephei stars, but not all of them, have radial modes so far been detected. Since the nonradial modes are low-degree p-modes, the β Cephei stars not only form a compact group in spectral type (B0.5-B2), but are isolated from other pulsating OB stars also in a parameter space defined by period and mode indices. In careful analyses of significant numbers of high-quality spectra of several stars, Aerts and collaborators (e.g., 1995) have re-classified the types of known modes and also detected additional ones. Recent mainly photometrically oriented reviews are by Sterken & Jerzykiewicz (1993) and Sterken (1993).

The knowledge about β Cephei stars has also benefitted much from the use of array detectors for the photometry of young stellar clusters in Galaxy (Balona et al. 1997), LMC (e.g., Balona 1992), and SMC (Kjeldsen & Baade, in preparation). The Galactic-cluster work showed that with the new opacities, the theoretical position of the β Cephei instability strip matches the observed one quite well. Balona et al. (1997) also develop an interesting additional application of the study of β Cephei pulsators: Because during their evolution B stars of different masses enter the instability strip at different ages, but also have different eigenfrequencies, the pulsation periods of β Cephei stars turn out to be surprisingly good cluster age indicators.

At one-half solar metallicity, Balona (1992, 1993) found no candidate β Cephei stars in LMC clusters. This is consistent with the need of the κ-mechanism for sufficient opacity to drive oscillations (cf. Dziembowski, these proceedings). However, in younger low-metallicity clusters with more massive stars in the instability strip, β Cephei stars are not excluded by theory. The tentative identification of a 4.3-σ candidate variable star (and several other $\geq 3\,\sigma$ stars) in the probably even more metal-poor SMC cluster NGC 330 by Kjeldsen & Baade (in preparation) is, therefore, not necessarily in conflict with theory. However, the period of somewhat less than 2 hours would be very short. Furthermore, the peak-to-peak b-band amplitude is only 0.003 mag, i.e. well below Balona's detection threshold of ~0.01 mag.

In the name-giving β Cephei itself a manifold of pecularities has been found which is so amazing (for references see Hadrava & Harmanec 1996 and Telting et al. 1997) that either it must serve as a warning how ill-understood β Cephei stars are in general or β Cephei actually is the proto-type of a quite different class of variables. After the discovery many years ago of a presumably cyclic modulation of the wind and of a magnetic field of several 100 G, new polarimetry and UV-spectroscopy have been used to suggest an oblique magnetic rotator model with period 12 days.

However, the evidence of genuine periodicity is weak. In particular, since only very few cycles are covered, the apparent double-wave variability may only be due to a variable amplitude. In this case the true wind and magnetic-field period, if any, would be 6 days. Furthermore, β Cep is the primary in an eccentric binary with 90-year period, and there are extended phases of Hα line emission. The wings of the Hα emission line vary in radial velocity with the period of the single radial mode.

Telting et al. (1997) find in addition to that radial mode four more frequencies. Three of these four and the radial mode are separated by 1/6 c/d corresponding to one half of the claimed double-wave period of 12 days. This is perhaps the strongest evidence to date of such a period in β Cephei, and there is a weak, but not (yet?) significant hint at a fifth frequency fitting that pattern. Shibahashi & Aerts (these proceedings) present a magnetic pulsator model in which the magnetic field distorts the radial eigenmode such as to add a quadrupole component. The latter would reveal itself by an evenly spaced frequency quintuplet. The model requires that the frequency spacing is identical to the rotation frequency, i.e. the double wave behaviour inferred from polarization and stellar wind data would be wrong.

The study of the line-profile variability associated with the sixth mode, which probably is nonradial (Telting et al. 1997), could cast light on a putative oblique nonradial pulsator model. If the pulsation axis is not aligned with the rotation axis, the line-profile variability depends on the rotational phase. Constraints on the angle between rotation and magnetic axis and on i can be derived from such a dependency.

Using *EUVE* and exploiting the particularly EUV-transparent line of sight to β Canis Majoris, Cassinelli et al. (1996) could for the first time determine the temperature variations in the Lyman continuum. They are about 110 K for the primary mode and about 25 K for the other two. Comparison of the 3 velocity and temperature amplitudes and monitoring of the beating effects should enable a detailed analysis of the atmospheric response to (nonradial) β Cephei pulsation to be made.

The disentangling of two almost resonating modes in 16 Lac with a beat period of ≤ 2 years by Jerzykiewicz & Pigulski (1996) is perhaps the most impressive example to date as to how important long and homogeneously analysed time series are.

3. 53 Per, ϵ Per and Slowly Pulsating B Stars

Various names are in use for the pulsating B stars that are neither β Cephei stars nor supergiants nor Be stars. But no systematic survey of the pulsational properties of this potentially very large group has been made, and the application of the nomenclature appears relatively arbitrary and not well tuned to the true pulsational heterogeneity of the stars in question which indeed seems to be emerging. The most comprehensive report on Slowly Pulsating B Stars is by Waelkens (1991); the 53 Per stars and ϵ Per were described by Smith (1981) and Gies (1991), respectively.

In the Hipparcos photometry, a large number of new Slowly Pulsating B Stars was detected (Waelkens et al. 1997; see also Aerts et al., these proceedings). All of them appear to fall on the red side of the β Cephei instability strip *and* have a $v \sin i$ of less than 150 km/s. Since there are numerous B stars with larger projected rotational velocity (in the Bright Star Catalogue their proportion is about 25% of which only one-third are Be stars), this limit appears to be highly significant. This is all the more so as for the usually invoked sectoral NRP modes the amplitude is the largest on the equator, thus maximizing the detection probability at large i. Since photometry can only detect NRP modes with $\ell \approx m \leq 4$, this result is consistent with unpublished observations by Penrod (Smith & Penrod 1984) and Baade (1987) of broad-lined ($v \sin i \geq 200$ km/s) non-emission line B stars (Bn stars). In both studies large-scale line-profile variability was not detected whereas small-scale moving bumps are nearly ubiquitous.

The moving-bump phenomenon is also commonly found in OB stars within and blueward of the β Cephei stability strip, one of the best studied cases being ϵ Per (B0.5 IV-III, Gies 1991; see also Howarth & Reid 1993 for the O9.5 V star HD 93521). Its detection over a very wide range in effective temperature would furthermore render significant the restriction of the Hipparcos discoveries to effective temperatures lower than those of the β Cephei stars. It would imply that in non-Be stars low-ℓ g-modes mainly occur redward of the β Cephei strip. On the other hand, low-degree p-modes seem to be present also in the hotter B stars. This follows from Smith's (1981) work on the 53 Per stars as candidates of which he preferentially selected narrow-lined early B-type stars. In fact, the lowest $|m|$ value found in ϵ Per, namely 3, (Gies 1991) makes this star a border-line case.

Obviously, T_{eff} and $v \sin i$ appear to be governing some basic pulsation properties:
- $T_{\text{eff}} \leq T_{\text{eff},\beta\text{Cep}}$ and $v \sin i < 150$ km/s: high-order g-modes with $\ell \approx 2$. These are the Slowly Pulsating B Stars. The 53 Per stars are largely identical to them but the nature of the hotter candidate 53 Per stars deserves to be checked.
- $T_{\text{eff}} \leq T_{\text{eff},\beta\text{Cep}}$ and $v \sin i \geq 200$ km/s: moving bumps without demonstrated general genuine periodicity; however, if due to NRP, at least some of their short periods would probably be indicative of p-modes. There is no obvious proto-type after which these stars could be named, and it is possible that they only form the hotter part of the following category.
- $T_{\text{eff}} \geq T_{\text{eff},\beta\text{Cep}}$: moving bumps and/or p modes with $\ell \geq 4$. ϵ Per could plausibly be called the proto-type of this category.

This enumeration is complemented by the fact that among the stars with $v \sin i$ above 200 km/s only the Be stars, which are found from late O to late B spectral types, seem to show low-ℓ line profile variability (which is probably due to high-n g-modes, see Sect. 5). It should be noted that all numerical thresholds given are crude estimates and only serve to illustrate the order of magnitude.

4. Evolved, Luminous Stars

Hipparcos contributed to the record of photometric variability also of OB supergiants (van Leeuwen et al. 1997, Waelkens et al. 1998). However, the diagnostic power of single-channel photometry for the understanding of the variability of these objects is limited and, perhaps, already exhausted. In particular, the variations in the wind and the photosphere are difficult to disentangle. Kaufer et al. (1996, 1997) have for several years closely monitored a few BA supergiants at high spectral resolution and with wide wavelength coverage. The time scales of both variabilities can exceed the one of the radial fundamental mode considerably. The wind-related variability probably largely exhibits rotationally induced aspect changes of a non-axisymmetric structure.

At the photospheric level, equivalent widths and the spectral energy distribution are constant (Kaufer et al. 1997). Moreover, the radial velocity variations do not depend on line strength, i.e. depth of formation. In some stars, even the moving-bump phenomenon is seen and maintains phase coherence for at least a few cycles; often the acceleration of the bumps is too high to be due to rotation only. But the variable superposition of different variabilities makes it difficult to draw a more certain

conclusion. This variability extends to some of the hottest not-subluminous stars
known (e.g., ζ Pup: $T_{\text{eff}} = 42{,}500$ K, Reid & Howarth 1996).

All available observations are compatible with the kinematical effects of nonradial
pulsation. However, contrary to low-luminosity stars, the inclusion of more data usually does not lead to a refinement of the temporal power spectrum. For as long as a
numbner of years, the power spectra of many stars retain a certain overall signature
which crudely distinguishes them from other stars. But the variations of individual
features can already after few cycles be so pronounced that a single period is no longer
sufficient to explain this behaviour.

5. Be Stars

Virtually all Be stars earlier than ∼B8 show low-order line-profile and/or photometric
variability; attempts to associate these two variabilities with one another have at least
in some cases been painful. Especially the spectroscopic variability seems to be the
second criterion that distinguishes Be stars from other rapidly rotating B stars (cf.
Sect. 3). The only well-known exception seems to be ζ Oph (Kambe et al. 1997) in
which only the higher-order line-profile variability is seen that is common to nearly
all broad-lined OB stars.

The periods are 0.5-2 days, remain stable over many years, and are statistically
indistinguishable from the expected rotation periods. But a single, truly multiperiodic
star would terminate the dispute (Baade & Balona 1994) whether the variability of
Be stars is due to some corotating stellar or circumstellar features or due to nonradial
pulsation. However, so far only tentative, albeit intriguing, identifications of no more
than two persistent periods have been possible in very few stars (e.g., ζ Oph: Kambe
et al. 1997). Similarly, hopes that nonradial low-ℓ pulsation could be responsible for
the seemingly irregular outbursts of many Be stars did not up to now materialize.

Recently, Rivinius et al. (1998) discovered in μ Cen six discrete frequencies in two
groups: $f_1/f_2/f_3/f_4 = 1.9884 / 1.9704 / 2.0222 / 1.9366$ c/d and $f_5/f_6 = 3.5536 / 3.5825$
c/d. Their phase proved coherent in four annual seasons spanning 5 years. The most
fascinating result is that the beat periods of f_1 with f_2 and f_3 are also the mean intervals between line emission outbursts, i.e. discrete mass loss events. Since these three
modes have the largest amplitudes in their group, apparently the vectorial co-addition
of the two velocity fields must exceed a certain threshold for an outburst to occur.
This explains that beating of, e.g., f_1 and f_5 is not important since for the $f_1 - f_4$
group $\ell \approx 2$ whereas $\ell_5 \approx \ell_6 \approx 3$.

Outbursts observed within a decade of the data used by Rivinius et al. are *with
the same periods* properly accounted for by this simple clock. Since this circumstellar
activity can be described so well by using the photospheric periods only, the trustworthiness of the latter is underligned because the measurements of the photospheric
radial velocities are independent of the circumstellar emission lines. It also suggests
that there are no hidden $\ell \approx 2$ modes with amplitudes larger than the one of f_3.

In a Be star with a typical rotation frequency of 1 c/d, the spacing of the $f_1 - f_4$
group is too small by an order of magnitude to be due to m-mode splitting (besides, all
four angular eigenfunctions look the same). Because the implied frequencies in the corotating frame are quite low, it is likely that these frequencies belong to g-modes with
similar but different radial orders, n, and identical angular indices ℓ and m. Since the
radial orders may be of order 100 and $|\Delta f_{2,4}| \approx |\Delta f_{1,3}| \approx 2 \times |\Delta f_{2,1}|$, n_1 and n_2 differ
by one-half of the differences between n_1 and n_3 and between n_2 and n_4. One may,

therefore, wonder whether between f_1 and f_3 as well as between f_2 and f_4 one mode each (more if n_1 and n_2 differ by more than unity) is either not excited or not detected. If the rotation period can be properly constrained, these frequency differences and the corotating frequencies should enable the first crude asteroseismological experiments with a rapidly rotating star or firmly rule out the NRP model for Be stars.

The difference between the $f_1 - f_4$ and the $f_5 - f_6$ frequencies, ~ 1.56 c/d, could plausibly be the rotation frequency. For a difference of unity in m between the two groups, which inspection of the line profile variations does not rule out, rotational m-mode splitting is, therefore, a possibility. For the same reason, but if also $m = +\ell$ is adopted together with the above ℓ values, i.e. retrograde modes are assumed, all 6 eigenfrequencies would in the corotating frame be nearly identical. This would raise the question whether rotation plays a role in the mode selection.

The large number of variable Be stars in the LMC cluster NGC 330, which seem to have the same period distribution as Galactic Be stars (Balona 1992), would suggest that the driving of g-modes in these rapidly rotating stars is less sensitive to reduced metallicity than is the driving of radial and p-modes in β Cephei stars.

6. Physical Parameters

The description in the previous sections of the pulsations of OB stars by types of objects, can tentatively be summarized in terms of physical parameters as follows:

Temperature: Outside the domain of the evolved, very luminous stars, incidence and/or amplitude of pulsations drop steeply at spectral types B8/B9. The pulsating OB stars are, therefore, clearly separated from the δ Scuti (Breger, these proceedings) and γ Doradus (Krisciunas, these proceedings) variables. By contrast, a blue limit to this instability domain has not yet been established. It is not even clear whether there is any major population of Galactic OB stars in which also high-resolution spectroscopy would not find the signature of nonradial pulsations. It would be worthwhile investigating whether the chemically peculiar stars form such an exception. There is evidence that blueward of the β Cephei instability strip low-ℓ NRP modes are less common than on the red side. The β Cephei strip continues to be the only locus of radial and low-degree p-modes.

Metallicity: The incidence of β Cephei pulsation apparently decreases strongly with metallicity. In LMC clusters, β Cephei stars seem to be very much rarer than in the Galaxy. On the other hand, even in the SMC traces of radial pulsations may possibly occur. In marked contrast to this theoretically expected dependency, periodically variable Be stars are in some LMC clusters at least as common as in Galactic clusters and seem to have the same period distribution.

Age: Since more massive B stars reach the β Cephei instability strip earlier, more massive β Cephei stars are younger. The increase in the pulsation frequency of the radial fundamental mode with mass therefore leads to the more readily observable correlation between cluster age and pulsation period.

Rotation: Among rapid rotators ($v \sin i \geq 200$ km/s) low-ℓ NRP modes seem to be common in Be stars only; their rotation and inertial-frame pulsation periods are very similar. Virtually all broad-lined B stars display higher-order line-profile variability. The latter *looks* similar to high-ℓ NRP modes but (the few) attempts to verify this identification more firmly have failed so far. With current observing techniques it is not possible to find out whether high-order line-profile variability occurs in narrow-line stars, too. It is an exciting possibility that the difficult subject of the theroretical treatment of rapid rotation may soon benefit from the incipient detection of multi-mode NRP in rapidly rotating stars. Because young clusters in LMC and SMC abound in periodically variable Be stars whereas β Cephei stars are rare, it may be possible that rapid rotation helps (but is not sufficient) to destabilize high-order low-degree g-modes.

Luminosity: The variability of luminous OB stars is in most cases characterized by an apparently much more pronounced irregularity. However, evidence is mounting that the variability is due to multi-mode nonradial pulsation in high-n g-modes with highly variable amplitudes. This would indicate that the ill-understood but very effective mode-selection mechanism, which apparently is acting in other candidate g-mode pulsators (Be and SPB stars), plays a much lesser role in supergiants.

Mass loss: In high-amplitude radial pulsators, IUE has not found significant alterations of the mass loss rate by the associated outward moving shock fronts. In μ Cen, the clock of its Be-star typical discrete mass loss events may finally have been identified: If the positive superposition of the velocity fields of two or more quadrupole modes exceeds a certain amplitude threshold, an outbursts occurs. In the case of more than two such modes also the long-term activity cycles of some Be stars may become reproducible. The physics of the outburst process, however, still remains unexplained.

Magnetic fields: The example of β Cephei proves that a significant magnetic dipole field and *bona fide* β Cephei pulsation are not incompatible with each other. If diffusion has created chemically distinct patches in the photosphere, the above conjecture of the CP stars being the only major exception to pulsational instability in OB stars would be challenged. The question of an oblique magnetic pulsator à la roAp stars should be answerable on the basis of the dependence or not of the line-profile variability on rotational phase. Perhaps more interestingly, β Cephei is the only OB star known to date in which the effect of a strong magnetic field on the radial eigenmode can be studied.

It is important to remember that to most of the above 'rules' exceptions are known. But as guidelines, maybe even incentives, for further research they should be useful. Areas which observers should be curious to see commented on by theorists include:

- The occurrence of higher-ℓ modes towards higher temperatures than is the case for low-degree modes which are quite well reproduced by theory (cf. Dziembowski, these proceedings).
- The instability of the temporal power spectra of supergiants.
- The possibly inverse question of the very effective mode selection in lower-luminosity stars.
- The roughly 50:50 dichotomy (evolutionary bifurcation or consecutive stages?) among rapid rotators between Bn stars that do not show low-degree g-mode pulsations and Be stars that (i) do so and (ii), perhaps, only as a result of multi-mode beating eject matter, thereby possibly acquiring the circumstellar disk which actually makes them become Be stars.
- The unabated persistence of this phenomenon even at LMC metallicities.
- The confrontation of the observed frequencies of μ Cen with model calculations.
- The physics of the apparently pulsationally clocked outburst mechanism of μ Cen.

Acknowledgement: I thank Petr Harmanec, Don Kurtz, Thomas Rivinius, Myron Smith, and Stanislav Štefl for valuable suggestions for improvements.

References

Aerts, C., Mathias, P., Van Hools, T., De Mey, K., Sterken, C. and Gillet, D.: 1995, *Astron. Astrophys.* **301**, 781
Baade, D.: 1987, in A. Slettebak and Th.P. Snow (eds.): Physics of Be Stars, Proc. IAU Coll. No. 92, Cambridge Univ. Press, Cambridge, 361
Baade, D. and Balona, L.A.: 1994, in L.A. Balona, H.F. Henrichs, and J.M. Le Contel (eds.): Pulsation, Rotation, and Mass Loss in Early-type Stars, Proc. IAU Symp. No. 162, Kluwer, Dordrecht, 311
Balona, L.A.: 1992, *Mon. Not. R. Astron. Soc.* **256**, 425
Balona, L.A.: 1993, *Mon. Not. R. Astron. Soc.* **260**, 795
Balona, L.A., Dziembowski, W.A. and Pamyatnykh, A.A.: 1997, *Mon. Not. R. Astron. Soc.* **289**, 25
Cassinelli, J.P., Cohen, D.H., MacFarlane, J.J., Drew, J.E., Lynas-Gray, A.E., Hubeny, I., Vallerga, J.V., Welsh, B.Y., Hoare, G.M.: *Astrophys. J.* **460**, 949
Gies, D.R.: 1991, in D. Baade (ed.): Rapid Variability of OB-Stars: Nature and Diagnostic Value, ESO Workshop and Conf. Proc. No. 36, Europ. South. Observ. Garching, 229
Hadrava, P. and Harmanec, P.: 1996, *Astron. Astrophys.* **315**, L401
Howarth, I.D. and Reid, A.H.N.: 1993, *Astron. Astrophys.* **279**, 148
Jerzykiewicz, M. and Pigulski, A.: 1996, *Mon. Not. R. Astron. Soc.* **282**, 853
Kambe, E., Ando, H., Hirata, R., Walker, G.A.H., Kennelly, E.J. and Matthews, J.M.: 1993, *Publ. Astron. Soc. Pacific* **105**, 1222
Kambe, E., Hirata, R., Ando, H., Cuypers, J., Katoh, M., Kennelly, E.J., QWalker, G.A.H., Štefl, S. and Tarasov, A.E.: 1997, *Astrophys. J.* **481**, 406
Kaufer, A., Stahl, O., Wolf, B., Gäng, Th., Gummersbach, C.A., Kovács, J., Mandel, H. and Szeifert, Th.: 1996, *Astron. Astrophys.* **305**, 887
Kaufer, A., Stahl, O., Wolf, B., Fullerton, A.W., Gäng, Th., Gummersbach, C.A., Jankovics, I., Kovács, J., Mandel, H. Peitz, J., Rivinius, Th. and Szeifert, Th.: 1996, *Astron. Astrophys.* **320**, 273
Reid, A.H.N. and Howarth, I.D.: 1996, *Astron. Astrophys.* **311**, 616
Rivinius, Th., Baade, D., Štefl, S., Stahl, O., Wolf, B., Kaufer, A.: 1998, *Astron. Astrophys.*, submitted
Smith, M.A.: 1981, in G.E.V.O.N. and C. Sterken (eds.): Workshop on Pulsating B Stars, Nice Obs., Nice, 317
Smith, M.A. and Penrod, D.G.: 1984, in R. Stalio and J. Zirker (eds.): Relat. between Chromosph.-coronal Heating and Mass Loss in Stars, Trieste, 394
Sterken, C.: 1993, in J.M. Nemec and J.M. Matthews (eds.): IAU Coll. No. 139, New Perspectives on Stellar Pulsation and Pulsating Variable Stars, Cambridge University Press, Cambridge, 171
Sterken, C. Jerzykiewicz, M.: 1993, β Cephei Stars from a Photometric Point of View, Space Science Reviews **62**, 95
Telting, J.H., Aerts, C. and Mathias, P.: 1997, *Astron. Astrophys.* **322**, 493
van Leeuwen, F., van Genderen, A.M., Zegelaar, I.: 1998, submitted to *Astron. Astrophys.*
Waelkens, C.: 1991, *Astron. Astrophys.* **246**, 453
Waelkens, C., Aerts, C., Kestens, E., Grenon, M., Eyer, L.: 1998, submitted to *Astron. Astrophys.*

B STAR PULSATION – THEORY AND SEISMOLOGICAL PROSPECTS

W.A. DZIEMBOWSKI
Copernicus Astronomical Center
ul. Bartycka 18, 00-716, Warsaw, Poland

1. Introduction

Progress in understanding B-type pulsators has been reviewed at several recent meetings (e.g. Dziembowski, 1993, 1995; Moskalik, 1995). Not much happened afterwards. Owing to further improvement in the stellar opacity calculations (Iglesias and Rogers 1996), the theoretical pulsation-instability domain in the upper Main Sequence is now more precisely determined. In the next section I will compare the predicted domain with the positions of various B-type pulsators in the H-R diagram.

There has been no progress in nonlinear modeling of B-type stars and therefore our understanding of pulsation is still limited to identification of the driving mechanism for the observed modes. However, we do not understand, in particular, how amplitudes of the unstable mode are determined. In the last section I will discuss needs for going beyond the linear approximation.

Most of this review is devoted to applications of B star pulsators to probing stellar systems and to testing stellar evolution. Some progress has been achieved but much remains to be done.

2. B star instability strip

The opacity mechanism acting in the metal-bump zone at $T \approx 2 \times 10^5$ K is the cause of pulsation excitation in the whole range of Main Sequence B stars. In stellar models corresponding to the B3–B9 spectral type range the instability is found only for high order g-modes. Periods of the unstable dipole ($\ell = 1$) modes extend from about 1 to 5 days. For the $\ell = 2$ mode the range is 0.5 - 3 days. Both the spectral type and pulsation period ranges agree with those of Slowly Pulsating B (SPB or 53 Per) stars.

In hotter stars the instability occurs also in low order p- and g-modes with periods in the 0.1-0.3 day range. Excitation of such modes is responsible for the existence of β Cep type pulsators. The two instability domains for modes with $\ell \leq 2$ are shown in Fig. 1.

As expected, the extent of the instability domains is very sensitive to the metal abundance, Z. It should be noted that even at $Z = 0.01$ there is a sizable SPB domain. The corresponding β Cep domain is limited to the high luminosity range.

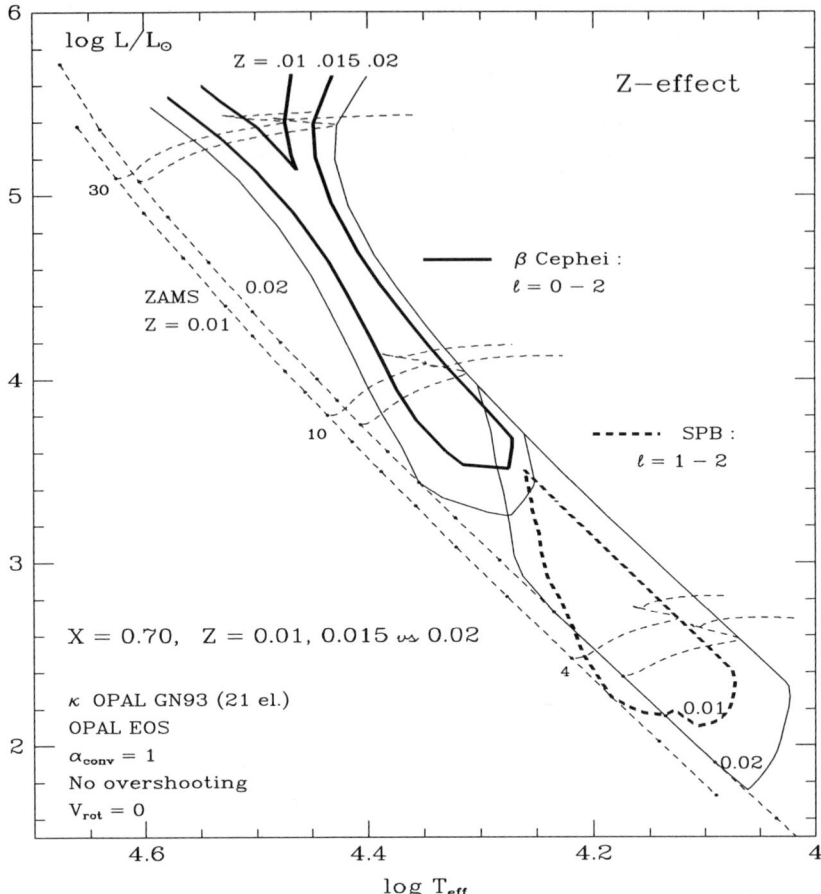

Figure 1. Domains of the pulsation instability in the upper part of the Main Sequence band as determined for models with the three indicated values of the metal abundance parameter, Z. The red boundary of the β Cep domain is identified with the Terminal-Age Main Sequence line. The instability continues into the Hertzsprung gap. For the SPB stars the instability terminates at the TAMS. Models were calculated with a standard stellar evolution code ignoring effects of overshooting and rotation. For $Z = 0.01$ and 0.02, the ZAMS lines and evolutionary tracks for the three indicated values of $M/M_\odot$ are shown (from Pamyatnykh, 1997).

This area, however, is apparently avoided by real objects. At $Z = 0.02$ there is a small overlap of the two instability domains. An object with the two types of modes excited simultaneously has not yet been discovered.

The instability to high order g-modes is indeed more common and more SPB than β Cep variables have been predicted. However, due to their long periods and lower amplitudes the former stars are much more difficult to discover, which explains why, until recently, we knew only 8 such stars. The situation changed after analysis of Hipparcos data which led to discovery of 72 [1] new SPB stars and only 4 β Cep stars (Waelkens et al., 1997). Now the numbers of known objects of the two types are comparable. The newly discovered SPB stars fill up the theoretical instability domain for $Z = 0.02$.

There is an agreement between observations and the theory that β Cep type pulsation is an episode in life of massive stars at the end of their Main Sequence evolutionary phase. The instability, in fact, extends beyond the Main Sequence phase. However, because the post-MS evolution is very fast we do not expect to find objects past the Terminal Age Main Sequence (TAMS) line in the H-R diagram. The position of this line is sensitive to the assumed extent of the convective overshooting from the stellar core. A small amount of overshooting helps to accommodate photometric data on some β Cep stars, however this may be accomplished also by taking into account the effects of rotation and/or by adjustments in the chemical composition parameters. More stringent constraints on the overshooting distance are expected from construction of seismic models of multimode β Cep stars.

The apparent paucity of β Cep type pulsations in stars with $\log L/L_\odot > 5$, in spite of the fact that the instability strip widens at high luminosities, calls for an explanation. Perhaps the instability has different consequences at the highest luminosities. The common low amplitude variability in blue supergiants is probably connected with the same driving mechanism (Kiriakidis et al., 1993).

3. Applications to probing stellar systems

The strong sensitivity of the instability domain boundaries to the Z-value suggests that the occurrence of B type pulsators in a stellar system may be used to determine the lower limit of metal abundance. This potential application remains to be explored. So far there is no confirmed discovery of a β Cep or SPB star outside our Galaxy, and within the Galaxy the inventory of B-type pulsators is incomplete.

Another application is in dating stellar systems. We may expect β Cep stars in a specified period range only in a certain time interval of the stellar system's life. In Fig. 1 we see that for $Z = 0.02$ at zero age we may have only low mass and, therefore, short period objects. The next group to enter the instability domain are massive, long-period objects, which follows from the shape of the blue boundary of the domain and the speed of stellar evolution. Then come the objects with typical β Cep masses of 10 to $15 M_\odot$. After some 15 Myr all but stars with masses less than $10 M_\odot$ leave the instability domain. Thus, by studying properties of β Cep stars in a stellar system we may determine its age.

Recently, Balona et al. (1997) analyzed the data on β Cep star in three young clusters (NGC 329s, NGC 4755, NGC 6231) which are abundant in this type of star. These authors determined the ages by isochrone fitting and by making use of pulsational data; they concluded that the latter method is more accurate.

4. Seismic models of β Cep stars

Applications of β Cep stars to testing massive star evolution were attempted soon after the cause of their pulsation had been identified. Two problems are of particular interest. The first (the extent of the overshooting from the core) was already mentioned

---

[1]This number was recently increased to 100 (Aerts, de Cat and Waelkens, this volume).

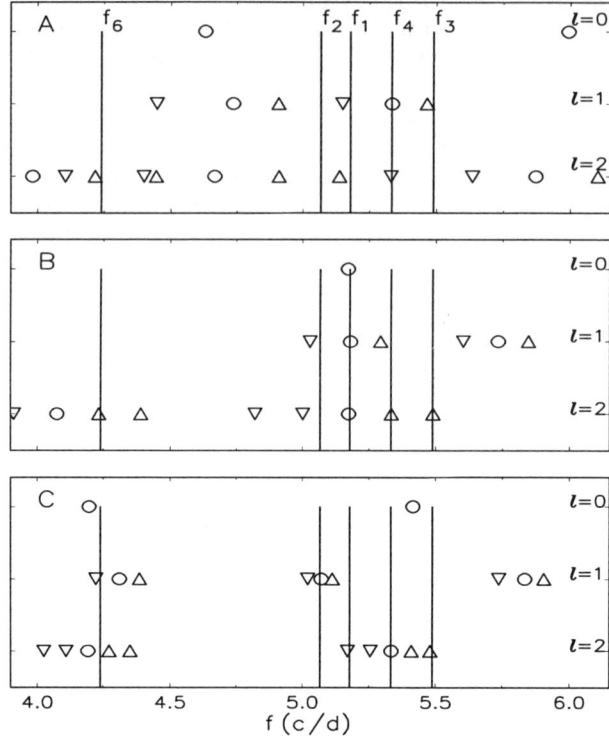

Figure 2. Measured frequencies for DD Lac (Jerzykiewicz, 1978) (vertical lines), compared with calculated frequencies of low degree modes ($\ell \leq 2$) in three models. The $f_5 = f_1 + f_4$ frequency is not shown. Circles denote centroid modes, upward and downward triangles denote prograde and retrograde modes, respectively. The three panels correspond to the following different identifications of the observed equidistant triplet (f_1, f_4, f_3) and to different stellar parameters. In model A it is the $\ell = 1$, p_1 triplet; in model B it is a part of the $\ell = 2$, g_1 quintuplet and in C a part of the $\ell = 2$, p_0 quintuplet. Models are characterized by the following parameters. A: $M/M_\odot = 11.84$, $\log T_{\text{eff}} = 4.37$, $\log L/L_\odot = 4.21$, $V_{\text{rot}} = 114$ km/s; B: $M/M_\odot = 10.44$, $\log T_{\text{eff}} = 4.35$, $\log L/L_\odot = 4.03$, $V_{\text{rot}} = 67$ km/s; C: $M/M_\odot = 12.24$, $\log T_{\text{eff}} = 4.37$, $\log L/L_\odot = 4.28$, $V_{\text{rot}} = 40$ km/s.

in Section 2. We do not have a credible theory of this phenomenon, nor adequate numerical models. The overshooting causes element mixing beyond the convective core boundary which has far-reaching consequences for stellar evolution. The second problem is the evolution of angular momentum. Also, in this case we lack good theory. In particular, we do not know how nonuniform may be the rotation rate in stellar interiors. Amongst modes that may be driven in β Cep stars there are ones with large amplitude in the chemically inhomogeneous zone surrounding the convective core. These modes are of particular interest here.

For several multimode β Cep stars, Mike Jerzykiewicz and I undertook a program of constructing seismic models, i.e., models reproducing all measured frequencies. Our first work (Dziembowski and Jerzykiewicz, 1996) was devoted to EN Lacertae, a star with four mode frequencies measured. The star is also an eclipsing and single-lined spectroscopic binary and this yields additional constraints on its model. We could reproduce all measured frequencies only upon assuming rather substantial increase of the rotation rate toward the center. We plan to revisit the star with our improved treatment of effects of rotation and with use of one additional frequency that has been recently determined (Jerzykiewicz and Pigulski, 1997).

Here, I will consider in greater detail another object from our program, DD Lac (Dziembowski and Jerzykiewicz, in preparation, see also our poster paper in this volume). The most conspicuous feature of the amplitude spectrum is the equidistant frequency triplet (f_1, f_4, f_3). There are conflicting conclusions regarding the ℓ and m values of the three modes. However, the observers agree that the f_1 mode is nonradial. Therefore, we assumed only that the three frequencies represent either a rotationally split $\ell = 1$ triplet or a part of an $\ell = 2$ quintuplet. With the observational constraint on T_{eff} we still had to consider 11 distinct identifications for the modes in the observed triplet. Three examples are shown in Fig. 2. Other possibilities involve the $\ell = 1$, g_1 triplet and four different choices of subsets of the $\ell = 2$ quintuplets for two possible radial orders.

For each identification we may find a model which best fits all measured frequencies. In the cases of the three models in Fig. 2 the fit was only preliminary as may be easily noticed. The only stellar parameters adjusted were mass M and initial angular momentum – the quantity translated to the current equatorial velocity of rotation $V_{\text{rot}} = R\Omega$. The fit may be improved by adjusting T_{eff} and chemical composition parameters, X and Z. The models were calculated with the standard Population I values: $X = 0.7$, $Z = 0.02$. There should be also room for parameters describing nonuniform rotation and convective overshooting.

Model C seems quite close to a good fit. However, prospects for achieving the frequency fit at level of the observational frequency errors (10^{-4} c/d) are in fact quite distant. One problem is the accuracy of the perturbation treatment of the rotation effects on oscillations. Here we rely on the second order perturbation theory in Ω, and our calculated frequencies are reliable only to 10^{-3} c/d.

There is another, more interesting, problem. The data yield for frequency separation $\Delta f = f_4 - (f_1 + f_3)/2 = -0.0001 \pm 0.0003$ c/d. The frequency splitting is equidistant within the errors. This is not predicted by model calculations. If the $(f_3 - f_1)$ difference is used to assess the value of Ω, then model values of Δf range, depending on identification, from 0.002 to 0.05. The lowest value corresponds to the $\ell = 2$, p_0, $m = 0$ identification. It is possible that in this case an agreement may be achieved with very specific model parameters. However, there is an alternative possibility which implies a limitation for use of frequencies calculated with the linear theory. Buchler et al. (1995) showed that a nonlinear interaction between modes in a rotationally split triplet may lead to a limit cycle with constant (not equal) amplitudes and the phase lock causing the frequencies to appear exactly equidistant while the linear frequencies are only approximately equidistant.

5. The need for nonlinear theory

The fact that the linear approximation may not be sufficient for the high precision frequency calculation is bad news for asteroseismic applications. Nonlinear calculations are far more complicated than the linear ones. Furthermore, if we do not have credible determination of the ℓ's and m's for the modes, we cannot be even sure that what we are observing is a rotationally split triplet. The phase lock leading to exactly equidistant frequency separations may occur for any three modes whose linear frequencies satisfy an approximate relation $\nu_2 \approx (\nu_1 + \nu_3)/2$, providing that the integral of $|Y_1|^2 Y_2 Y_3$, where Y_k's denote spherical harmonics of the respective modes, is nonzero.

The nature of the triplet in DD Lac must be explained before we may construct the seismic model of the star. A reliable determination of the ℓ's and m's would be of great help. However, development in nonlinear theory seems a necessity. This effort is important for interpretation of the oscillation for this and other β Cep stars, as well as for many multimode pulsators of different types. For variable white dwarfs we have abundant data which may be interpreted only within the framework of nonlinear theory of nonradial oscillations (S.O. Kepler, D.E. Winget. private communications).

Perhaps the most interesting application of β Cep stars is for testing nonlinear theories of multimode stellar pulsation. These stars are the best for starters because they are the simplest amongst stars showing this type of pulsational behavior. The observed pulsation spectra do not exhibit the complexity encountered in variable white dwarfs. Furthermore, the effects of convection and magnetic fields, which always present formidable problems for the theory are, most likely, unimportant in β Cep pulsations.

I am grateful to Mike Jerzykiewicz for allowing me to use our unpublished results and for his useful comments, and to Alosha Pamyatnykh for Fig. 1. This work was supported by KBN-2P304-013-07.

References

Balona, L.A., Pamyatnykh, A., and Dziembowski, W.A. , 1997, *MNRAS*, **289**, 35
Buchler, J.R., Goupil, M.J., and Serre, T. , 1995, *Astron. Astrophys.*, **296**, 405
Dziembowski, W.A. , 1993, in L.A. Balona, H.F. Henrichs and J.M. Le Contel (eds), *IAU Symposium No. 162 – Pulsation, Rotation and Mass Loss in Early-Type Stars*, 55
Dziembowski, W.A., 1995, in S.J. Adelman and W.L. Wiese (eds), *Astrophysical Applications of Powerful New Databases*, (ASP Conf. Ser.), 275
Dziembowski, W.A. and Jerzykiewicz, M. , 1996, *Astron. Astrophys.*, **306**, 436
Iglesias, C.R. and Rogers, F.J. , 1996, *Astroph.J*, **464**, 943
Jerzykiewicz, M. , 1978, *Acta Astronomica*, **28**, 465
Jerzykiewicz, M. and Pigulski, A. , 1996, *MNRAS*, **282**, 853
Kiriakidis, M., Fricke, K.J., and Glatzel, W. , 1993, *MNRAS*, **264**, 50
Moskalik, P. , 1995, in R.S. Stobie and P.A. Whitelock (eds), *Astrophysical Application of Stellar Pulsation*, (ASP Conf. Ser), 44
Pamyatnykh, A.A. , 1997, in P.A. Bradley and J.A. Guzik (eds), *A Half Century of Stellar Pulsation Interpretation*, (ASP Conf. Ser), in press
Waelkens, C., Aerts, C., Grenon, M., and Eyer, L. , 1997, *Astron. Astrophys.*, in press

THE EC14026 STARS – PULSATING HOT SUBDWARFS

C. KOEN, D. O'DONOGHUE, D. KILKENNY AND R.S. STOBIE
South African Astronomical Observatory, PO Box 9, Observatory 7935, Cape, South Africa

1. Introduction

The Edinburgh-Cape Blue Object Survey (Stobie et al. 1997a) is a southern hemisphere survey to discover hot, blue stellar objects brighter than B=18 at galactic latitudes more than 30° from the galactic plane. The main categories of object detected are hot subdwarfs, white dwarfs, blue horizontal branch stars, apparently normal B stars, cataclysmic variables and (stellar-like) active galactic nuclei. Over 50% of the EC Survey comprises hot subdwarfs.

In the course of high speed photometry runs on selected EC objects (e.g. candidate pulsating white dwarfs) a whole new class of rapidly pulsating stars was discovered serendipitously. These stars are now recognised to be pulsating sdB stars, dubbed the EC14026 stars after the prototype. Since the discovery of the pulsations and the realisation that this was a new class of rapid pulsators, a major search was initiated for other pulsators. Over 300 sdB stars have now been searched using the techniques of high speed photometry and frequency analysis, and to date (July 1997) twelve pulsators have been discovered.

This paper will summarise the results of the first four pulsators (Kilkenny et al. 1997 - paper I; Koen et al. 1997 - paper II; Stobie et al. 1997b - paper III; O'Donoghue et al. 1997a - paper IV), present photometry on two new pulsators, and preliminary results on three other pulsators.

2. Photometric Analysis

To discover new pulsators high speed photometry in white light with a time resolution of 10 s is obtained for 1-2 hours. Both photoelectric photometers and a high time-resolution CCD camera have been used with 0.5-m to 1-m class telescopes. The data are analysed for coherent frequencies by periodogram techniques. For stars with B<15 it is normally possible to detect pulsations with semi-amplitude > 3 mmag. Once pulsations are confirmed the star is

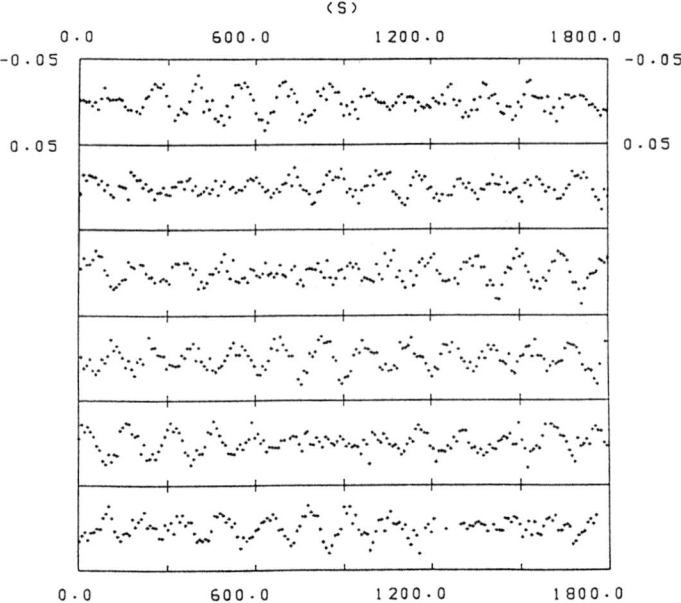

Figure 1. Light curve of pulsating sdB star PG1047+003. The ordinate units are differential magnitudes.

followed intensively for a number of nights. The details of the photometric techniques are described in papers I-IV.

Fig. 1 shows the time series photometry of a new pulsator PG1047+003, (O'Donoghue et al. 1998). Oscillations with a period of ~140 s and a peak-to-peak amplitude of several hundredths of a magnitude are evident. The multiperiodic nature of the oscillations is evident from the 'beating' phenomenon. The periodogram of this photometric run is shown in Fig. 2 revealing five frequencies well above the noise (the peak at 120 s is spurious, caused by the drive error of the 0.75-m telescope). Frequency analysis of all the nights reveals at least 9 periods in the range 104-162 s.

Other EC14026 stars have similar pulsation properties, though mostly not with such a large peak-to-peak amplitude. In Table 1 we list the properties of eight of the twelve pulsators that have been studied to date. The three dominant pulsation periods of each pulsator are listed together with their semi-amplitudes. Note that many of these semi-amplitudes are variable in time, possibly as a result of beating between close frequencies. In some cases (PB8783, PG1047+003) many more than three periods are known. The EC14026 stars listed are observed to be multiperiodic with periods in the range $90\ s < P < 220\ s$ and semi-amplitudes of component frequencies normally < 10 mmag. These pulsating hot subdwarfs are amongst the most rapid pulsators known, comparable to the shortest-period white dwarf pulsators.

High speed photometry of the largest amplitude pulsator, PG1605+072, (not listed in Table 1) is illustrated in Fig. 3. This pulsator is unusual in

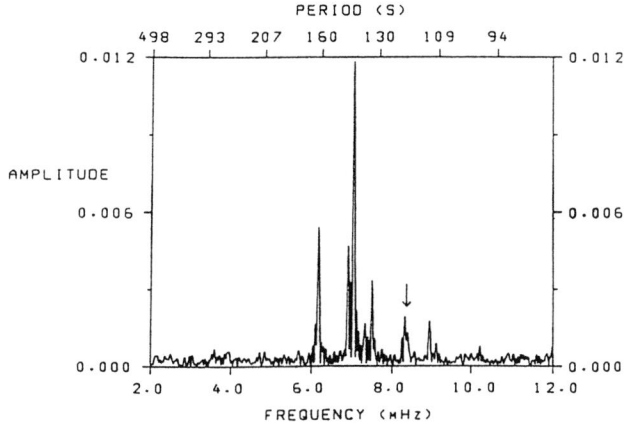

Figure 2. Fourier amplitude spectrum of 5.7 h high speed photometry of PG1047+003. The arrow indicates the frequency of the SAAO 0.75-m telescope drive.

comparison to the properties of the other pulsators because of its much larger amplitude (∼0.3 mag peak-to-peak), the length of the periods of oscillation

TABLE 1. Spectroscopic and pulsation properties of EC14026 stars

star	spectral type	T_{eff}	log g	P_1 (s)	a_1 mmag	P_2 (s)	a_2 mmag	P_3 (s)	a_3 mmag
EC14026-2647	sdB+G2V	34700	6.1	144	12	134	4	-	-
PB8783	sdB+F3V	35700	5.6	122	9	123	4	134	2
EC10228-0905	sdB+G0V	35100	6.0	140	7	152	5	147	4
EC20117-4014	sdB+F6V	34800	5.9	137	4	142	1	157	1
PG1047+003	sdB	35000	5.9	142	10	162	5	145	4
PG1336-018	sdB+M5V	33000	5.7	184	10	141	5	-	-
EC05217-3914	sdB	-	-	217	5	214	3	-	-
EC11583-2708	sdB+	-	-	149	5	143	3	-	-

∼480 s, and the large number of frequencies identifiable on a single night (at least 15). As this pulsator presents one of the best opportunities for mode identification, a multisite campaign was recently organised to maximise the number of frequencies identified and to minimise the alias problems. More than 30 frequencies have been identified in the preliminary analysis.

The potential of the pulsations of the EC14026 stars as diagnostic probes of their interiors can only be realized fully by intensive campaigns of multisite time series photometry. To this end a campaign spanning 2 weeks and involving data from 5 different sites was carried out on PB8783 (O'Donoghue et

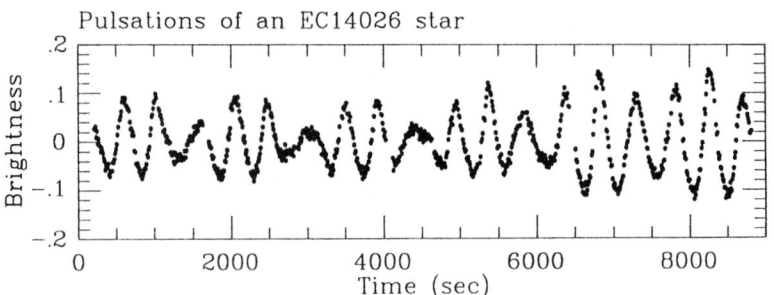

Figure 3. Light curve of largest amplitude sdB pulsator. The ordinate units are fractional changes in intensity.

al. 1997b - Paper V). Frequency analysis yielded at least 12 pulsation periods in 7 discrete groups between 94 and 136 s. Three of the periods in one specific group are equally-spaced in frequency with a spacing of 0.96 μHz, implying a rotation period of ~12 d. The periods of a detailed model of an evolved extreme horizontal branch sdB star give good agreement with the observed periods (table 3 of paper V). Although the period agreement is good, we caution against this being the only possible mode identification. The problem is that in pulsation models of sdB stars, the frequency spacing between modes with different n and/or ℓ is occasionally very small (e.g. for $n=3$, $\ell=0,2$ in table 3 of paper V) – whereas most observed spacings are much larger. Although the exact values of n and ℓ must await detailed modelling, small values are implied.

3. Spectroscopic Analysis

Low (~3.5 Å) and medium (~1.2 Å) resolution spectra have been obtained of the EC14026 pulsators using the SAAO 1.9-m telescope equipped with an intensified Reticon spectrograph. The spectra all show broad, strong Balmer lines superimposed on a blue continuum, characteristics of sdB stars. However, some spectra also show (unexpectedly for sdB stars) the CaII K line and sometimes the G-band. These features are indicative of a composite spectrum comprising a hot sdB star and a cool main sequence star. The details of the fitting technique to derive the best estimate of the spectral types of the two components and the values of T_{eff} and log g for the sdB stars are given in paper IV and the results listed in Table 1.

Optical spectroscopy will only detect a main sequence companion to an sdB star if the companion is of spectral type earlier than ~G5. For later K and M spectral types infrared photometry is required to detect the presence of a companion. In Table 1 the fraction of composite systems is 5 out of 8. This is a minimum and it is conceivable that all EC14026 stars are binary in nature.

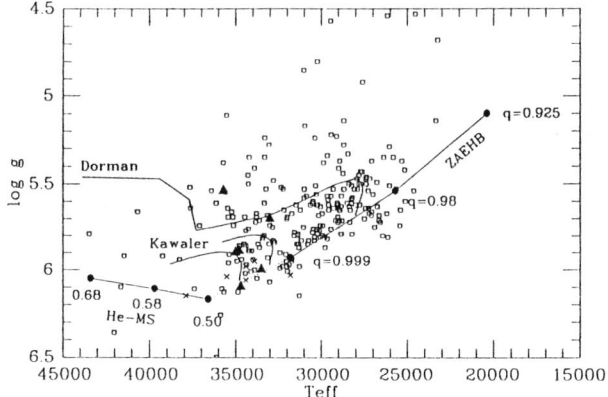

Figure 4. The log g - T_{eff} plane showing sdB field subdwarfs (open symbols) and EC14026 pulsators (filled triangles)

It should be noted that the M5 spectral type of the companion in PG1336-018 was estimated not from the spectroscopic analysis but by the reflection effect in this close binary with P_{orb} = 0.1010174 d (Kilkenny et al. 1998).

The position of the sdB oscillators in the log g - T_{eff} plane is shown in Fig. 4 by the filled triangles. The open symbols are log g and T_{eff} determinations for a large sample of field sdB subdwarfs by Saffer et al. (1994). The zero age extreme horizontal branch, the helium main sequence and evolutionary tracks are shown. With the exception of PG1605+072, the range of log g (5.6 to 6.1) and T_{eff} (33 000 to 35 700 K) exhibited by the EC14026 stars occupies a small fraction of the subdwarf parameter space, suggestive of an instability strip. The sdB stars lie on the extended blue horizontal branch and as such are core helium burning stars comprised of a ~0.5 $M_\odot$ helium core surrounded by a thin (M_{env} < 0.01 $M_\odot$) hydrogen envelope. The pulsators at the hot end of the sdB star range contain even thinner (M_{env} < 0.0005 $M_\odot$) hydrogen envelopes.

4. Discussion

Pulsation models by Kawaler (see paper III) with 'standard' zero-age horizontal branch (ZAHB) models have been constructed to compare the observed and theoretical pulsation periods. The models have a helium core mass of 0.485 $M_\odot$ with a hydrogen layer thickness of only ~0.00015 $M_\odot$ to bring their temperatures into agreement with observation. The periods of the low order (ℓ=0,1,2; n=0,1,2) p-modes are in the observed range. It is not expected that higher order ℓ modes would be easily visible because of geometric cancellation effects. Furthermore, it is clear from the stars with three or more observed periods that not all the modes can be radial.

Although the periods of the models agree with the observed range, none of the above models was pulsationally unstable. However, independently of the observational discovery of these pulsators, Charpinet et al. (1996) demonstrated that significant driving in sdB stars could be caused by the opacity bump associated with heavy element ionization (similar to the instability mechanism in β Cephei stars). This could lead to pulsational instability if the heavy element abundance in the relevant zone was high enough ($Z \geq 0.04$). In a later paper, Charpinet et al. (1997) presented improved models with nonuniform iron abundance distributions obtained through the condition of diffusive equilibrium between gravitational settling and radiative levitation. The enhanced iron abundance led to unstable models for low order radial and nonradial pulsation modes in the range $36\ 500 > T_{\text{eff}} > 29\ 000$ K. These and other results are presented in more detail by Fontaine in this volume.

Once modes have been identified these stars represent an excellent laboratory for elucidating the structure of sdB stars. This could lead to a better understanding of how sdB stars evolve into this state from their red giant progenitors. Furthermore, with the short periods and ability to measure period change within a few years it should be possible to determine the evolutionary time scale of core helium burning on the horizontal branch.

References

Charpinet, S., Fontaine, G., Brassard, P. and Dorman, B. (1996) The potential of asteroseismology for hot, subdwarf B stars: A new class of pulsating stars?, *Astrophys. J.* **471**, pp L103-106

Charpinet, S., Fontaine, G., Brassard, P., Chayer, P., Rogers, F.J., Iglesias, C.A. and Dorman, B. (1997) A Driving Mechanism for the Newly Discovered Class of Pulsating Subdwarf B Stars, *Astrophys. J.*, in press

Kilkenny, D., Koen, C., O'Donoghue, D. and Stobie, R.S. (1997) A new class of rapidly pulsating star – I. EC14026-2647, the class prototype, *Mon. Not. Roy. astr. Soc.* **285**, pp 640-644 (Paper I)

Kilkenny, D., O'Donoghue, D., Koen, C. and van Wyk, F. (1998) The EC14026 stars - VII. PG1336-018 – a pulsating sdB star in an HW Vir-type eclipsing binary, preprint (Paper VII)

Koen, C., Kilkenny, D., O'Donoghue, D., van Wyk, F. and Stobie, R.S. (1997) A new class of rapidly pulsating star – II. PB8783, *Mon. Not. Roy. astr. Soc.* **285**, pp 645-650 (Paper II)

O'Donoghue, D., Lynas-Gray, A.E., Kilkenny D., Stobie, R.S. and Koen, C. (1997a) A new class of rapidly pulsating star – IV. Oscillations in EC20117-4014 and atmospheric analysis, *Mon. Not. Roy. astr. Soc.* **285**, pp 657-672 (Paper IV)

O'Donoghue, D., Koen, C., Solheim, J.-E., Barstow, M.A., Dobbie, P.D., O'Brien, M.S., Clemens, J.C., Sullivan, D.J. and Kawaler, S.D. (1997b) The EC14026 stars –V. The pulsation periods of PB8783, preprint (Paper V)

O'Donoghue, D., Koen, C., Lynas-Gray, A.E., Kilkenny, D. and van Wyk, F. (1998) The EC14026 stars – VI. PG1047+003, preprint (Paper VI)

Saffer, R.A., Bergeron, J., Koester, D. and Liebert, J. (1994) Atmospheric parameters of field subdwarf B stars, *Astrophys. J.* **432**, pp 351-366

Stobie, R.S. et al. (1997a) The Edinburgh-Cape Blue Object Survey - I. Description of the survey, *Mon. Not. Roy. astr. Soc.* **287**, pp 848-866

Stobie, R.S., Kawaler, S.D., Kilkenny, D., O'Donoghue, D. and Koen, C. (1997b) A new class of rapidly pulsating star – III. Oscillations in EC10228-0905 and pulsation analysis, *Mon. Not. Roy. astr. Soc.* **285**, pp 651-656 (Paper III)

EC14026 STARS: THEORETICAL CONSIDERATIONS

G. FONTAINE, S. CHARPINET, P. BRASSARD
Département de Physique, Université de Montréal
Montréal, Québec, Canada H3C 3J7

P. CHAYER
Center for EUV Astrophysics, University of California
Berkeley, California 94720, U.S.A.

F.J. ROGERS, C.A. IGLESIAS
Lawrence Livermore National Laboratory
Livermore, California 94550, U.S.A.

AND

B. DORMAN
Laboratory for Astronomy and Solar Physics, NASA/GSFC
Greenbelt, Maryland 20771, U.S.A.

1. Introduction

The recent discovery of pulsating subdwarf B (sdB) stars at the South African Astronomical Observatory (Kilkenny et al. 1997; Koen et al. 1997; Stobie et al. 1997; O'Donoghue et al. 1997) has opened a brand new avenue in the field of asteroseismology. The study of these pulsators (dubbed EC14026 stars, after the prototype) offers the exciting possibility of exploiting the full power of asteroseismology to investigate the sdB phase of stellar evolution, one of the last frontiers in our general understanding of the history of stars.

The existence of pulsating sdB stars was predicted theoretically by Charpinet et al. (1996) as part of the first systematic investigation of the asteroseismological potential of stellar models on the extreme horizontal branch (EHB) and beyond. This was made possible thanks to significant progress in our ability to compute increasingly sophisticated and realistic models for this relatively neglected phase of stellar evolution (see Dorman 1995 for a review). Charpinet et al. (1996) uncovered an efficient driving mechanism due to an opacity bump associated with iron ionization in such models.

Oddly enough, both observational and theoretical efforts were carried out totally independently of each other at about the same time. The discovery of the first EC14026 stars in South Africa, however, gave our group added confidence in the basic validity of the approach of Charpinet et al. (1996). This also led us to further refinements of the physical description of the iron bump mechanism responsible for driving pulsation modes in models of sdB stars. Thus, in Charpinet et al. (1997), we demonstrated the existence of a theoretical instability strip in which all the currently known EC14026 stars fall, and obtained an excellent qualitative agreement between the periods of the expected driven modes and the observed periods. Low order radial, and low order and low degree nonradial (p and f) modes are involved. In the following, we summarize these and still more recent theoretical developments.

2. The nature of the sdB stars

It is now well established that sdB stars are evolved, compact, EHB and post–EHB objects. They are certainly evolved from the RGB, but we still do not know exactly how they are formed. The sdB stars are quite abundant in terms of surface density in shallow photographic surveys such as the Palomar–Green, Montréal–Cambridge–Tololo, and Edinburgh–Cape surveys. For instance, they dominate the population of blue objects for $V \lesssim 16$. About 300 sdB stars are brighter than $V \sim 14.3$ in the complete Palomar–Green survey.

Quantitative studies of the spectra of sdB stars (see Saffer et al. 1994 and references therein) have revealed that their atmospheric parameters are found in the ranges 40,000 K $\gtrsim T_{\rm eff} \gtrsim$ 24,000 K and 6.2 $\gtrsim \log g \gtrsim$ 5.1. The average values in the Saffer sample of 213 stars are $< T_{\rm eff} > \simeq$ 30,740 K and $< \log g > \simeq$ 5.68. In addition, they are all chemically peculiar. Their atmospheres are dominated by hydrogen, with helium typically underabundant by more than one order of magnitude (in number). Heavier elements (particularly C, N, and Si which have been well studied) show abundance anomalies that can be quite large (see, e.g., Heber 1991). It is believed that diffusion processes (mostly gravitational settling and radiative levitation) are at work in these stars, possibly in competition with weak stellar winds (Michaud et al. 1985).

The positions of the real sdB stars in the $T_{\rm eff}$–$\log g$ plane (or, equivalently, in the HR diagram) force their identification with low–mass ($M \lesssim 0.5~M_\odot$), helium core–burning models with outer H–rich envelopes that are too thin to sustain appreciable hydrogen shell–burning on the ZAEHB. Consequently, after core helium exhaustion, these objects do not ascend the AGB; they instead turn left during their evolution in the HR diagram and, ultimately, collapse as low–mass white dwarfs. A typical evolutionary timescale for a star in the sdB phase is of order $\sim 10^8$ yr.

3. The driving mechanism in EC14026 star models

In a first effort, Charpinet et al. (1996) investigated the pulsation properties of detailed, full evolutionary models of sdB stars with uniform solar composition in the outer H–rich envelope. Although all the modes investigated turned out to be globally stable, this led to the discovery of a region of strong local driving associated with a maximum in the Rosseland opacity in several of these models. This bump in opacity, sometimes referred to as the "Z–bump", is essentially due to the ionization of an electronic shell of iron contained in the solar mixture assumed in the stellar models. This opacity feature is sufficiently important and well located in sdB star models to produce significant local driving through the κ–effect.

Because diffusion processes are known to be at work in sdB stars (causing local overabundances –and underabundances– of a given element, depending on the location in the star), Charpinet et al. (1996) reasoned that iron could plausibly be overabundant in the driving region, thus further boosting the local opacity there. This was justified by rough radiative levitation calculations indicating that iron should indeed be overabundant in the critical region. They investigated the question with relatively crude envelope models in which the (uniform) metallicity was artificially increased beyond the solar value. Models with $Z \geq 0.04$ were found to be unstable and, on this basis, Charpinet et al. (1996) made the prediction that a subclass of sdB stars should show luminosity variations resulting from pulsational instabilities.

In a second effort, we reinvestigated the stability problem by constructing more sophisticated models in which the crude assumption of uniform metallicity has been replaced by the more realistic condition of diffusive equilibrium between gravitational settling and radiative levitation on the metal –iron– responsible for the driving process (see Charpinet et al. 1997). This necessitated detailed calculations of radiative forces on iron in the models (see, e.g., Chayer, Fontaine, & Wesemael 1995) and the computations of special OPAL opacity tables taking into account the large variations of the iron abundance about the cosmic value predicted by equilibrium radiative levitation theory.

Figure 1 illustrates some properties of these more sophisticated stratified envelope models. The solid curve in each panel gives the iron abundance as a function of fractional mass depth $\log q$. The panels correspond to a series of representative sdB models with $\log g = 5.8$ and with effective temperatures ranging from $\sim$ 22,000 K to about $\sim$ 42,000 K. Note that large overabundances of iron can be supported by radiative levitation in certain regions of the envelope of the stars, especially for the hotter models.

Associated with the nonuniform profile of iron is the run of the Rosseland opacity (dotted curve in each panel). The location and shape of the iron opacity peak are the critical factors in the efficiency of the driving process. It turns

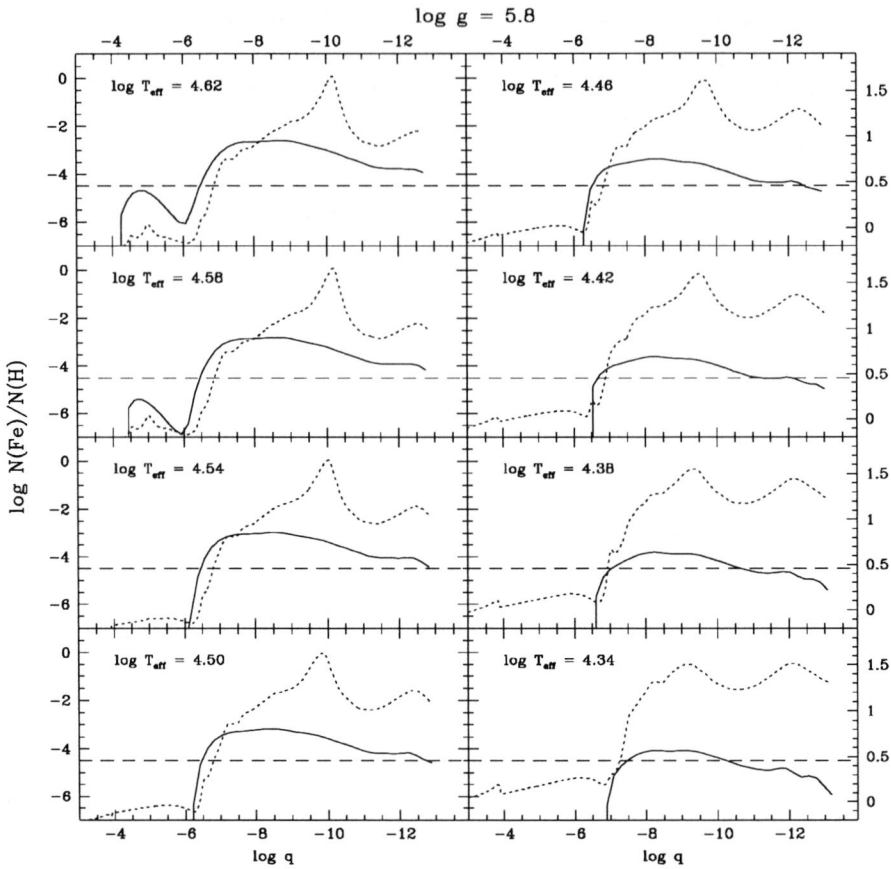

Figure 1. Equilibrium abundance of iron (solid curve) as a function of the fractional mass depth $\log q$ ($= \log(1 - M(r)/M_*)$) for a series of representative models of sdB stars with $M = 0.48\ M_\odot$, $\log g = 5.8$, $\log q(H) = -3.70$, and $\log T_{\rm eff}$ from 4.34 to 4.62 in steps of 0.04. In each panel the tip of the solid curve on the right hand side corresponds to the location of the Rosseland photosphere. The dashed horizontal line gives the normal value of the Fe/H number ratio. Also shown is the profile of the Rosseland opacity (dotted curve) whose logarithmic value can be read on the right axis.

out that, at low effective temperatures, there is not enough iron supported by radiative levitation, the iron opacity peak is rather broad and located relatively deep, and no overall driving is possible. By increasing the effective temperature, the iron opacity peak gets narrower, stronger, and moves toward the surface (the latter effect being caused by increasing overall ionization). In effect, the peak ultimately moves through the critical region for efficient driving. At too high effective temperatures, the peak has moved too far up in the envelope, past the region of efficient driving. This leads naturally to the formation of a broad instability strip.

4. Some sample results

Figure 2 shows how the theoretical period spectrum is affected when changing the effective temperature, the 3 other free parameters of the models being kept fixed at typical values. Here, all modes with $l = 0, 1, 2$, and 3 are considered, and the period window illustrated covers the range from 80 to 280 s. The various curves refer to the fundamental modes (except for the case $l = 1$ and $k = 1$), and divide the diagram between the domain of the p–modes (below the curves) and that of the g–modes (above the curves). The excited modes define a broad instability strip in the range 39,000 K $\gtrsim T_{\text{eff}} \gtrsim$ 27,000 K (for these $\log g = 5.8$ models). Only some of the fundamental and some of the p–modes are excited; the g–modes are not driven in the present models.

Except for the f–modes and g–modes, the period of a given (p) mode is not very sensitive to T_{eff}. However, if we focus only on the *excited* modes, the figure reveals that the iron–bump mechanism can only drive high order modes near the blue edge of the instability strip, and the lowest order modes near the red edge. Hence, we find a correlation between T_{eff} and excited periods such that the higher the effective temperature, the shorter the periods of the excited modes (prediction #1). In addition, the figure reveals that the largest growth rates are found for models with $T_{\text{eff}} \sim$ 33,000–36,000 K. Hence, we would expect the pulsators with representative surface gravities near $\log g \sim$ 5.8 to be clustered primarily around these effective temperatures (prediction #2).

Figure 3 is similar, except that, this time, the effects of varying the surface gravity are illustrated. The effective temperature, total mass, and H–rich envelope mass are now kept fixed at typical values in this second batch of models. As before, the curves divide the diagram into the domain of g–modes and that of the p–modes. Except for the odd g–mode in the lowest gravity models, the iron bump mechanism is seen to be able to excite only low order p–(and f–)modes. Physically, this is due to the fact that the p–modes are primarily envelope modes in sdB star models, while g–modes are primarily core modes with important amplitudes only in regions located much deeper than the driving region. (This distinction becomes more fuzzy in the lowest gravity models.) On the basis of our calculations, we do not expect to observe g–modes in pulsating sdB stars (prediction #3).

The figure also illustrates the large effects on the period spectrum produced by changing the surface gravity while keeping the effective temperature constant. Of course, this is no surprise here as the mechanical structure is strongly affected. The period of a given mode (i.e., for fixed values of l and k) increases substantially with decreasing surface gravity, and this is why we were forced to consider the wider range of periods, 80–800 s, in Fig. 3. The periods of the *excited* modes follow the same trend, so we expect that the lower

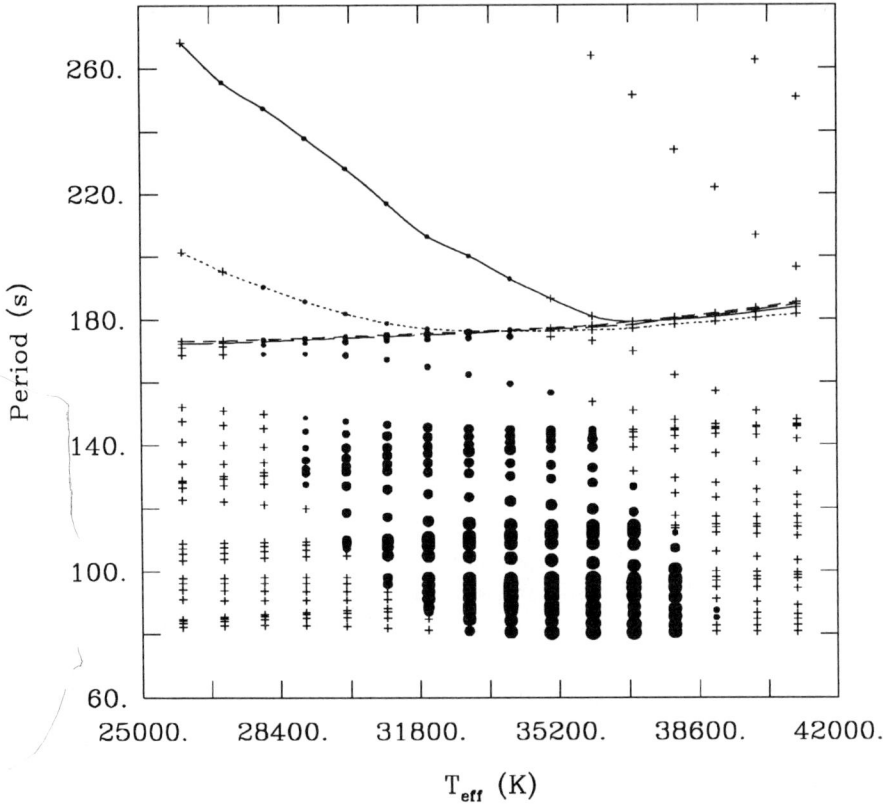

Figure 2. Period spectrum as a function of effective temperature for representative sdB models with $M = 0.48\ M_\odot$, $\log g = 5.8$, and a H–rich envelope mass $\log q(H) = -3.70$. All modes with $l = 0, 1, 2,$ and 3, and with periods in the range 80–280 s are illustrated. A stable mode is represented by a small cross, while an excited mode is represented by a filled circle. There are 7 different sizes of filled circles, each size representing a range of values of the modulus of the imaginary part of the complex frequency, σ_i. In order of increasing circle size, the relevant values of σ_i (in Hz) are $< 10^{-7}$, $10^{-7} \leq \sigma_i < 5 \times 10^{-7}$, $5 \times 10^{-7} \leq \sigma_i < 10^{-6}$, $10^{-6} \leq \sigma_i < 5 \times 10^{-6}$, $5 \times 10^{-6} \leq \sigma_i < 10^{-5}$, $10^{-5} \leq \sigma_i < 5 \times 10^{-5}$, and $\geq 5 \times 10^{-5}$. The most unstable modes are thus those represented by the largest filled circles. The curves join together the fundamental modes and separate the p–modes from the g–modes in the diagram. The dashed (long–dashed, solid, dotted) curve joins together the modes with $l = 0$ and $k = 0$ ($l = 1$ and $k = 1$; $l = 2$ and $k = 0$; $l = 3$ and $k = 0$).

gravity EC14026 stars should show the longer periods (prediction #4).

Finally, one can also observe in the figure that the largest growth rates are found in the lowest gravity models. Insofar as linear theory may be relied upon, this suggests that the lower gravity EC14026 stars should show the larger amplitudes (prediction #5).

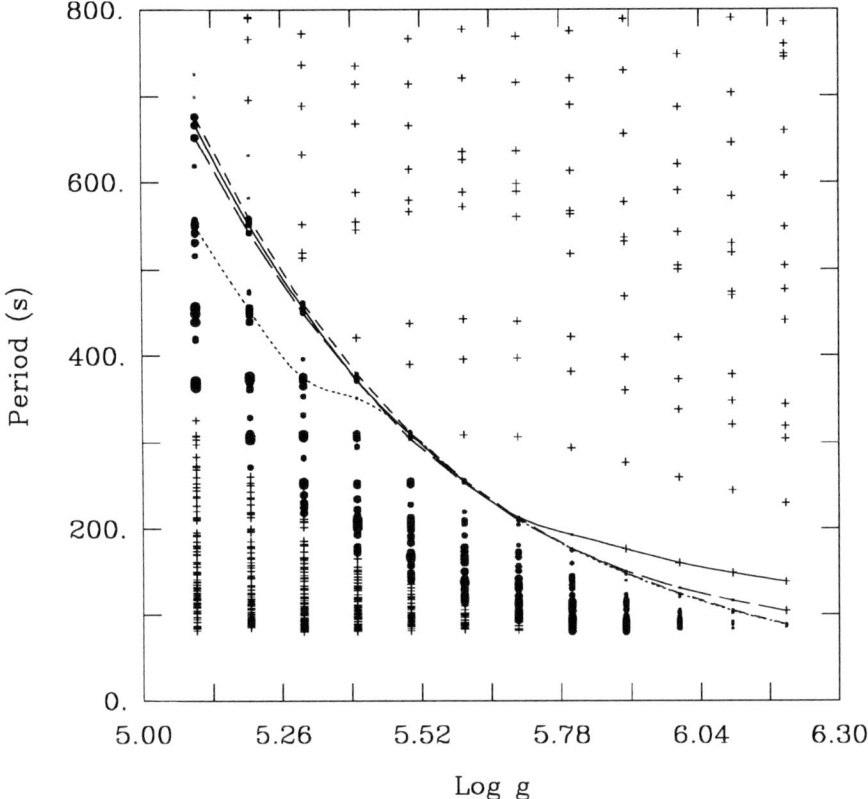

Figure 3. Period spectrum as a function of surface gravity for representative sdB models with $T_{\text{eff}} = 34{,}000$ K, $M = 0.48\ M_\odot$, and a H–rich envelope mass $\log q(H) = -3.70$. All modes with $l = 0,\ 1,\ 2,$ and 3, and with periods in the range 80–800 s are illustrated. As in Fig. 2, a stable mode is represented by a small cross, while an excited mode is represented by a filled circle. There are 5 different sizes of filled circles, each size representing a range of values of the modulus of the imaginary part of the complex frequency, σ_i. In order of increasing circle size, the relevant values of σ_i (in Hz) are $< 10^{-7}$, $10^{-7} \leq \sigma_i < 10^{-6}$, $10^{-6} \leq \sigma_i < 10^{-5}$, $10^{-5} \leq \sigma_i < 10^{-4}$, and $\geq 10^{-4}$. As in the previous figure, the most unstable modes are thus those represented by the largest filled circles. The curves join together the fundamental modes and separate the p–modes from the g–modes in the diagram. The dashed (dotted, solid, long–dashed) curve joins together the modes with $l = 0$ and $k = 0$ ($l = 1$ and $k = 1$; $l = 2$ and $k = 0$; $l = 3$ and $k = 0$).

5. Discussion and conclusion

The results of our nonadiabatic studies of models of sdB stars lead to definite predictions which can, in principle, be tested against the statistics of EC14026 stars. Although not enough statistics have accumulated on these objects yet, the few results available are very encouraging and tend to support strongly the reality of the destabilization mechanism we have uncovered in our models. For instance, the values of the periods of the modes discovered so far in EC14026 stars are *all* consistent with p–modes (prediction #3). Also, it turns out that the first 6 EC14026 stars discovered have very similar atmospheric parameters: $T_{\text{eff}} \simeq 34{,}450$ K (450 K, rms deviation) and $\log g \simeq 5.85$ (0.09, rms deviation). This is the clustering of pulsators expected in the range 33,000–36,000 K for stars with these gravities (prediction #2). Their pulsation properties are also similar with periods found in the range 104–184 s and amplitudes of the individual modes ranging from a fraction of a millimag up to 14 millimag. We know of the existence of 2 other EC14026 stars: the recently discovered pulsator KPD2109+4401 (Billères et al. 1997), a star with $\log g \simeq 5.83$, but with a somewhat lower value $T_{\text{eff}} \simeq 31{,}160$ K . In that case, the period range is pushed to somewhat higher values, 182–198 s, as expected for a cooler star (prediction #1). The case of PG1605+072 (Kilkenny, private communication) is even more interesting. This star currently stands out as an EC14026 object with a relatively low effective temperature, $T_{\text{eff}} \simeq 30{,}000$ K, and, even more significantly, with a rather low surface gravity, $\log g \simeq 5.20$. As expected (prediction #4), the range of excited periods is broader and reaches values of up to 601 s. Also, the amplitude of the dominant mode at 482 s reaches a value of more than 24 millimag, the largest so far observed in EC14026 stars (prediction #5). Quite clearly, these results are rather encouraging, but must be seen as the first attempt to elucidate the asteroseismology of EC14026 stars, a field still in its infancy. For an update of the observational situation, the reader is referred to the interesting paper of Stobie et al. in these Proceedings.

References

Billères, M., Fontaine, G., Brassard, P., Charpinet, S., Liebert, J., Saffer, R.A., Bergeron, P., & Vauclair, G. 1997, *Ap.J.*, in press
Charpinet, S., Fontaine, G., Brassard, P., & Dorman, B. 1996, *Ap.J.*, **471**, L103
Charpinet, S., Fontaine, G., Brassard, P., Chayer, P., Rogers, F.J., Iglesias, C.A., & Dorman, B. 1997, *Ap.J.*, **483**, L123
Chayer, P., Fontaine, G., & Wesemael, F. 1995, *Ap.J.Suppl.*, **99**, 189
Dorman, B. 1995, in Proc. 32nd Liège Astrophysical Colloq., Stellar Evolution: What Should Be Done, ed. A. Noels, D. Fraipont-Caro, N. Grevesse, & P. Demarque (Liège: Institut d'Astrophysique), 291
Heber, U. 1991, in IAU Symposium 145, Evolution of Stars: The Photospheric Abundance Connection, ed. G. Michaud & A. Tutukov (Dordrecht: Kluwer), 363
Kilkenny, D., Koen, C., O'Donoghue, D., & Stobie, R.S. 1997, *M.N.R.A.S.*, **285**, 640
Koen, C., Kilkenny, D., O'Donoghue, D., Van Wyk, F., & Stobie, R.S. 1997, *M.N.R.A.S.*, **285**, 645
Michaud, G., Bergeron, P., Wesemael, F., & Fontaine, G. 1985, *Ap.J.*, **299**, 741
O'Donoghue, D., Lynas-Gray, A.E., Kilkenny, D., Stobie, R.S., & Koen, C. 1997, *M.N.R.A.S.*, **285**, 657
Saffer, R.A., Bergeron, P., Koester, D., & Liebert, J. 1994, *Ap.J.*, **432**, 351
Stobie, R.S., Kawaler, S.D., Kilkenny, D., O'Donoghue, D., & Koen, C. 1997, *M.N.R.A.S.*, **285**, 651

ASTEROSEISMOLOGY FROM EQUIVALENT WIDTHS: A TEST OF THE SUN

C.U. KELLER, J.W. HARVEY, S.C. BARDEN,
M.S. GIAMPAPA, F. HILL AND C.A. PILACHOWSKI
National Optical Astronomy Observatories
P.O.Box 26732, Tucson, AZ 86726-6732, USA

1. Introduction

Kjeldsen et al. (1995) reported a probable detection of solar-like, low-amplitude p-mode oscillations in η Bootes using equivalent width measurements from low-resolution spectra of hydrogen Balmer lines. However, this detection has not been confirmed so far. Indeed, there is no confirmed detection of p-mode oscillations in a solar-like star using the equivalent width technique or any other approach (Bedding 1998).

An important test of the equivalent width method consists in applying it to integrated solar light. This has been done by Kjeldsen et al. (1995), but they did not detect an obvious excess in the power spectrum at 3 mHz at their noise level of 5.1 ppm. While techniques such as accurate Doppler measurements have been tested using solar light, there has been no successful test of the equivalent width technique using Balmer lines on the Sun.

We recently obtained six continuous days of solar integrated light measurements with a low-resolution spectrograph, but have failed so far to see clear evidence for p modes. Hence, we recorded resolved solar observations where we integrated only over a small part of the solar disk to enhance the apparent amplitude of the oscillations. This allows us to test the equivalent width technique, understand how it works, estimate the expected amplitude for integrated sunlight measurements, and compare it to theoretical amplitude predictions for the Sun.

2. Observations and Data Reduction

The time series was obtained with the 1.5-m McMath-Pierce main telescope and the stellar spectrograph. The telescope was out of focus so that the effective area observed on the Sun covered about 1 arcmin. One camera of the Zurich Imaging Stokes Polarimeter ZIMPOL I (e.g. Keller 1992, Povel 1995) with its fast read-out of 38 ms per frame and 12-bit sampling was used to obtain spectra at a very high signal-to-noise ratio. We observed for 4.3 hours with a sampling rate of 1/15 Hz and a dispersion of 0.27 Å/pixel. An effective signal-to-noise ratio of 2.6×10^4 per pixel was achieved by adding 128 frames in 15 s and averaging along the spectral lines.

We subtracted the dark current and spectrograph stray-light, but no flatfield correction was applied due to the lack of a suitable lamp. However, if the flatfield is constant in time, its influence on the final power spectra should be negligible.

The Doppler velocity was determined by cross-correlating the individual spectra with the average spectrum and subtracting spectrograph drifts as measured with a fiducial mark in the optical system.

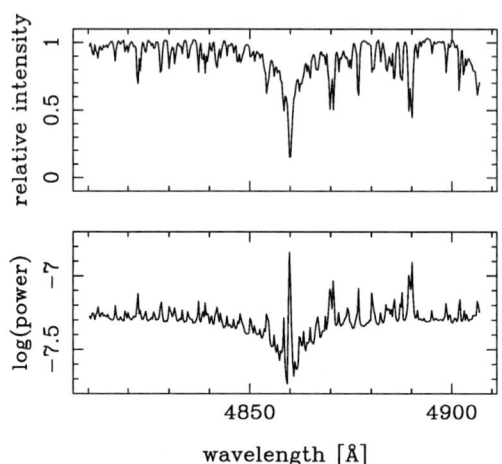

Figure 1. Power in the 2 to 4 mHz band as a function of wavelength around the Hβ line.

The following steps were applied to the time series of each spectral pixel: normalize and remove long-term trends to obtain the fractional variation independent of the intensity; decorrelate against apparent Doppler shift signal to remove the influence of intensity variations due to Doppler velocities and spectrograph drifts; and calculate the power spectrum with two-point filtering (e.g. Duvall et al. 1993) and apodization to minimize the leakage of large-amplitude low-frequency variations into the 5-minute band. Finally, the power spectra were integrated between 2 and 4 mHz to estimate the power in the 5-minute band as a function of wavelength (see Figure 1).

The Hβ equivalent width was determined by measuring the intensity in Gaussian windows centered on Hβ and two continuum locations. The equivalent width is then given by $1 - I_{H\beta}/I_{continuum}$, and the fractional variation is determined by normalizing with a fifth-order polynomial to remove long-term variations.

Power spectra of the Doppler velocity, the intensity in the continuum windows, the Hβ window, and the Hβ equivalent width were calculated using two-point filtering and apodization.

3. Results

The 5 minute oscillations are reduced in Hβ to about 70% of the continuum signal as shown in Figure 1. A similar effect has been seen in Hγ and Hδ by Ronan et al. (1991). Therefore, at least for the sun, the equivalent width technique relies on the suppression of the oscillation signal in the Balmer lines as compared to the continuum. The measured signal corresponds to the difference in amplitude between the Hβ line and the continuum, which is only about a third of the absolute amplitude in the continuum. However, the fractional variation in the equivalent width is similar in amplitude to the fractional variation in the continuum.

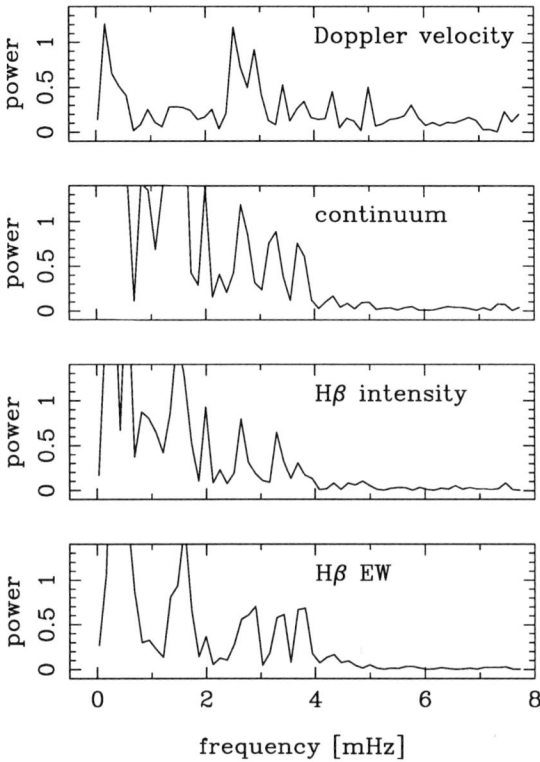

Figure 2. Power spectra of four quantities of resolved solar light (arbitrary units).

Figure 2 shows the power spectra for Doppler velocity, and the fractional variations of the continuum intensity, the Hβ intensity, and the Hβ equivalent width. The intensity measurements show considerable low-frequency power due to atmospheric transparency variations. The Hβ equivalent width shows a clear power excess around 3 mHz.

TABLE 1. Equivalent sine-wave amplitudes of solar integrated-light observations estimated from resolved solar observations.

quantity	resolved	integrated light	reference
velocity amplitude	60 m/s	0.23 m/s	Kjeldsen & Bedding (1995)
continuum amplitude	1.2×10^{-3}	4.6 ppm	estimated here
Hβ equivalent width	1.0×10^{-3}	3.8 ppm	estimated here

For asteroseismology, the amplitudes determined with these resolved observations are not of interest. What is of interest are the amplitudes we would expect for integrated light. They can indeed be estimated by assuming that the scaling factor between the velocity amplitude around 5 mHz for resolved and integrated light can also be applied to other quantities (see Table 1). This extrapolation of the oscillation signal seen in the spatially resolved data suggests an amplitude of about 3.8 ppm for integrated light measurements of the Hβ equivalent width. As usual, we quote the amplitude of a sine wave that matches the observed power spectrum.

4. Discussion

Spatially resolved observations of the Hβ equivalent width at solar disk center reveal that the oscillation signal is suppressed in the wings of Hβ as compared to the continuum. This leads to a clear signal around periods of 5 minutes in the Hβ equivalent width. The amplitude of the absolute (in contrast to the fractional) variations is about three times smaller in the equivalent width as compared to the continuum. The amplitudes of the fractional variations are of similar magnitudes. However, since the absolute amplitude is important for the signal-to-noise ratio, continuum measurements from space should be more sensitive than equivalent width observations.

Our estimate for the integrated-light continuum amplitude is in good agreement with theoretical predictions and other observations, both giving amplitudes of 4.7 ppm (Kjeldsen & Bedding 1995). Our estimate of 3.8 ppm for the amplitude of the fractional variation of the Hβ equivalent width is substantially smaller than the theoretically predicted amplitude of 6 ppm (Bedding et al. 1996). However, the accuracy of our extrapolation is limited and cannot rule out the predicted, higher amplitude.

The Hβ equivalent width power spectrum in Figure 2 shows considerable power at frequencies below 2 mHz, which is not expected since the equivalent width should be independent of atmospheric transmission variations. These variations at low temporal frequencies are not correlated with intensity fluctuations or the position of the spectrum on the detector. They might be due to wavelength dependencies of scintillation and transparency variations, tiny water vapor blends, chromospheric activity, or the evolution of solar features.

Acknowledgments

The National Optical Astronomy Observatories are operated by the Association of Universities for Research in Astronomy Inc. under cooperative agreement with the National Science Foundation. T. Bedding and D. Kurtz provided helpful comments.

References

Bedding T.R., 1998, these proceedings
Bedding T.R., Kjeldsen H., Reetz J., Barbuy B., 1996, MNRAS 280, 1155
Duvall T.L., Jefferies S.M., Harvey J.W., Osaki Y., Pomerantz M.A., 1993, APJ 410, 829
Keller C.U., Aebersold F., Egger U., Povel H.P., Steiner P., Stenflo J.O., 1992, LEST Technical Report 53
Kjeldsen H., Bedding T.R., 1995, A&A 293, 87
Kjeldsen H., Bedding T.R., Viskum M., Frandsen S., 1995, AJ 109, 1313
Povel H., 1995, Optical Engineering, 34, 1870
Ronan R.S., Harvey J.W., Duvall T.L., 1991, APJ 369, 549

ASTEROSEISMOLOGY OF THE β CEPHEI STAR 12 (DD) LACERTAE

W.A. DZIEMBOWSKI
Copernicus Astronomical Center
ul. Bartycka 18, 00-716 Warsaw, Poland

AND

M. JERZYKIEWICZ
Wrocław University Observatory
ul. Kopernika 11, 51-622 Wrocław, Poland

1. The observed frequencies

The frequency spectrum in Fig. 1 shows that at least five pulsation modes are excited in DD Lac. Three frequencies, f_1, f_4, and f_3, form an equidistant triplet. In addition to the value of the central frequency, f_4, the triplet can be characterized by the mean separation, $S = (f_3 - f_1)/2$, and the asymmetry, $\Delta f = f_4 - (f_1 + f_3)/2$. Taking the values of the frequencies from a recent analysis of all available data (Pigulski 1994), we get $S = 0.15544 \pm 0.00021$ and $\Delta f = -0.00014 \pm 0.00029$ d^{-1}.

2. The discrepant spherical harmonic degrees

For the f_1 mode, the l value has been derived from the observed light, color and radial-velocity amplitudes by Cugier et al. (1994). These authors showed that modes with different $l \leq 2$ are particularly well resolved when the color to light amplitude ratio, A_{U-V}/A_V, is used as abscissa, and the radial-velocity to light amplitude ratio, K/A_V, as ordinate. In such a diagram, the f_1 term falls on the $l = 1$ sequence.

For the f_2 mode, the UBV photometry of Sato (1973, 1977, 1979) and all available radial velocities (Pigulski 1994) yield $A_{U-V}/A_V = 0.30 \pm 0.10$ and $K/A_V = 460 \pm 35$ km s^{-1}/mag. In the above-mentioned diagram of Cugier et al. (1994) these coordinates indicate $l = 1$ or 2.

The above identifications can be compared with the results of the line-profile observations. Unfortunately, the three modern line-profile studies of DD Lac yield conflicting results. Smith (1980) maintains that the f_1, f_4, f_3 triplet should be identified with the $l = 2$, $m = 0, -1, -2$ states, and that the f_2 mode is radial. On the other hand, Mathias et al. (1994) conclude that the f_1 mode is either sectoral, or tesseral with $l - 1 = |m|$. More recently, the line-profile data of Mathias et al. (1994) have been re-analyzed by Aerts (1996). She finds $l = 2$, $m = -1$ for f_1, and $l = 2$, $m = -2$

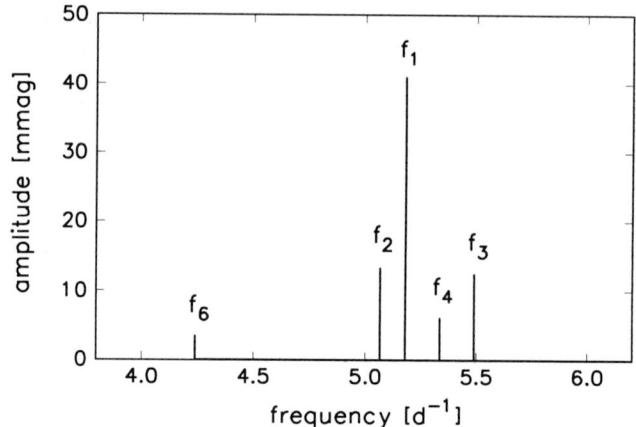

Figure 1. The frequency spectrum of DD Lac, derived by Jerzykiewicz (1978) from the yellow-magnitude data of Abrami et al. (1957). The combination frequency $f_5 = f_1 + f_4$ is not shown.

for f_3, but $l = 3$, $m = +1$ for f_4. Thus, according to Aerts (1996), the triplet f_1, f_4, f_3 does not consist of three m states of the same l. For f_2, Aerts (1996) gets $l = 2$ or 3 and $m = 0$.

The only constraint which follows from these discrepant results is that the f_1 mode is nonradial, with l equal to either 1 or 2.

3. The effective temperature and surface gravity of DD Lac

From the star's MK spectral type of B 1.5 III and the OAO-2 empirical effective temperatures (Code et al. 1976), we derived $\log T_{\text{eff}} = 4.369$. Another value can be obtained from the Strömgren indices. Using the observed c_0 and β, and the recent calibration of Napiwotzki et al. (1993), we got $\log T_{\text{eff}} = 4.380$. The mean of these two numbers, 4.374, we adopt as $\log T_{\text{eff}}$ of DD Lac. Taking into account uncertainties of the calibrations and the mean errors of the data, we estimate the standard deviation of this value to be 0.020.

We also derived $\log g$ of DD Lac. Using the observed value of β and Smalley and Dworetsky's (1995) grid of synthetic β indices, we obtained $\log g = 3.76 \pm 0.15$.

In the $\log T_{\text{eff}} - \log g$ plane, the values we obtained place DD Lac close to the 12 $M_\odot$ evolutionary track, at $\log L/L_\odot = 4.3$, that is, in the advanced phases of core hydrogen-burning. The masses allowed by the mean error of $\log T_{\text{eff}}$ span the range from 10 to 13.5 $M_\odot$.

4. The triplet

We shall now assume that f_1, f_4, and f_3 arise as a result of rotational splitting of an $l > 0$ mode. According to the conclusion reached at the end of Sect. 2, the l in question must be 1 or 2. Since for a given l there are $2l + 1$ spherical harmonic orders, m, there is only one possibility of accounting for the frequency triplet if $l = 1$, and four possibilities if $l = 2$. If $l = 1$, the order of frequencies implies the following m-

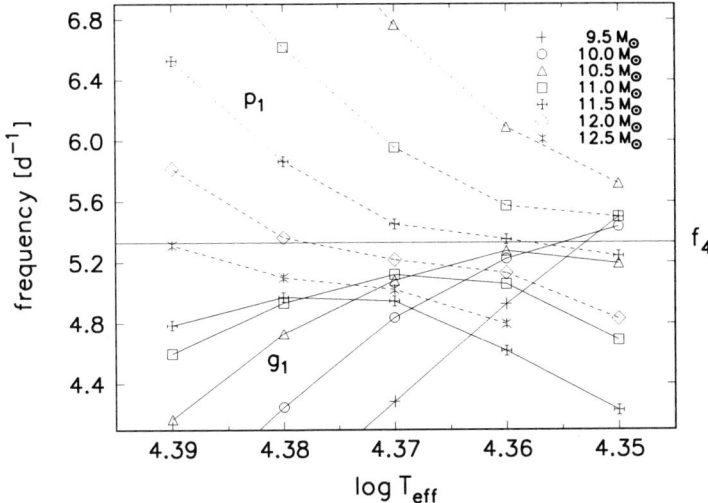

Figure 2. Fitting the observed $f_4 = 5.33427$ d^{-1} to the $l = 1$, $m = 0$ modes.

triplet: $m = +1$ (retrograde) for f_1, 0 for f_4, and -1 (prograde) for f_3. If $l = 2$, the four possible m-triplets are: $(0, -1, -2)$, $(1, 0, -1)$, $(2, 1, 0)$, and $(2, 0, -2)$.

5. The radial order of the f_4 mode

A number of modes of $l = 1$ or 2, and $m = -1$, 0 or 1, with frequencies close to f_4 are unstable in models of DD Lac in the range of effective temperature and mass allowed by the position of the star in the $\log T_{\rm eff} - \log g$ plane. An example is presented in Fig. 2, where computed frequencies of the two $l = 1$, $m = 0$ modes are plotted together with f_4 (horizontal solid line). The designations p_1 and g_1 refer to properties of the modes on ZAMS, where the p- and g-mode spectra are separated in frequency. The models were computed with the updated OPAL opacities (Iglesias and Rogers 1996) for $X = 0.7$ and $Z = 0.02$.

As can be seen from Fig. 2, f_4 can be fitted with the frequency of the p_1 mode in the whole range of $\log T_{\rm eff}$. On the other hand, for the g_1 mode such fit is possible only at the lower limit of the temperature range and for the lowest-mass models which have $\log g$ still consistent with the observed one. In addition, an $l = 1$, $m = 0$, p_2 mode in models with $\log T_{\rm eff} \approx 4.39$ and $M > 13\,M_\odot$ (not shown in Fig. 2) also has its frequency close to f_4.

6. The mean separation and asymmetry of the triplet

Having identified the modes that could reproduce f_4 (those shown in Fig. 2 and many others), we varied the rotation frequency, Ω, until computed frequencies of the outlying members of the triplet, f_1 and f_3, were also fitted. The rule we adopted was that the observed mean separation of the triplet, S, should be reproduced exactly. The interaction between rotation and pulsation was treated as in Soufi et al. (1997). However, only terms up to quadratic ones in rotation frequency were kept in the perturbation treatment. The effect of the near degeneracy of modes differing in l by 2 was taken into account.

Out of a large number of models computed in this way, we selected those which had the smallest asymmetry, Δf, and reproduced – at least approximately – the two frequencies outside the triplet, f_2 and f_6. In all cases we encountered the following two problems: (1) the number of unstable modes was much greater than observed; this problem is common to all linear pulsation analyses (see, e.g., Dziembowski 1997), and (2) the computed Δf was greater than the observed one. A discussion of three representative models has been already provided by one of us (WAD, this volume).

This work was supported by KBN grants 2 P304 013 07 and 2 P03D 016 11.

References

Abrami, A., Bakos, G., Broglia, P. et al., 1957, *Nature*, **180**, 1112
Aerts, C., 1996, *Astron. Astrophys.*, **314**, 115
Code, A.D., Davis, J., Bless, R.C., and Hanbury Brown, R., 1976, *Astrophys. J.*, **203**, 417
Cugier, H., Dziembowski, W.A., and Pamyatnykh, A.A., 1994, *Astron. Astrophys.*, **291**, 143
Dziembowski, W.A., 1997, in J. Provost and F.X. Schmider (eds) *IAU Symposium No. 181 – Sounding Solar and Stellar Interiors*, in press
Iglesias, C.A. and Rogers, F.J., 1996, *Astrophys. J.*, **464**, 943
Jerzykiewicz, M., 1978, *Acta Astronomica*, **28**, 465
Mathias, P., Aerts, C., Gillet, D., and Waelkens, C., 1994, *Astron. Astrophys.*, **289**, 875
Napiwotzki, R., Schönberner, D., and Wenske, V., 1993, *Astron. Astrophys.*, **268**, 653
Pigulski, A., 1994, *Astron. Astrophys.*, **292**, 183
Sato, N., 1973, *Astrophys. Space Sci.*, **24**, 215
Sato, N., 1977, *Astrophys. Space Sci.*, **48**, 453
Sato, N., 1979, *Astrophys. Space Sci.*, **66**, 309
Smalley, B. and Dworetsky, M.M., 1995, *Astron. Astrophys.*, **293**, 446
Smith, M.A., 1980, *Astrophys. J.*, **240**, 149
Soufi, F., Goupil, M.J., and Dziembowski, W.A., 1997, *Astron. Astrophys.*, in press

RECOVERING THE PULSATION VELOCITY DISTRIBUTION ON STELLAR SURFACE

J. HAO
Beijing Astronomical Observatory
Beijing 100080, CHINA

1. Introduction

The analytical expression between the line profile and its corresponding pulsation velocity field is derived by the assumption of Doppler Imaging (DI). Based on this approach, numerical experiments of the recovery of the one dimensional nonradial pulsation velocity distribution from the residual line profiles are presented.

2. Motivation

Since the present methods of line profile analysis of nonradial stellar pulsation based on the DI principle employ the residual line profile series as the input data instead of the velocity field itself, they are more or less empirical and lack of firm mathematical foundation. Here we find that there exists the analytical relation between the line profiles and their corresponding velocity field in the case where the DI assumption is satisfied. Based on this relation, we give a method of reconstructing the one dimensional velocity distribution from the residual line profiles. We hope this effort can lead the way to approach more accurate mode identification of nonradial pulsation.

3. Basic relations

The mathematic form of a line profile can be expressed as

$$p(v) = \int\int_S b(x,y) f(v - v_{\rm rot}(x,y) - v_{\rm pul}(x,y)) dS,$$

where $b(x,y) = 1 - u_\lambda + u_\lambda\sqrt{1-(x^2+y^2)}$, is the limb dark law, and $f(v)$ is the intrinsic profile. The Fourier transform is expressed as: $P(\omega) = \int_{-\infty}^{\infty} p(v) e^{-i\omega v} dv$. When $\omega V_{\rm pul} \ll 1$, the DI assumption is satisfied and we can obtain $P(\omega) = F(\omega)(R(\omega) - i\omega V(\omega))$, where $F(\omega)$, $R(\omega)$ and $V(\omega)$ are the Fourier transforms of $f(v), r(x) = 2(1-u_\lambda)\sqrt{1-x^2} + \frac{\pi u_\lambda}{2}(1-x^2)$, and $\overline{v_{\rm pul}}(x) \equiv \int_{-\sqrt{1-x^2}}^{\sqrt{1-x^2}} b(x,y) v_{\rm pul}(x,y) dy$, respec-

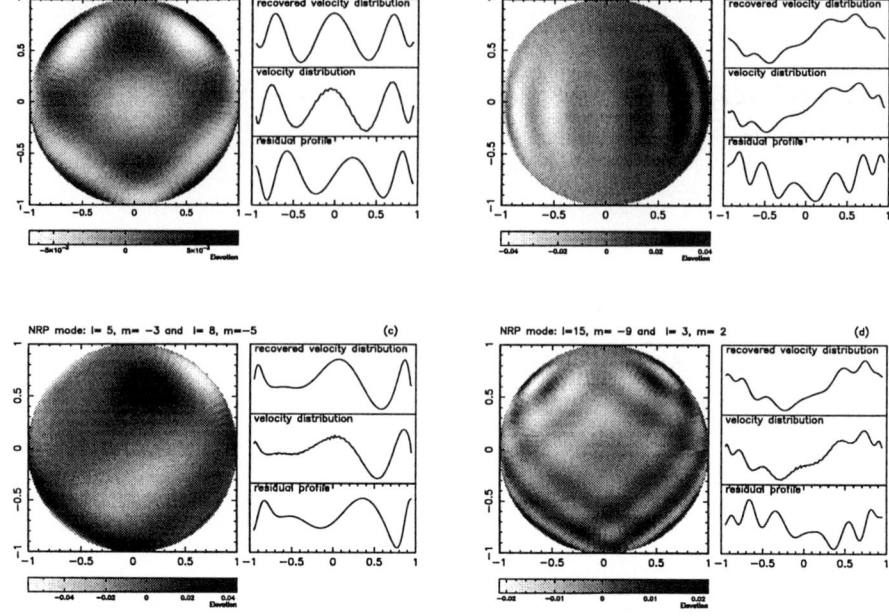

Figure 1. The results of numerical experiments for some modes. The parameters used in the calculation are as follows: (a) $l = 8$, $m = -5$ with amplitude of 0.01 $v \sin i$, $k = 0.25$, $i = 75°$, $v \sin i$ is 17 times the width of the intrinsic profile. (b) $l = 14$, $m = -14$ and $l = 2$, $m = 2$ with amplitudes of 0.01 and 0.1 $v \sin i$, respectively, $k = 0.25$, $i = 85°$, $v \sin i$ is 20 times the the width of intrinsic profile. (c) $l = 5$, $m = -3$ and $l = 8$, $m = -5$ with amplitudes of 0.09 and 0.03 $v \sin i$, respectively, $k = 0.25$, $i = 75°$, $v \sin i$ is 17 times the the width of intrinsic profile. (d) $l = 15$, $m = -9$ and $l = 3$, $m = 2$ with amplitudes of 0.01 and 0.03 $v \sin i$, respectively, $k = 0.25$, $i = 75°$, $v \sin i$ is 17 times the width of the intrinsic profile.

tively. Considering $F(\omega)R(\omega)$ is the Fourier transform of non-pulsation profile, $p^*(v)$, we obtain the relation:

$$\mathcal{F}[p(v) - p^*(v)] = -i\omega F(\omega)V(\omega), \quad \text{or} \quad p(v) - p^*(v) = -i\mathcal{F}^{-1}[\omega] * f(x) * \overline{v_{\text{pul}}(x)},$$

where $\mathcal{F}$ and $\mathcal{F}^{-1}$ denote the Fourier transform and the reverse transform, respectively.

4. How to apply the technique

We use the observed residual profiles, $p(v) - p^*(v)$, and perform the deconvolution to them with $\mathcal{F}^{-1}(\omega)$, then the intrinsic profile, $f(x)$. Finally, the surface velocity distribution, $\overline{v_{\text{pul}}(x)}$, can be recovered. The numerical experiments are made and the results are shown in Fig.1.

A HIGH SPEED PHOTOMETRIC SURVEY OF NORMAL AND PECULIAR A-TYPE STARS

O.M. KURTANIDZE AND M.G.NIKOLASHVILI
Astronomical Observatory, 383762 Abastumani,
Republic of GEORGIA
okur@abao.kheta.ge, maria@abao.kheta.ge

Abstract. The programme of long-term high-speed photometric survey about two hundred Normal and Peculiar A-type stars with 125cm RC telescope equipped by Two-Star Photometer is presented.

1. INTRODUCTION

Rapidly oscillating Ap stars form a unique group of objects having pulsations with frequencies close to those seen in the Sun, but with amplitudes several times larger. They are chemically peculiar which pulsate in low-degree ($l < 3$), high-overtone ($n >> l$) non-radial p-modes with periods in the range of 4-15min. The peak-to-peak amplitudes are $\leq$ 16mmag. The searches for pulsations in Ap stars were summarized in the recent reviews (Kurtz, 1990; Martinez, 1993; Martinez and Kurtz, 1995).

2. ABASTUMANI SURVEY

Prior beginning of the Abastumani survey, only 27 roAp stars were discovered since 1978. Most of roAp stars discovered (25) are located in the southern hemisphere. Two roAp located in the northern hemisphere 10 Aql and γ Equ are very bright, brighter than 6mag which definitely indicates that all northern sky photometric surveys of roAp stars were very ineffective. In order to ameliorate this situation we have initiated a long-term high-speed photometric survey of the peculiar and normal A-F stars in the northern hemisphere. The scientific goal of the survey are as follows.

1. To increase the number of known roAp in the norhtern hemisphere, so that the extent of the roAp phenomenon can be better determined.
2. To apply the techniques of asteroseismology to these new roAp stars.
3. To study the relation between the roAp stars and other pulsating stars in the same region of the H-R diagram, such as δ Sct stars.

The candidates in our survey are drawn from the Catalogue General des Etoiles Ap et Am (Renson 1991) and the Michigan Spectral Survey. The instrument used was the Two-Star Photometer attached to the 125cm RC telescope (1/13) of Abastumani Astrophysical Observatory. It permits to carry out simultaneous measurements of two objects in the field 22x22 sq.min with exposure of 10^{-3} to 10^2. Photometer operates on line with PC/AT computer.

In principle, the use of Two-Star photometers in high-speed photometry should make it possible to eliminate sky transparency and sky background variations to produce scintillation limited amplitude spectra at all frequencies. But it is usually difficult to acquire a comparison star of similar brightness in the limited field of second channel. Hence, if the program star data are normalized to the comparison star data on a point-by-point basis, the large photon noise of the comparison star data degrades the quality of the program star data. We smoothed the comparison star data over half the period of interest in the program star and then normalized the program star data with the comparison star data, assuming that the photon noise for comparison star is less than sky transparency noise when integrated over half the period of interest in the program star (see Kurtz, 1984). On nights of significantly good quality the individual channels can be analysed separetely. This allows more thorough interpretation of the data since each star produces its own result that can be used along with the combined channels, making identification and interpretation of significant periodicities more straightforward.

The duration of a typical data set is 2.5-3h. Data set significantly shorter than these do not produce periodograms of sufficient quality or resolution for the periods of interest here. On the other hand, to produce an extremely high resolution periodogram, much more than 3h of observations are required. If multiple frequenes are present, time series longer than any beat frequencies are required to resolve all components. The 20 or 30 arcsec diameter apertures and integration times 2-10 sec were used. The observations are interrupted at regular intervals for measurements of the sky background and dark. The software for data processing includes the Fourier filtering and fine spectral scanning methods for periodic fast phenomena and the correlation analysis for non-periodic variations of light curves.

Up to now we have searched nine SrCrEu stars for high-overtone pulsations. Seven of them were observed on more than one night. Unfortunately none of these stars show variations, although we have precision required to reveal the presence of oscillations even for lower amplitude roAp stars. Several roAp stars may lie undiscovered in these null results because we happend to observe them near quadrature (amplitude minimum) or during a time when several oscillation frequencies in the star interfered destructively.

Unfortunately at present due to very seldom electricity supply of the Observatory we are unable to carry out the Programme effectivelly.

References

Kurtz, D.W., 1984, *NASA CP-2350*, 56.
Kurtz, D.W., 1990, *ARA&A*, v.28, 607.
Martinez, P., Kurtz, D.W., 1995, *ASP CS* v.83, 58.
Martinez, P., 1993a, *Ph.D.Thesis*, Univ. of Cape Town.
Renson, P., 1991, *Catalogue General des Etoiles Ap et Am*.

LINE PROFILE VARIATIONS IN THE SPECTRA OF THE γ DOR STAR HR 2740

E. PORETTI AND L. MANTEGAZZA
Osservatorio Astronomico di Brera
Via E. Bianchi 46, 22055 Merate, Italy

AND

C. KOEN, P. MARTINEZ, F. BREUER, D. DE ALWIS AND H. HAUPT
South African Astronomical Observatory
P.O. Box 9, Observatory 7935, Cape Town, South Africa

1. Photometric results

HR 2740 was a target for a photometric campaign carried out at La Silla (ESO) and Sutherland (SAAO) from 1997 January 14 to 1997 February 11 (Poretti et al., 1997). The campaign revealed that HR 2740 is one of the brightest γ Dor stars, a class of variable stars located near the cool border of the instability strip, and in which gravity pulsation modes are excited. Four frequencies were identified (f_1=1.0434, f_2=0.9951, f_3=1.1088, f_4=0.9019 c d^{-1}), which together yield a satisfactory solution to the observed light curve. The frequency analysis was not simple, but thanks to the large coverage in longitude we could separate the effect of aliasing on the two terms f_3=1.1088 c d^{-1} and f_4=0.9019 c d^{-1}, linked by the relationship $f_4 = 2 - f_3$. Moreover, only the long time baseline allowed us to resolve the two close terms f_1=1.0434 c d^{-1} and f_2=0.9951 c d^{-1}. See Poretti et al. (1997) for a detailed discussion.

2. Spectroscopic observations

HR 2740 is also reported by Slettebak et al. (1975) as a standard for rotational velocity measurements; we used it for this purpose in two observing runs on the 1.4-m Coudé Auxiliary Telescope (CAT; ESO, La Silla) in 1992 November and 1994 October. Inspection of the normalized spectra revealed clear line profile variations. This prompted us to monitor HR 2740 photometrically, obtaining the results described in the previous section. Figure 1 shows a good example of the line profile variations we observed. In the bottom spectrum the line minima are shifted redward; in the middle spectrum the lines are not as deep as in the other two cases; in the top spectrum the asymmetry is again well marked, but this time towards the bluer wavelengths. A $v \sin i$ value of 60±5 km s^{-1} was also obtained, which renders the rotational splitting again less suitable to match the observed frequency spectra of HR 2740 (see par. 5.5

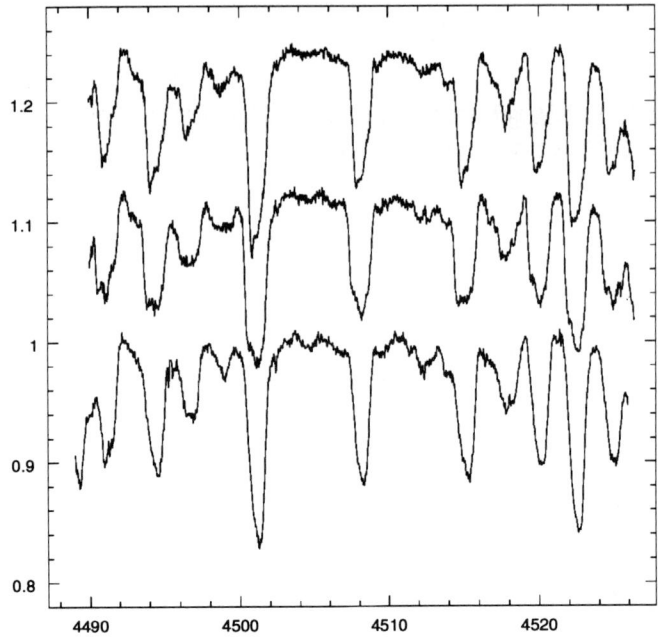

Figure 1. Three high–resolution spectra of HR 2740 (Bottom: JD 2448955.84; middle: JD 2449651.83; top: 2449654.84). The star was originally observed as a standard for rotational velocity measurements, but it exhibits line profile variations indicative of pulsation.

in Poretti *et al.*, 1997). At present, the photometric variability of γ Dor variables has been well studied in a number of cases, but a spectroscopic survey is still lacking (Krisciunas 1997). A combined photometric and spectroscopic approach to the study of the γ Dor stars should produce the same synergistic benefits noted from the combined applications of these techniques to the δ Sct stars (Mantegazza *et al.*, 1997).

References

Krisciunas, K. (1997), The discovery of non–radial gravity mode pulsations in γ Doradus–type stars, this volume

Mantegazza, L., Poretti, E., Bossi, M., Nunez, N.S., Zerbi, F. (1997), Photometry and spectroscopy as a synergic approach to sound the interiors of δ Scuti stars, in *A half century of stellar pulsations interpretations*, Los Alamos, 1997 June, in press

Poretti, E., Koen, C., Martinez, P., Breuer, F., de Alwis, D., Haupt, H. (1997), Discovery and analysis of Gamma Doradus type pulsations in the F0 IV star HR 2740$\equiv$QW Pup, *Monthly Notices Royal Astron. Soc.*, in press

Slettebak, A., Collins, G.W., Boyce, P.B., White, N.M., Parkinson, T.D. (1978), A system of standard stars for rotational velocity determinations, *Astrophys. J. Suppl.*, **Vol. no. 29**, pp. 137–159

THE LIGHT CURVES OF DOUBLE–MODE CEPHEIDS: THE CO AUR CASE

E. PORETTI AND I. PARDO
Osservatorio Astronomico di Brera
Via E. Bianchi 46, 22055 Merate, Italy

1. Fourier decomposition

In two recent papers (Pardo & Poretti 1997; Poretti & Pardo 1997) we analyzed all the available photometry of galactic double–mode Cepheids (DMCs) with the aim of detecting in each case the importance of the harmonics and of the cross coupling terms. We found that no a priori fit can be reliably applied to the measurements of a DMC, but a careful frequency analysis must be done to evaluate the importance of each term. As a further application of this technique, we obtained very precise indications about the properties of the Fourier parameters. When discussing the generalized phase differences $G_{i,j}$ we demonstrated that plotting them as a function of the order $|i|+|j|$, there are well–defined regions where they are confined: the second order terms have $\pi < G_{i,j} < 3\pi/2$; the third order terms have $\pi/2 < G_{i,j} < \pi$; the fourth order terms cluster around 2π.

2. CO Aurigae

By performing the Fourier decomposition of the light curves we found a close similarity between the parameters of the classical Cepheids and those of the fundamental mode of the DMCs and also between the parameters of the s–Cepheids and those of the 1st overtone mode of the DMCs. This fact is an independent confirmation of the different pulsation mode in Classical and s–Cepheids, as first suggested by Antonello et al. (1990) on the basis of the different progressions in the $\phi_{21} - P$ diagrams. Moreover, it seems that in the DMCs the light curve shape in one mode is not influenced by the simultaneous excitation of the other one.

The importance of this result can be appreciated if we consider the case of CO Aur, the only galactic DMC pulsating in the 1st and 2nd overtone. There is no clear indication of the presence of single–mode 2nd overtone pulsators in the Galaxy and we do not know their light curve shape; for an approach to the subject, see Antonello & Kanbur (1997). If we consider the similarity discussed above, we can infer that the shape of a 2nd overtone light curve should be similar to the 2nd overtone component of CO Aur. Figure 1 shows the light curves of the two periods of CO Aur; as can be seen, the f_2 curve is symmetrical, with no appreciable deviation from a sinusoid.

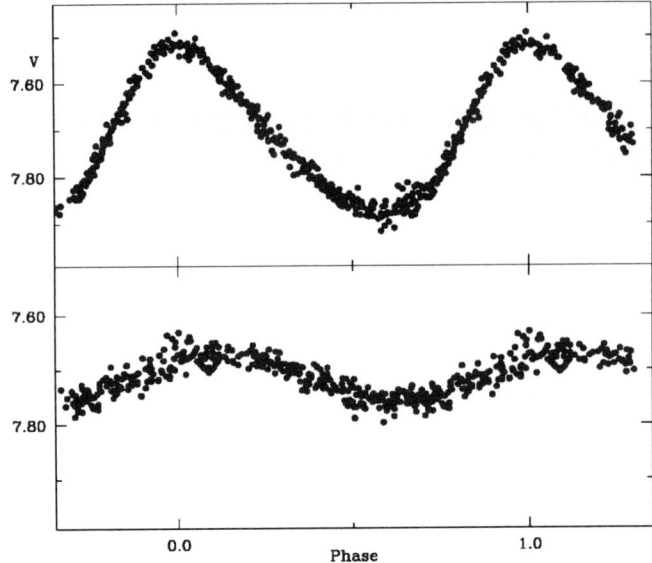

Figure 1. Light curves of the two independent frequencies $f_1=0.560844$ c d^{-1} (1st overtone; top panel) and $f_2=0.700390$ c d^{-1} (2nd overtone; bottom panel) as obtained from the all available V photometry of CO Aur

This result deserves further analysis, but it is clear that very accurate measurements should be performed to detect the $2f$ contribution in the 2nd overtone candidates. On the other hand, the f_1 curve is asymmetrical, since the $2f_1$ term has an amplitude of 0.031 mag; this fact confirms the need of calculating the harmonic content for each independent frequency, without applying an a priori fit, since in this case insignificant terms will be included, distorting the results.

References

Antonello, E., Kanbur, S.M. (1997), The characteristics of second–overtone mode Cepheids predicted by non–linear pulsation modes, *Monthly Notices Royal Astron. Soc.*, **Vol. no. 286**, pp. L33–L37

Antonello, E., Poretti, L., Reduzzi, L. (1990), The separation of s–Cepheids from classical Cepheids and a new definition of the class, *Astron. Astrophys.* **Vol. no. 236**, pp. 138–148

Pardo, I., Poretti, E. (1997), The galactic double–mode Cepheids. I. Frequency analysis of the light curves and comparison with single–mode Cepheids, *Astron. Astrophys.* **Vol. no. 324**, pp. 121–132

Poretti, E., Pardo, I. (1997), The galactic double–mode Cepheids. II. Properties of the generalized phase differences, *Astron. Astrophys.* **Vol. no. 324**, pp. 133–136

ASYMPTOTIC DESCRIPTION AND THE DIAGNOSTIC PROPERTIES OF LOW-DEGREE STELLAR P-MODES

I.W. ROXBURGH[1] AND S.V. VORONTSOV[1,2]
[1] *Astronomy Unit, Queen Mary and Westfield College*
Mile End Road, London E1 4NS, UK;
[2] *Institute of Physics of the Earth*
B.Gruzinskaya 10, Moscow 123810, Russia

Abstract. Standard asymptotic descriptions of stellar p-modes use $1/\omega$ as a small parameter; either the degree ℓ is kept constant in the asymptotic expansions (e.g. Tassoul 1980), or the parameter $\tilde{w} = (\ell + 1/2)/\omega$, which specifies the position of the inner turning point (e.g. Vorontsov 1991). At low degree, due to the strong effects of gravity perturbations, these expansions are known to produce rather poor results, even when developed to higher order (Roxburgh and Vorontsov 1994). Here, we employ an alternative asymptotic expansion, capable of providing much better accuracy at low degree—with $\tilde{w}$ as a small parameter instead of $1/\omega$, and test its diagnostic capabilities.

The asymptotic eigenfrequency equation, which includes terms of order $\tilde{w}^2$, is

$$F_0(\omega) + \tilde{w}^2 F_2(\omega) \simeq \frac{\pi}{\omega}\left[n + \ell/2 + 1/4 + \alpha(\omega)\right],$$

where n is radial order and $\alpha(\omega)$ is the surface phase shift. When terms of order $\tilde{w}^2$ are neglected, the frequencies degenerate with respect to $n + \ell/2$; the function $F_0(\omega)$ is an effective acoustic radius. Small frequency separations are described by $F_2(\omega)$:

$$F_2(\omega_{\ell,n}) \simeq \frac{\pi}{2}\left(n + \frac{\ell}{2} + \frac{1}{4}\right)\frac{\omega_{\ell,n} - \omega_{\ell+2,n-1}}{2\ell + 3}.$$

After $F_2(\omega)$ has been measured and approximated by linear regression $F_2(\omega) \simeq A + B\omega$, the surface term $\alpha(\omega)$ in its combination with $F_0(\omega)$ can be inferred as

$$\beta^*(\omega) = \omega^2 \frac{d}{d\omega}\left[\frac{F_0(\omega)}{\pi} - \frac{\alpha(\omega)}{\omega}\right] \simeq \frac{\bar{\omega}}{\omega_{\ell,n+1} - \omega_{\ell,n}} - n - \frac{\ell}{2} - \frac{3}{4} + \frac{(\ell + \frac{1}{2})^2}{\pi\bar{\omega}}(2A + B\bar{\omega}),$$

at frequency $\bar{\omega} = (\omega_{\ell,n+1} + \omega_{\ell,n})/2$. Numerical results are illustrated by Figs 1 and 2.

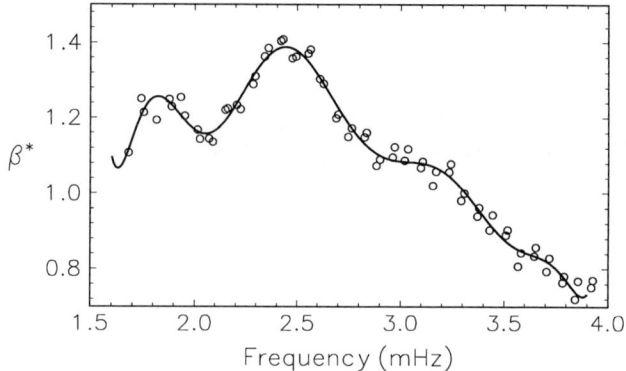

Figure 1. Function $\beta^*(\omega)$ inferred from the solar eigenfrequencies of degree ℓ from 0 to 3. The solid line shows a best-fit polynomial of degree 10. The prominent quasi-periodic variation with frequency is produced by the HeII ionization region

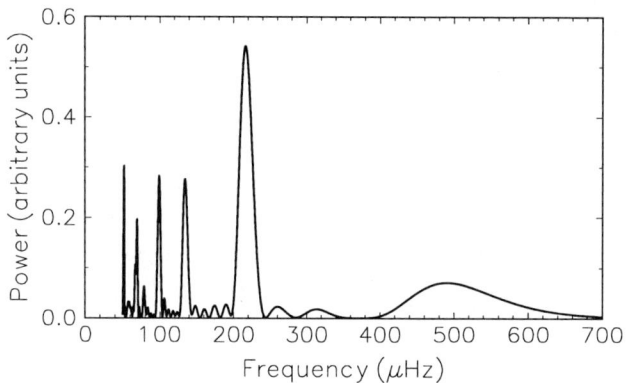

Figure 2. Fourier transform of the approximation residuals seen in Fig.1. The prominent peak at about 220 μHz is produced by the rapid variation of the sound-speed gradient at the base of the solar convection zone

References

Tassoul, M. (1980), *ApJS* **43**, p.469
Roxburgh, I. W. and Vorontsov, S. V. (1994), *Mon. Not. R. astr. Soc.* **268**, p.143
Vorontsov, S. V. (1991), *SvA* **35**, p.400

SPECTRAL LINE-PROFILE VARIABILITY AS A PROBE FOR L AND M

C. SCHRIJVERS
Anton Pannekoek Instituut, University of Amsterdam
Kruislaan 403, 1098 SJ Amsterdam, Netherlands

AND

J.H. TELTING
Isaac Newton Group of Telescopes, ASTRON/NFRA
Apartado 321, 38780 Santa Cruz de La Palma, Spain

We investigate the observable spectroscopic characteristics of non-radial pulsation for stars with rotation rates large enough to resolve the stellar surface by Doppler imaging. We show that the intensity variations in time series of theoretical spectra, at each position in the line profile, cannot be described by a single sinusoid: at least one harmonic sinusoid needs to be included to describe the data. Across the line profile the relative amplitudes and phases of both these sinusoids vary independently.

We use a model of adiabatic pulsations that accounts for stellar rotation effects and includes parameters to simulate non-adiabatic temperature effects. From extensive Monte-Carlo simulations of time series of line-profiles and the subsequent Fourier analysis we derive the following. The blue-to-red phase difference found at the main pulsation frequency turns out to be an indicator of the degree ℓ, rather than the azimuthal order m; the phase difference of the variations with the first harmonic frequency is in many cases (especially for p-modes) an indicator of m. Hence, the evaluation of the variability at the harmonic frequency can improve the results derived from an analysis of observed line profiles.

We conclude that it is possible to derive the ℓ value of the mode(s), if the rotational velocity is high enough to resolve the stellar surface. For p-modes an m-value can be derived as well, if the line-profile variations are non-sinusoidal due to the contribution of the surface-velocity fields. For detailed description of the model, the analysis, and some applications to stellar data we refer to Schrijvers et al. (1997), Telting & Schrijvers (1997ab) and Schrijvers and Telting (1998).

References

Schrijvers C., Telting J.H., Aerts C., Ruymaekers E., Henrichs H.F., 1997, A&AS 121, 343
Telting J.H., Schrijvers C. 1997a, A&A 317, 723
Telting J.H., Schrijvers C. 1997b, A&A 317, 742
Schrijvers C., Telting J.H. 1998, A&A, in prep

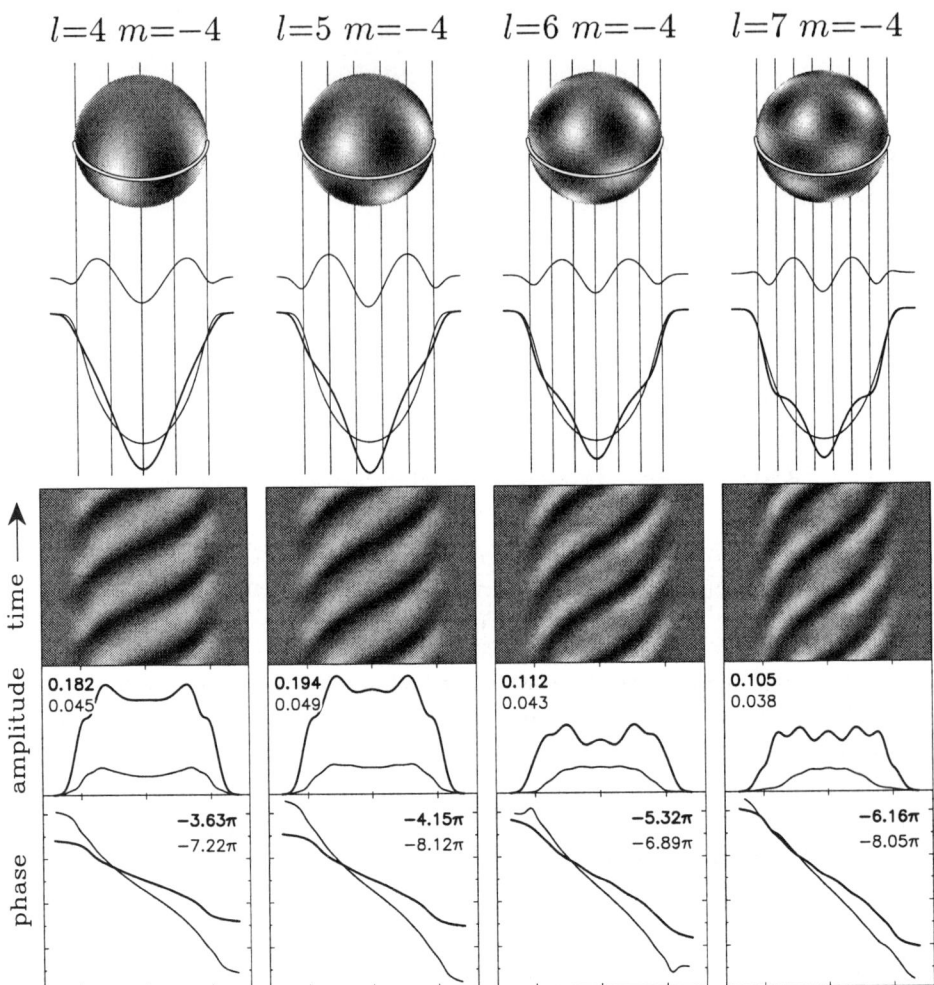

Figure 1. Line-profile variations due to Doppler shifts caused by the three dimensional velocity field of non-radial pulsations. For different values of the pulsation degree ℓ we show (from top to bottom): radial part of the eigenfunction V_r — difference of line profiles of pulsating and non-pulsating case — superposed line profiles of pulsating and non-pulsating case — gray scale representation of residual spectra of 3 pulsation cycles — distribution of the amplitudes of the variations with observed pulsation frequency (thick line) and first harmonic (thin line); the numbers express the maximum values in units of average central line depth — distribution of the phase of the variations with input pulsation frequency (thick line) and first harmonic of the pulsation frequency (thin line); numbers refer to the blue-to-red phase differences.

In the top part of the figure $V_e \sin i$ is indicated by the outer vertical lines, and in the bottom part by the tick marks on the horizontal axis. The stellar and pulsation parameters for this example are: inclination $i=55°$, intrinsic line width $W=0.15*V_e \sin i$, order $m=-4$, ratio of horizontal to vertical pulsation amplitudes $k^{(0)}=0.3$ and the pulsation amplitudes were chosen such that the maximum vector velocity due to the pulsation $V_{\max}=0.15*V_e \sin i$. Note that at any instant the line profile has more bumps for higher values of ℓ. The main phase distribution changes as a function of ℓ, while the slope of the harmonic phase distribution stays rather constant

ASTEROSEISMOLOGY OF β CEPHEI

H. SHIBAHASHI
Department of Astronomy, University of Tokyo, Japan

AND

C. AERTS
Instituut voor Sterrenkunde, Katholieke Univ. Leuven, Belgium

Abstract. We discuss the oscillation features of β Cephei, which is a magnetic star in which the magnetic axis seems to be oblique to the rotation axis. We interpret the observed equi-distant fine structure of the frequency spectrum as a manifestation of a magnetic effect on an eigenmode, which would be a radial mode in the absence of the magnetic field. Besides these frequency components, we interpret another peak in the frequency spectrum as an independent quadrupole mode. By this mode identification, we deduce the mass, the evolutionary stage, the rotational frequency, the magnetic field strength, and the geometrical configuration of β Cephei.

Since many of the β Cephei-type stars show multi-periodic pulsations, it is expected that a lot of information can be extracted from the pulsations of these stars. Telting et al. (1997) investigated the line-profile variability of β Cephei in detail by means of an extensive data set of high-resolution, high S/N spectra obtained at the Observatoire de Haute Provence during one month, and deduced that the variations show periodicity with at least five frequencies. Three frequencies among them are separated from the main frequency $f_1 = 5.250$ day^{-1} by $-2/6$ day^{-1}, $-1/6$ day^{-1}, and $+1/6$ day^{-1}, respectively. Hence, we regard the four peaks (and the small peak at $f_1 + 2/6$ day^{-1}) as part of a frequency quintuplet with equal spacing. The central frequency f_1 has been identified as the radial pulsation mode. Besides the variations in the luminosity and in the velocity field, a magnetic field has been detected and the field strength seems to vary with a period of 6 or 12 days (Rudy & Kemp 1978). Furthermore, the UV-wind lines of this star reveal a periodic variation with a 6 or 12 day period (Henrichs et al. 1993). Hence we conclude that the magnetic field of β Cephei is oblique to the rotation axis, and that the star rotates with a period of 6 or 12 days. We interpret the quintuplet fine structure found by Telting et al. (1997) as a manifestation of a magnetic perturbation of a radial mode. We assume that the dominant component of the magnetic field is dipolar, and that the effect of the magnetic field on the oscillations is stronger than that of the rotation of the star. Being influenced by such a strong magnetic field, the eigenmode, which would be a pure radial mode in the absence of a magnetic field, is deformed to have an axially symmetric quadrupole component, whose symmetry axis coincides with the magnetic axis. Hence, the eigenfunction is characterized by means of a superposition of the spherical harmonic with $\ell = m = 0$ and that of $\ell = 2$ and $m = 0$ with respect to the magnetic axis (Shibahashi 1994). The aspect angle of the pulsation axis varies with the rotation of the star. Therefore, the contribution of the quadrupole component of the eigenfunction to the apparent variation changes with time and produces a quintuplet fine structure with an equal spacing of the rotation frequency in the power spectrum. Hence we conclude that the star is rotating with a period of 6 days rather than 12 days. The relative ratios of the side-peak amplitudes to the central peak amplitude depend on the strength of the magnetic field. On the other hand, the relative ratios among the side-peak amplitudes are dependent on the geometrical configuration of the star, — that is, the angle between the magnetic axis and the rotation axis, β, and the angle between the rotation axis and the line of sight, i. By analyzing the amplitude ratios, we estimate $\tan\beta \tan i \simeq 3$. The factor $\tan\beta \tan i$ can be independently estimated from the variation in the magnetic field strength, and the result is consistent with the estimation from the pulsation power spectrum.

The second highest peak $f_2 = 5.380$ day^{-1} in the power spectrum has been identified by Telting et al. (1997) as a mode with $\ell = 2$. We assume that f_2 belongs to a quadrupole axisymmetric mode with respect to the magnetic axis. The magnetically induced fine structure of this mode is undetectable, since the amplitude of the f_2 component is much smaller than the one of the radial component. Since we have two independent frequencies, f_1 and f_2, we can identify the evolutionary stage of the star. The candidates are (i) a $\sim 18 M_\odot$ star at the middle of the hydrogen core-burning phase, (ii) a $\sim 12 M_\odot$ star near the turning point, (iii) a $\sim 9 M_\odot$ star at the late stage of the hydrogen core-burning phase. The oscillation mode f_1 is identified as the radial fundamental mode in the case (i) or (iii). On the other hand, it is identified as the radial first harmonic in the case (ii). Case (i) seems unlikely because of the high mass required. Though case (ii) cannot be ruled out, we think case (iii) is more likely because the radial fundamental mode is more easily excited. The radius of the star for case (iii) is estimated from the stellar evolution calculation as $R \simeq 6.5 R_\odot$. Since $2\pi/\Omega = 6$ day and $V_e \sin i \simeq 25$ km/s, this means $i = 30°$. Combining this with $\tan\beta \tan i \simeq 3$, we obtain $\beta \simeq 80°$. We have calculated the theoretically expected power spectra and compared them with the observations. A magnetic field strength of 800 G and the geometrical configuration described in the above lead to a reasonable fit.

References

Henrichs, H. F., Bauer, F., Hill, G. M., Kaper, L., Nichols-Bohlin, J. S., & Veen, P. M. 1993, in Proc. IAU Colloq. 139, New Perspectives on Stellar Pulsation and Pulsating Variable Stars, ed. J. M. Nemec & J. M. Matthews (Cambridge: Cambridge Univ. Press), 186.

Rudy, R. J., & Kemp, J. C. 1978, MNRAS, 183, 595.

Shibahashi, H. 1994, in ASP Conf. 76, GONG '94: Helio- and Asteroseismology from the Earth and Space, ed. R. K. Ulrich, E. J. Rhodes, Jr., & W. Däppen (San Francisco: ASP), 618.

Telting, J. H., Aerts, C., & Mathias, P. 1997, A&A, 322, 493.

PULSATION OF PMS 1.8 $M_\odot$ STARS: A THEORETICAL INVESTIGATION

DORU MARIAN SURAN

ASTRONOMICAL INSTITUTE OF ROMANIAN ACADEMY,
Str. Cutitul de Argint, No.5, 75212 Bucharest 28, ROMANIA,
e-mail: suran@roastro.astro.ro

Herbig Ae stars are pre-main sequence of masses $\sim 2M_\odot$. As they trace their course toward the ZAMS, they cross the classical instability strip. At this stage, their evolution has considerably slowed down. Their structure differs from that of stars evolving after the ZAMS (δ Scuti stars) in the very deep layers where nuclear burning has only recently started. But the outer layers for stars before and after the ZAMS are expected to be similar. As these layers drive the pulsation of δ Scuti stars, it is reasonable to expect that PMS stars in the instability strip also are destabilized by the κ mechanism with a similar range of unstable modes. Indeed, recently a Herbig Ae star has been reported as a variable (Kurtz and Marang 1995).

To study more precisely the internal structure of a PMS star of 1.8 $M_\odot$, we use two different theoretical stellar models: *CESAM* code (Morel et al., 1992) and *HENYEY* code (Suran 1993), and check the consistency of the calculations (limited by possible differences in the physical estimations for the EOS, opacities, energy release, convection, overshooting effect,...). We studied also post-MS 1.8-2 $M_\odot$ stars, which lie in the region of the HR diagram as the PMS 1.8 $M_\odot$ star.

For our purpose, it is enough to assume that by the time the proto-star has arrived in the region of the HR diagram of relevance here (classical instability strip, t>5.9 Myr), complex dynamical and thermal phenomena that take place in the very early stages of PMS evolution have long disappeared and have generated an internal structure which can be computed according to a classical treatment for this kind of star.

Details of the structure of PMS models which can be of interest are the *convection* (l_{MLT} =1, 1.5 for the Henyey model, l_{MLT} =1.5 for the CESAM model), and the *gravitational and nuclear energy* release: $\epsilon = \epsilon_g + \epsilon_n$. The difference between a PMS star and a post-MS star at a given position in the HR diagram lies in the fact that the PMS star is still contracting, in a mechanical and thermal imbalance due to the release of gravitational energy. The internal structure, and therefore the pulsation characteristics, such as frequencies and growth rates, evolve on a much more rapid (Kelvin-Helmholtz) time-scale ($\sim 10^6$ yr). Because of the release of contraction (or expansion), an extra energy, due to the ϵ_g, is supplied to or retrieved from the oscillation and affects the stability of the modes. This extra energy is, however, very small and confined in the deep (or respectively outer) adiabatic regions and therefore can be neglected.

Numerical computations (for $l=0-3$, $n<10$) have been done with a *fully nonadiabatic code* (Suran, 1993) and compared with the *Dziembowski's code* (Goupil 1995, private communication). In our analysis we found a wide ranges of excited modes (g,f,p types and also mixed g-p modes).

Due to a similar mean density and similar outer layers, frequencies of pure p modes of given (n,l) are similar. Mixed modes and g modes, trapped in the interior, however differ ($l=2$, p_2, g_2 *mixed modes)*. If we suppose we have:

$$W = W(\epsilon_n) + W(\epsilon_g) + W(\kappa), \qquad (1)$$

the results indicate negligible contribution from nuclear energy and contraction (or expansion) energy. Only for high-overtone g modes does the influence of ϵ_n become of importance. These kind of modes are only quasi-excited modes ($-0.05 < \eta < 0$) in the δ Scuti region (and represent in fact a prolongement of the excited high-overtone g region of the SPB stars). From a seismological point of view: mixed modes would discriminate as they are very sensitive to the structure of the inner layers which differ between PMS and MS stars. This can be done differentially

$$\nu_i(pms) - \nu_i(postMS) \qquad (2)$$

at same position in the HR diagram. This could be seen particularly in the vicinity of models for which post-MS stars have developed a large μ gradient, whereas PMS stars have almost non-existent convective cores with no μ gradient .

References

Dziembowski, W. (1977), Oscilations of giants and supergiants, *Acta Astr.* **27**, pp.95.
Kurtz, D.W., Marang, F. (1995), The discovery of a δ Sct pulsational variability in the premain-sequence Herbig Ae star HR 5999, and the discovery of a rotational light variability in the remarkable He-weak Bp star, HR 6000, *MNRAS* **276**, pp.191.
Morel, P., Berthomieu, G., Provost, J., Lebreton, Y. (1993), CESAM solar model, *IAU Coll. 137 Inside the Stars*, eds. A.Baglin, W.Weiss, pp.54.
Suran, M.D. (1993), Asteroseismologic models with rotation, *IAU Coll 137 Inside the Stars*, eds. A.Baglin, W.Weiss, pp.560.

OSCILLATIONS OF LONG-PERIOD VARIABLES

D. R. XIONG
Purple Mountain Observatory, Academia Sinica, Nanjing 210008, P.R. China

L. DENG
Beijing Astronomical Observatory, Beijing 100080, P.R. China

AND

Q. L. CHENG
Purple Mountain Observatory, Nanjing 210008, P.R. China

Abstract. We have performed a linear pulsational stability survey of long period variable models. The dynamic and thermodynamic couplings between convection and oscillations are treated by using a statistical theory of nonlocal and time-dependent convection. The results show that the fundamental and all the low overtones are always pulsationally unstable for the low-temperature models when the coupling between convection and oscillations is ignored. When the coupling is considered, there is indeed a "Mira" pulsationally instability region outside of the Cepheid instability strip on the H-R diagram. The coolest models near the Hayashi track are pulsationally stable. Towards high temperature the fundamental mode first becomes unstable, and then the first overtone. Some of the 2nd -4th overtones may become unstable for the hotter models. All the modes higher than 4th ($n > 4$) are pulsationally stable. The position and the width of such an instability region on the H-R diagram critically depends on the mass, luminosity and metal abundance of the star.

1. Conclusions and Discussion

By using a nonlocal and time-dependent theory of convection, we have very carefully treated the coupling between convection and oscillations. The linear stability analysis for luminous red stars gives us the following results:

1. The coupling between convection and oscillations is the dominant factor for the pulsational instability, which favours estabilishing a Mira instability strip outside the Cepheid instability strip;
2. Except for some slightly hotter models which may have pulsationally unstable second to fourth overtone modes, luminous red variable stars pulsate in the fundamental mode or the first overtone. All the modes higher than 4 are pulsationally stable;
3. For the low-temperature red stars, the dynamic coupling between convection and oscillations is in the same order of magnitude as the thermodynamic coupling, and may even overtake the later;
4. The effect of turbulent viscosity grows very quickly towards high overtones for the luminous red stars. For high overtones, it becomes the main damping mechanism.

As for the long disputed pulsational mode for o Ceti, some authors think that o Ceti is pulsating in the first overtone, while some others believe that it oscillates in the fundamental mode. Following our numerical results, there is no model for which the first overtone pulsation constant exceeds 0.045 days. Normally, $Q_1 \leq 0.041$. Assuming for o Ceti, $M \approx M_\odot$, $T_e = 2900K$ (Wood P.R. 1990a, 1990b) and that it pulsates at the first overtone, then we have $L \approx 12500 L_\odot$ or $M_{bol} \approx -5.5$. This would make it 0.7^m more luminous than the value of $M_{bol} = -4.8$ given by the P-L relation for Miras in the Large Magellanic Cloud (Glass et al 1987, Hughes & Wood 1990). This is unlikely to be real. Wood(1990b) proposed that if Mira's P-L relation depends on the metal abundance, then the discrepancy between the observed and the theoretical Q values could be removed. Our results of linear pulsational stability survey support Wood's guess. The instability strip moves indeed towards low temperature as the metal abundance increases. Following our linear stability analysis, it appears more likely for o Ceti pulsates in the fundamental mode than in the first overtone.

The details of the present study are given in our recent papers (Xiong et al 1997a, 1997b).

References

Hughes S.M.G. & Wood P.R., (1990), (1990), *AJ*, 99, 784
Glass I.S., Catchpole R.M., Feast M.W., Whitelock P.A. & Reid I.N., (1987), in *In the Late Stages of Stellar Evolution*, eds. Kwok, S. and Pottasch, S. R., (Reidel), p. 51
Wood P.R. (1990a), in *From Miras to Planetary Nubulae* ed. Mennesier, M.O. and Omont, A. (Editions Frontieres), p. 67
Wood P.R. (1990b), in *A.S.P. Conf. Ser.* Vol 11, eds. Cacciari, C. and Clementini, G. (San Francisco, CA), p355
Xiong D.R., Deng L. & Cheng Q.L. (1997a), *submitted to ApJ*
Xiong D.R., Cheng Q.L. & Deng L. (1997b), *submitted to ApJ*

OSCILLATIONS OF HB RED VARIABLE STARS

D. R. XIONG AND Q. L. CHENG

Purple Mountain Observatory, Nanjing 210008, P.R. China

AND

L. DENG

Beijing Astronomical Observatory, Beijing 100080, P. R. China

Abstract. Using a nonlocal time-dependent theory of convection, we have calculated the linear non-adiabatic oscillations of the Horizontal Branch (HB) stars, with both the dynamic and thermodynamic coupling between convection and oscillations been carefully treated. Turbulent pressure and turbulent viscosity have been included consistently in our equations of non-adiabatic pulsation. When the coupling between convection and oscillations is ignored, for all models with $T_e \leq 7350K$, the fundamental through the second overtone are pulsationally unstable; while for $T_e \leq 6200K$ all the models are unstable up to (at least) the 9th overtone. When the coupling between convection and oscillations is included, the RR Lyrae instability strip is very well predicted. Within the strip most models are pulsationally unstable only for the fundamental and the first few overtones. Turbulent viscosity is an important damping mechanism. Being exclusively distinct from the luminous red variables (long period variables), the HB stars to the right of the RR Lyrae strip are pulsationally stable for the fundamental and low-order overtones, but become unstable for some of the high-order overtones. This may provide a valuable clue for the short period, low amplitude red variables found outside the red edge of the RR Lyrae strip on the H-R diagram of globular clusters. Moreover, we present a new radiation modulated excitation mechanism functioning in radiation flux gradient regions. The effects of nonlocal convection and the dynamic coupling between convection and oscillations are discussed. The spatial oscillations of the thermal variables in the pulsational calculations have been effectively suppressed.

1. Discussion and Conclusions

We have reported the detailed study of HB variables in our recent work in Xiong et al (1997a, 1997b), in which the physical justifications and our numerical efforts can be found. The main results of our work can be summarized as the following,

1. When not considering the coupling between convection and oscillations, it is not possible to explain the red edge of the RR Lyrae instability strip. The true reason for the existence of the red edge of the RR instability strip is because of the coupling convection and oscillations. Our numerical results show that, when taking the coupling into consideration, all low temperature HB models with $T_e \leq 5940K$ have their fundamental and low-order overtones ($n \leq 3$) pulsationally stable. The turbulent viscosity is an important damping mechanism of oscillations.
2. For the low temperature red HB star models some of the high-order overtones ($n \geq 4$) become pulsationally unstable. This might be the clue for the red variables found outside the RR Lyrae instability strip in globular clusters by Yao and his collaborators.
3. Radiation Modulated Excitation exists in radiation flux gradient regions at the top and bottom of convection zone.
4. The spatial oscillations are due to the local treatment of convection. They can be suppressed effectively by non-local treatment of convection.

The effects of convection on the instability can not be omitted for the blue HB stars ($T_e \geq 7000K$), although the convective envelope is already very shallow and convective energy transport becomes negligible ($L_c/L \ll 1$). We took a numerical test, in which the dynamic coupling was omitted. The results shows that all the low-order overtones become pulsationally stable. We are reluctant to draw any further conclusion on the pulsational properties of the blue HB stars because there are some undetermined factors. This problem deserves future investigation. High-precision photometry of variables is available for us to determine whether there are low-order unstable modes in the blue HB stars.

References

Xiong D.R., Deng L. & Cheng Q.L. (1997a), *submitted to ApJ*
Xiong D.R., Cheng Q.L. & Deng L. (1997b), *submitted to ApJ*

CARBON MONOXIDE AND THE TEMPERATURE STRUCTURE OF THE SOLAR ATMOSPHERE

T.R. AYRES
Center for Astrophysics and Space Astronomy
University of Colorado, Boulder, CO 80309 USA

Abstract. The themes woven into this review are: (1) the solar outer atmosphere is highly inhomogeneous; (2) anomalies in the behavior of particular wavelength intervals can signal key inconsistencies in an accepted model; and (3) the observations still hold surprises, particularly in the thermal infrared.

1. Introduction

The surface layers of the Sun serve as a crucial boundary condition for processes of the deep interior; for example, the reflector for trapped acoustic modes, undeniably relevant to the present Symposium. The stratification of the solar atmosphere once was thought to be well understood, based on a wealth of ultraviolet and optical diagnostics (Vernazza, Avrett, & Loeser 1973).

Over the past two decades, however, studies of infrared molecular spectra have unearthed puzzling anomalies. Strong lines of the $\Delta v = 1$ (fundamental) vibration-rotation bands of carbon monoxide near 4.7 μm display very cool temperatures at the extreme limb (Noyes & Hall 1972), and remarkable off-limb emissions protruding hundreds of kilometers into the supposedly hot chromosphere (Solanki, Livingston, & Ayres 1994). One proposal is that the low chromosphere is not hot at all, but instead is permeated by CO-cooled gas (Ayres 1981)—a "COmosphere" (Wiedemann et al. 1995), if you will.

The conflicting pictures of hot chromosphere versus cold COmosphere have raised suspicions that those "layers" of the atmosphere are considerably more inhomogeneous than previously suspected. Here, I will describe the infrared molecular evidence in favor of the COmosphere, drawing mainly on recent work at the National Solar Observatory's McMath-Pierce telescope.

I will begin at the end. After briefly describing the classical picture, I will present the punchline; a modern view of the solar atmosphere based on a

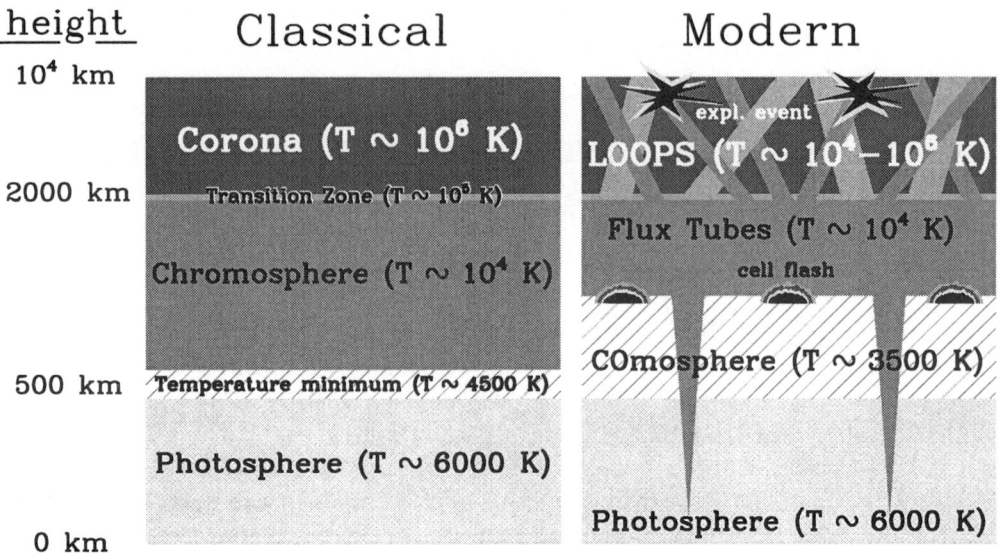

Figure 1. Classical and modern views of solar atmosphere. Height "0" is at $\tau_{5000} = 1$.

synthesis of ultraviolet and infrared diagnostics. I then will focus on aspects of the model which are illuminated by infrared work, and discuss areas of controversy. The ultraviolet side of the story is just as important. Timely discussions can be found in recent reviews by Avrett (1997) and Ayres (1997).

2. Atmospheric Paradigms

2.1. THE CLASSICAL VIEW: LAYERS

The classical model, illustrated in the lefthand panel of Figure 1, is a plane-parallel stratification. The outer atmosphere is divided into a hierarchy of distinct **layers**, controlled by different energy budgets, and through which the temperature runs smoothly. The layered atmosphere represents the idealized response of a hydrostatically stratified gas in which the mechanical heating varies slowly with height. The lower portion of the model is the **photosphere**, where the energy flow is dominated in the deepest levels by convection and in the higher levels by radiation. As the density falls with altitude, the radiative equilibrium becomes increasingly disrupted by mechanical heating. At a critical height the outwardly falling temperature profile bottoms out and begins to rise. That **temperature minimum** marks the interface between the radiative-heating dominated lower atmosphere, and the mechanical-heating dominated higher layers.

The temperature continues rising slowly through a region, many scale heights thick, deemed the **chromosphere**. There, the progressive ionization of hydrogen—with its large reservoir of bound electrons—can fuel the collisionally-excited line cooling side of the energy budget, roughly independent of the outwardly plummeting hydrogen density. After the pressure has fallen several orders of magnitude through the extensive chromospheric temperature "plateau," the hydrogen is mostly ionized, the reservoir of bound electrons depleted, and the gas no longer can maintain a high n_e—and the associated radiative cooling. A thermal instability ensues, and the gas is driven to **coronal** temperatures ($T \sim 10^6$ K) through a thin **transition zone** (TZ).

The cooling of the coronal layers is accomplished partly by thermal conduction back to the chromosphere and partly by high-temperature optically-thin line and continuum emission (Athay 1976). In this picture, the location of the second key interface, the TZ, is determined by the overall level of heating. The TZ will occur at relatively low altitudes (high densities) if the mechanical heating is large (as in active regions), and at higher altitudes if the heating is less (as in the quiet Sun).

Within the context of the simple layered view it is possible to derive scaling laws that connect, for example, the thickness of the chromospheric plateau and its mean electron density, with stellar surface gravity and the average level of nonradiative heating (Ayres 1979). Furthermore, the layered view naturally is compatible with 1 D numerical simulations of line and continuum formation. Sophisticated NLTE spectral synthesis algorithms have been devised to infer optimum plane-parallel thermal profiles from a wide range of multiwavelength solar data (Vernazza, Avrett, & Loeser 1973), including the possibility of treating horizontal spatial inhomogeneities by distributions of mildly perturbed 1 D models (Vernazza, Avrett, & Loeser 1981; Fontenla, Avrett, & Loeser 1993). The layered 1 D paradigm also is a natural starting point for investigations of late-type coronal stars, lacking all the fine detail of routine solar observations.

2.2. THE MODERN PICTURE: STRUCTURES

The new paradigm, illustrated on the righthand side of Fig. 1, is not particularly new at all, but has been around in one form or another ever since the first high spatial resolution pictures of the chromosphere were taken decades ago. The lower part—the photosphere—is identical to that of the layered model. In the middle photosphere, above the convective overshoot region, the horizontal variation in temperature probably is rather mild compared with that at higher altitudes, and any inhomogeneities can be treated successfully by a perturbed standard model. The large departures between the classical and modern views are confined to the heights above the photosphere, dominated by **structures**.

For example, the fact that the corona exists not as a layer of uniform temperature and density, but rather as a collection of distinct **loop**-like structures, dates back at least to the *Skylab* era (e.g., Vaiana & Rosner 1978). Contemporary daily images from *SOHO* EIT and *Yohkoh* SXT have reinforced the view, emphasizing the wide range of temperatures and densities in coronal magnetic structures, particularly in active regions (Gurman 1997).

We now think of the "transition zone" not only in terms of conductive interfaces at the footpoints of hot coronal loops, but also as a separate population of usually small-scale "warm" ($T \sim 10^5$ K) structures mostly associated with the chromospheric network, although larger-scale features occasionally are seen, such as "plumes" over sunspots (Foukal 1975). Even cooler structures ($T \sim 10^4$ K) exist in the "corona:" prominences and spicules (Athay 1976; and more recently—Hirayama 1985; Suematsu, Wang, & Zirin 1995).

The jumble of magnetic fields which constitutes the corona is rooted in a **canopy** (e.g., Solanki & Steiner 1990), occupying what previously we would have called the upper chromosphere. The canopy itself is further rooted mainly in the network, by slender funnel-shaped **flux tubes** that penetrate down through the photosphere into the convection zone, ultimately connecting to the deep source region of the magnetic flux. Interactions between the tangled fields in and above the canopy can give rise to localized explosions traced by high-velocity jets in TZ-temperature lines (the so-called TZ **explosive events**; e.g., Dere, Bartoe, & Brueckner 1989; and more recently—Moses et al. 1994; Innes et al. 1997; cf., the **nanoflares** of Parker 1983).

The pervasive canopy zone in the middle chromosphere is threaded by warm gas with large horizontal density contrasts. Some of the "chromospheric" emission we normally associate with that zone in the layered view arises in the canopy. Additional emission comes from the subcanopy extensions of the flux tubes, which evidently are hotter than the surrounding "quiet" medium down into the middle photosphere. A third component of chromospheric emission arises from **cell flashes** at the base of the canopy (the acoustic disturbances described by M. Carlsson at this Symposium; see, also, Carlsson & Stein 1995, 1997; Carlsson, Judge, & Wilhelm 1997). The canopy emission is ubiquitous and its density striations are traced in Hα filtergrams (e.g., Zirin 1988). The cell flash and low-altitude flux tube components are compact, giving rise to transient and persistent bright points in, for example, Ca K and CN spectroheliograms (Sheeley 1969).

Between the upper photosphere ($h \sim 500$ km) and the base of the canopy ($h \sim 1000$ km) lies a region of controversy, the **COmosphere** mentioned previously. For reasons that I will describe in more detail later, the cool zone likely is a pervasive component of the outer atmosphere, replacing what previously was designated the low chromosphere. The "cold clouds" surrounding the hot flux tubes probably owe their existence to strong radiative cooling by the CO molecules themselves (Ayres 1981), in what for all intents and purposes is

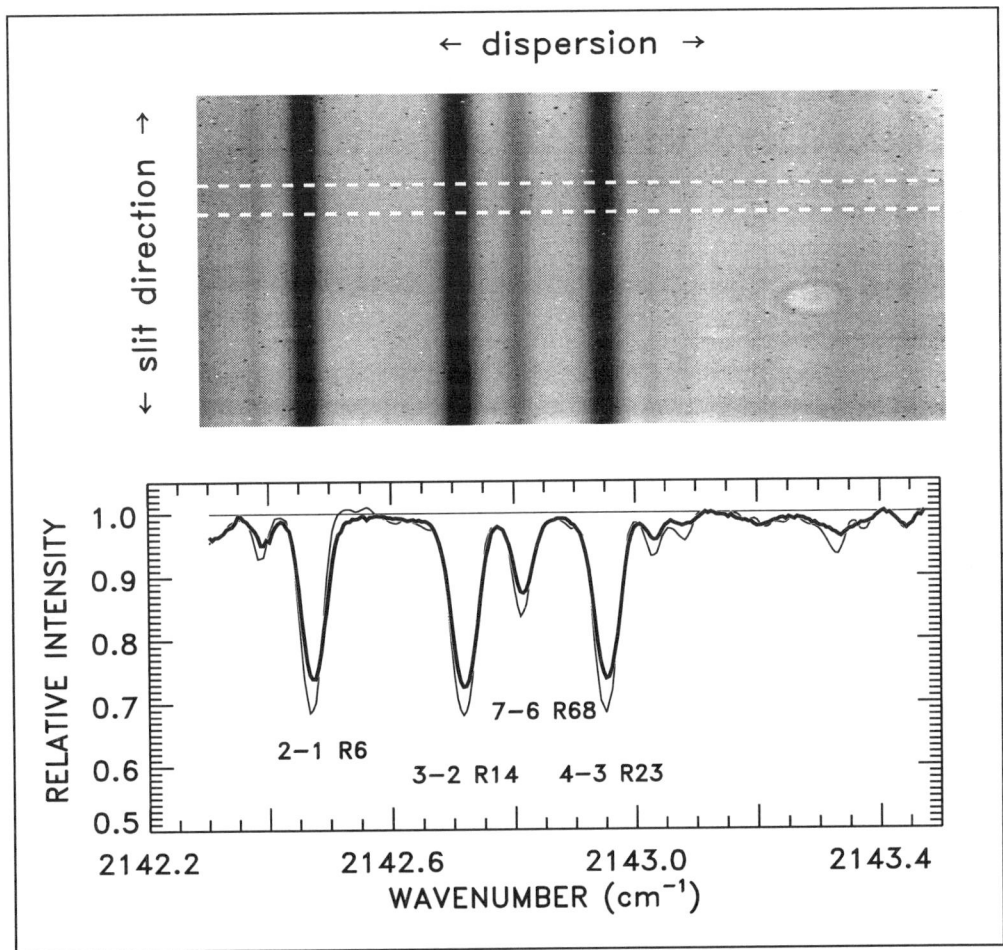

Figure 2. Long-slit double-pass spectrum from McMath-Pierce Main spectrograph.

merely an upward extension of the radiation-dominated photosphere.

I personally view the COmosphere as a relatively passive zone of the outer atmosphere, although some degree of mechanical heating can be tolerated before the low-temperature CO-controlled equilibrium jumps to the high-temperature hot-line dominated branch (Ayres 1981; Anderson & Athay 1989). The importance of the COmosphere comes in how one attributes the radiative activity of different levels of the atmosphere in a global assessment of the energy budget. The gas density is large in the 500–900 km range compared with the higher-lying canopy zone. Because the radiative cooling depends on $\sim n^2$, the intermediate altitude range is a key player in the energy budget. One can imagine that a calculation of total radiative losses (which must balance the input heating, and therefore are a proxy for it; see Vernazza, Avrett, & Loeser 1981) will yield very different answers if there is a temperature *rise* in those layers (classical chromosphere) versus a temperature *drop* (COmosphere). If the thermal model is biased, gross errors in the inferred mechanical energy input are possible.

Now I will describe some of the evidence in favor of the COmosphere view.

3. Observations of Solar CO in the Thermal Infrared

In the late 1970s and early 1980s the instrument of choice for recording the solar 4.7 μm CO fundamental bands was the 1 m Fourier transform spectrometer (FTS) on the McMath-Pierce telescope at Kitt Peak (Ayres & Testerman 1981; Ayres, Testerman, & Brault 1986; Ayres & Brault 1990). Recent work, however, has turned to long-slit stigmatic imaging with the 13.5 m Main spectrograph, its newly-installed large (419×320 mm^2) IR grating, and a 256×256 Amber Engineering InSb camera commissioned in 1993 (see Uitenbroek, Noyes, & Rabin 1994; Ayres & Rabin 1996 [hereafter AR]). The all-reflecting unobscured design of the McMath-Pierce is vital in the thermal infrared, and the large (1.5 m diameter) primary provides diffraction-limited performance at ∼1″, compatible with the best seeing routinely encountered at the telescope. The solar image can be stepped across the spectrograph slit to build up 2 D surface maps, using a control system developed for NSO's Near Infrared Magnetograph (Rabin 1994).

Figure 2 illustrates a typical long-slit spectrum of the quiet Sun obtained at disk center on 9 May 1996. A tracing (*thick curve, lower panel*) of the portion highlighted by white dashed lines is compared with the NSO FTS photospheric atlas (Livingston & Wallace 1991; *thin curve, lower panel*). During that run, we operated the spectrograph in double-pass mode. One obtains double the spectral resolution, at the expense of free spectral range. Oversampling the spectral image along both the dispersion and slit (spatial) directions renders the derived line parameters—central depth, width, and Doppler shift—less susceptible to flatfield errors. Double pass also suppresses, to some extent, the far wings of the instrumental profile responsible for scattered light, a particular nuisance in the grating spectrometer (and absent in the FTS). The scattering fills in the absorption cores of strong CO lines (Fig. 2), affecting the derived brightness temperatures (proxies for the kinetic temperatures at $\tau_L \sim 1$).

In classical scanning double-pass systems an intermediate slit clips the far profile wings, and a reading of diffuse light can be taken when the internal slit is closed. A narrow intermediate slit, however, would eliminate the spectral multiplex advantage. We therefore work around the scattered light by focussing on differential behavior (i.e., temperature *changes*), rather than ab-

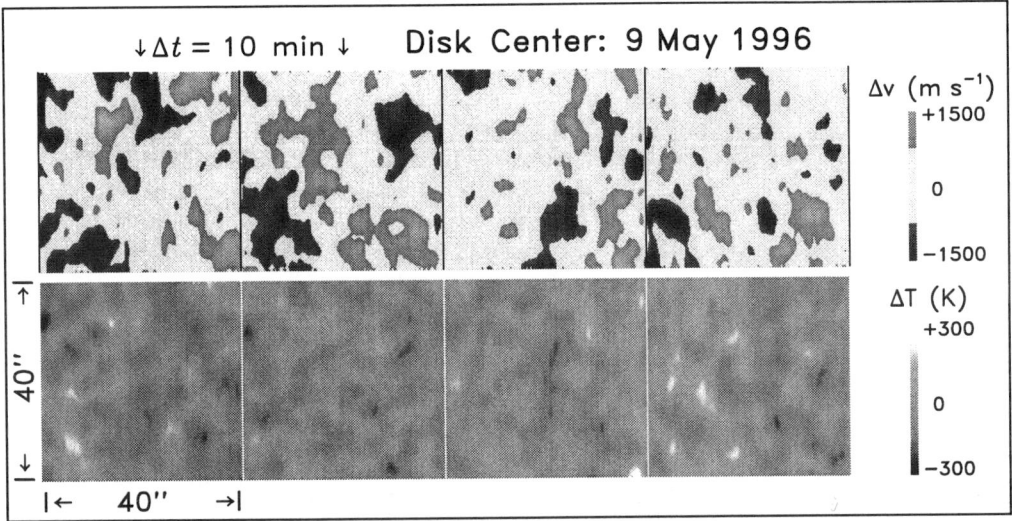

Figure 3. Velocity (*upper*) and temperature (*lower*) maps derived from strong CO lines.

solute quantities, and by incorporating a model of the scattering in our spectral simulations.

3.1. THERMAL AND VELOCITY MAPS OF THE SOLAR SURFACE

Figure 3 illustrates a sequence of four thermal and velocity maps derived from the three strong CO lines of the 2143 cm^{-1} interval shown in Fig. 2. The scans were obtained near disk center on 9 May 1996 under good observing conditions. Each frame is 40"×40" and required 10 minutes to complete. The color scales emphasize excursions beyond the normal fluctuations of the photospheric p-mode pattern at high altitudes (see AR). The velocity maps show p-mode interference "wavepackets." The thermal scenes are relatively bland, however, with no obvious very dark or very bright structures (say, with $|\Delta T| > 500$ K), at least at the ∼2"–3" seeing-limited resolution of these frames. In larger-scale maps, one often can recognize persistent bright points coincident with Ca K network fragments (e.g., AR), although none are evident in the sequence here.

3.2. CO OFF-LIMB EMISSIONS

The off-limb emissions of CO are a fundamental signature of cool gas in the altitude range corresponding to the low chromosphere of conventional models. 2 D spherical transfer simulations suggest that molecular gas is present up to

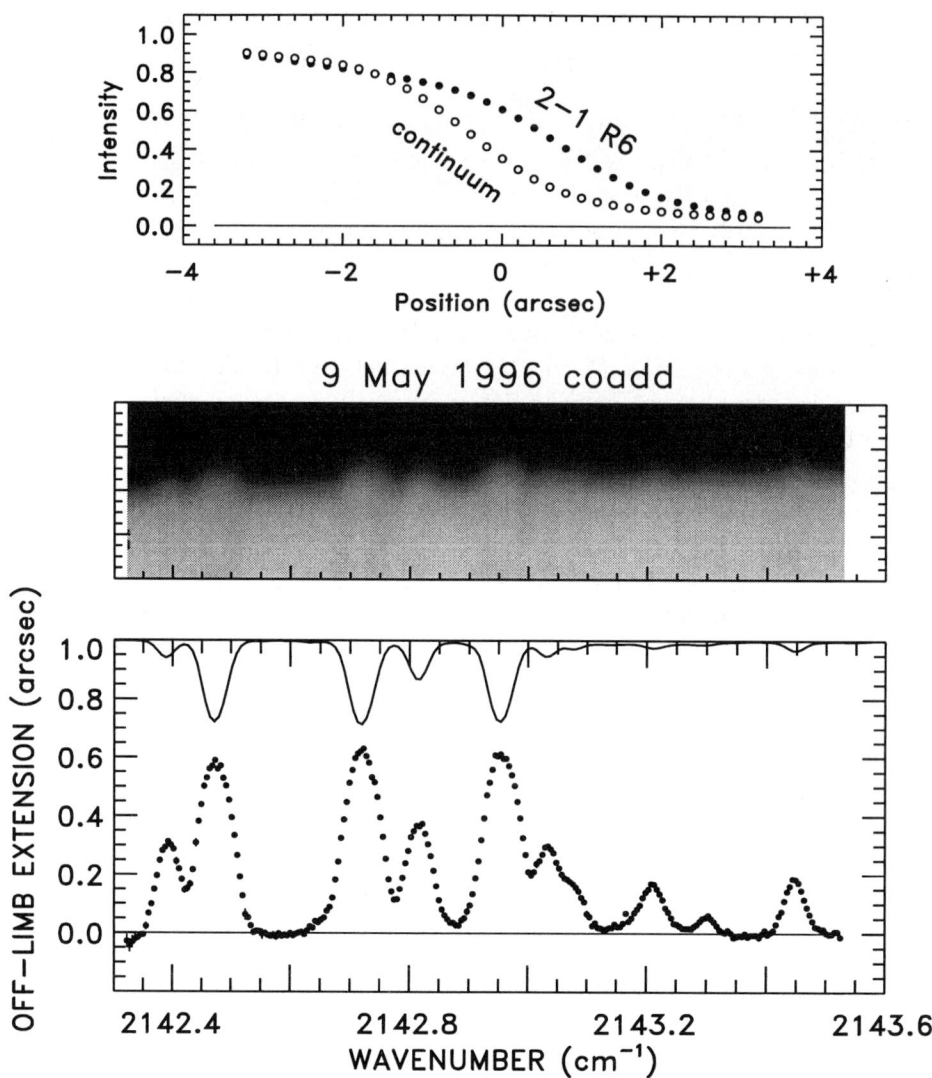

Figure 4. Off-limb emissions of CO lines in 2143 cm^{-1} region.

at least 900 km. The emissions have a relatively low brightness temperature, ∼3500 K, but are seen with high contrast against the dark sky (beyond the continuum edge). Figure 4 depicts the off-limb emissions in the 2143 cm^{-1} region. The double-pass observations were obtained during the morning of 9 May 1996, under clear skies with moderately good seeing. The middle panel illustrates the coaddition of the 9 best spectrograms out of a run of 1000, selected on the basis of limb sharpness (AR). The CO lines are the distinct mounds of emission extending beyond the bright continuum edge. The individual frames were registered by cross-correlation, and divided by an average on-disk spectral trace to normalize the intensities and suppress telluric absorptions.

The upper panel compares the translimb intensity profile of the 2142.6 cm^{-1} continuum with that of the core of the adjacent CO 2–1 R6 feature. The continuum edge should be sharp on a scale of 1″; the apparent smoother roll-off is caused by seeing and diffraction. Nevertheless, the excess displacements of the CO emissions relative to the limb can be measured accurately as a differential effect. The lower trace in the bottom panel (*dots*) shows the spectrally-resolved off-limb extensions of the 2143 cm^{-1} interval relative to the continuum. The upper trace in the bottom panel (*thin curve*) illustrates a disk-center FTS relative intensity spectrum for comparison.

The strong CO absorptions at disk center show prominent extensions above the limb; the weaker features show larger extensions at the limb than one might anticipate from their relative strengths at disk center. The weaker lines have low opacities in the 500–900 km zone, but the tangential raypaths above the limb are tens of thousands of kilometers long, allowing the weak transitions to build up significant optical depths. On the other hand, the strong CO lines might achieve $\tau_L = 1$—and thereby saturate—over only the first few hundred kilometers of the sightline.

4. Interpretation

4.1. HOW PERVASIVE IS THE COOL GAS?

The off-limb CO emissions are accumulated through an enormous tangential pathlength. They tell us directly that there is some cool gas present somewhere along the sightline, but not where or how much. The key question is how widespread the cool gas truly is.

To explain the extreme-limb darking of the CO line cores requires either: (1) cool material with a high filling factor (say, ∼80%) at high altitudes; or (2) a smaller covering fraction (say, ∼20%) of much colder structures. In the first case, the intensity behavior is dominated everywhere by the pervasive cool component. In the second case, the isolated cold, but transversely-opaque, "pillars" dominate only near the limb where they effectively "shadow" any hot structures (which naturally are transparent in CO).

One consequence of the small filling factor model is that the cold regions (with $\Delta T \sim 1000$ K) should show up as small-scale "dark spots" with the ∼20% filling factor in disk-center thermal maps (AR). So far, within the sustained ∼2″–3″ resolution typically encountered under good conditions at Kitt Peak, we have not seen anything remotely like the anticipated sprinkling of dark

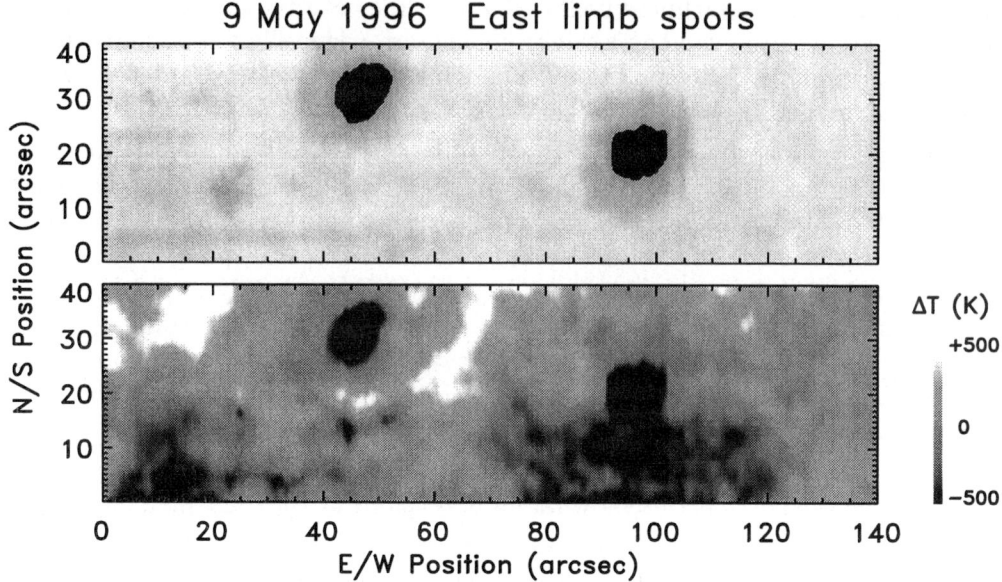

Figure 5. Long spatial scan through small spot group near east limb.

spots. We conclude that the opposite model—pervasive COmosphere—is more promising.

We would feel more confident in our conclusion, of course, if we had access to subarcsecond spatial resolution (i.e., via a large-aperture all-reflecting solar telescope with active-optic image stabilization) to eliminate the (remote) possibility of ultra–fine-scale dark points, below the resolution of the present observations (see discussion in AR).

4.2. DYNAMICAL COMOSPHERE?

An alternative view of the COmosphere must be mentioned at this point (Carlsson & Stein 1995; Avrett et al. 1996). 1 D radiation-hydrodynamics simulations suggest that the 500-1000 km altitude range might be in a continually evolving thermal state owing to the passage of steepening acoustic wavetrains from below. At those low densities, the CO chemical timescales are long (see AR). The molecular population cannot respond to the instantaneous temperature perturbations, but rather attains an equilibrium based on the time-average thermal profile (rather cool in the Carlsson & Stein model). Such a steady-state background population of CO, existing in a fluctuating thermal environment, might be capable of explaining the off-limb emissions and extreme-limb darkening of the infrared $\Delta v = 1$ lines. At the same time one would obtain significant time-averaged ultraviolet signatures from the same spatial volume (mostly from the extreme temperature peaks). The observational consequences should be pursued further; as well as the incorporation of the CO chemistry (and cooling) into the dynamical simulations.

4.3. THERMAL SHADOWS IN ACTIVE REGIONS?

What I have presented is mostly in the way of review. I will conclude with something new, a brief discussion of an intriguing observation of the CO lines in an active region.

Figure 5 displays a thermal scan across a small spot group, near the east limb, which D. Rabin and I followed for four days of our observing run in early May 1996. The upper panel shows a 2142.6 cm^{-1} continuum map. We see the two dark sunspots and, faintly, the surrounding penumbral regions. In the lower panel, the CO thermal map again displays the cold umbrae, although at the heights of the photosphere probed by the $\Delta v = 1$ lines, the penumbrae show little if any contrast. We also see bright plage regions, which have no counterpart in the continuum images. The warm, low-density flux tubes that comprise the plages act as low-opacity windows through the molecular zone of the high photosphere, revealing the hot continuum below.

The fact that sunspots are dark in CO, and plages are bright is not new. What is different in these maps are the regions of $\sim$500 K cooler temperatures to the south of the westward spot, and to the south east of the eastward spot. These areas are much cooler than anything seen in the quiet Sun (e.g., Fig. 3), but are not connected with any obvious spot-like structures in the continuum map. Possibly, they are associated with near-surface horizontal fields which partially suppress convection and thereby steepen the temperature gradient (yielding cooler temperatures higher up, since the absolute temperature T near $\tau_C \sim 1$ is pegged by the constraint to radiate away the energy flux from below: $\mathcal{F} \sim T^4$).

Such "thermal shadows" might provide insight into the subsurface character of active regions, and possibly allow one to anticipate the future evolution of a growing or decaying spot group; relevant to predicting solar activity as well as to understanding the dynamo at its heart (Parker 1970).

I offer this preliminary untested result as an indication of how much we have yet to learn in the exciting rapidly-developing arena of infrared solar physics.

Acknowledgments. I thank D. Rabin and D. Jaksha for their help with the project. This work was supported by grant AST-9618505 from the National Science Foundation to the University of Colorado. I also thank the International Astronomical Union for a partial travel grant to the Kyoto General Assembly (and this Symposium).

References

Anderson, L. S., & Athay, R. G. 1989, ApJ, 346, 1010
Athay, R. G. 1976, The Solar Chromosphere and Corona: Quiet Sun, (Dordrecht: D. Reidel)
Avrett, E. H. 1997, in Synoptic Solar Physics, 18th NSO/Sacramento Peak Summer Workshop, ed. K. Balasubramaniam, J. Harvey, & D. Rabin, (Tucson: NSO), in press
Avrett, E. H., Höflich, Uitenbroek, H., & Ulmschneider, P. 1996, in Cool Stars, Stellar Systems, and the Sun, 9th Cambridge Workshop, ed. R. Pallavicini & A. K. Dupree, ASP Conf. Ser., Vol. 109, p. 105
Ayres, T. R. 1979, ApJ, 228, 509
_____. 1981, ApJ, 244, 1064
_____. 1997, in Synoptic Solar Physics, 18th NSO/Sacramento Peak Summer Workshop, ed. K. Balasubramaniam, J. Harvey, & D. Rabin, (Tucson: NSO), in press
Ayres, T. R., & Brault, J. W. 1990, ApJ, 363, 705
Ayres, T. R., & Rabin, D. 1996, ApJ, 460, 1042 (AR)
Ayres, T. R., & Testerman, L. 1981,
Ayres, T. R., Testerman, L., & Brault, J. W. 1986, ApJ, 304, 542
Carlsson, M., Judge, P. G., & Wilhelm, K. 1997, ApJ, 486, L63
Carlsson, M., & Stein, R. F. 1995, ApJ, 440, L29
_____. 1997, ApJ, 481, 500
Dere, K. P., Bartoe, J.-D. F., & Brueckner, G. E. 1989, Solar Phys., 123, 41
Fontenla, J. M., Avrett, E. H., & Loeser, R. 1993, ApJ, 406, 319
Foukal, P. V. 1975, Solar Phys., 43, 327
Hirayama, T. 1985, Solar Phys., 100, 415
Innes, D. E., Inhester, B., Axford, W. I., & Wilhelm, K. 1997, Nature, 386, 811
Gurman, J. B. 1997, in Cool Stars, Stellar Systems, and the Sun, 10th Cambridge Workshop, ed. R. A. Donahue & J. A. Bookbinder, ASP Conf. Ser., in press
Livingston, W., & Wallace, L. 1991, An Atlas of the Solar Spectrum in the Infrared from 1850 to 9000 cm^{-1} (1.1 to 5.4 μm), NSO Tech. Rept. No. 91-001
Moses, D., et al. 1994, ApJ, 430, 913
Noyes, R. W., & Hall, D. N. B. 1972, BAAS, 4, 389
Parker, E. N. 1970, ARA&A, 8, 1
_____. 1983, ApJ, 264, 635
Rabin, D. M. 1994, in IAU Symp. 154, Infrared Solar Physics, ed. D. M. Rabin, J. T. Jefferies, & C. Lindsey (Dordrecht: Kluwer), 449
Sheeley, N. R. 1969, Solar Phys., 9, 347
Solanki, S. K., Livingston, W., & Ayres, T. 1994, Science, 263, 64
Solanki, S. K., & Steiner, O. 1990, A&A, 234, 519
Suematsu, Y., Wang, H., & Zirin, H. 1995, ApJ, 450, 411
Uitenbroek, H., Noyes, R. W., & Rabin, D. 1994, ApJ, 432, L67
Vaiana, G. S., & Rosner, R. 1978, ARA&A, 16, 393
Vernazza, J. E., Avrett, E. H., and Loeser, R. 1973, ApJ, 184, 605
_____. 1981, ApJS, 45, 635
Wiedemann, G., Ayres, T. R., Jennings, D. E., & Saar, S. H. 1994, ApJ, 423, 806
Zirin, H. 1988, Astrophysics of the Sun, (Cambridge: Cambridge Univ. Press)

HIGH-FREQUENCY SOLAR OSCILLATIONS

STUART M. JEFFERIES
National Solar Observatory
Tucson, AZ, USA

Abstract. This paper reviews the properties of the solar oscillation power spectrum at frequencies near and above the acoustic cut-off frequency for the atmosphere. This region of the spectrum contains over 50% of the total peak structure observed and is a source of information not only on the outer layers of the sun and the origin of solar oscillations, but also on the physics of the solar core and chromosphere.

1. Introduction

Solar acoustic waves with frequencies less than the acoustic cut-off frequency for the atmosphere, $\omega_{ac} \simeq 0.033$ rad s^{-1}, are trapped in resonant cavities beneath the photosphere. The boundaries for each cavity are defined by the positions where the outward propagating wave is reflected by the sharp transition in the density gradient near the solar surface, and where refraction of the inward propagating wave — due to the rapid rise in sound speed in the solar interior — renders the wave propagation horizontal. The resulting modes of oscillation, called p-modes, are governed by a dispersion relation and are manifest as discrete peaks in the $k - \omega$ power spectrum. As the acoustic wave frequency approaches and passes through ω_{ac}, the cavities develop "leaky" upper boundaries that become increasingly transparent. Eventually, for $\omega \gg \omega_{ac}$, the waves are no longer trapped and are able to propagate through the chromosphere to the base of the corona. Despite the traveling wave nature of the high-frequency waves, the oscillation power spectrum shows peak structure that extends well beyond ω_{ac} (see Figure 1). Three explanations have been put forward for the preservation of "mode-like" structure at high-frequencies: (1) The analysis procedure of viewing the sun in terms of spherical harmonic functions results in a "geometrical" interference of direct and indirect traveling waves emitted from a sub-photospheric source (Kumar et al., 1990; Kumar and Lu, 1991). (2) Wave reflection from the chromosphere-corona transition region results in an "extended" acoustic cavity (Balmforth and Gough, 1990). (3) Chromospheric magnetism modifies the acoustic cut-off frequency such that waves with $\omega > \omega_{ac}$ are reflected at the photosphere (Jain and Roberts, 1996). Although all three mechanisms may be contributing to the observed signal, the ability of Kumar's model to reproduce the general characteristics of the high-frequency spectrum,

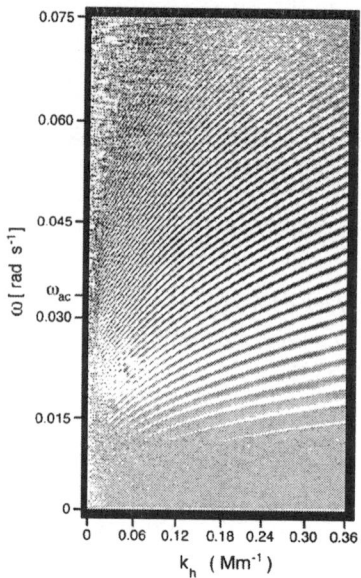

Figure 1. The oscillation power spectrum of 19 days of full-disk, Ca II K-line intensity observations acquired at South Pole over the period December 12, 1994 to January 1, 1995.

suggests that geometric interference accounts for the majority of the high-frequency structure (known as high-frequency interference peaks: HIPs). Unlike the p-modes, which have been extensively studied owing to their diagnostic capability for the solar interior, the HIPs have not been well studied. This is probably due, in part, to their intrinsically large widths (Duvall *et al.*, 1991) which prevents accurate frequency measurement (when compared to the p-modes). In this article I hope to demonstrate that the $k - \omega$ spectrum near and above ω_{ac} should receive more attention.

2. A Brief Overview of Kumar's HIP Model

Consider an acoustic noise source that is located just beneath the base of the photosphere and is emitting waves with frequency $\omega \gg \omega_{ac}$. When the solar surface is viewed in terms of spherical harmonics, the observed velocity signal of waves with spherical harmonic degree ℓ is given by

$$v(t) = A(\omega, \ell) e^{i\omega t} \left[1 + D e^{-i2\theta^*} \right] , \qquad (1)$$

where the first term represents a wave of amplitude $A(\omega, \ell)$ emitted upwards from the source (i.e. towards the solar surface), and the second term represents a downward emitted wave that returns to the source level, with amplitude $DA(\omega, \ell)$ and phase

$$2\theta^* = 2 \int_{r_1(w)}^{r^*} k_r(\omega, r) dr - \pi/2 + \pi\delta , \qquad (2)$$

after reflection at the lower turning point. Here k_r is the radial wavenumber, r^* and $r_1(w)$ are the radial locations of the excitation source and inner turning point, respectively, $w = \omega/(\ell + \frac{1}{2})$, the factor $\pi/2$ is the phase change that the inward wave undergoes on reflection at r_1 (Keller and Rubinow, 1960), and δ is a constant whose value depends on the type of radiation emitted by the source (0 for monopole and quadrupole radiation, 1 for dipole radiation). The power spectrum of $v(t)$,

$$P(\omega, \ell) = \mid A(\omega, \ell) \mid^2 \left[(1-D)^2 + 4D\cos^2\theta^*\right], \qquad (3)$$

has maxima when $\theta^* = n\pi$. The frequencies of the HIPs thus depend on the properties of the solar interior and the oscillation source. Since

$$\int_{r_1(w)}^{r^*} k_r(\omega, r) dr \simeq \omega F^*(w), \qquad (4)$$

where

$$F^*(w) \equiv \int_{r_1(w)}^{r^*} \left[1 - \frac{c^2}{r^2 w^2}\right]^{1/2} \frac{dr}{c} \qquad (5)$$

(Jefferies et al., 1994), and $\theta^* = n\pi$, then along with equation (2) we can see that

$$\omega F^*(w) = \pi \left(n + \frac{1}{4} - \frac{\delta}{2}\right). \qquad (6)$$

This is the dispersion relation for HIPs. The HIPs can be made to obey the dispersion relation for the trapped p-modes, (Jefferies et al., 1994; Roxburgh and Vorontsov, 1995; Vorontsov et al., 1997), i.e.

$$\omega F(w) = \pi[n + \alpha(\omega)], \qquad (7)$$

by defining the phase function

$$\alpha(\omega) \equiv \frac{\omega \Delta S}{\pi} + \frac{1}{4} - \frac{\delta}{2}. \qquad (8)$$

Here

$$F(w) \equiv \int_{r_1(w)}^{R_\odot} \left[1 - \frac{c^2}{r^2 w^2}\right]^{1/2} \frac{dr}{c}, \qquad (9)$$

$R_\odot$ is photospheric radius, c is the sound speed, n is the radial order, and

$$\Delta S \simeq \int_{r^*}^{R_\odot} \frac{dr}{c} \qquad (10)$$

is the sound travel time between r^* and $R_\odot$. The phase function is, in general, sensitive to both the physics of the near-surface layers and the source properties. From equation (8) it can be seen that, in the limit of zero reflectivity,

$$\pi \frac{d\alpha(\omega)}{d\omega} = \Delta S \qquad (11)$$

provides a measure of the source location (with respect to $R_\odot$), while

$$\beta = -\omega^2 \frac{d}{d\omega}\left(\frac{\alpha}{\omega}\right) = \pm\frac{1}{4} \qquad (12)$$

provides information on the source type ($+\frac{1}{4}$ for monopole or quadrupole radiation, and $-\frac{1}{4}$ for dipole radiation).

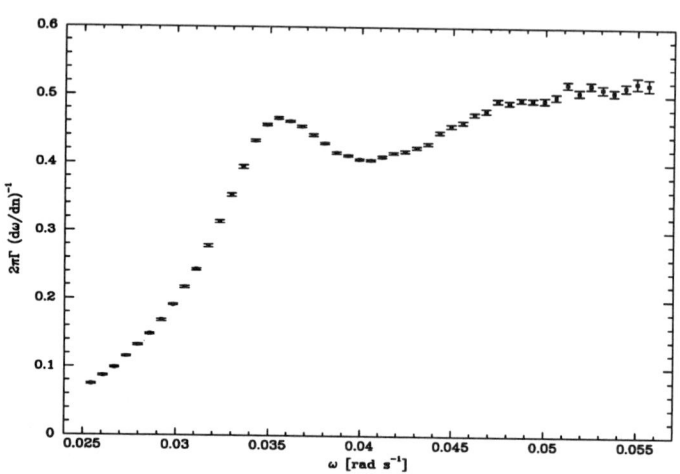

Figure 2. The quantity $2\pi\Gamma(\frac{\partial\omega}{\partial n})_\ell^{-1}$ as measured from the high-frequency region of the power spectrum shown in Figure 1, and averaged over the range $100 \leq \ell \leq 250$. The widths were measured using a single Lorentzian profile and then corrected, to first order, for the additional width introduced by the spatial leaks. Kumar's HIP model predicts a value of $\frac{1}{2}$ for $\omega \gg \omega_{ac}$ (see equation (13)).

Another property of the HIP model is that the frequency separation between two successive peaks, $\partial\omega/\partial n$, and the peak line width, Γ (FWHM), are determined entirely by the wave travel-time, $S^*(w)$, between the source and the inner turning point: i.e.

$$2\pi\Gamma = \frac{1}{2}\left(\frac{\partial\omega}{\partial n}\right) = \frac{\pi}{2}\left[\int_{r_1(w)}^{r^*}\frac{\partial k_r}{\partial\omega}dr\right]^{-1} = \frac{\pi}{2S^*(w)} \qquad (13)$$

3. Determining the Properties of the Oscillation Source and the Near-Surface Layers

By varying the radial location and type of the acoustic source in a solar model, Kumar (1997) is able to produce a theoretical power spectrum which closely matches the observed power spectrum in Figure 1 when the source is located 140 +/- 60 km below the base of the photosphere and emits quadrupole radiation.

Based on the high-frequency behavior of $(d\alpha/d\omega)$ and β for the same data, Vorontsov et al. (1997) estimate the oscillation source to be located at the base of the photosphere (within a pressure scale height), and to emit both positive and negative parity radiation (i.e. monopole/quadrupole and dipole radiation respectively). The discrepancy with Kumar's result for the source type is not understood.

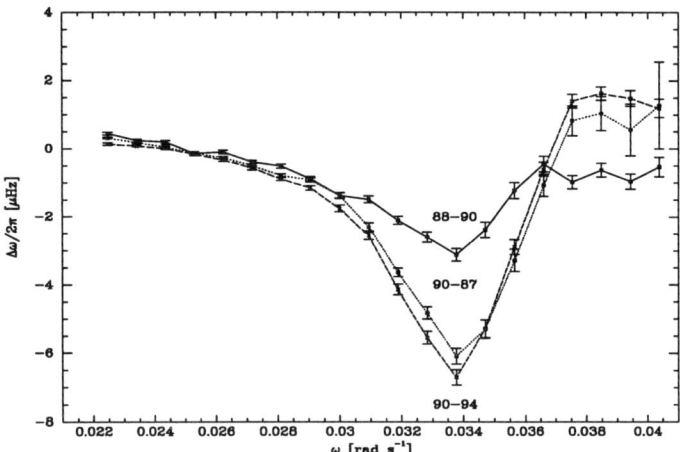

Figure 3. The frequency differences $[\omega(x) - \omega(y)]/2\pi$, measured for $100 \le \ell \le 250$ using a single Lorentzian profile: x and y are the years in which the observations were obtained. To first order, the spectral leakage is symmetric about each target peak. Thus the systematic error in ω incurred by fitting a multiplet with a single peak is small. The differences are plotted such that the mean solar activity level for the observations obtained in year x is higher than for year y. The data were all obtained from observations made at South Pole.

The observed frequency separation, $\frac{1}{2\pi}(\frac{\partial \omega}{\partial n})_\ell$ (Kumar et al., 1994), wave travel time (Jefferies et al., 1994), phase function (Duvall et al., 1991; Vorontsov et al., 1997), and line width (see Figure 2) all show a resonant-like behavior near ω_{ac}. This phenomenon has been interpreted as the manifestation of an acoustic resonance between the oscillation source and the effective level of partial reflection for the waves (Vorontsov et al., 1997). The properties of this "source resonance" are sensitive to the acoustic reflectivity of the solar atmosphere, and to the location and parity of the excitation source. The observed frequency of the resonance for the data shown in Figure 2 is $\omega_r \simeq 0.035$ rad s^{-1}. For this value of ω_r to be consistent with the source being located near the base of the photosphere, the source needs to have composite parity. This agrees with the inference based on the behavior of β.

Based on the solar-cycle variations in ω (Libbrecht and Woodard, 1991; Ronan et al., 1994) and $\alpha(\omega)$ (Vorontsov et al., 1997), Vorontsov et al. argue that the acoustic reflectivity increases with increasing solar activity. As the reflectivity of the atmosphere increases, the acoustic waves are repelled more and penetrate the atmosphere to a lesser extent: i.e. the effective acoustic cavity is smaller. Hence both the cavity eigenfrequencies and the source resonance frequency increase. Oscillations with frequencies well below ω_r, i.e. the p-modes, therefore show a positive difference for ω(high-activity) $- \omega$(low-activity) as is well known. Near ω_r, however, the frequency differences have the opposite sense (see Figure 3). They are also much larger than those observed for the p-modes. Although the observed frequency differences can be

qualitatively explained by simultaneous magnetic field and temperature changes in
the sun's chromosphere (Jain and Roberts, 1996), the temperature changes required
are quite large. In addition, a large part of the frequency differences near ω_r may be
caused by fitting symmetric profiles to asymmetric spectral peaks (Vorontsov et al.,
1997). Using a simple model for wave interference, it can be shown (Vorontsov et
al., 1997) that the spectral peaks in the immediate vicinity of ω_r are asymmetric
with the amount of asymmetry decreasing as $|\omega - \omega_r|$ increases. Increasing the
atmospheric reflectivity affects the line profiles in two ways. First, ω_r moves to higher
frequencies thus moving the most asymmetrical peaks to higher frequencies. Second,
the magnitude of the asymmetry increases. Under these circumstances, the systematic
frequency offset produced by fitting a symmetric profile to an asymmetric spectral line
can be much larger than (and in the opposite sense to) the change in the cavity eigenfrequency. The large asymmetries in the spectral profiles near the source resonance
frequency are also responsible for some of the resonant-like behavior in the measured
linewidths [c.f. Figure 2 with Figure 3 of Roxburgh and Vorontsov (1995)]. In summary, waves with $\omega \simeq \omega_{ac}$ appear to have an increased sensitivity to changes in the
near surface layers of the sun (Guenther, 1991). However, careful modeling of the
spectrum in this region is required before any inferences can be made about changing
chromospheric temperatures, magnetic fields, etc.

4. Probing the Solar Core

HIP-like structure with a spacing of $4.4\ 10^{-4}$ rad s^{-1}, has been detected in the unimaged velocity observations from the GOLF experiment (García et al., 1997). At first
sight, this is a surprising result (Fossat et al., 1992) as the GOLF experiment is only
sensitive to partial waves with $\ell \leq 5$. (Acoustic waves with ray paths that pass close
to the solar center can be considered as the superposition of low-ℓ partial waves.)
Based on Kumar's HIP model, the expected velocity signal is given by

$$v(t) = \sum_{\ell=0}^{5} a_\ell A(\omega, \ell) e^{i\omega t} \left[1 + De^{-i2\theta_\ell^*}\right] \quad (14)$$

where a_ℓ is the sensitivity of the observation to a given ℓ-value. Now for small ℓ,
equation (7) reduces to

$$F(w) \simeq \int_0^{R_\odot} \frac{dr}{c} - \frac{\pi}{2} \frac{\ell + 1/2}{\omega} \quad (15)$$

where the first term corresponds to the acoustic radius of the sun (~ 3518 seconds
(Gough, 1990)), and the second term is independent of the solar model. Assuming
that the excitation source is located close to $R_\odot$, then $F^*(w) \simeq F(w)$ and using
equations (2), (4) and (15) we have

$$\theta_\ell^* \simeq \omega \int_0^{R_\odot} \frac{dr}{c} + \frac{\pi}{2}(\delta - \ell - 1) = \theta_0^* - \frac{\pi\ell}{2}\ . \quad (16)$$

If we now assume that the low-ℓ modes are excited to equal amplitudes, then

$$v(t) \simeq A' e^{i\omega t}\left[1 + \eta De^{-i2\theta_0^*}\right] \quad (17)$$

with $A' = A \sum_\ell a_\ell$ and

$$\eta = \frac{\sum_\ell a_\ell(\ell = \text{even}) - \sum_\ell a_\ell(\ell = \text{odd})}{\sum_\ell a_\ell} . \tag{18}$$

Now for unimaged, full-disk observations, the numerator in equation(18) is much smaller than the denominator (Christensen-Dalsgaard and Gough, 1979) so that $\eta D \ll D$. Thus we can see from the power spectrum,

$$P(\omega) \simeq |A'|^2 \left[(1 - \eta D)^2 + 4\eta D \cos^2 \theta_0^*\right] , \tag{19}$$

that the amplitude of any HIP structure will be extremely small. Moreover, the expected frequency separation between adjacent HIPs will be

$$\frac{d\omega}{dn} = \pi \left[\int_0^{R_\odot} \frac{dr}{c}\right]^{-1} \simeq 8.8 \ 10^{-4} \text{rad s}^{-1} , \tag{20}$$

and not the observed value of $4.4 \ 10^{-4}$rad s^{-1}.

To explain the observations requires that the high-frequency waves undergo reflection from a layer above the source (García et al., 1997b). When there is wave reflection, the observed velocity signal is given by

$$v(t) = e^{i\omega t} \sum_{\ell=0}^{5} a_\ell A(\omega, \ell) \left[\frac{1 + De^{-i2\theta_\ell^*}}{1 - Re^{-i2\theta_\ell}}\right] . \tag{21}$$

If R is small, so that $(1 - Re^{-i2\theta_\ell})^{-1} \simeq 1 + Re^{-i2\theta_\ell}$, and if $\eta \simeq 0$, then the power spectrum is given by

$$P(\omega) \simeq |A'|^2 \left[(1 - RD)^2 + 4RD \cos^2(\theta + \theta^*)\right] \tag{22}$$

and it is easy to see that the expected peak separation is close to $4.4 \ 10^{-4}$rad s^{-1}. However, high-frequency acoustic waves are not expected to be significantly reflected at the solar surface (Kumar, 1997), so where does the reflection come from? Recent time-distance measurements of the high-frequency waves (Jefferies et al., 1997) suggest that a small amount of wave reflection ($\sim 6\%$) may be occurring in the sun's chromosphere. If the frequency dependence of the reflectivity for this atmospheric "boundary" is weak, then it may provide the reflection necessary to explain the GOLF spectrum at high ω. The detection of high-frequency structure in the GOLF spectrum is exciting as it represents waves which pass very close to the solar core and therefore provide information about the rotation and sound speed in this region. Unfortunately, the precision with which the peak frequencies can be measured is limited owing to the broad nature of the profiles. Despite this, the data are important as they cover a range of w which is otherwise poorly sampled.

5. Determining the Horizontal Structure of the Chromosphere

Since high-frequency acoustic waves are essentially vertically propagating in the solar atmosphere, it should be possible to directly measure the wave travel-time across the atmosphere. This could be accomplished by observing the acoustic wavefield simultaneously at two heights in the atmosphere (e.g. using photospheric and chromospheric lines) and using the cross-correlation techniques that have been developed for time-distance analyses (Duvall et al., 1993). The wave travel-times could then be inverted to produce maps of the horizontal structure of the chromosphere, averaged over the vertical extent between the two observing heights, similar to the tomographic maps recently produced for the solar interior (Duvall et al., 1996).

Acknowledgements

I would like to thank Y. Osaki, H. Shibahashi and S. Vorontsov for many enlightening conversations about the high-frequency oscillations, and for the "5-star" hospitality extended to me during my visits to the Universities of Tokyo and London. This material is based upon work supported by the National Science Foundation under grant numbers INT-9417091 and OPP-9219515.

References

Balmforth, N. J., and Gough, D. O.: 1990, *Astrophys. J.* **256**, 266.
Christensen-Dalsgaard, J., and Gough, D. O.: 1979, in H.A. Hill and W. A. Dziembowski (eds.), *Nonradial and Non-linear Stellar Pulsation*, Lecture Notes in Physics 125, 184.
Duvall, T. L. Jr., Harvey, J. W., Jefferies, S. M., and Pomerantz, M. A.: 1991, *Astrophys. J.* **373**, 308.
Duvall, T. L. Jr., Jefferies, S. M., Harvey, J. W., and Pomerantz, M. A.: 1993, *Nature* **362**, 430.
Duvall, T. L. Jr., D'Silva, S., Jefferies, S. M., Harvey, J. W., and Schou, J.: 1996, *Nature* **379**, 235.
Fossat E., Régulo C., Roca Cortés T., Ehgamberdiev S., Gelly B., Khalikov S., Khamitov I., Lazrek M., Pallé P.L. and Sánchez Duarte L.: 1992, *Astron. Astrophys.* **266**, 532.
García, R. A., Pallé, P. L., and the GOLF team: 1997, these proceedings.
García, R. A. et al.: 1997b, in preparation.
Gough, D. O.: 1990, in Y. Osaki and H. Shibahashi (eds.), *Progress of Seismology of the Sun and Stars*, Lecture Notes in Physics 367, 283.
Guenther, D. B.: 1991, *Astrophys. J.* **369**, 274.
Jain, R., and Roberts, B.: 1996, *Astrophys. J.* **456**, 399.
Jefferies, S. M., Osaki, Y., Shibahashi, H., Duvall, T. L. Jr., Harvey, J. W., and Pomerantz, M. A.: 1994, *Astrophys. J.* **434**, 795.
Jefferies, S. M., Osaki, Y., Shibahashi, H., Harvey, J. W., D'Silva, S., and Duvall, T. L. Jr.: 1997, *Astrophys. J.* **485**, L49.
Keller, J. B., and Rubinow, S. I.: 1960, *Ann. Phys.*, **9**, 24.
Kumar, P., Duvall, T. L. Jr., Harvey, J. W., Jefferies, S. M., Pomerantz, M. A., and Thompson, M. J.: 1990, in Y. Osaki and H. Shibahashi (eds.), *Progress of Seismology of the Sun and Stars*, Lecture Notes in Physics 367, 87.
Kumar, P., and Lu, E.: 1991, *Astrophys. J.* **375**, L35.
Kumar, P., Fardal, M. A., Jefferies, S. M., Duvall, T. L. Jr., Harvey, J. W., and Pomerantz, M. A.: 1994, *Astrophys. J.* **422**, L29.
Kumar, P.: 1997, in J. Provost and F.-X. Schmider (eds.), *Sounding Solar and Stellar Interiors*, IAU Symposium 181, (Dordrect: Kluwer), in press.
Libbrecht, K. G., and Woodard, M. F.: 1991, *Science* **253**, 152.
Ronan, R. S., Cadora, K., and LaBonte, B. J.: 1994, *Solar Phys.* **150**, 389.
Roxburgh, I. W., and Vorontsov, S. V.: 1995, *Monthly Notices Roy. Astron. Soc.* **272**, 850.
Vorontsov, S. V., Jefferies, S. M., Duvall, T. L. Jr., and Harvey, J. W.: 1997, *Monthly Notices Roy. Astron. Soc.*, submitted.

THE INFLUENCE OF A MAGNETIC FIELD ON RADIATIVE DAMPING OF MAGNETOATMOSPHERIC OSCILLATIONS

DIPANKAR BANERJEE
Indian Institute of Astrophysics, Bangalore 560034, India.
Armagh Observatory, Armagh, BT61 9DG, N. Ireland [†]

S. S. HASAN
Indian Institute of Astrophysics, Bangalore 560034, India.

AND

J. CHRISTENSEN-DALSGAARD
Teoretisk Astrofysik Center, Danmarks Grundforskningsfond, and Institut for Fysik og Astronomi, Aarhus Universitet, DK-8000 Aarhus C, Denmark

Abstract. We investigate the influence of a magnetic field on the radiative damping of magnetoatmospheric waves, extending our previous work on the adiabatic modes of an isothermal stratified atmosphere with a uniform vertical magnetic field. Banerjee, Hasan & Christensen-Dalsgaard (1996, 1997) generalized this work to include radiative effects using Newton's law of cooling for a weak magnetic field. The present study examines the variation of the mode damping rate with increasing magnetic field strength. We find that a moderate field suppresses radiative damping.

Key words: MHD, magnetic fields, atmosphere, oscillations

1. Introduction

It is now generally accepted on the basis of observations (Stenflo 1989) that the solar magnetic field is highly structured, with most of the magnetic flux in the photosphere being clumped into intense flux tubes with kilogauss field strengths. Flux tube diameters vary from a few hundred kilometers in small-scale tubes to several thousand

---

[†]Present address

kilometers in sunspots. Oscillations in a fairly broad range of frequencies have been reported in magnetic flux tubes (Moore & Rabin 1985). More recently MHD oscillations have been observed in the solar photosphere (Norton et al. 1997) and active regions (Horn & Staude 1997). The purpose of this contribution is to examine whether radiative damping is important in intense flux tubes.

The present investigation is a continuation of earlier work by Hasan & Christensen-Dalsgaard (1992) and Banerjee, Hasan & Christensen-Dalsgaard (1995), who examined the effects of a vertical magnetic field on the normal adiabatic modes of an isothermal stratified atmosphere by combining a semi-analytic approach, based on asymptotic dispersion relations, with numerical solutions. These results were extended by Banerjee et al. (1997) to include radiative dissipation using Newton's law of cooling for a weak magnetic field. In the present work we examine the variation of the mode damping rate with magnetic field strength.

2. Magnetic modes in an isothermal radiative atmosphere

The radiativly damped normal modes of an isothermal atmosphere with a vertical magnetic field have been discussed in detail by Banerjee (1997). We consider a cavity of vertical extent d and look for standing-wave solutions. We introduce a dimensionless wave number $K = k/H$ and dimensionless frequency $\Omega = \omega H/c_s$, where H is the scale height of the atmosphere and c_s is the sound speed. We characterize the radiative losses by a dimensionless constant radiative cooling time scale

$$\tilde{\tau}_R = \frac{c_s \rho c_V}{16 H \chi \sigma T^3}, \tag{1}$$

where χ is the mean absorption coefficient per unit length, c_V is the specific heat at constant volume and σ is the Stefan-Boltzman constant. Banerjee (1997) derived a dispersion relation which allows the effect of a weak magnetic field on the modes to be studied. It was shown that the normal-mode frequency can be expressed as a sum of an elementary mode frequency and a first-order correction, due to coupling between different modes. Asymptotic analysis revealed that in the weak- and strong-field limits, the modes are essentially decoupled into magnetic and p-type modes.

We focus on the magnetic modes and study the radiative damping of these modes as the field strength is increased. For the adiabatic case, the magnetic modes have frequencies (in the weak-field limit) $\Omega_m \propto n\epsilon$, where n denotes the mode order and $\epsilon = v_{A,0}/c_s$, ($v_{A,0}$ being the photospheric Alfven speed at the lower boundary).

To illustrate the effect of the magnetic field strength on the damping of the magnetic modes, Fig. 1 shows the variation of the imaginary part of frequency (which is a measure of the damping) with ϵ for fixed $K = 0.3$, $\tilde{\tau}_R = 0.5$ and $D = d/H = 1.0$. These results were generated by solving the MHD equations numerically assuming rigid boundary conditions at the top and bottom boundaries. The various line styles corresponds to different order magnetic modes. It is evident that as we increase the value of ϵ (increasing magnetic field strength) the imaginary part of Ω decreases, indicating a reduction in the damping of the wave modes. In the limit of a very strong field, the damping rate becomes negligibly small. The non-monotonic behaviour of the curves in Fig. 1 is due to mode interaction, which leads to the occurrence of avoided crossings when the frequencies of different modes become almost equal. Also, the damping of the modes is enhanced at these crossings (Banerjee et al. 1996, 1997). Our conclusion that the damping rate decreases with field strength is in qualitative agreement with the work of Bogdan & Knölker (1989), who examined a similar

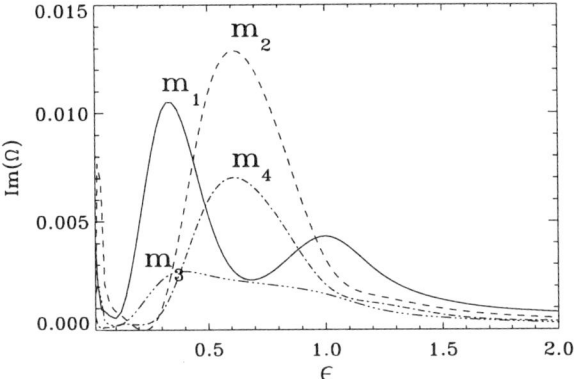

Figure 1. Variation of the imaginary part of Ω with ϵ for $K = 0.3$, $\tilde{\tau}_R = 0.5$ and $D = 1.0$ for various order magnetic modes, shown in different line styles.

problem for an unstratified atmosphere. They argued that a uniform magnetic field reduces the temperature perturbations associated with MHD waves and therefore suppresses the radiative damping of these disturbances. A similar result was found for a horizontal magnetic field by Bünte & Bogdan (1994).

3. Umbral oscillations

We now consider the importance of radiative effects in damping umbral oscillations in sunpots. For simplicity we assume that the umbral atmosphere is isothermal and extends over several scale heights. It is well known that periods of oscillations above a sunspot's umbra range between 100 and 200 s. Extensive measurements in different spectral lines have been made (Thomas *et al.* 1984, 1987; Alisandrakis *et al.* 1992). Kentischer & Mattig (1995) recently observed umbral oscillations in the middle and upper chromosphere, with pronounced power at frequencies between 5 and 9 mHz. They concluded that the first peak at 5.9 mHz consists mainly of upward propagating waves traveling at the sound speed. There are indications that these waves have their origin in the photospheric resonator. The second distribution is located at 7.5 mHz and involve standing waves which are confined to the chromosphere (chromospheric resonator). Previous studies (Hasan 1991; Wood 1990, 1997) have shown that the the waves in the chromosphere are essentially slow magnetoacoustic modes or p−modes. The frequencies of the slow magnetoacoustic modes or p−modes with radiative damping for $K = 0$ are given by (Banerjee 1997),

$$\Omega_{p,n} = \sqrt{\tilde{\gamma}\left(\frac{n^2\pi^2}{D^2} + \frac{1}{4}\right)}, \qquad (2)$$

where n is the order and $\tilde{\gamma} = (1+i\Omega\tilde{\tau}_R\gamma)/(\gamma+i\Omega\tilde{\tau}_R\gamma)$, γ being the adiabatic exponent.

Table 1 gives the eigenfrequencies of different order p−modes for an isothermal atmosphere with $D = 10$, $\tilde{\tau}_R = 0.5$ and $\epsilon = 0.84$ ($B \sim 2$ kG). The relation between the horizontal wave number and the radius of the flux tube was discussed by Banerjee (1997). It is evident from Table 1 that radiative cooling introduces a positive imaginary

	Nearly adiabatic case ($\tilde{\tau}_R = 100$)			Radiative case ($\tilde{\tau}_R = 0.5$)			Nearly isothermal case ($\tilde{\tau}_R = 0.05$)		
Mode	Re(Ω)	Im(Ω)	P(s)	Re(Ω)	Im(Ω)	P(s)	Re(Ω)	Im(Ω)	P(s)
p_1	0.5903	0.0007	164	0.5203	0.0658	186	0.458	0.0125	211
p_2	0.8030	0.0009	120	0.7075	0.0895	136	0.624	0.0170	155
p_3	1.0700	0.0013	90	0.9400	0.1190	103	0.828	0.0227	117

TABLE 1. Eigenfrequencies of different order p-modes for a model atmosphere with $D = 10$, $B = 2$ kG, $\tilde{\tau}_R = 0.5$ and $K = 0.1$.

component of the eigenfrequencies, corresponding to comparatively small damping. The computed frequencies match very well with those calculated from expression (2). Furthermore, the p_2-mode has a period of 136 s in the presence of radiative losses, which is very close to the observed peak at 7.5 mHz (period 134 s) by Kentischer et al. (1995). This observation was reported at layers near the temperature minimum which correspond to about 1000 km over the photosphere ($D = 10$, in our notation).

In summary, we have examined the influence of the magnetic field on wave damping in an isothermal stratified atmosphere. We have shown that as the field strength is raised, the damping rate decreases. We have applied our calculations to umbral oscillations in sunspots and demonstrated that they exhibit modest damping, even when the radiative cooling time is comparable with the pulsation period.

This work was partly supported by the Danish National Research Foundation through the establishment of the Theoretical Astrophysics Center. and the Smithsonian Institution through a foreign currency grant.

References

Alisandrakis, T. E., Georgakilas, A. A., Dialetis, D., 1992, Solar Phys., **138**, 93
Banerjee, D., 1997, Ph.D Thesis, Bangalore University, India
Banerjee, D., Hasan, S. S., Christensen-Dalsgaard, J., 1995, Astrophys. J. **451**, 825
Banerjee, D., Hasan, S. S., Christensen-Dalsgaard, J., 1996, Bull. Astron. Soc. India **24**, 325
Banerjee, D., Hasan, S. S., Christensen-Dalsgaard, J., 1997, Solar Phys., **172**, 53
Bünte, M., Bogdan, T. J., 1994, Astron. Astrophys. **283**, 642
Bogdan, T. J., Knölker, M., 1989, Astrophys. J. **339**, 579
Hasan, S. S., 1991, Astrophys. J. **366**, 328
Hasan, S. S. Christensen-Dalsgaard, J., 1992, Astrophys. J. **396**, 311
Horn, T., Staude, J., 1997, these proceedings
Kentischer, T. J., Mattig, W., 1995, Astron. Astrophys. **300**, 539
Moore, R. L., Rabin, D., 1985, Ann. Rev. Astron. Astrophys. **23**, 239
Norton, A., Ulrich, R. K., Bogart, R. S., Bush, R. I., Hoeksema, J. T., 1997, these proceedings
Stenflo, J. O., 1989, Astron. Astrophys. Rev **1**, 3
Thomas, J. H., Cram, L. E., Nye, A. H., 1984, Astrophys. J. **285**, 368
Thomas, J. H., Lites, B. W., Gurman, J. B., Ladd, E. F., 1987, Astrophys. J. **312**, 457
Wood, W. P., 1990, Mon. Not. Rol. Astr. Soc., **251**, 427
Wood, W. P., 1997, Sol. Phys., **173**, 259

PHASE RELATIONS AND OTHER DIAGNOSTICS OF SOLAR ATMOSPHERIC STRUCTURE AND DYNAMICS

FRANZ - LUDWIG DEUBNER
Astronomisches Institut der Universität Würzburg
D - 97074 Würzburg, Germany

Abstract. Based on spectral observations with high spatial and temporal resolution, we discuss in this brief review the dynamic behaviour of the upper layers of the quiet solar atmosphere. A proper description of this behaviour is fundamental for the understanding (and simulation) of the temperature structure and the heating of the chromosphere, and the interpretation of such conspicuous phenomena like the chromospheric "grains". The evidence is drawn from waveforms, power spectra, and phase spectra of velocity and brightness fluctuations in numerous Fraunhofer lines formed over an extended radial range of the atmosphere. A comparison with recent predictions of simulation calculations documents the strong need to extend the simulation codes to 3-D, and also to incorporate a realistic chromosphere/coronal transition layer, to better reproduce the observations.

1. Introduction

This paper focusses on the dynamics of the quiet (i.e. non-active) solar atmosphere, and in particular on the diagnostics available to examine some current models and simulations of solar atmospheric phenomena. Only a few of the observations I am going to discuss are new, most of them have been shown before; in any case it appears useful to provide some elementary background to introduce the following discussion.

It is a widespread creed that the wakes of sub-photospheric p-modes and convective perturbations pass through the photosphere as evanescent or running waves (depending on frequency) into the upper chromospheric layers, where running waves form shocks, and dissipate an amount of mechanical energy that is still being debated. The shocks become visible e.g. as K_{2v} bright points (or "grains") which populate the chromospheric internetwork regions with an average density of 1 grain in (5 - 10 arcsec)^2 at any time. These grains have the tendency to brighten repeatedly at intervals of 3 min for 10 to 30 sec with variable amplitude (Rutten & Uitenbroek, 1991; cf. also Fig. 5 in Gaizauskas, 1985).

It has been suggested (see the following review by Carlsson & Stein, these proceedings) that the temperature rise (in the internetwork regions) of semi-empirical models of the chromosphere is a result of the non-linear response of the *intensity* radiated by the shock region to the local temperature perturbation ΔT. In reality, therefore, the mean temperature gradient with height may be flat, or even negative. Highly advanced calculations (Carlsson & Stein, 1994) appear to support this picture. Certain details, like the observed periodicity of the grains, the asymmetric shape of the wave forms, and the typical evolution of the chromospheric line profiles, during the appearance of a grain in the field of view, can be reproduced with impressive similarity.

Yet, we believe that the *similarity* of these simulations with observation is misleading. Our doubts are chiefly based on a class of diagnostics that have not yet been applied rigorously to the simulations, and that appear to invalidate some of the conclusions reported there.

2. Observations and Diagnostics

2.1. WAVEFORMS

Evidently, typical waveforms of radial velocity wavetrains with amplitudes > 2 km/s deduced from measurements of Dopplershift in chromospheric lines are asymmetric (cf. e.g. Fleck et al., 1994; Hofmann et al., 1996). The slope from maximum downward velocity to maximum upward velocity is steeper than the complementary part of the cycle. However, only in exceptional cases can the asymmetry be characterized by a down-to-up ratio exceeding the value of 2:1. In most events with strong asymmetry, the observed Ca K core displacements correspond to velocities of typically ± 5000 m/s, and are accompanied by a brightening of the K_{2v} feature ("grain") by a factor of 2 to 3 (see Fig. 2 in Fleck et al. 1994, or Fig. 3 and 4 in Hofmann et al., 1996).

In contrast, calculations (Carlsson & Stein, 1993) predict extended wavetrains with almost discontinuous wavefronts, formed periodically as low as at the level of the infrared Ca 8542 line. The average core displacement amplitudes in the simulated 5 cycle wavetrain are about ± 4000 m/s in Ca 8542, and ± 8000 m/s in Ca H_3, i.e. even much larger then those extreme observed values quoted above. Also, steep wavefronts like in the simulated wavetrain have never been observed. One might conclude already from this comparison between observed and predicted waveforms and amplitudes that the formation of shocks in the upper chromosphere is less important than theory suggests. However, we need to be aware of the effects of *seeing* in groundbased observations, that degrade the spatial resolution. Even the temporal resolution may suffer potentially from the transfer effects caused by the finite width of the velocity contribution function on the line shape.

2.2. WAVE PROPAGATION

A much more powerful diagnostic can be obtained from an observational analysis of the characteristics of the vertical propagation of wave packets in the solar atmosphere (e.g. Deubner, 1995). The calculation of phase difference spectra comparing the variations of Doppler shift at different levels in the atmosphere as function of frequency (V−V spectra) permits a conclusive characterisation of the wavetrain in terms of standing or running waves, direction of propagation, etc. Observations of this kind have been obtained to date by many authors for a large number of spectral lines covering the whole vertical extent of the visible atmosphere. Observations including also the transition region and lower corona accessible by the SUMER, EIT, and CDS experiments onboard SOHO are presently being explored to extend the range of wave diagnostics (see e.g. Steffens et al., 1997a).

The observations confirmed that below a level of about 1000 km above $\tau_5 = 1$ waves with frequencies less than the acoustic cutoff frequency $\nu_a \approx 5.3$ mHz are evanescent, and begin propagating (upward) only at higher frequencies, in accordance with the theory of linear waves in a stratified atmosphere. However, if one analyses pairs of lines which are both formed higher than this level (Ca 8542, Ca K, He 10830), nearly uniform phase spectra bear ample evidence that there the velocity field is dominated by non-propagating, i.e. evanescent or standing waves, independent of frequency (Lites, 1994; Bocchialini & Koutchmy, 1995; Fleck et al., 1994; for theory, cf. Skartlien et al., 1994, Fig. 4).

The evidence is further strengthened by phase difference spectra deduced from fluctuations of the intensity signal in a selected range of the line profile in correlation with the Doppler shift of the bisector of that range (V–I spectra). Again, chromospheric V–I spectra deviate significantly from the theoretical prediction, in displaying almost uniform phase distributions with phases in the range of $-90°$ at frequencies far above the acoustic cutoff frequency, where the phase lag (V–I) is expected to converge to zero, in LTE (Fleck & Deubner, 1989).

Two arguments are frequently brought forward challenging the conclusions based on the phase spectra: 1) The phase lag obtained from ground based observations is subject to the effects of seeing (especially by image motion), at high frequencies. The phase values tend towards zero or $\pm 180°$, depending on the sign of the average *correlation* of the two signals compared (Endler & Deubner, 1983). Evidently, this argument does not apply to the observed V–I spectra, since the phases remain close to 90° or 100° at *all* frequencies. Neither can these phases be altered by poor spatial resolution alone. On the other hand, the characteristic frequency dependence of those V–V spectra which connect to at least one photospheric line differs strongly from those obtained with the combination of two chromospheric lines. Note that all the time series used in this comparison (Fleck et al., 1994) were taken strictly simultaneously with the same instrument. 2) The other argument applies primarily to the V–I spectra, in pointing out correctly that intensity (brightness) is a very poor if not misleading proxy of temperature under the non-LTE conditions of the solar chromosphere. At the present state of the discussion we note, however, that the V–I phase lag in the infrared Ca line simulations (Skartlien et al., 1994, Fig.2), deviates systematically by 40° to 50° from the observed values.

From the evidence presented so far we infer that we observe, in the chromosphere, a regime of standing (and partly evanescent) waves with strong, near sonic velocity amplitudes in conjunction with non-linear brightness changes induced by the wave motion. Is this scenario compatible with the formation of shock waves? *How can shocks evolve in a non-propagating wave system, where all fluid elements oscillate in phase, as observations seem to imply? Or is there a mechanism, which mitigates the observable consequences of the generation of shock waves in the upper atmosphere?*

In this context it is worth noting that observed radial velocity amplitudes reach maximum r.m.s. values of the order of 2.0 to 2.5 km/s at approximately the level of formation of Ca K_3. It decreases by about $\frac{1}{2}$ in the He I 10830 line core (Fleck et al., 1994; Bocchialini & Koutchmy, 1995; cf. however Rutten, 1995). Further up, in transition region and low coronal lines, shocks or wavetrains appear to be not regularly connected with dynamic events observed in the upper chromosphere (Steffens et al., 1997b).

2.3. REFLECTION

Apart from the observational evidence mentioned already, other observations naturally lead to the discussion of the effects of reflection in the context of wave dynamics of the solar atmosphere.

2-D phase spectra obtained recently (Straus et al., these proceedings) from SOHO MDI data confirm and enhance important details of the diagnostic diagram first published by Deubner (1990), and discussed by Marmolino et al. (1993), which have led to the interpretation of the previously neglected V–I phases observed *in between* the p-mode ridges in terms of wave scattering, and which stress the importance of coherent partial reflection (even of evanescent waves) in this context. The effects of reflection in certain parts of the atmosphere are even more manifest in observations reported by Harvey et al. (1993), Steffens et al. (1995), and now by Jefferies et al. (this conference), which seem to imply at least one atmospheric resonant frequency (or "mode"). The time–distance analysis applied by Jefferies et al. made it possible to estimate both the reflectivity and the height of the upper reflecting layer, the latter one being close to the top of the chromosphere (2000 km).

In parallel to these beautiful demonstrations of *oblique reflection* at the chromosphere/coronal interface, now covering ℓ-values from about 80 to several thousand, we have continued exploiting the diagnostic power of the V–I and V–V phase spectra. Long after the serendipitous detection of a 180° phase jump in phase spectra of the NaD_2 and Mgb_2 line (Staiger et al., 1984; Deubner et al., 1984) we have begun a systematic search for similar features, as a function of height in the atmosphere (Deubner et al., 1996). By climbing the profile of a spectral line (NaD_2) from the bottom of the core to the outer wings, it was possible to bridge a certain range of the atmosphere (about 750 km) in steps of arbitrary finesse.

This systematic search has not only reproduced the previously observed discontinuity in the V–I spectra; another sharp discontinuity stands out in the V–V phase difference spectra of velocity signals from the core of the line compared against signals measured at a well defined position in the line wing.

These phase discontinuities occur each only at a certain wave frequency and at one distinct height of the atmosphere approached by selecting the corresponding level in the line profile. There is a rapid transition from the spectrum with the 180° discontinuity, to spectra with a much smoother phase transition generated with adjacent levels in the line profile. No discontinuities are found at other levels. We infer from these observations the presence of discrete nodal layers determined by the structural properties of the atmosphere. One must not confuse the nodal layers of the eigenfunctions of velocity, or pressure in the atmosphere with the levels where the phase discontinuities of V–V or V–I signals are detected spectroscopically. Also, in our interpretation, we take into account that in spatially averaged power spectra the observed resonances are expected to be broadened, because horizontal inhomogeneities of sound speed and/or vertical scale introduce some scatter of the nodal layer heights. This has been discussed in Deubner et al. (1996), and we shall not expand on the details here.

We recall that, to produce the distinct discontinuities, the presence of only *one* reflecting layer (on top of the atmosphere, say) is not sufficient. It would generate a continuous spectrum of standing waves with nodal layers at all heights, depending on frequency only. However, the characteristic modulation of the oscillatory power of Doppler shifts and brightness fluctuations as a function of both frequency and height as derived from the observations (Fig. 1) clearly testify to the presence of vertical resonances in the solar atmosphere.

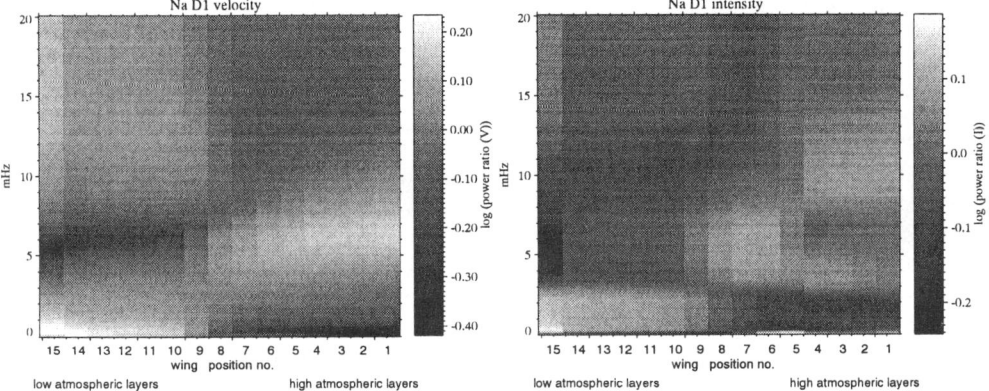

Figure 1. This diagram presents the most intuitive evidence of the vertical resonances of the solar atmosphere. Power spectra of velocity (left) and intensity (right) fluctuations in the Na D_2 line as function of *frequency*, and *height* in the atmosphere corresponding to the mean level in the line profile at which the fluctuations are measured. Note the spatial (viz. vertical) phase shifts of intensity *vs* velocity "nodes" (power minima) e.g. at 6 and 9 mHz (cf. also Fig. 8 in Deubner et al. 1996).

Worrall (1997, priv. comm.) has mentioned the possibility that in chromospheric lines like NaD$_2$ non-LTE effects could affect the response of spectral intensity to changes of temperature in such a way that the sign of dI/dT changes as a function of the position in the line profile. Even though it would seem difficult to explain the distinct *frequency* dependence in the I-signal, such an effect might influence its variation with *height* in the atmosphere. Severino (1997, priv. comm.) argues, on the basis of non-LTE simulations of the NaD$_2$ line in static atmospheres, that within the range of the inner Doppler core a reversal of the temperature response with a maximum effect of the order of $2\,10^{-3}$ of the mean spectral intensity is expected to be observed for a 5% increase in temperature. I doubt, however, that such a tiny effect could be detected at all. Regarding V–V phase discontinuities, a reversal of Doppler shift, in response to a given velocity perturbation, with height in the atmosphere is not easily visualized, in particular in those parts of the line profile (formed in the lower photosperic layers), where the lower level contribution to the V–V phase difference is discontinuous.

3. Discussion

3.1. THE TEMPERATURE STRUCTURE OF THE ATMOSPHERE

Understanding the structure of the solar atmosphere, transition region, and corona is fundamental for a proper understanding of the physical processes in those layers of the sun that are most directly involved in solar-terrestrial relations. Since the very same layers can be directly inspected whenever the sun shines, the problem of modelling the underlying physics should appear an easy task. In reality, it is not. Rather, it has become one of the major challenges in solar physics, resisting even the most advanced theoretical tools available.

We have already laid out the two lines of approach which are presently being pursued in this field: 1) Multi-line observations with high spatial and spectral resolution, with sufficient temporal resolution to capture the relevant frequency range of the dynamics involved, and with sufficient spatial coverage: to preserve the 3-D aspects of the observed phenomena; to distinguish various sources of dynamic phenomena, or different conditions in the atmosphere; and to enhance statistics. 2) Non-adiabatic, non-LTE computer simulations of wavetrains in a realistic solar atmosphere, based on an excitation model guided by observations. The agreement between observation and theoretical prediction is not satisfactory, as we have seen above. In fact, it is poor; the conclusions from the two lines of approach differ fundamentally with regard to basic behaviour of the solar atmosphere.

We have mentioned above certain deficiencies of the data: limited vertical and horizontal resolution due to seeing and radiation tranfer effects, and inadequacy of the spectral *intensity* as a proxy for *temperature* for the interpretation of the chromospheric V–I phase diagrams. The simulation results of Carlsson and Stein, on the other hand, suffer from several constraints to the computer code which could easily bias the conclusions (cf. however Cheng & Yi, 1996): 1-D geometry, transmitting upper boundary (without transition layer), and a source of perturbations which is inconsistent with the geometry, because it is derived from the observed 3-D superposition of a large number of non-radial wavemodes. The first two constraints are severe because they modify drastically the conditions for reflection at the upper boundary, and, therefore, the character and amplitude of the waves in the wavefields below the boundary, and above. With regard to the drastic physical consequences of such unrealistic conditions, it is highly desirable to explore other model calculations, which permit to relax the limitations of the Carlsson/Stein code. Since computer power presently limits all attempts to complete such simulations with sufficiently realistic models, Steffens et al. (1997c) have solved numerically in 3-D the linear adiabatic wave equations for a VAL atmosphere, including the convection zone down to 10^4 km below $\tau_5 = 1$, and a realistic transition zone and corona up to $1.7\,10^4$ km. Several chromospheric temperature distributions were chosen to account for various modifications of the VAL atmosphere, recently suggested, and to study their effects on properties of the wavefield.

The results of this work are rather encouraging. The $k-\omega$ diagrams derived from this 2-D study reveal significant resonances in the range close to and above the acoustic cut-off frequency, particularly near 6 and 8 mHz. These resonances resemble closely in position and in shape the "chromospheric mode" detected by Harvey et al. (1993), and, seen as a "ridge", by Steffens et al. (1995). They are also strangely reminiscent of the secondary peaks in high resolution power spectra of photosperic and chromospheric Fraunhofer lines in earlier studies (Deubner, 1976), attributed erroneously to *optical* filtering effects of the actual velocity or intensity contribution functions. The frequency of these resonances is only weakly dependent on wavenumber as expected, since they are thought to reside chiefly in the atmosphere; the slight upward bend of the ridges at higher wavenumbers is first noticeable at $k \approx 1.5$ rad/Mm ($\ell \approx 1000$). Generally, however, the occurrence of the resonances appears to depend only moderately on the details of the model of the atmosphere employed by the authors (Steffens et al., 1997c).

These results suggest strongly that a potential well for acoustic waves, formed by the transition region together with the convection zone and atmospheric structure, succeedes in trapping sufficient acoustic power to give rise to well defined resonant features in spectra and $\nu - \ell$ diagrams of photospheric and chromospheric oscillations. These resonances also provide the most compelling explanation of the phase discontinuities discussed in Sec. 2.3.

3.2. CONCLUSIONS

In the debate over the energy balance of the quiet solar atmosphere, an assessment of the net amount of energy flux carried by acoustic waves, and how it is distributed, is of importance. It is yet uncertain, owing to intrinsic difficulties in correcting the measured amplitudes for insufficient temporal or spatial resolution, how much power the sun actually feeds into high frequency ($\geq$10 mHz) perturbations of the mean atmosphere. However, the observations seem to suggest that, in the chromosphere, highly non-radial waves within a range of periods rather closer to the acoustic cut-off frequency, and with a degree typically of the order of $\ell = 1000$ (viz. the K_{2v} bright points) are closely linked with the process of converting mechanical into radiative energy. These conditions call for a very careful scrutiny of the chromospheric dynamics.

We have seen no observed phase relations yet which support the notion of travelling acoustic waves in the chromosphere. On the contrary, the observed phase spectra derived from Doppler shifts and brightness fluctuations (with due caution regarding the limited value of brightness as a proxy for temperature) deviate significantly and systematically from predictions of advanced calculations with running acoustic waves and ensueing shocks. The observed waveforms do *not* regularly exhibit the discontinuous wavefronts as shown in these simulations, nor do they reach amplitudes nearly as high as predicted. Rather, the observational diagnostics that we have discussed here imply a coherent scenario with a significant if not dominant contribution to the wavefield of resonant (i.e. standing) wave trains maintained by the resonant structure naturally provided with the transition region as upper boundary. It should be noted that the time–distance analysis of Jefferies et al. (1997, and these proceedings), yielding a reflection coefficient $R \leq 9\%$, is based on spatially filtered data with $\ell \approx 125$, whereas the effective degree of the waves seen in high resolution observations used to study the Ca K grains is rather higher than $\ell \approx 1000$, which would seem to entail much better (if not total) reflection.

In conclusion, it appears that at some distance from supergranular lanes or compact magnetic fields the character of the chromospheric waves is less dramatic than what is implied from most 1-D simulations. Even observed radial velocities of ~ 6000 m/s are of little concern in a standing wave system, where all fluid elements move in phase, and *return* in phase after half a period, as evidenced in the observations. The conspicuous spatial and temporal intermittency of the Ca K grains can still be understood as a consequence of interference and compression at nodal layer crossings, enhanced by the non-linear response of the source function to temperature changes, as suggested earlier (Deubner, 1991, cf. also Rutten & Uitenbroek, 1991). The V–V phase spectra are in good, and the V–I spectra in reasonable, agreement with inferences based on the resonant standing wave model.

All this accumulating observational evidence of the 3-D structure and dynamics of the quiet solar atmosphere, in conjunction with further elaborated techniques employed for simulation of what is possibly not quite so rough a medium, should hopefully in the near future lead to an improved understanding of this important boundary of the solar body that appears so easily accessible.

References

Bocchialini, K. and Koutchmy, S.: 1995, in J.T. Hoeksema, V. Domingo, B. Fleck, and B. Battrick (eds.), *Fourth SOHO Workshop: Helioseismology*, ESA, p. 499.
Carlsson, M. and Stein, R.F.: 1993, in P. Maltby and H.L. Pécsely (eds.) *Wave Phenomena in Solar Terrestrial Plasmas* Univ. of Oslo, p. 21.
Carlsson, M. and Stein, R.F.: 1994, in M. Carlsson (ed.), *Chromospheric Dynamics*, Univ. of Oslo, p. 47.
Carlsson, M. and Stein, R.F.: *these proceedings*
Cheng, Q.Q. and Yi, Z.: 1996, *Astron. Astrophys.* **313**, 971.
Deubner, F.-L.: 1976, *Astron. Astrophys.* **51**, 189.
Deubner, F.-L.: 1990, in J.O. Stenflo (ed.) *Solar Photosphere: Structure, Convection, and Magnetic Fields* IAU Symp. **138**, 217.
Deubner, F.-L.: 1991, in P. Ulmschneider, E.R. Priest, and R. Rosner (eds.) *Mechanisms of Chromospheric and Coronal Heating*, Springer, p. 6.
Deubner, F.-L.: 1995, in R.K. Ulrich, E. Rhodes, Jr., and W. Däppen (eds.), *Helio- and Asteroseismology from the Earth and Space*, ASP Conf. Series **76**, 303.
Deubner, F.-L., Endler, F., and Staiger, J.: 1984, *Mem. S. A. Ital.* **55**, 135.
Deubner, F.-L., Waldschick, Th., and Steffens, S.: 1996, *Astron. Astrophys.* **307**, 936.
Endler, F. and Deubner, F.-L.: 1983, *Astron. Astrophys.* **121**, 291.
Fleck, B. and Deubner, F.-L.: 1989, *Astron.Astrophys.* **224**, 245.
Fleck, B., Deubner, F.-L., Hofmann, J., and Steffens, S.: 1994, in M. Carlsson (ed.), *Chromospheric Dynamics*, Univ. of Oslo, p. 103.
Gaizauskas, V.: 1985, in B. W. Lites (ed.), *Chromospheric Diagnostics and Modelling*, Sunspot, N.M., p. 25.
Harvey, J.W., Duvall, T.L., Jefferies, S.M., Pomerantz., M.A.: 1993, in T.M. Brown (ed.) *GONG 1992: Seismic Investigation of the Sun and Stars*, ASP Conf. Series **42**, 111.
Hofmann, J., Steffens, S., and Deubner, F.-L.: 1996, *Astron. Astrophys.* **308**, 192.
Jefferies, S.: *these proceedings*
Jefferies, S., Osaki, Y., Shibahashi, H., Harvey, J.W., D'Silva, S., and Duvall, T.L., Jr.: 1997, *Astrophys. J.* **485**, L 49.
Lites, B.: 1994, in M. Carlsson (ed.), *Chromospheric Dynamics*, Univ. of Oslo, p. 1.
Marmolino, C., Severino, G., Deubner, F.-L., and Fleck, B.: 1993, *Astron. Astrophys.* **278**, 617.
Rutten, R.J.: 1995, in J.T. Hoeksema, V. Domingo, B. Fleck, and B. Battrick (eds.) *Fourth SOHO Workshop: Helioseismology*, ESA, p. 151.
Rutten, R.J. and Uitenbroek, H.: 1991, *Solar Phys.* **134**, 15.
Severino, G.: 1997, *private communication*
Skartlien, R., Carlsson, M., and Stein, R.F.: 1994, in M. Carlsson (ed.), *Chromospheric Dynamics*, Univ. of Oslo, p. 79.
Staiger, J., Schmieder, B., Deubner, F.-L., and Mattig, W.: 1984, *Mem. S. A. Ital.* **55**, 147.
Steffens, S., Deubner, F.-L., Fleck, B., and Wilhelm, K.: 1997a, in A. Wilson (ed.), *The Corona and Solar Wind Near Minimum Activity*, ESA, p. 685.
Steffens, S., Deubner, F.-L., Fleck, B., Wilhelm, K., Harrison, R., and Gurman, J.: 1997b, in A. Wilson (ed.), *The Corona and Solar Wind Near Minimum Activity*, ESA, p. 679.
Steffens, S., Deubner, F.-L., Hofmann, J., and Fleck, B.: 1995, *Astro. Astrophys.* **302**, 277.
Steffens, S., Schmitz, F., and Deubner, F.-L.: 1997c, *Solar Phys.* **172**, 85.
Straus, Th., Deubner, F.-L., Fleck, B., Marmolino, C., Severino, G., and Tarbell, T.; *these proceedings*
Worrall, G.: 1997, *private communication*

THE NEW CHROMOSPHERE

MATS CARLSSON
Institute of Theoretical Astrophysics
P.O. Box 1029 Blindern, N-0315 Oslo, Norway

AND

ROBERT F. STEIN
Department of Physics & Astronomy, Michigan State University
East Lansing, MI 48824, USA

Abstract. The natural state of the Solar chromosphere is very dynamic. Any photospheric disturbance will grow and naturally form shocks over the twenty scale-heights in density between the photosphere and the corona. Observations in the resonance lines from singly ionized calcium and recently in the ultraviolet region of the spectrum observed with the Solar and Heliospheric Observatory satellite also show a very dynamic chromosphere. This dynamic picture is further supported by numerical simulations. Static and dynamic pictures of the chromosphere are fundamentally different. The simulations also show that time variations are crucial for our understanding of the chromosphere itself and for the spectral features formed there.

1. Introduction

The existence of a million degree corona has been known since Edlén's (1939) identification of coronal emission lines as arising from highly ionized atoms. Initially, the existence of a chromosphere was inferred from the sharp increase in the emission scale height of lines and continua just above the limb of the sun. Starting with the analysis of the 1952 eclipse data by Thomas & Athay (1961) and others, the chromosphere was inferred from the increase in radiation temperature with increasing opacity in lines and continua. One dimensional models were constructed of the temperature and density that reproduced the observed spatially and temporally averaged continua and line intensities.

Avrett and co-workers (*e.g.*, Vernazza et al. 1981, Maltby et al. 1986, Fontenla et al. 1993) constructed spatially resolved models representing regions of different Lyman continuum intensity corresponding to dark cell interior to very bright network locations. These static models show a temperature increasing with height from a minimum value at about 500 km height above where the optical depth is unity at 500 nm.

Observations with higher spatial and temporal resolution have been made of CO and Ca II emission (*e.g.*, Solanki et al. 1994, Uitenbroek et al. 1994, and Lites et al. 1993). These results are consistent with a large fraction of the internetwork solar surface having no temperature rise in the first 500 km above the location of the temperature minimum in the VAL models. The very dynamic behaviour seen in observations of the Ca II resonance lines even raises the question whether the chromosphere exists as a quasi-static state or if it is wholly dynamic in nature (see *e.g.*, Lites 1985).

The layout of this paper is as follows. In Section 2 we argue that considerations of basic physics leads one to conclude that the natural state of the Solar chromosphere is dynamic and not static, in Section 3 we discuss observations that also imply a very dynamic chromosphere. Section 4 shows that this dynamic picture is also very much supported by numerical simulations. In Section 5 we explain that a static picture and a dynamic picture of the chromosphere are fundamentally different. Section 6 gives some examples of how the temporal variations are crucial for our understanding of spectral features formed in the Solar chromosphere. We conclude in Section 7 with a short summary.

2. Basic physics

In the photosphere we have temperatures on the order of 5×10^3 K and a gas pressure much larger than the magnetic pressure. In the corona we have temperatures on the order of 10^6 K, at heights more than 2 Mm above the photosphere, and a magnetic pressure that is much larger than the gas pressure. This corresponds to about 20 scale heights in density and 15 scale heights in gas pressure. The large drop in density and gas pressure has two important consequences:

- perturbations in the photosphere will grow in amplitude with height to conserve wave energy. The growth over 20 scale heights will cause formation of shocks unless we have a very strong damping of the wave.
- while the magnetic field is pushed around by the gas in the photosphere the opposite is true in the corona.

The natural state of the region between the photosphere and the corona is therefore a *dynamic* state with large amplitude variations. The variations around a mean can be expected to be so large that it is questionable whether a mean atmosphere makes any sense at all. Furthermore, because the magnetic field will start to dominate over the gas somewhere in the chromosphere we expect a one-dimensional non-magnetic model to break down at some height. A static, one-component model can thus almost be ruled out as a meaningful picture already from very basic physical considerations.

3. Observations

It is not altogether straightforward to diagnose the Solar chromosphere from observations. In order for a spectral feature to carry information about the chromosphere the opacity has to be high enough for the radiation to be formed there. In the optical region of the spectrum the continuum opacity is much too low with optical depth unity in the photosphere. We thus need a large line-opacity. Since the opacity is proportional to the population density of the lower level, this condition translates to lines from the ground state of the dominant ionization stage of an abundant element. Unfortunately, most such lines are found in the ultraviolet part of the spectrum with the notable exception of the resonance lines from singly ionized calcium, Ca II.

These two lines, named the H and K lines by Fraunhofer, constitute almost the only available diagnostic of the Solar chromosphere in the optical part of the spectrum and they have therefore received a lot of attention: (e.g., Rutten and Uitenbroek 1991 and references therein; Harvey et al. 1992; Lites et al. 1993; Bocchialini et al. 1994; Von Uexkuell and Kneer 1995; Hofmann et al. 1996; Steffens et al. 1996).

In regions void of strong photospheric magnetic fields (called the inter-network) the Ca II resonance lines are strongly assymetric and vary in time. A brightening in the line-wing propagates in time towards line-center and ends up as a strong brightening blue-ward of line-center. There is no corresponding brightening on the red side. The line-center absorption is at the same time shifted slowly more and more to the red. Just after the brightening the line-center absorption is abruptly shifted to the blue. In a spectro-heliogram the brightenings of the violet peak can be seen as roundish areas a few megameters wide. They are called Ca II K_{2V} (and H_{2V}) bright grains. A brightening in one place often repeats three-four times with about 3 minutes period.

The interpretation of the observations is both simple and complicated. The complication comes from the fact that the line-formation is optically thick and Local Thermodynamic Equilibrium (LTE) is a bad approximation. The good news is that the opacity in the line is a simple function. Almost all calcium is singly ionized in the Solar chromosphere so the opacity is set by the density which is basically given by the hydrostatic stratification. The excitation is collisionally dominated up to rather large heights so the source function does not decouple completely from the Planck function until about 1 Mm above the photosphere. We can thus expect the lines to give information about local conditions in the chromosphere even though modelling is necessary for a detailed understanding. We will return to the interpretation of the observations in Section 4 but it is important to note that even without sophisticated modelling we can conclude from the observations alone that the chromosphere must be very dynamic.

If we also include the ultraviolet part of the spectrum we get access to many more spectral features formed in the chromosphere. All continua shortward of the silicon edge at 152 nm are formed at heights above 500 km. Variations in the continuum intensity at various wavelengths will give information on the variations in temperature in the chromosphere. Unfortunately, there is no simple inversion to get the physical conditions directly from the observations. The reason is that the source function has decoupled from the Planck function so the intensity is not a good measure of the local temperature. The ionization/recombination may also be out of equilibrium from what the instantaneous values of temperature, density and electron density would imply because of slow recombination rates.

Observations from the SUMER spectrograph on-board the Solar and Heliospheric Observatory (SOHO) satellite show that, indeed, the UV continuum intensities do vary enormously with time (*e.g.*, Carlsson et al. 1997).

4. Simulations

To properly model the dynamic Solar chromosphere one has to take into account the important physical processes. Radiative losses are important in continua and lines of hydrogen and calcium. The population densities can not be calculated from LTE but has to be solved for using the full rate equations including time-derivative and advection terms because of the slow recombination rates. Shock formation also has to be allowed.

Such a self-consistent radiation-hydrodynamic modelling of the solar chromosphere was performed by Carlsson & Stein (1992, 1994, 1995, 1997) and we will here summarize some of the results.

A schematic representation of the computational scheme is shown in Fig. 1. We solve the one-dimensional equations of mass, momentum, energy and charge conservation together with the non-LTE radiative transfer and population rate equations, implicitly on an adaptive mesh. We employ 6 level model atoms for hydrogen and singly ionized calcium. We include in detail all transitions between these levels, thus including four Lyman lines (α, β, γ, δ), three Balmer lines (α, β, γ), Paschen α, β, Bracket α, the Lyman-, Balmer-, Paschen-, Bracket- and Pfundt continua of hydrogen, the H and K resonance lines, the infrared triplet and the photoionization continua from the five lowest levels of singly ionized calcium. Other continua are treated as background continua in LTE, using the Uppsala atmospheres program (Gustafsson 1973). Microturbulence broadening was set to a constant 2 km/s throughout the atmosphere.

Our initial atmosphere is in radiative equilibrium above the convection zone (for the processes we consider) without line blanketing and extends 100 km into the convection zone, with a time constant divergence of the convective energy flux (on a column mass scale) calculated with the Uppsala code without line blanketing.

Waves are driven through the atmosphere by a piston located at the bottom of the computational domain (100 km below $\tau_{500} = 1$). The piston velocity is chosen to reproduce a 3750 second sequence of Doppler-shift observations in an Fe I line at $\lambda 396.68$ nm in the wing of the Ca H-line (Lites et al. 1993).

The velocity spectrum as a function of frequency changes with height, both in amplitude and phase. The velocity amplitude of propagating waves increases with height in a stratified atmosphere to maintain a constant flux as the density decreases. Damping reduces this amplitude increase. Propagating modes also show a phase shift due to their finite phase speed. Evanescent modes are attenuated but have no change in phase as a function of height.

We calculate this change in amplitude and phase between the piston height and 260 km as a function of wave frequency and multiply the observed Doppler-shifts with the inverse of this transfer function to obtain our piston velocities. Comparing the simulated velocities at 260 km with the observed Doppler-shift in the iron line provides a check of this procedure (Fig.1). We applied this procedure to the observed Doppler-shifts at five different slit positions. At the top of the computational domain there is a transmitting boundary condition.

In the dynamic calculation only hydrogen and calcium are treated self-consistently in non-LTE. With the time-variation of hydrodynamic variables (density, temperature, electron density, velocity) taken from this calculation, the statistical equilibrium equations were solved for an additional set of "minority species" (solution assumed not to influence the hydrodynamics or energetics of the atmosphere): neutral aluminium, magnesium, silicon and carbon.

Figure 2 shows that waves are continually traversing the chromosphere. The waves increase in amplitude and form shocks above 1 Mm height. A shock traveling in the cool downflow of a preceding shock may be overtaken by the next shock forming a

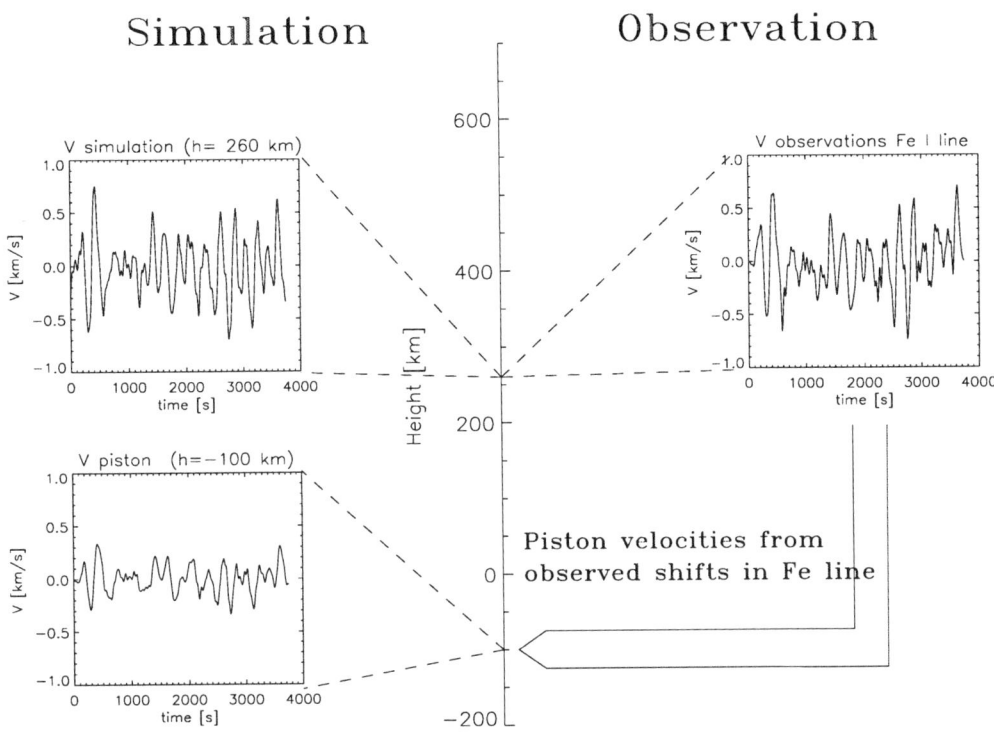

Figure 1. A schematic representation of the computational method. Observed Doppler shifts in an iron line are transformed in amplitude and phase to get a piston velocity that gives a simulated velocity at the iron line formation height (260km) close to the observed Doppler-shifts. The equations of conservation of mass, momentum, energy and charge are solved together with the non-LTE rate equations for 6 levels of hydrogen and 6 levels of ionized calcium implicitly on an adaptive mesh.

merged, high-amplitude shock. The temperature difference across a shock can be more than 10^4 K.

The simulations thus also show that the Solar chromosphere is very dynamic. What can we learn from these simulations? We present one example of how static models can be very misleading and several examples of dynamic diagnostics.

5. Failure of Static models in a Dynamic Chromosphere

Classical static models of the Solar chromosphere are based on temporal and spatial averages of intensity — either continuum intensities shortward of the silicon edge at 152 nm (like in the models by Avrett and co-workers, *e.g.*, Vernazza et al. 1981, Maltby et al. 1986, Fontenla et al. 1993) or line profiles of resonance lines from ionized calcium and magnesium. The crucial factor here is that the mean is taken in the ultraviolet part of the spectrum where the temperature dependence of the Planck function is more exponential than linear. The source function is quite decoupled from

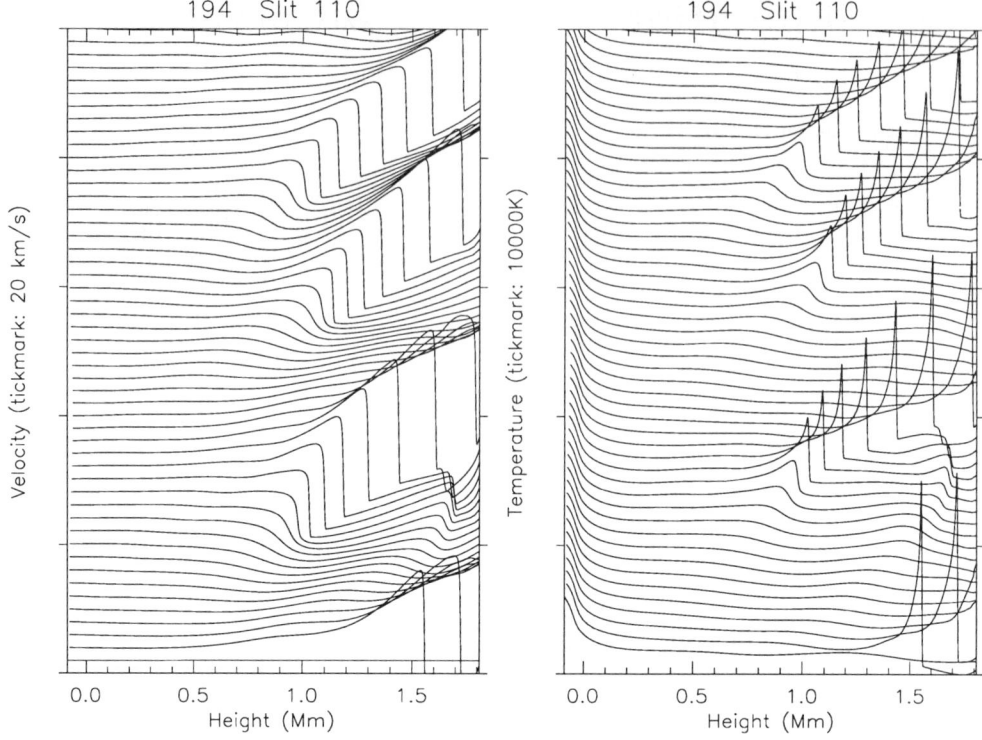

Figure 2. The velocity and temperature as functions of height and time. The curves are 10s apart in time and offset in the vertical direction. The amplitudes can be seen from the tickmarks that are 20 km/s apart in velocity and 10^4 K apart in temperature. Waves propagate and form shocks above 1 Mm height. Shocks may overtake an earlier shock and merge.

the Planck function and has a smaller amplitude than the local Planck function but the fact remains that a mean preferentially samples the high temperatures. This was dramatically illustrated in Carlsson & Stein (1995) where the best semi-empirical fit to the mean intensities in the simulation showed a chromospheric temperature rise while the mean temperature showed no increase, see Fig. 3.

This result does not mean that there is no chromosphere. The simulations have enhanced chromospheric emission much above the radiative equilibrium values and not too different from the observed values (see also Section 6.2). This emission is, however, not caused by an increase in the mean temperature — the mechanical work goes directly into radiation instead of first causing a temperature increase.

The effects of waves with frequencies above 20 mHz and of absorption in the helium continuum of radiation from the corona have been neglected in the simulations. Both of these effects would favour a mean temperature rise. The Solar chromosphere may thus well have a mean temperature rise but the essential conclusions remain:

- a mean temperature structure can *not* be deduced from mean intensities in the blue part of the spectrum.

– in a dynamic chromosphere the mean temperature is not only difficult to deduce but is also a *meaningless* and even *misleading* quantity. The energy balance can only be deduced from a dynamical model and not from any such mean model.

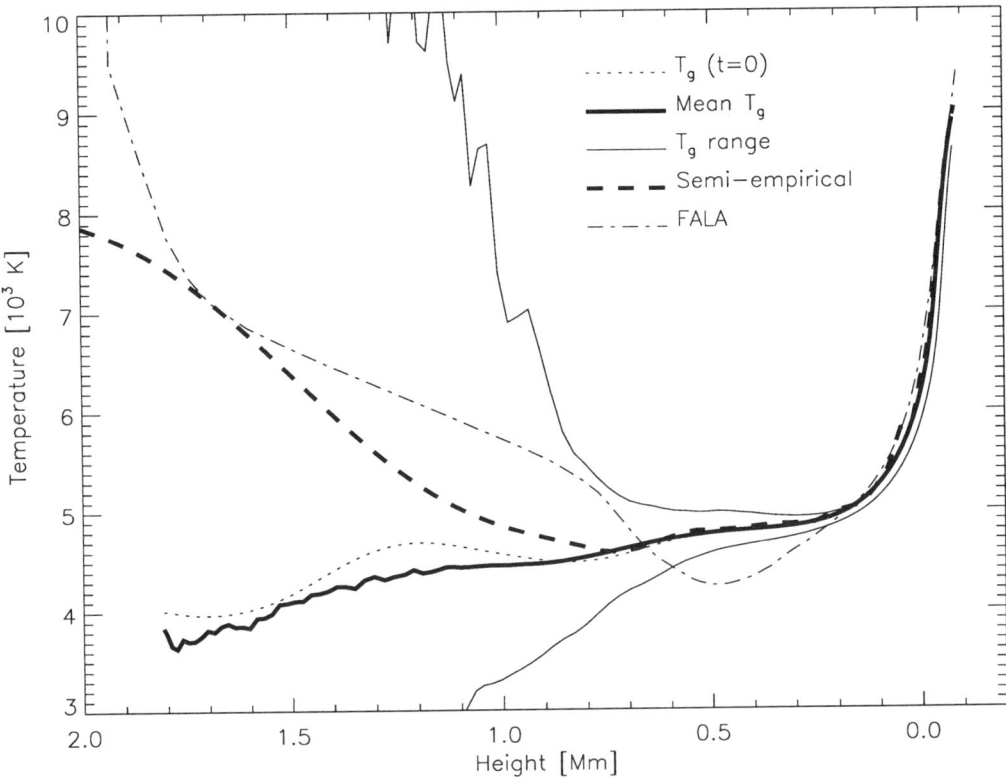

Figure 3. Time average of the temperature in the dynamical simulation, the range of temperatures in the simulation, the semi-empirical model that gives the best fit to the time average of the intensity as a function of wavelength calculated from the dynamical simulation, the starting model for the dynamical simulation and the semi-empirical model FAL-A. The maximum temperatures are only reached in narrow shock spikes of short duration. The semi-empirical model giving the same intensities as the dynamical simulation shows a chromospheric temperature rise while the mean temperature in the simulation does not. From Carlsson & Stein (1995).

6. Dynamic Diagnostics

In a dynamic atmosphere one must be cautious in applying traditional methods used to analyze static atmospheres:

– the height of formation of a spectral feature varies in time. Increased temperature may lead to increased opacity and a formation higher in the atmosphere where the temperature is lower. Increased temperature may thus lead to a *decrease* in intensity.

- the Doppler-shift of a spectral line typically has a response height higher than the intensity. This will cause phase differences between Doppler-shift and intensity to be different than the phase differences between fluid velocity and temperature.
- slow rates may cause the ionization/excitation equilibrium to depend on the previous history of the atmosphere and deviate from the equilibrium of the instantaneous atmosphere. This is the case for hydrogen where the ionization/recombination time-scale is longer than the dynamic time-scale in the Solar chromosphere (see Carlsson & Stein 1992).
- contribution functions may be multi-peaked with one peak close to optical depth unity and another at a temperature spike in a shock (see Carlsson & Stein 1995).

In the following sections we present two examples of dynamic diagnostics.

6.1. DETERMINING VELOCITY AND TEMPERATURE FROM LINE OBSERVATIONS

As pointed out above the Doppler-shift and intensity of a line may be bad proxies for the fluid velocity and temperature. In a dynamic model we have access to all quantities and may test different diagnostic methods. One such method is to use the Doppler-shift and intensity of the line core to diagnose the velocity and temperature at the height of formation. Different spectral lines with different opacities will give information from different heights. Stein & Carlsson (1996) compared line center Doppler-shifts and temperatures with actual fluid values for several sodium lines concluding that the fluid velocity and line core Doppler-shift are tightly correlated in the photospheric lines but more loosely correlated in the chromospheric D-line. The correlations between line core radiation temperature and fluid temperatures is much looser because of decoupling of the source function from the Planck function.

Observationally it may be difficult to simultaneously observe a large number of spectral lines in order to sample different formation heights. Across one spectral line one has in principle information from all heights between the formation height of the line core and the continuum. The difficulty lies in disentangling the effects of Doppler-shifts and intensity changes. One method that attempts this decoupling is the lambda-meter method. Rigid "rods" of given lengths in wavelength are fitted to an observed line-profile. The mid-point of the rod is taken as the Doppler-shift and the intensity value as a proxy for local temperature. The shorter the length of the rod the higher the formation height. An example with 15 different length rods fitted to the profile of the D-line of neutral sodium is given in Fig. 4. The length of the rods and the choice of the sodium D-line was motivated by the work of Deubner et al. (1996) (see also Deubner 1998, these proceedings). They used this technique to study propagation properties of waves in the Solar upper photosphere and lower chromosphere. An important assumption in their work is that the Doppler-shift of the rod is a good proxy for the fluid velocity at a fixed height in the atmosphere. For the interpretation of their intensity-Doppler-shift phase diagrams the additional assumption is that the rod intensity is a good measure of the fluid temperature.

From Fig. 4 it is evident that the Doppler-shift of the rod is a good proxy for the velocity at the mean response height for the line-wing, which is formed in the deep photosphere where the wave amplitude is small. At line-center the correlation is still quite good but with a large scatter. The intensity, however, is a rather bad proxy for the local temperature even deep down. The radiation temperature is systematically lower than the fluid temperature. At line-center there is almost no correlation with

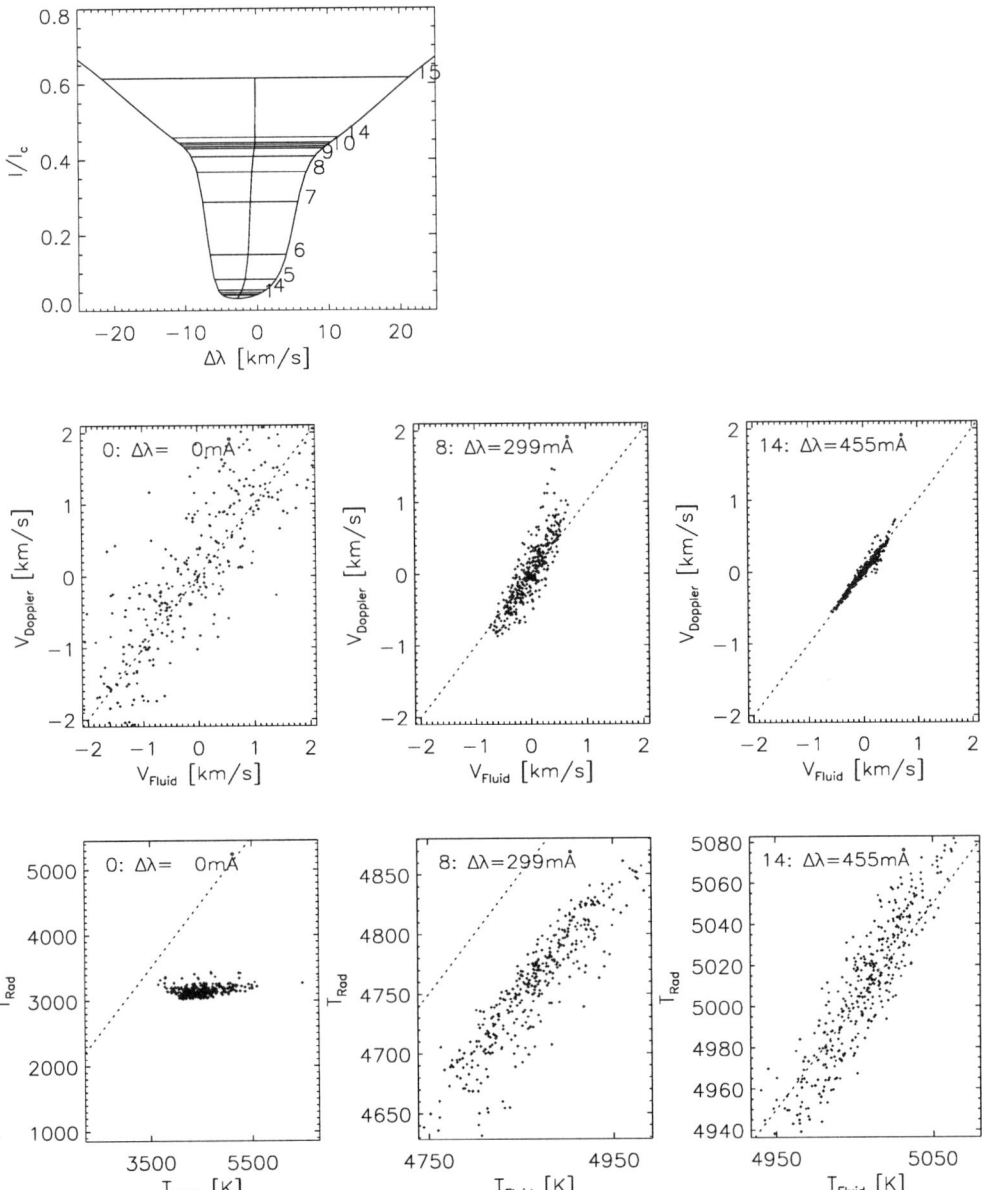

Figure 4. Lambda-meter method. The top panel shows the Na I line profile from one snapshot in the simulation with the 15 diagnostic rods of given lengths shown. The middle panels show the correlation between the fluid velocity at the static formation height and the Doppler-shift of a given rod (line-center, number 8 and number 14 from left to right). The bottom panels show the correlation between the fluid temperature at these heights and the radiation temperature of the intensity at the position of the rod. The dotted lines show $V_{Doppler} = V_{Fluid}$ and $T_{Rad} = T_{Fluid}$.

the fluid temperature. Doppler-shift – intensity phase relations can thus *not* be interpreted in terms of a velocity – temperature phase difference. Even Doppler-shift – Doppler-shift phase relations have to be treated with caution. Phase diagrams constructed from the measured Doppler-shifts in the simulation show similarities with the phase diagrams of Deubner et al. (1996) with a phase jump close to 7 mHz while phase diagrams constructed from the fluid velocities look more like what is expected from propagating waves. The simulations correspond to only a few spatial points and the phase diagrams are therefore very noisy. For a proper comparison with the observations a much larger simulation would be necessary, preferably in 3D. Such work is in progress.

6.2. CaII H & K GRAIN FORMATION

The H and K resonance lines of singly ionized calcium constitute the best available diagnostics of the Solar chromosphere available from the ground. The observations show complex temporal variations with marked line assymetries, see Section 3. The simulations by Carlsson & Stein (1997) show how these lines are formed. Since the bottom boundary condition is taken from observed Doppler-shifts of a photospheric line it is possible to compare the computed H-line intensities with the observations. This is done in Fig. 5. The correspondence is remarkable. Slight alterations of the piston velocities give different intensity behaviour. The sensitivity of the results to the driving velocity field together with the good agreement between simulation and observations give confidence that we can extract physics from the simulations that has bearing on the Sun.

The bright grains are produced by shocks near 1 Mm above $\tau_{500}=1$. Shocks in the mid chromosphere produce a large source function (and therefore high emissivity) because the density is high enough for collisions to couple the Ca II populations to the local conditions. The asymmetry of the line profile is due to velocity gradients near 1Mm. Material motion Doppler-shifts the frequency where atoms emit and absorb photons, so the maximum opacity is located at – and the absorption profile is symmetric about – the local fluid velocity, which is shifted to the blue behind shocks. The optical depth depends on the velocity structure higher up. Shocks propagate generally into downflowing material, so there is little matter above to absorb the blue Doppler-shifted radiation. The corresponding red peak is absent because of small opacity at the source function maximum and large optical depth due to overlying material. The brightness of the violet peak depends on the height of shock formation. The lower the shock, the higher the density and the larger the source function. The position in wavelength of the bright violet peak depends on the bulk velocity at the shock peak and the width of the atomic absorption profile (described with the microturbulence fudge parameter).

The bright grains are produced primarily by waves near and slightly above the acoustic cutoff frequency. The precise time and strength of a grain depends on the interference between these waves at the acoustic cutoff frequency and higher frequency waves. When waves near the acoustic cutoff frequency are weak, then higher frequency waves produce grains. The "five-minute" trapped p-mode oscillations are not the source of the grains, although they can modify the behavior of higher frequency waves. The wave pattern that exists at the solar surface is due to the interference of many trapped and propagating modes, so that the grain pattern has a stochastic nature.

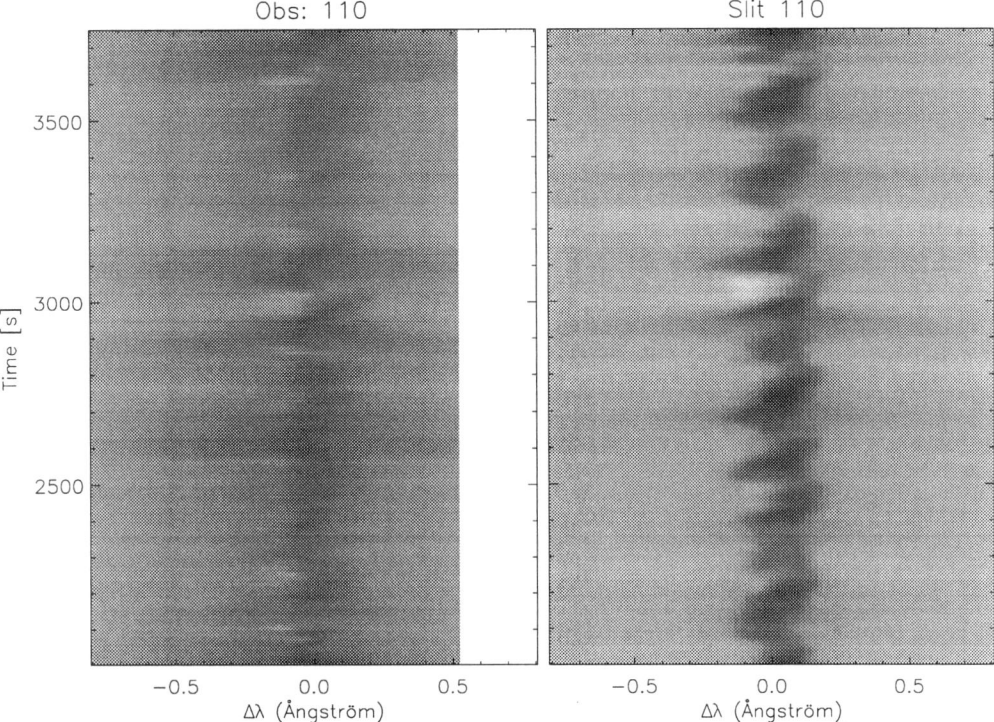

Figure 5. The computed Ca II H line intensity as a function of wavelength and time compared with observations.

7. Summary

Considerations of basic physics, observations and simulations all end up with the same conclusion: the Solar chromosphere is very dynamic. The very essence of the chromosphere lies in the dynamics. To study the Solar chromosphere with a mean model is as meaningful as taking the mean of Beethoven's 9th symphony (http://www.astro.uio.no/~matsc/iau97/beethoven_mean.au).

References

Bocchialini, K., Vial, J.-C., Koutchmy, S. 1994, ApJ, 423, L67
Carlsson, M., Judge, P. G., Wilhelm, K. 1997, ApJ, 486, L63
Carlsson, M., Stein, R. F. 1992, ApJ, 397, L59
Carlsson, M., Stein, R. F. 1994, in M. Carlsson (ed.), Proc. Mini-Workshop on Chromospheric Dynamics, Institute of Theoretical Astrophysics, Oslo, p. 47
Carlsson, M., Stein, R. F. 1995, ApJ, 440, L29
Carlsson, M., Stein, R. F. 1997, ApJ, 481, 500
Deubner, F. L., Waldschik, T., Steffens, S. 1996, A&A, 307, 936
Fontenla, J. M., Avrett, E. H., Loeser, R. 1993, ApJ, 406, 319
Gustafsson, B. 1973, Uppsala Astr. Obs. Ann., 5, No. 6

Harvey, J., Jefferies, S., Pomerantz, M., Duvall, T., J. 1992, BAAS, 180, 1705
Hofmann, J., Steffens, S., Deubner, F. L. 1996, A&A, 308, 192
Lites, B. W. 1985, in H. U. Schmidt (ed.), Theoretical Problems in High Resolution Solar Physics, MPA/LPARL Workshop, Max-Planck-Institut für Physik und Astrophysik MPA 212, München, p. 273
Lites, B. W., Rutten, R. J., Kalkofen, W. 1993, ApJ, 414, 345
Maltby, P., Avrett, E. H., Carlsson, M., Kjeldseth-Moe, O., Kurucz, R. L., Loeser, R. 1986, ApJ, 306, 284
Rutten, R. J., Uitenbroek, H. 1991, Solar Phys., 134, 15
Solanki, S. K., Livingston, W., Ayres, T. 1994, Science, 263, 64
Steffens, S., Hofmann, J., Deubner, F. L. 1996, A&A, 307, 288
Stein, R. F., Carlsson, M. 1996, in J. Christensen-Dalsgaard, F. Pijpers (eds.), Solar Convection and Oscillations and their Relationship, Proc. of AArhus Workshop
Thomas, R. N., Athay, R. G. 1961, Physics of the Solar Chromosphere, Interscience, New York
Uitenbroek, H., R.W.Noyes, Rabin, D. 1994, ApJ, 432, L67
Vernazza, J. E., Avrett, E. H., Loeser, R. 1981, ApJS, 45, 635
Von Uexkuell, M., Kneer, F. 1995, A&A, 294, 252

HIGH FREQUENCY SIGNAL IN GOLF DATA

R. A. GARCÍA[1], P. L. PALLÉ[2] AND THE GOLF TEAM[†]
[1] SAp, DAPNIA/DSM, CE Saclay, 91191 France
[2] Instituto de Astrofisica de Canarias, E-38205, Tenerife, Spain
[†] GOLF Team: P. Boumier, J. Charra, A.H. Gabriel, G. Grec, J.M. Robillot, T. Roca Cortés, S. Turck-Chièze, R.K. Ulrich

Abstract.
The analysis of ≈ 460 days of high quality data provided by the GOLF experiment on board SOHO, has unambiguously revealed the presence of signal in the power spectrum in the region 5.8 to 7.5 mHz, well above the p-modes cut-off frequency. In this contribution, the observed structure of these full-disk observations is presented. High frequency peaks (hereafter HFPs) well above the acoustic cutoff frequency of the solar p modes (≈ 5.5 mHz) have been already observed in intermediate and high spatial resolution oscillation data (Duvall et al. 1991). The different interpretations of the observed pattern (Kumar 1994 and references therein; Rast and Gough 1995) agree that it provides direct information on the location and some characteristics of the source of the acoustic oscillations (see also Vorontsov et al. 1997).

1. Observations

The high frequency range of the GOLF velocity power spectrum has been analyzed using 460 days beginning on April 11, 1996. GOLF is a disk-integrated sunlight resonant scattering spectrophotometer flying on board the SOHO mission. In its actual configuration mode, it obtains the line of sight velocity of the integrated visible solar surface by measuring the Doppler shift in the blue line wings of the sodium doublet.

As the high frequency structures are not expected to be coherent over a time longer than several hours, 115 subseries of 4 days have been extracted from the GOLF time series and their power spectrum has been averaged. Individual spectra were calculated using average of FFTs and crossed-spectra between the two photomultipliers.

2. Results

- Providing enough statistics, HFPs are clearly visible and independent of the spectral

technics (average of FFTs or crossed-spectra), the number of days averaged (4-10), degree of smoothing, and method of velocity calibration.
- 3 different regions have been observed (see figure 1):
  1. < 5.2 mHz. Unresolved high n p-mode envelopes (odd and even degree), separated $\sim$ 60 to 80 μHz depending on the frequency interval.
  2. $\sim$ 5.4-5.8 mHz. No signal. Acoustic cut off frequency region.
  3. > 5.8 mHz. HFPs pattern with characteristics of pseudo-modes.

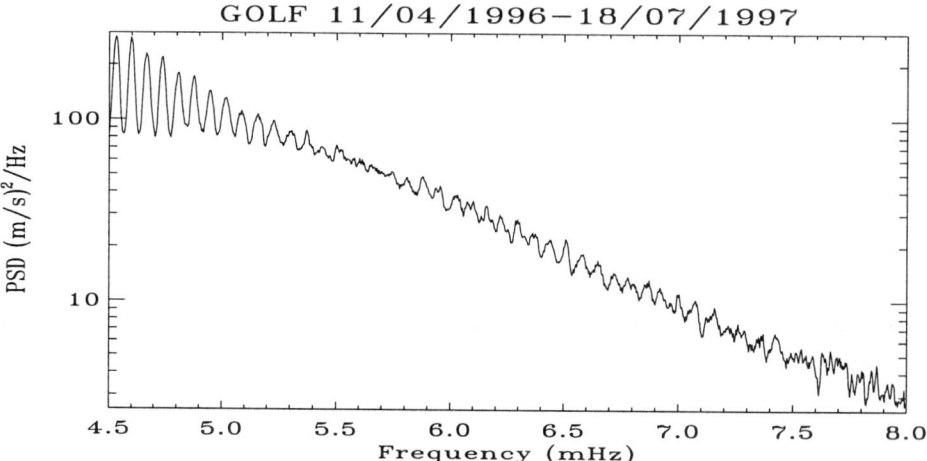

Figure 1. Average GOLF power spectrum of 115 Subseries of 4 days each smoothed with a boxcar of 5 points. A clear transition region between the p modes and the HFPs can be observed $\approx$ 5.5 − 5.8 mHz. The HFPs pattern are visible between 5.8 to 7.5 mHz.

- HFPs pattern consists on a unique series of broad (FWHM $\sim$ 28.89 $\pm$ 2.00μHz) equally spaced peaks ($\sim$ 70.38 $\pm$ 1.83μHz) that can be modelled with a simple damped sinusoid.
- Global Sun observations do not allow to see visible variations of the width or spacing of the peaks as a function of frequency.
- Neither is there a phase shift seen in the HFPs pattern between intensity and velocity, nor between the two intensities at different height of the solar atmosphere.
- Above 7.5 mHz, more data are required.

References

Duvall, T.L.Jr., Harvey, J.W., Jefferies, S.M. and Pomerantz M.A. (1991), *ApJ* **Vol. 373**, p. 308

Kumar, P.(1994), *ApJ* **Vol. 428**, p. 827

Rast, M.P. and Gough, D.O. (1995), *GONG'94: Helio-and Astroseismology from the Earth and Space. A.S.P. Conference Series*, R.K. Ulrich, E.J. Rhodes Jr. and W. Dappen. Ed., **Vol. 76**, p. 322

Vorontsov, S.V., Jefferies, S.M., Duvall, T.L.Jr. and Harvey, H.(1997), *MNRAS* Submitted

OSCILLATION OF THE MAGNETIC FIELD IN AN ACTIVE REGION

T. HORN AND J. STAUDE
Astrophysikalisches Institut Potsdam
Telegrafenberg A27; D-14473 Potsdam - Germany

Observations. The Vacuum Tower Telescope (VTT) at Tenerife has been used together with a polarimeter in front of a two-dimensional imaging spectrometer with a Fabry-Perot interferometer. The spectrometer has been described by Bendlin et al. (1992) and Bendlin et al. (1995), the remaining instrumentation, calibration, and data reduction by Horn et al. (1996). Some parameters of the setup were: 384x286 pixel CCD-camera (0.2" per pixel); resolution for the images better than 0.8"; filtergrams in the photospheric spectral line Fe I 6173.4 Å with $\Delta\lambda = 10.9$ mÅ; 45 filtergrams for each scan; 127 scans within 114 min. $\Delta t = 54$ s. White-light images from the same field of view were taken strictly simultaneously to all narrow-band pictures to correct for image blurring and motion.

The present observations were obtained on July 20, 1994, and were focused on the main spot of the active region NOAA 7757, 30° NE from the disk center.

Results. From the $(I+V)$ and $(I-V)$ images the Doppler shifts (velocity v) and the Zeeman splittings of the observed line were derived by fitting a polynomial to the line cores of the σ-components. The measured splittings are influenced by the total field strength B and to a minor extent by the angle of inclination between the line-of-sight and the vector **B**, if the splitting is incomplete. The maps show the power of v and δB in two bands of frequency ν (around $P \approx 3$ min and 5 min, respectively), moreover, the coherence and phase difference $\Delta\Phi$ between both types of oscillations are given. The power spectra of $v(t)$ in the umbra show the known features of strong power in bands of periods around 3 min (strengthened) and 5 min (weakened with respect to the quiet Sun). The data show significant power of oscillations of $\delta B(t)$ as well, which is concentrated in the same frequency bands as the power of $\delta v(t)$.

To exclude a possible influence of the jitter of the telescope the intensity fluctuation of a small area at the umbra-penumbra boundary has been checked: significant power is observed at frequencies $\nu < 1.5$ mHz only. An inspection of the observed umbral contrast ϕ shows a large amount of stray light, but it is very stable in time: the power density of the fluctuations of ϕ is very weak and equally distributed over all ν.

The spatial distribution of the power of $v(t)$ and $\delta B(t)$ is inhomogeneous across the umbra and shows marked spatial fine structures. Maximum power is measured in those parts of the umbra (close to the umbra-penumbra boundary) where we are looking along the lines of force of **B**, thus demonstrating the longitudinal character of the oscillations with respect to the direction of **B**. In the same parts of the umbra

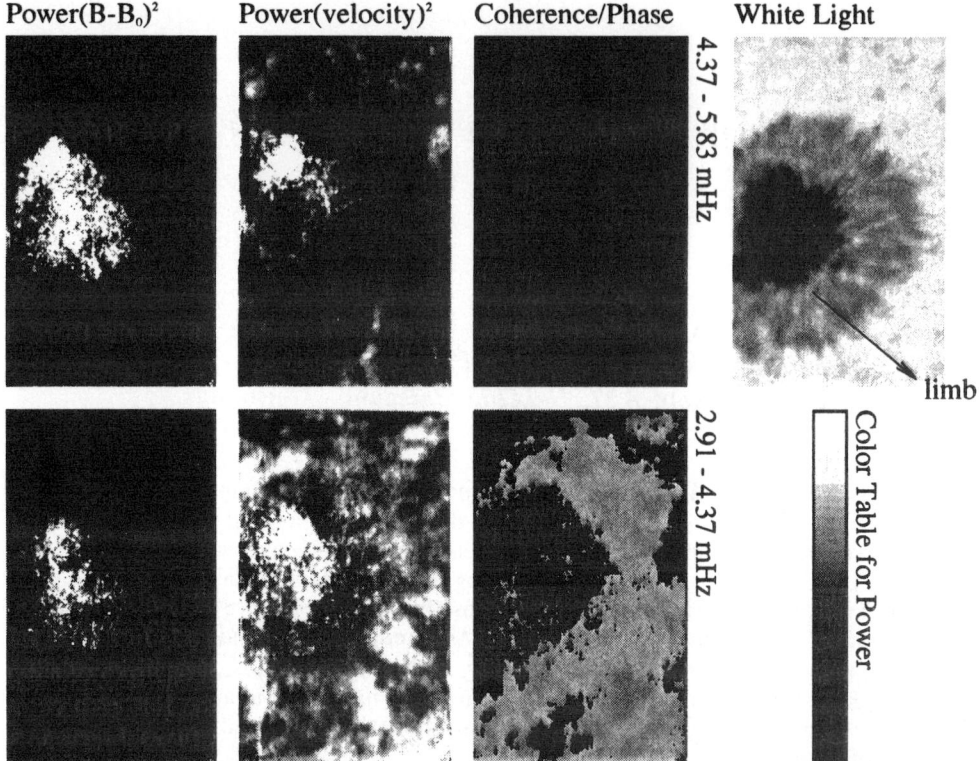

Figure 1. Spatial distribution of power of the oscillations of $B(t)$ (left), $v(t)$ and phase difference $\Delta\Phi$ (middle) for two selected ν bands; the right column shows the direction towards the limb, the power gray scale (maximum power is brightest), and an I_c picture of the same region. The gray coding for the $\Delta\Phi$ map is: black for a coherence < 0.85, elsewhere gray for $\Delta\Phi \approx \pi/2$ and white for $\Delta\Phi \approx 0$

we found also the largest correlation and $\Delta\Phi \approx 0$ between both types of oscillations, in the 3-min band in particular. The penumbral regions show rather an quadrature in the 5-min band.

References

Balthasar, H., Schleicher, H., Bendlin, C., and Volkmer, R. (1996) *Astron. Astrophys.* **315**, 603
Bendlin, C., and Volkmer, R. (1995) *Astron. Astrophys.* **112**, 371
Bendlin, C., Volkmer, R., and Kneer, F. (1992) *Astron. Astrophys.* **257**, 817
Horn, T., Hofmann, A., and Balthasar, H. (1996) *Solar Phys.* **164**, 321
Horn, T., Hofmann, A., and Landgraf, V. (1997) *Solar Phys.* **172**, 69

DYNAMICS OF THE DEEP SOLAR PHOTOSPHERE AT SUBGRANULAR SCALES

A. NESIS, R. HAMMER, M. KIEFER, AND H. SCHLEICHER
Kiepenheuer-Institut für Sonnenphysik
Schöneckstr. 6, D-79104 Freiburg, F.R.G.

1. Introduction

Extending our previous studies of the dynamics of solar granulation (Nesis *et al.*, 1997) we investigated the relationship between granular flow and the emergence of turbulence in the deep photosphere. Our main goal is to explore if such a relationship exists, and if so, to define it quantitatively. To this end we take advantage of the excellent signal approximation property of wavelets. The material for the present work is a series of spectrograms of high spatial resolution covering a time span of 12 min. They were taken at the center of the solar disk with the German Vacuum Tower Telescope in Izaña (Tenerife, Spain) in 1994, and include several absorption lines of different strengths; for more details see Nesis *et al.* (1997). The spectrograms were digitized and processed with wavelet techniques and regression analysis, in order to investigate the granular convective flow, the associated turbulence, and their mutual connection.

2. Results and Conclusions

Figure 1. shows the traces of the Doppler velocity (full line) and the turbulent velocity (dotted line) along the spectrograph slit. The Doppler velocity corresponds to the Doppler shift of the line core, whereas the turbulent velocity is reflected by the *fwhm* of the line. We found that granular flow speed and turbulence cannot be related by a regression line; rather the convective flow and the turbulence appear to be related by an attractor in the convective flow speed-turbulence phase space. The behavior of the regression analysis follows closely the history of the granulation dynamics over the entire 12 min time span, which corresponds roughly to a mean turn-over time of a granule (Mehltretter *et al.*, 1978). Thus, we tend to assert that convective flow and turbulence can be interpreted in terms of a dynamical system, which is shown

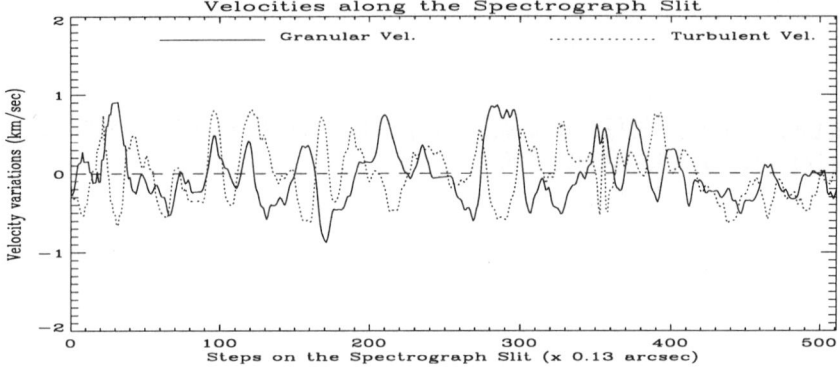

Figure 1. Doppler velocity and turbulent velocity.

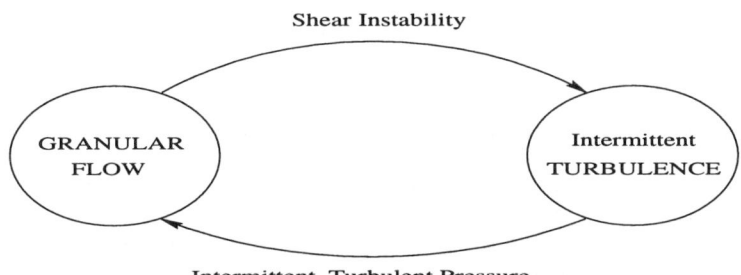

At locations of steep gradients in the granular flow patterns, shear instability produces turbulence. The associated turbulent pressure will backreact on the convective flow.

Figure 2. The dynamical system.

in Fig. 2. The calculated turbulent pressure shown in Table 1 supports this assertion.

TABLE 1. Hydrodynamical implications of our observations.

Re	ν_{turb}	P_{turb}	P_{turb}/P_{gas}	all values at
10^9	$10^8 \, m^2 s^{-1}$	$\leq 10^4 \, Pa$	≤ 0.5	$\approx \tau_{5000} = 1$

References

Nesis, A., Hammer, R., Hanslmeier, H., Schleicher, H., Sigwarth, M., and Staiger, J. (1997), *Astronomy and Astrophysics* **Vol. 326**, p. 851

Mehltretter, P. (1978), *Astronomy and Astrophysics* **Vol. 62**, p. 3

MHD OSCILLATIONS OBSERVED IN THE SOLAR PHOTOSPHERE WITH THE MICHELSON DOPPLER IMAGER

A.A. NORTON AND R.K. ULRICH
Dept. of Phys. and Astron., Univ. of California at Los Angeles

AND

R.S. BOGART, R.I. BUSH AND J.T. HOEKSEMA
Hansen Experimental Phys. Laboratory, Stanford Univ.

Magnetohydrodynamic (MHD) oscillations are observed in the solar photosphere with the Michelson Doppler Imager (MDI). The absorption of acoustic waves by sunspots (Braun *et al.*, 1988) and the possible conversion of acoustic waves into MHD waves (Spruit *et al.*, 1992) provide a plausible basis for the existence of MHD waves in the solar atmosphere and motivate this attempt at observation.

Images of surface velocity, V, and magnetic flux, B, with 4" spatial and 60 second temporal resolution are analyzed. A 2-D gaussian aperture with 10" FWHM is applied to the data in regions of sunspot and plage and the averaged signal is returned each minute. Sunspot passage is followed using a modified Carrington rate. Representative signals are shown in Figure 1. Significant power is observed in the B oscillations with 5 minute periods, as seen in Figure 2. It is assumed that the large amplitude acoustic waves with 5 minute periods are the driving mechanism behind the B oscillations.

To determine if the B oscillations are instrumental or solar in origin, the effect of misregistration between left and right circularly polarized (LCP, RCP) images is investigated and found not to be the cause of the observed B oscillations, see Figure 3. If the B signal were due to misregistration, the sunspot signal would decrease to the level of noise at a given misregistration, as found in quiet sun.

Ulrich (1996) proposed that phase relations between B and V contain the signatures of Alfénic and/or magnetoacoustic waves, which have 0 and ± 60 second time lags, respectively. Phase angles between V and B oscillations are calculated for 19 sunspot and plage regions, see Figure 4. An average weighted time lag of -67 seconds is consistent with magnetoacoustic waves.

References

Braun DC, LaBonte BG, Duvall TL: 1988, 'The absorption of high-degree p-mode oscillations in and around sunspots,' *ApJ.* **335**, 1015

Spruit HC, Bogdan TJ: 1992, 'The conversion of p-modes to slow modes and the absorption of acoustic waves by sunspots,' *ApJ.* **391L**, 109

Ulrich, RK: 1996, 'Observations of magnetohydrodynamic oscillations in the solar atmosphere with properties of Alfvén waves,' *ApJ.* **465**, 436

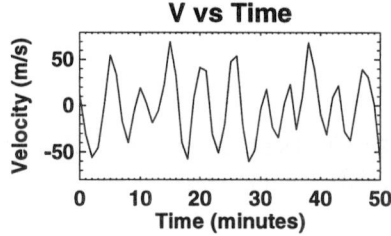
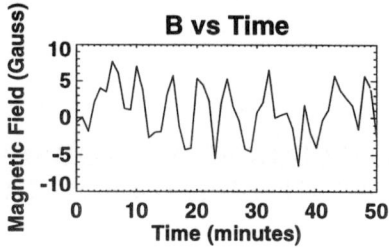

Figure 1. Representative V and B signals from an Aug 30, 1996 sunspot region. Signals are detrended with a 30 minute boxcar average. Average V and B values are subtracted.

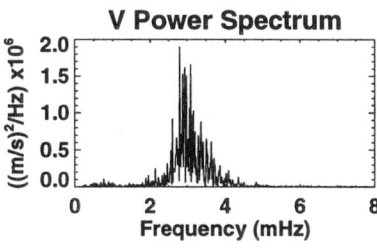
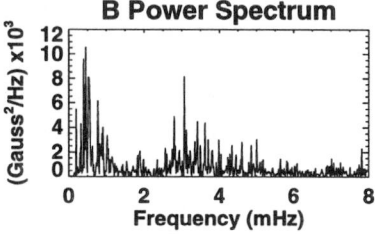

Figure 2. Velocity and magnetic power spectra from an August 30, 1996 sunspot region.

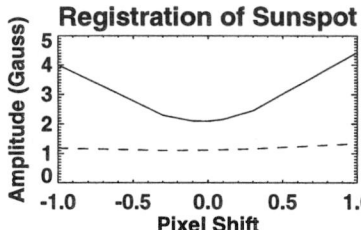
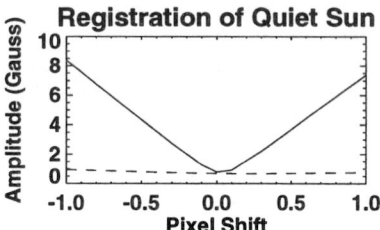

Figure 3. LCP and RCP images are shifted against each other to generate misregistered datasets. Resulting B signal (solid) and noise (dashed) amplitude are plotted vs N-S misregistration for sunspot (left) and quiet sun (right).

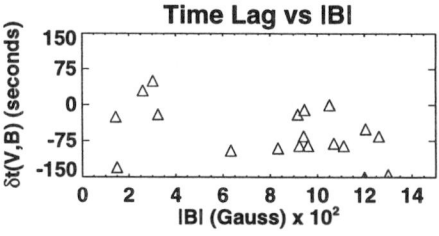

Figure 4. The time lag between V and B is plotted vs absolute magnetic field strength. V is cross correlated with B and the phase is calculated as the time lag, δt, corresponding to the largest cross correlation coefficient within a 300 second period.

PHASE SPECTRA SEEN FROM SPACE

TH. STRAUS
Osservatorio Astronomico di Capodimonte, Napoli, Italy

F.-L. DEUBNER
Institut für Astronomie und Astrophysik, Würzburg, Germany

B. FLECK
ESA Space Science Department, NASA/GSFC, Greenbelt, USA

C. MARMOLINO
Università "Federico II", Napoli, Italy

G. SEVERINO
Osservatorio Astronomico di Capodimonte, Napoli, Italy

AND

T. TARBELL
Lockheed-Martin Advanced Technology Center, Palo Alto, USA

A 15 hour time series of simultaneous spectroscopic measurements of the continuum intensity, Doppler shift, and line depth in the Ni I 6768Å line at disk center of the quiet sun has been obtained during a guest investigator campaign with the Michelson Doppler Imager (MDI, see Scherrer et al. 1995). In addition, the line intensity has been calculated from the continuum intensity and the line depth data. The spatial shifts due to solar rotation have been approximately removed during the observation run by shifting the CCD read-out area by 1 pixel every 4 minutes. Small residual shifts have been removed during the data analysis. Before applying the Fourier analysis, a (temporal) low pass filtered mean velocity (line of sight component of solar rotation) and mean intensity (center-limb variation) has been removed from each frame. Finally, the continuum intensity (C), line intensity (I), and velocity (V) time series have been subjected to a 3D Fourier transform, yielding $k-\omega$ power, cross power, phase difference, and coherence spectra. The data can now be compared to recent ground-based observations (Deubner, 1990; Deubner et al., 1992; Straus & Bonaccini,1997).

The data from space conclusively prove the existence of different phase regimes in the $k-\omega$ diagram (see Fig. 1), as indicated from ground-based observations. These different regimes, in part theoretically understood (see Fig. 2-6 of Beckers 1981 for a summary of phenomena in the $k-\omega$ diagram), in part to be investigated in more detail, are:

(1) the p-modes showing negative $V-I$ phase values and an upward progressive component; they can now be followed up to 8.3 mHz and even to higher frequencies

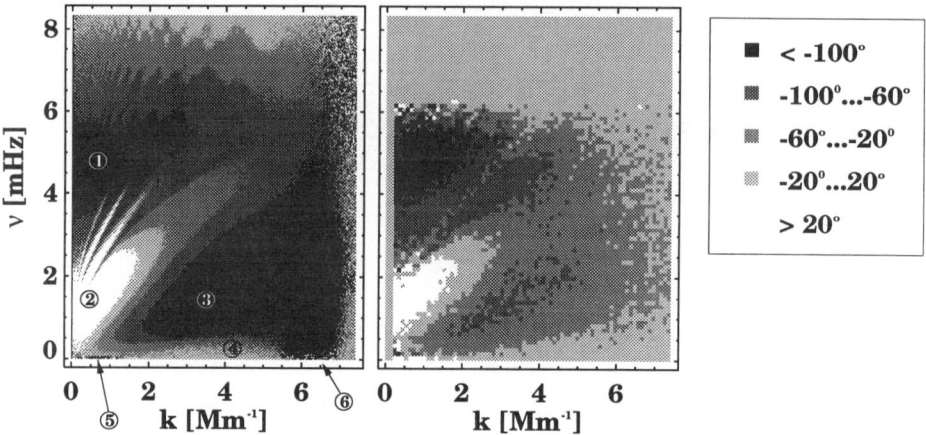

Figure 1. Comparison between $k - \omega$ phase difference spectra between the velocity and the intensity fluctuations, as observed from space (left panel) and from ground (right panel, Straus & Bonaccini, 1997); the labels refer to the text.

(see Jefferies, this conference, for a review on high frequency pseudo modes). (2) a low power, intermediate coherence plateau showing positive $V - I$ phase, which continues to higher frequencies into the space between the ridges. This regime has been attributed to downward propagating waves (Marmolino et al., 1993), which can now clearly be demonstrated by the $I - C$ phase spectrum. (3) A distinct regime of negative $V - I$ phase, which has been attributed to internal gravity waves in the atmosphere (Deubner et al., 1992), as indicated by the downward propagating character. (4) A nearly zero $V - I$ phase regime corresponding to the convective overshoot; this regime is limited to spatial wavenumbers less than $5.5 \, \text{Mm}^{-1}$. Furthermore, two other regimes can be identified which merit further investigations: (5) a regime at very long periods and spatial wavenumbers up to $1.5 \, \text{Mm}^{-1}$, which shows anticorrelation between V and I, recently surmised as being due to the chromospheric network (Straus & Bonaccini, 1997); and (6) for the first time visible, a regime at very high spatial wavenumbers which also shows a clear anticorrelation between V and I. Fine scale magnetic field structures might be a candidate for the latter phenomenon.

Acknowledgements

This work has been partially funded by ASI contract ASI 95-RS-69.

References

Beckers, J. M.: 1981, in Jordan, S. (ed.), *The Sun as a Star*, NASA SP 450.
Deubner, F.-L.: 1990, in Stenflo, J. O. (ed.),*Solar Photosphere: Structure, Convection, and Magnetic Fields*, IAU Symp. 138, p. 217.
Deubner, F.-L., Fleck, B., Schmitz, F., Straus, Th.: 1992, *Astron. Astrophys.* **266**, 560.
Jefferies, S.: 1997, this proceedings.
Marmolino, C., Severino, G., Deubner, F.-L., Fleck, B.: 1993, *Astron. Astrophys.* **278**, 617.
Scherrer, P.H., Bogart, R.I., Bush, J.T., et al.: 1995, *Solar Phys.* **162**, 129.
Straus, Th., Bonaccini, D.: 1997, *Astron. Astrophys* **324**, 704.

OBSERVATIONAL RELATIONSHIP BETWEEN MESO-SIZED CONVECTION AND 5-MIN OSCILLATION IN THE SOLAR ATMOSPHERE

S. UENO
Kwasan Observatory, Kyoto University,
Yamashina, Kyoto, Japan 607

AND

R. KITAI
Hida Observatory, Kyoto University,
Yoshiki-gun, Gifu, Japan 506-13

1. Introduction

Solar convection has been studied theoretically and observationally for very long time. At present, three kinds of solar surface structures are supposed as manifestations of solar convection; granulation, mesogranulation, supergranulation, in increasing order of spatial size. Among them, observations of mesogranulation, however, do not have long history, yet, since November's discovery (1981). Their basic characteristics such as lifetime, velocity field, radial size, and relation to other kinds of convection, have not been precisely determined, yet.

By the way, currently, there are important discussions about where the source of the excitation of the solar oscillation is situated. The small sized oscillating cell compared with the granulation scale had been considered as beat phenomena between larger scale oscillations for a long time (White & Cha 1971; Wolff 1973; Fossat & Ricort 1975; Deubner 1975). There are even opinions that meso-granulation is also the apparent phenomenon by superposition of oscillating cells. Quite recently, however, a few observations showed very important results. Rimmele et al. (1995) and Espagnet et al. (1996) showed that strong solar oscillations (having large amplitudes) often seem to be closely associated, both temporally and spatially, with the dark lanes of granulation and down flowing regions.

Then, in this time, we observationally observed time variations of granulation, meso-granulation, and 5-min oscillation and 3-dimensional velocity components of meso-granulation, in order to investigated features of the mesogranulation and relationship between solar convections and 5-min oscillation cells.

2. Observation, Analysis and Results

For the purpose described in section 1, we performed two different kinds of observation simultaneously with the Domeless Solar Telescope (DST) of Hida Observatory. One is 2D imaging observation of solar quiet photosphere in wavelengths around $\lambda 4308$ Å, whose duration is 90 min. The other one is slit spectroscopy of the neutral iron line $\lambda 6302.5$ Å at the central region of the field-of-view of the 2D imaging observation; its duration is only 30 min. Thus the horizontal velocity field is obtained from the former observation by local correlation tracking method (November & Simon 1988; Kitai et al. 1997; UeNo & Kitai 1997) using granulation patterns as tracers, while the radial velocity field is obtained from the latter one at the same region.

From the map of horizontal velocity field, we can find meso-sized velocity divergence structures which distribute all over the photosphere. By calculating the spatial power spectrum, we could know that the size of this structure is about 15". Moreover, we made a histogram of the meso-sized structures' lifetime from the time series of velocity field maps for 90 min. It shows a clear peak of lifetime at 40 min, and the frequency gradually decreases toward long lifetime, though this obtained lifetime is rather shorter than ones reported in the past studies.

Next, we separated the radial velocity field into four components, i.e. granulation, mesogranulation, oscillation, and high frequency noises, by Fourier filtering. We also calculated the power spectrum of the meso-granular radial velocity field, and obtained the same size as horizontal velocity structure. This clearly suggests the existence of a three-dimensional meso-sized structure . In fact, both patterns of the meso-sized radial and horizontal velocity fields are well consistent, and the values of velocities satisfy the continuity equation for convection.

Further, we investigated the temporal and spatial relationship between the 5-min oscillations and convection by using the data of the radial velocity field. At first, we investigated the temporal development of the velocity fields of granulation and oscillation. At the location where amplitudes of 5-min oscillation increase suddenly, or where large phase changes occur, we noted a strong downflow in the granular field, i.e. a granular collapse immediately before a change of the 5-min oscillation. This finding corroborates reports by Rimmele et al. and Espagnet et al.. Moreover, we checked the spatial relation between the meso-granular velocity field and the 5-min oscillation field. We found that regions, where oscillations are strongly excited for more than 30 min, correspond to downflow regions in the meso-granulation field.

References

Deubner, F. L. (1975) *A&A* **44, 371**
Espagnet, O., Muller, R., Roudier, Th., Mein, P., Mein, N., and Malherbe, J. M. (1996) *A&A* **313, 297**
Fossat, E., Ricort, G. (1973) *Sol. Phys.* **28, 311**
Kitai, R., Funakoshi, Y., UeNo, S., Sano, S., and Ichimoto, K. (1997) *PASJ* **49, No. 4**
November, L. J., Toomre, J., Gebbie, K. B., and Simon, G. W. (1981) *ApJ letter* **245, L123**
November, L. J., Simon, G. W. (1988) *ApJ* **333, 427**
Rimmele, T. R., Goode, P. R., Harold, E., Stebbins, R. T. (1995) *ApJ letter* **444, L119**
UeNo, S., Kitai, R. (1997) *submitted to PASJ*
White, O.R., Cha, M.Y. (1973) *Sol. Phys.* **31, 23**
Wolff, C.L. (1973) *Sol. Phys.* **32, 31**

ON THE CHROMOSPHERIC BEHAVIOUR OF PHOTOSPHERIC Mn 539.47 nm SPECTRAL LINE

I. VINCE AND S. ERKAPIĆ

Astronomical Observatory, Volgina 7, 11050 Belgrade, Yugoslavia

1. Introduction

The measured equivalent width of the Mn 539.47 nm line irradiance shows a cycle variability of about 2% (Livingston, 1992). In this paper we analyse the unusually large variation of this line.

2. Observations, Data Analysis and Hypotheses

Reduced equivalent widths (EW) and central depths (r_c) of Fe (539.32, 539.52 nm) and Mn (539.47 nm) lines obtained on Kitt Peak Observatory (1979-1996; KP data) and used for this analysis with kind permission of Livingston(1997). From 1987 on, these lines have also been observed at Belgrade Astronomical Observatory (BAO data).

The BAO data were divided into two samples: the high activity sample (HAS; 1989-1992) and the low activity sample (LAS; 1987, 1995, 1996). The relative changes of EW and r_c of the Mn line between HAS and LAS are found by dividing these parameters by the corresponding parameters of the comparison Fe 539.32 nm line (this line has no cycle response (Livingston (1992))). A comparison of the HAS and LAS parameters gives the following results (HAS/LAS): $(\frac{\Delta EW}{W})_{rel} = 1.2\% \pm 4\%$; $(\frac{\Delta r_c}{r_c})_{rel} = 2.6\% \pm 2\%$. Both the EW and r_c increase with activity. The same procedure of reduction was applied to the KP data yielding the following results: $-1.6\%\pm0.2\%$ for EW and $-1.2\% \pm 0.1\%$ for r_c. The EW and r_c decrease with activity. This discrepancy between KP and BAO data is probably due to systematic errors in BAO data (e.g., water vapour blends or an instrumental change in 1994). Because of that our further discussions will be based on KP data, and a special care will be taken in future for elimination of systematic errors in BAO data. The Fourier spectrum of the KP data shows a significant peak at frequency $3.0\times10^{-4} day^{-1}$ with a relative variation of 1.12% for EW and with 1.05% for r_c.

The observed relative changes EW of about 1% can be explained by a temperature gradient variation of about 5 K or by a temperature change of about 3 K (Erkapić and Vince, 1995). According to Gray and Livingston (1996) the amplitude of temperature variation (derived from CI 538.03 nm line) during a solar cycle is only 1.5 K.

Using the VAL models (Vernazza et al., 1981) and a plage model (Kučera and Baranovsky, 1992) we calculate the EW and r_c of Mn line for different bright network elements (BNE), and for plages. To explain the observed variations of EW and r_c during the cycle we need to assume filling factors changes that are too large in comparison to the observed values.

Strong chromospheric emission spectral lines could influence the electron population of Mn atom energy levels, but we did not find any important Mn bound-bound transitions that could be influenced by those lines.

There is also a possibility that a chromospheric component of the Mn 539.47 nm line causes the observed variations. Our synthetic profile calculations, truncating step by step the upper layers of the VALC model, show that layers over 450 km contribute to EW about -3% and to r_c about -6%. Consequently, the variation of the chromospheric Mn emission component from 20% to 50 % could explain the observed behaviour of the Mn line. We plan to check the existence of the Mn emission component by observing the transition of the photospheric into the chromospheric spectrum during the 1999 solar eclipse.

Since photons with $\lambda \leq 166.8$ nm can ionize the MnI atoms, and one of the largest transparency jumps of the solar atmosphere occurs at this wavelength region (such that chromospheric photons could penetrate to MnI line formation layers), we assume that photoionization is the main source of the MnI 539.47 nm line variability. Unfortunately, we have not relevant data at our disposal to check this assumption yet.

References

Erkapić S. and Vince I. (1995), Influence of photospheric parameters on solar spectral line parameters, *Publ. Obs. Astron. Belgrade* **49**, pp. 159–162

Gray D.F. and Livingston W.C. (1996), Monitoring the solar temperature: spectroscopic temperature variations of the sun (Preprint, accepted for pub. in *Astrophys. J.*)

Livingston W. (1992), Observations of solar spectral irradiance variations at visible wavelengths, in it Proc. of the Workshop on the Solar Electromagnetic Radiation Study for Solar Cycle 22 (ed. F. Donnelly), pp. 11-19

Livingston W. (1997), The equivalent width data, *Private communication*

Kariyappa R. and Sivaraman K.R. (1994), Variability of the solar chromospheric network over the solar cycle, *Solar Physics* **152**, pp. 139-144

Kučera A. and Baranovsky E.A. (1992), Solar plage model, (Preprint to appear in *Proc. of IAU Symp. 154*)

Vernazza J.E., Avrett E.H. and Loeser R. (1981), Structure of the solar chromosphere. III. Models of the EUV brightness components of the quiet sun, *Astrophys. J. Suppl.* **45**, pp. 635-725

A SPATIAL ANALYSIS OF LOCAL SOURCES OF OSCILLATION

Y. YAN[1,2], F.-L. DEUBNER[1] AND R. KLEINEISEL[1]
[1] Astronomisches Institut, Universität Würzburg, Am Hubland,
97074 Würzburg, Germany
[2] Beijing Astronomical Observatory, Chinese Academy of Sciences,
Beijing 100080, China

1. Introduction

Recent observational studies of the spatial and temporal relations between impulsive and periodic perturbations (Deubner and Kleineisel, 1997; Espagnet et al., 1996; see also Goode et al., these proceedings) suggest that oscillations in the solar atmosphere are excited locally and stochastically by turbulent convection. Proper characterization of the dynamical processes involved requires careful analysis of the observations in the spatial as well as in the temporal domain. Wavelet transforms (Combes et al., 1984) allow the characterization of waveforms together with the temporal delays that occur between the observed fluctuations. We have employed Morlet wavelets (Daubechies, 1992) in the present analysis. In addition to the phase relation between convective eddies and the atmospheric waves we study in particular the 2-D spatial evolution of the perturbations, to better distinguish different modes of oscillation. The observational material used is the same as in Deubner and Kleineisel (1997).

2. Methods of analysis

Spatio-temporal filters have been utilized to separate the "convective" and "oscillatory" components of the observed brightness fluctuations, corresponding to gravity waves, and to the evanescent and propagating p-mode regime in the solar atmosphere, respectively.
In contrast to Deubner and Kleineisel (1997), the present study is not limited to preselected "events". For 4 subcubes (Deubner and Kleineisel, 1997) as well as for the whole data cube, we have calculated the spatio-temporal cross-correlation functions between the subsonic and oscillatory components of the white light and CaK_2 data. Furthermore, the Morlet wavelet centered at a wave period of 200 seconds was used to obtain the amplitudes of the oscillatory wave packets. Then spatio-temporal cross-correlation functions between the subsonic components and wavelet transformed oscillations were calculated again. At the photospheric level, we find that the onset of a granule is followed by a decrease of oscillatory amplitude, whereas the opposite happens at the borders of the intergranular lanes, confirming the findings in Deubner

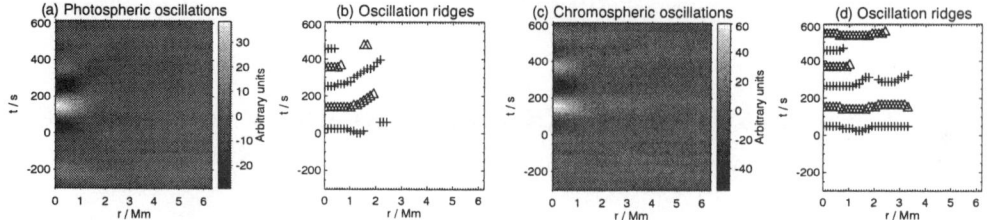

Figure 1. Oscillatory intensity at positions of intergranular lanes in the photosphere (a) and in the chromosphere (c). Corresponding ridges of minimal and maximal intensity (b) and (d)

and Kleineisel (1997).
Using our data we have defined a new criterion for the granular and intergranular "events" by searching automatically for the onset of a strong granule followed by a significant decrease of oscillatory power after a lapse of time as determined by the cross-correlation above, or respectively, for rapid darkening in intergranular lanes followed by an increase of oscillatory power. We also sharpen the criterion by specifying that no adjacent "event" center is within a subcube with 2" radius and 40 min duration of each "event" center. In this way, we have selected 1897 granular and 1762 intergranular events that are not located on the supergranular network.

3. Results and discussions

Centered around the positions defined above, we have averaged the evolution of the brightness distribution both azimuthally and over all 1897 granule, respectively 1762 intergranule events, covering a radial range of 7.6 Mm, and a time interval of 300 s before and 600 s after the event center. Figure 1 shows the temporal-spatial evolution of the oscillatory data at the intergranular event positions at both heights.

4. Conclusion

Our results corroborate strongly previous suggestions that the solar oscillations may indeed be excited by convective downflows.

Y. Y. is grateful to the Alexander-von-Humboldt Foundation for granting a Humboldt fellowship.

References

Combes, J. M., Grossmann, A., and Tchamitchian, Ph. (eds.): *Wavelets*, Springer-Verlag, Berlin, 1989.
Daubechies I.: *Ten Lectures on Wavelets*, SIAM, Philadelphia, 1992.
Deubner, F.-L., and Kleineisel, R. in: Provost, J., Schmider, F.-X. (eds.): *Sounding Solar and Stellar Interiors*, Dordrecht, 1997
Espagnet,O., Muller, R., Roudier, T., Mein, P., Mein, N., and Malherbe, J.M.: 1996, *Astron. Astrophys.*, **313**, 297.

DISTRIBUTIONS OF THE SOURCES OF THE MAGNETIC FIELD DURING THE DOUBLE MAGNETIC CYCLE

ELENA E. BENEVOLENSKAYA
Pulkovo Observatory
St. Petersburg, 196140, Russia

The longitudinal and latitudinal distributions of the solar magnetic field have been investigated by many authors see e.g. (Bumba and Howard 1969; Gaizauskas et. al. 1983). The main longitudinal structures such as active longitudes, sector boundaries in the solar wind, and coronal holes appear to be the consequence of non-symmetric magnetic modes (Stix 1971; Ivanova and Ruzmaikin 1977) and can be explained within the framework of dynamo models. According to these models, solar magnetic fields are generated by helicity and differential rotation in the solar convection zone (Parker 1979) and the linear dynamo process may generate a carrier frequency ω_c ($\omega_c = 2\pi/T$, $T = 22$ yr).

The 22-yr cycle of solar activity is a magnetic cycle that manifests itself as a change of polarity of the sunspots. In parallel with this, the poloidal magnetic field, or background field (B_r-component), shows a 22-yr periodicity (Howard and LaBonte 1981).

According to the studies (Stenflo 1994) the low-latitude and polar fields clearly show the 22-year periodicity too and there is an indication for low-amplitude power at higher frequencies, corresponding to periods of about 2 yr. In non-linear theory stable frequencies with a period shorter than that of the cycle period may occur (Hoyng 1990). All above mentioned allow to assume that cycle consists of two component: high frequency (Hale cycle) and low frequency (biennial cycle). The purpose of our investigation is to find longitudinal structure of the solar magnetic cycle.

Line-of-sight ($B_{\parallel}$) magnetograph data observed by the Wilcox Solar Observatory at Stanford are used in the present investigation. Synoptic observations of the Sun's photospheric magnetic field were obtained covering the interval from 1976 to 1996, spanning cycles 21 and 22.

In the analysis, the first harmonic (f_1) corresponds to $T_1 = 256$ Carrington rotations (being, approximately, 20 yr), and the tenth harmonic (f_{10}) corresponds to $T_{10} = T_1/10$ (being, approximately, 2 yr). The results are obtained as a power spectrum $P_{yy}(\mu, f)(\mu = cos\theta)$ for each longitude l_i. Finally, the power spectra $P_{yy}(\mu, f)$ were averaged over latitude, separately, for northern and southern hemispheres.

This investigation has revealed that the low-frequency (f_1) and high-frequency (f_{10}) components of longitudinal distribution are distributed non-uniformly on the solar surface and rotate approximately at the same rate. In particular, both the low-frequency (f_1) and high-frequency (f_{10}) components reach their maximum values in different longitudinal zones that are separated by 20°.

A natural question is whether differences between the latitudinal distribution of the low-frequency f_1 component and high-frequency f_{10} component take place for a given longitude.

The first harmonic f_1, having a period $T_1 = 20$ yr, appears dominant and has a maximum at latitudes greater than 40° in both hemispheres, as does the f_{10} component with $T_{10} = 2.0$ yr. These results coincide with those found in the distribution of high-frequency and low-frequency components of the magnetic field for the axisymmetrical case (Benevolenskaya 1994, 1995). Thus, it may be conceivable that these distributions correspond to the reality to the physical picture of a double magnetic cycle of which both components reach their maximum values at high latitude.

Therefore, we conclude that the poloidal magnetic field of the solar activity cycle consists of both axisymmetrical and non-axisymmetrical modes: a high-frequency (biennial) component and a low-frequency (Hale) component.

Acknowledgements

I am grateful to Dr. J. Todd Hoeksema (Stanford University) for providing the Wilcox Solar Observatory line-of-sight magnetograph data, Prof. Toshio Fukushima for a possibility to attend the GA IAU (LOC IAU grant) and Prof. Franz-L. Deubner for useful comments.

References

Benevolenskaya, E. E. 1994, Astron. Lett., 20, 468
Benevolenskaya, E. E. 1995, Solar Phys., 161, 1
Bumba, V. and Howard, R. 1969, Solar Phys., 7, 28
Gaizauskas, V., Harvey, K. L., Harvey, J. W. and Zwaan, C. 1983, Astrophys. J., 265, 1056
Hoyng, P. 1990, Solar Photosphere: Structure, Convection and Magnetic Fields, Stenflo, J. O., IAU Symp. 138, 359
Howard, R. and LaBonte, B. J. 1981, Solar Phys., 74, 131
Ivanova, T. S. and Ruzmaikin, A. A. 1977, Astron. Zh., 54, 846
Parker, E. N. 1979, Cosmical Magnetic Fields, Oxford University Press
Stenflo, J. O. 1994, Solar Surface Magnetism, Rutten, R. J. and Schrijver C. J., NATO Advanced Research Workshop, Kluwer
Stix, M., 1971, Astron. Astrophys., 13, 203

TIME VARIATION OF THE GLOBAL SOLAR MAGNETIC FIELD INFERRED FROM THE SUN'S SHADOW AS SEEN IN 10 TeV COSMIC RAYS

L.K. DING[1], M. NISHIZAWA[2], T. SASAKI[3], Y.H. TAN[1], Y. YAMAMOTO[3],
T. YUDA[4] AND THE OTHER MEMBERS OF THE TIBET ASγ COLLABORATION

[1] *Institute for High Energy Physics, Academia Sinica, Beijing, China*
[2] *National Center for Science Information Systems, Tokyo 112, Japan*
[3] *Department of Physics, Konan University, 8 Okamoto, Kobe 658, Japan*
[4] *Institute for Cosmic Ray Research, University of Tokyo, Tanashi, Tokyo 188, Japan*

1. Introduction

Air shower arrays with high counting rates at high altitude provide a unique means for the study of the time dependence of the Sun's shadow as seen in cosmic rays (Amenomori et al. 1992). With the Tibet-I array, operated from 1990 to 1993 at Yangbajing (4300m), we detected for the first time the influence of the solar and interplanetary magnetic fields (IMF) on the Sun's shadow. In this experiment the Sun's shadow seen by 10 TeV cosmic rays was found at a position $0.°7$ away from the position of the Sun. This large displacement is considered to be caused by IMF which changed considerably in 1990-1993, near maximum, and during the declining phase of solar activity (cycle 22). A new Tibet-II array, enlarged in 1994, with a seven times larger effective area than the Tibet-I, has been operating since 1995 and allows us to observe the Sun's shadow every 3-4 months. The solar activity, being in the most quiet phase now in 1995-1997, will return to more active phase in 1998. Here, we present some results obtained in 1996 with Tibet-II array.

2. Yearly Variation of the Sun's Shadow

In the previous papers (Amenomori et al. 1993, 1994, 1996), we showed that the position of the Sun's shadow is not only largely displaced from the apparent Sun's center but also varies with change of solar activity as shown in Fig. 1 in 1991-1993. The data in the "away" and "toward" sectors of IMF gave also fairly different positions of the Sun's shadow. The strength of the solar magnetic field by Stanford Mean Solar Magnetic Field (NOAA/USAF 1997), and the mean sunspot number reached the maximum in 1991 and decreased rapidly, although IMF measured by IMP-8 (NASA 1997) at the Earth's orbit was rather stable.

The first ten months data with the Tibet-II array in 1995-1996 gives a deep Sun's shadow whose position is almost at the apparent Sun's center (Fig. 1), consistent with a quiet phase of solar activity. No clear difference between the "away" and "toward" fields is seen in contrast with the results in 1991-1993. This may be interpreted by an ill-defined sector separation in 1995-1996.

3. Summary

Using the data obtained in 1995-1996 with the Tibet-II array, we found the Sun's shadow almost at the apparent Sun's position, although it is largely displaced, and moves from year to year, in 1991-1993.

The continuous observation of the Sun's shadow with the Tibet-II array can provide a new clue for understanding the three-dimensional configuration of the global solar magnetic field under the influence of the coming solar activity cycle 23 with a predicted maximum around the year 2000.

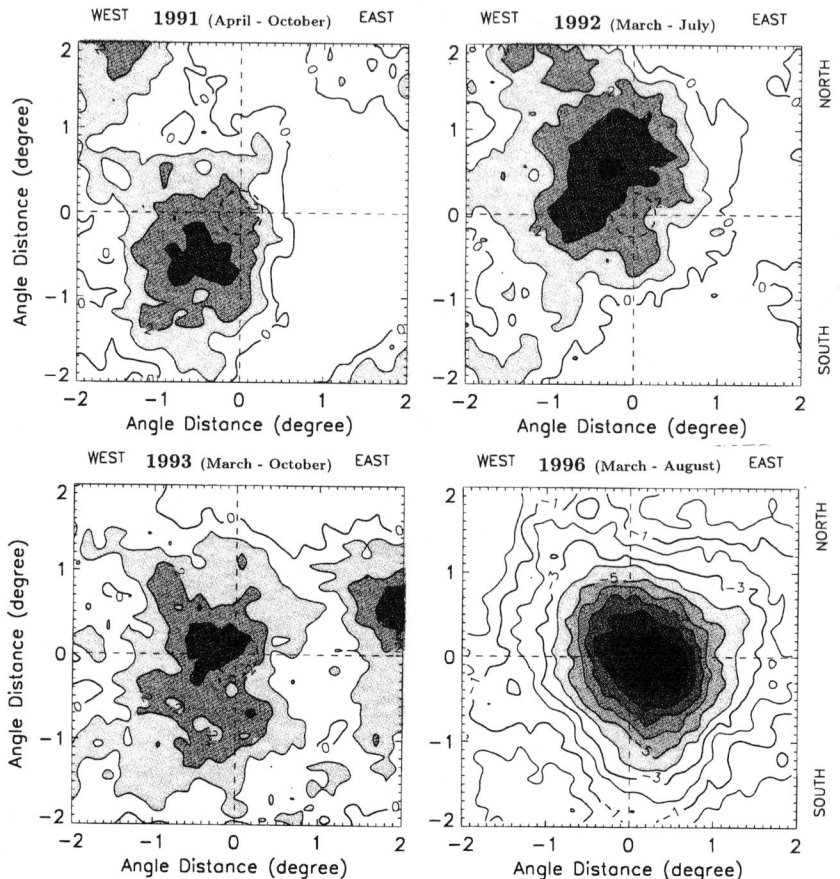

Fig. 1 Yearly variation of the position of the Sun's shadow from 1991 to 1996.

References

Amenomori M. et al. 1992, Phys. Rev. Lett., **69**, 2468.
Amenomori M. et al. 1993, Phys. Rev., **D47**, 2675.
Amenomori M. et al. 1994, ApJ, **415**, L147.
Amenomori M. et al. 1996, ApJ, **464**, 954.
NASA 1997, National Space Science Data Center, (URL http://nssdc.gsfc.nasa.gov/omniweb/).
NOAA/USAF 1997, Space Environmental Information Center, Solar Geophysical Data, (URL http://www.ngdc.noaa.gov/stp/stp.html).

NUMERICAL SIMULATION OF TWISTED SOLAR CORONA

S. PARHI[1], B. P. PANDEY[2], M. GOOSSENS[3] and G. S. LAKHINA[2]

[1] *Communications Research Laboratory, Tokyo 184, Japan*
[2] *Indian Institute of Geomagnetism, Bombay 400 005, India*
[3] *Katholieke Universiteit Leuven, Leuven, Belgium*

Introduction

The solar corona supports a variety of waves generated by convective upwelling motion in the photosphere. In order to explain the observed coronal temperature profile, resonant absorption of MHD waves by coronal plasma (Goossens et al., 1995) has been proposed as a possible candidate. The physical picture is that the footpoint motion in the photosphere constantly stirs the coronal plasma leading to the MHD wave generation which is then resonantly absorbed producing the enhanced heating of the corona. Here we consider the problem of MHD wave propagation in a twisted solar corona.

Governing Equations and Boundary Conditions

The coronal plasma in cartesian geometry obeys the usual compressible, time-dependent and resistive equations. The equations are solved in a rectangular spatial domain $D = [-5a, +5a] \times [-8a, +8a]$, using a 2.5D resistive MHD code. Rigid wall and free-slip boundary conditions are considered at the top and bottom of the simulation regions: $V_y = 0$ and $\partial f/\partial y = 0$ for $f = \rho, p, \mathbf{B}, V_x$, and V_z. At the remaining (x) boundaries open (zero gradient) conditions for all of the variables are considered (Parhi et al., 1996; Parhi et al., 1997).

We present all our numerical results for the Lundquist number $aV_a/\eta = 2000$ and plasma β varying from 5×10^{-3} to 0.95. The equilibrium density $\rho_0(x) = \rho_e + (\rho_i/\rho_e - 1)\rho_e \mathrm{sech}^3(|x| - a)$ for $|x| > a$ and $\rho_0(x) = \rho_i$ for $|x| \leq a$ with $\rho_i/\rho_e = 4$ and $V_{ae}/V_{ai} = 3$. $B_{oy}(x)$ has similar structure and $B_{0z}(x) = cxe^{-x^2}$, where c is the twist parameter. At $x = -8a$, V_z is perturbed by $F_z(x,y,t) = F_d(x)\sin(\omega_d t)$, where $F_d(x)$ has similar structure as $\rho_0(x)$. The following is considered: $F_d = 0.02\,\rho_e V_{ae}^2/a$, $c = 0.06\,V_{ae}\sqrt{\mu\rho_e}$ and $\omega_d = 0.5\,V_{ae}/a$.

Numerical Results

The driver excites Alfvén wave which in turn excites the magnetosonic waves. Consequently, these waves are resonantly absorbed in the corona leading to coronal heating. The evolution of plasma through sausage or kink modes depends on the amount of shear in the magnetic field (the following figure at $t = 200\ a/V_{ae}$). As plasma evolves the current sheets which provide the heating at the edges are distorted and fragment into two current sheets at each edge which in turn come closer when the twist is enhanced. The twist facilitates the formation of current sheet which produces the heating. Increase in the shear of the magnetic field causes the resonance layer to shift to the middle of the coronal tube. Steep gradients in slow wave at the slab edges which are signature of resonance layer where dissipation takes place are observed. It is observed that the thickness of the Alfvén resonance layer is more that that of the slow wave resonance layer. Effort is made to distinguish between slow and Alfvén wave resonance layers.

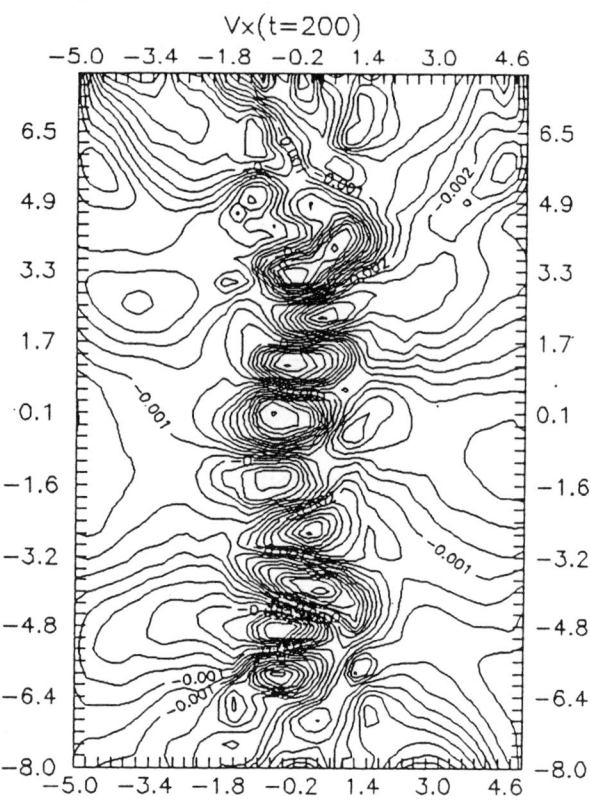

References

Goossens, M., Ruderman M. and Hollweg, J. V.: 1995, *Solar Phys.*, **157**, 75.

Parhi, S., De Bruyne P., Murawski, K., Goossens, M. and DeVore, C. R.: 1996, *Solar phys.* **167**, 181.

Parhi, S., Pandey, B. P., Goossens, M., Lakhina, G. S. and De Bruyne P. : 1997, *Astro. Space Sci.*, in press.

CORONAL HEATING MECHANISM IN THE PRESENCE OF A FLOW: A NUMERICAL APPROACH

S. PARHI and T. TANAKA

Communications Research Laboratory, Tokyo 184, Japan

Introduction

The solar corona is a hot, tenuous plasma permeated with the structured magnetic fields. A variety of waves is generated in the corona due to the convective upwelling motion in the photosphere. The excitation of MHD fluctuations is generated by the footpoint motion of the field lines in the photosphere. Resonant absorption of the Alfvén waves in an inhomogeneous plasma has been suggested as a means of driving current and plasma heating in the corona (Sakurai et al., 1991). We study this problem in the presence of flow.

Simulation Model

The coronal plasma in cartesian geometry obeys the usual compressible, time-dependent and resistive equations. The y direction is along the height of the loop, the z direction corresponds to the azimuthal direction and the inhomogeneity occurs in the x direction. Rigid wall and free-slip boundary conditions are considered at the top and bottom of the simulation regions: $V_y = 0$ and $\partial f/\partial y = 0$ for $f = \rho, p, \mathbf{B}, V_x$, and V_z. At the remaining (x) boundaries open (zero gradient) conditions for all of the variables are considered (Murawski et al., 1996; Parhi et al., 1996).

We present all our numerical results for the Lundquist number $aV_a/\eta = 10^4$ and plasma β of 0.04 far outside the slab. The equilibrium density $\rho_0(x) = 1 + sech^3(x/5)$ and the magnetic field $B_{0y}(x) = 1 - sech^3(x/5)$. The only flow $V_{0y}(x)$ has similar structure as density. The magnetic field $B_{0z}(x)$ along the z direction is taken as $B_{0z}(x) = cxe^{-x^2}$, where c is an amplitude factor to be chosen. The plasma undergoes twist due to this magnetic field. At the bottom of the simulation region we superimpose on V_z the perturbation $F_z(x,y,t) = F_d(x)sin(\omega_d t)$, where $F_d(x) = sech^3(x/5)$. The following is considered: $F_d = 0.01\ \rho_e V_{ae}^2/a$, $c = 0.05\ V_{ae}\sqrt{\mu\rho_e}$, $\omega_d = 0.9\ V_{ae}/a$ and $V_{0y} = 0.1\ V_{ae}$.

Simulation Results

We see some signature of kink instability and traces of logarithmic singularities at early stage of evolution. Further temporal evolution completely removes the singularity. We infer that the flow inhibits the development of kink instability. We see that the presence of flow facilitates heating. The inclusion of a small non-zero V_y has tremendous effect on resonance absorption manifested in the large change of the vortex structures. The center of the simulation loop appears consisting of elongated plasma vortices (the following figure at $t = 177 \ a/V_{ae}$) which break down on further evoluion. The microstructures thus formed are the possible signatures of the direct cascade of energy. The flow brings more or less symmetric distribution of heating. Attempt is made to correlate vortex formation and heating pattern.

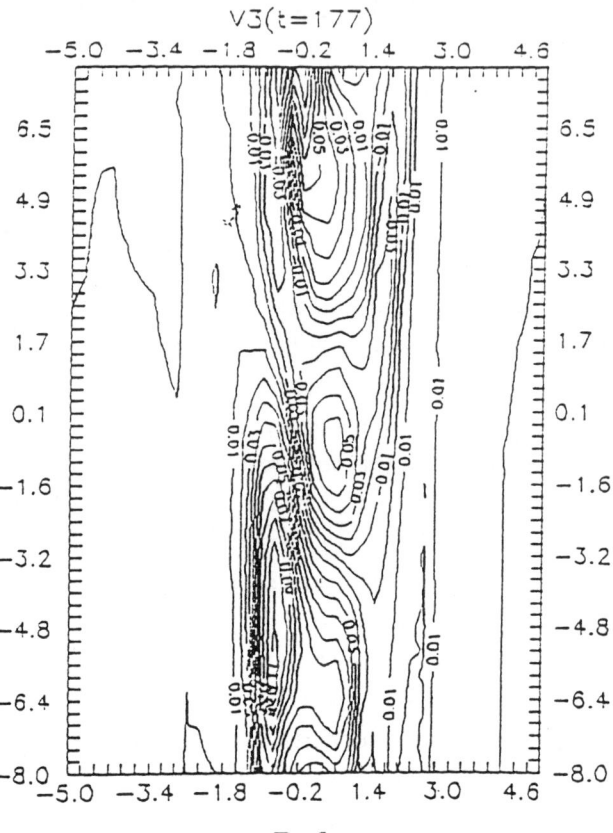

References

Murawski, K., DeVore, C. R., Parhi, S. and Goossens, M.: 1996, *Planet Space Sci.* **44**, 253.

Parhi, S., De Bruyne P., Murawski, K., Goossens, M. and DeVore, C. R.: 1996, *Solar phys.* **167**, 181.

Sakurai, T., Goossens, M. and Hollweg, J. V.: 1991, *Solar Phys.* **133**, 227.

SYNTHESIS OF SOLAR-STELLAR SEISMOLOGY — MEETING SUMMARY

S. M. CHITRE
Tata Institute of Fundamental Research
Homi Bhabha Road, Mumbai 400005, India; and
Queen Mary and Westfield College, Mile End Road, London, U.K.

1. Introduction

The Sun has generally been described as the Rosetta Stone of astronomy. This label is especially apt because our star is sufficiently close to Earth for a detailed scrutiny of its surface, a luxury that is not readily available for distant stars. It is our fond hope that the interior of the Sun and stars can be used as an astrophysical laboratory for testing theories of atomic and nuclear physics, high-temperature plasma physics, magnetohydrodynamics and neutrino physics.

The internal layers of the Sun are clearly not directly accessible to observations; nonetheless, it is possible to construct a reasonable picture of its interior. This can be accomplished with the help of a set of mathematical equations governing the mechanical and thermal equilibrium of the Sun, along with the boundary conditions provided by observations such as the mass, radius, luminosity and chemical composition.

The standard solar model is constructed with a number of simplifying assumptions, based on spherical symmetry without the inclusion of effects of rotational and magnetic forces on its structure. It adopts standard nuclear and neutrino physics, uniform initial chemical composition with the solar age of 4.6 billion years. A description of the convection zone is incorporated in the framework of local mixing-length theory with practically no overshoot into the adjoining stable layers. It also includes the gravitational settling of helium and heavy elements from the convection zone into the radiative interior.

The theoretical description of the Sun's interior is largely based on extensive numerical solutions of the structure equations, coupled with input physics pertaining to the opacity, equation of state and nuclear energy generation. This determines the physical conditions inside the Sun such as the profiles of the sound speed, density, temperature, chemical composition etc. The broad picture of the sun that has emerged is the following: The Sun is powered by thermonuclear reactions converting hydrogen into helium mainly by the proton-proton chain that operates in its energy-generating core which has a temperature upwards of 15 million K and a density close to 150 g cm^{-3}. The radiant energy slowly makes its way out of the central regions and in the outer third of the solar interior the energy transport is largely by turbulent convection.

The crucial issue is whether there is any way of checking the correctness of these numerically constructed models describing the internal constitution of our Sun!

2. Windows on the Sun's interior

There are at least two diagnostic probes available for looking inside the Sun:
a) Solar neutrinos,   b) Solar seismology

Valiant attempts have been made since the nineteen sixties to measure the flux of neutrinos released in the Sun's energy-generating core (cf., Bahcall 1989). This diagnostic probe designed for the exceedingly difficult measurement of neutrinos from the thermonuclear reaction network sustaining the solar luminosity, is expected to provide a handle on the temperature and chemical composition prevailing in the central regions of the Sun. Davis's radiochemical chlorine experiment was the first effort to make this exceedingly difficult experiment of neutrino fluxes where the ^{37}Cl nuclei in a tank containing liquid perchloroethylene act as solar neutrino absorbers. The measured solar neutrino counting rate by Davis is reported to be 2.55 ± 0.25 SNU (SNU = 10^{-36} captures per target atom per second) which clearly shows a puzzling deficit by about a factor of 3 over the predicted capture rate of 9.3 ± 1.2 SNU calculated by Bahcall and Pinsonneault (1995) for a standard solar model using OPAL opacities (Iglesias and Rogers 1996) and diffusion of heavy elements.

The Kamiokande experiment with its threshold of 7.5 MeV is sensitive to the ^{8}B neutrinos generated in the nuclear reaction network. The measured flux of these high energy neutrinos by Superkamiokande is reportedly $(2.44 \pm 0.25) \times 10^6$ cm^{-2} s^{-1}, while the expected flux according to the standard solar model is 6.62×10^6 cm^{-2} s^{-1}. Unlike the foregoing experiments which are insensitive to the low-energy pp neutrinos, two current radiochemical experiments which use gallium detectors are capable of measuring the lower end of the solar neutrino spectrum. The reported measurements of the neutrino counting rate are 69.7 ± 8 SNU for GALLEX and 69 ± 13 SNU for SAGE (cf. Takata, these proceedings), while the theoretically predicted capture rate is 137 ± 7 SNU (cf., Bahcall and Pinsonneault 1995). There is, thus, a clear and persistent discrepancy between the measured and predicted neutrino fluxes, assuming the neutrinos to have standard physical properties (i.e., no mass, no flavour-mixing, no magnetic moment). This has cast doubts on the reliability of standard solar models, and has prompted solar physicists to look for some independent, complementary probe to determine physical conditions in the solar interior.

The Sun's surface undergoes a series of mechanical vibrations which manifest as Doppler shifts oscillating with a period centred predominantly around five minutes (Leighton, Noyes and Simon 1962). A theoretical explanation of these oscillations was given by Ulrich (1970) and independently, by Leibacher and Stein (1971) emphasizing the role of the interior acting as an acoustic cavity for the waves excited and trapped inside the solar body. Later, Deubner (1979) made accurate measurements of the periods and horizontal wavelengths of the five-minute oscillations to derive a power spectrum showing narrow ridges. These global oscillations are superpositions of millions of resonant normal modes which sample different layers of the Sun and carry information about the hidden solar layers, in much the same way as geoseismology reveals the structure of the earth's interior. This seismic tool, provided by a rich spectrum of velocity fields observed at the solar surface, with frequencies determined to a precision better than 1 part in 10^5, has indeed revealed the physical make-up of the Sun's interior with tantalizing precision.

The progress and puzzles in helioseismology were outlined by Tim Brown who emphasized the need to study the line asymmetries, shapes and their shifts, particularly for locating the source of excitation for these oscillations. The observational input for helioseismic measurements comes from ground-based networks and space-based instruments:

RESOLVED SUN	SUN AS A STAR
GONG	BISON
LOWL	IRIS
Mt WILSON	SOUTH POLE
SOHO (MDI)	SOHO (GOLF & VIRGO)

Elsworth stressed the overwhelming advantages in having continuous data sets from these varied observations and recommended a search for low degree modes, preferably of low order, in order to gain more accurate information about the central regions of the Sun. Frank Hill described the strategies for data analysis and drew particular attention to the possible pitfalls in the reduction of massive datasets, highlighting the many steps that go into the data processing such as calibration, proper mode identification, image geometry determination, mode parameters (e.g. amplitudes, frequencies, linewidths) and special care for considering spatial leaks structure, cross-talk, multi-site merging, etc.

The accurately determined helioseismic data can then be analyzed in two ways:
i) Direct model fitting (Forward method)
ii) Inversion method.

In the Forward method, a set of solar models is constructed with one or more adjustable parameters which are perturbed to obtain the linear eigenfrequencies of oscillations for comparison with the observed p-mode frequencies. The fit cannot, in practice, be perfect, but such an approach indicates the depth of the outer convection zone to be $\simeq 200,000$ km, deeper than previous estimates and the helium abundance by mass in the envelope to be $\simeq 0.25$. An interesting outcome of such an approach is that the computed opacities near the base of the convection zone ($T \gtrsim 2 \times 10^6$ K) are found to be too low, a situation later modified with more up-to-date Livermore calculations. (Iglesias and Rogers 1996).

The inversion methods which have been effectively developed for inferring the physical conditions inside the Sun with the accurately available p-mode data, were described by Thompson. Most of these inversions are based on some kind of a linearization process and have resorted to one of the following methods: Regularized Least Squares, Optimally Localized Averaging Kernels, Asymptotic Representation, with some ad hoc trade-off parameter. A major accomplishment of the inversion technique has been a determination of the sound speed throughout much of the solar interior to a precision of better than 0.1% and the density to a somewhat lower accuracy (Gough et al. 1996). Fig. 1 shows the relative difference in the sound speed between the Sun and the standard solar model with gravitational settling of helium and heavy elements for the GONG, MDI and BBSO + BiSON datasets. The dominant features in the sound speed difference, which are shared by essentially all the data sets, occur in regions of the model with substantial composition gradients such as in the nuclear burning core and beneath the convection zone from which helium and heavy elements diffuse into the interior.

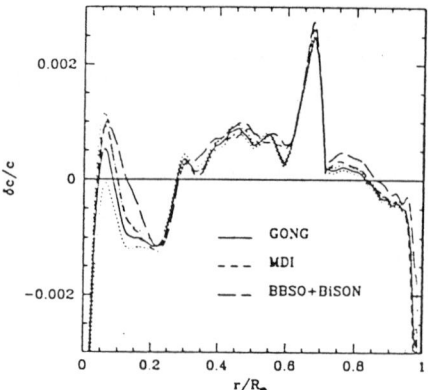

Figure 1. Relative difference in sound speed between the Sun and the standard solar model of Christensen-Dalsgaard et al. (1996) as inferred using various sets of observed frequencies.

3. Helioseismic constraints on solar structure

Vauclair summarized the principal constraints on the internal structure of the Sun inferred from helioseismic inversions.

1. The base of the convection zone: $r = (0.713 \pm 0.001) R_\odot$,
2. Extent of the overshoot beneath the convection zone: $d \lesssim 0.05 H_p$,
3. Helium abundance in the solar envelope, $Y = 0.24 - 0.25$. Evidence for gravitational settling of helium and heavy elements out of the envelope and resultant chemical segregation,
4. Revision of input microphysics: opacities and equation of state (OPAL), nuclear reaction rates and Canuto-Mazzitelli prescription for convection,
5. Detection of boundaries of convection zones and seismic signatures of the HeII ionization zone.

Chaboyer discussed the role of lithium as a tracer of mixing in stars, particularly in the context of the Sun where the surface abundance of lithium is lower by about a factor of 140 compared to the meteoritic value. This can be attributed to the lithium depletion due to a combined action of rotationally induced mixing and diffusion at the base of the convection zone. Such a mixing would result in a flattening of the composition gradient beneath the convection zone and this should lead to a reduction in the pronounced bump in $\delta c/c$ around $r = 0.67 R_\odot$ in Fig. 1.

The acoustic structure, namely, $c(r)$ and $\rho(r)$ obtained from the primary inversion of the accurate helioseismic data, can be profitably used to infer the thermal and composition profiles in the solar interior. For this purpose, of course, we need to employ the equations of thermal balance and energy transport, together with the auxiliary input provided by the opacity, nuclear energy generation rate and thermodynamic state of matter. Three approaches have been adopted in the literature for constructing seismic models (cf., Antia and Chitre 1995; Kosovichev 1996; Shibahashi and Takata 1996; Roxburgh 1996), and once the run of temperature, density and chemical composition is determined, it becomes possible to calculate the expected neutrino fluxes for these models. Remarkably, with the allowance of arbitrary variations in the input opacities and even relaxation of the thermal equilibrium condition, it turns out to be difficult to produce a seismic model that is simultaneously consistent with any two of the

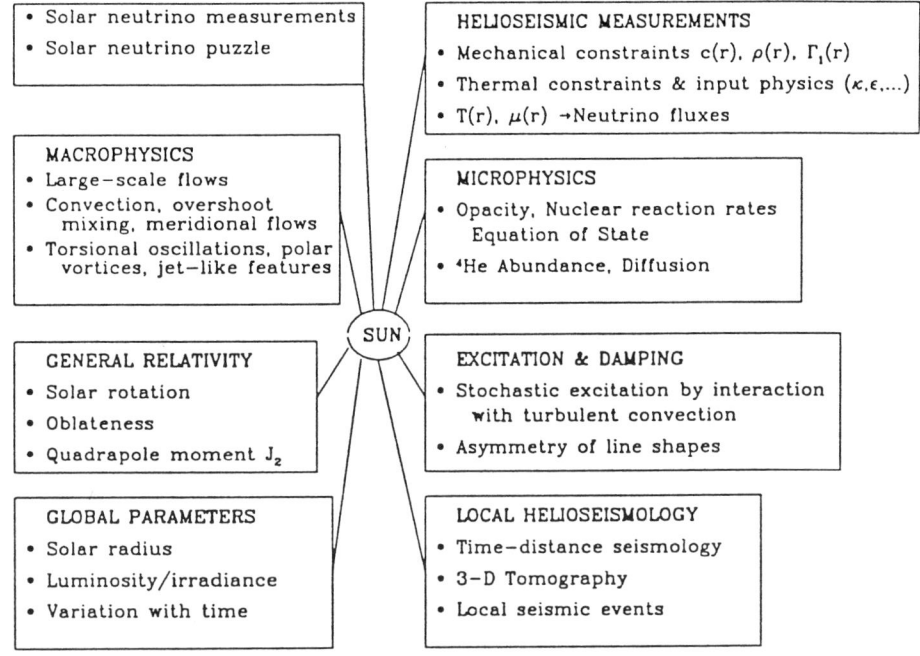

Figure 2.

existing solar neutrino experiments (Antia and Chitre 1997). A possible resolution of the solar neutrino puzzle should, therefore, be sought not in the astrophysical domain, but in the framework of non-standard neutrino physics, cf. the MSW effect (Bahcall and Bethe 1990). The various windows to look into the solar interior are schematically illustrated in Fig. 2.

4. Hydrodynamical flows and internal rotation

Convection is the primary mode of energy transport in the regions just below the solar surface which extends to a depth of $0.287 R_\odot$. Ulrich stressed the role of rotation in modifying the large-scale convective flow and redistributing the kinetic energy, and of the magnetic fields in redirecting the energy flow with the resultant changes in the output of solar radiation. Basically, the convection zone may be regarded as a reservoir for the energy storage through changes in the solar radius. From the MDI and the Mt Wilson supersynoptic chart data, Ulrich furnished evidence for long-duration convective patterns highlighting the striking east-west velocity flows that drift towards the equator. The suggestive existence of long-lifetime convective flows (duragranulation), intermediate in scale between supergranulation and giant cells, and their possible interaction with the rotation field was discussed by Ulrich for chanelling energy into east-west flows. Schou also reported a signature for such torsional oscillation patterns from the inversion of f-mode eigenfrequency splittings, and pointed out that magnetic fields do not seem to play much role in driving these flows.

Ayukov pointed out the importance of including the effects due to non-adiabaticity

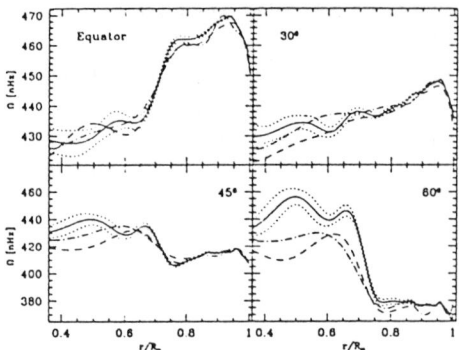

Figure 3. Rotation rate inside the Sun at different latitudes as inferred using various sets of observed frequencies. The continuous line represents the results using MDI data, the dashed line those using the GONG data while the dot-dashed line is for BBSO+BiSON combined data. The dotted lines represent the 1σ error limits on MDI results.

and contributions from the turbulent pressure for better agreement between theoretical treatment of the structure of the outer convection zone and observed oscillation frequencies. With the imposition of the helioseismic constraints (sound speed and state of matter) Shibahashi obtained the mechanical and thermal properties of the solar convective envelope for deducing the depth of the convection zone and the helium abundance. Roxburgh emphasized the importance of convective overshooting and mixing in the context of solar and stellar evolution, especially since penetration from convective cores into adjoining stable regions can have a non-negligible influence on the structure as well as the evolutionary properties.

It is now widely accepted that the observed solar p-mode oscillations are stochastically excited by turbulent convection in the sub-surface layers of the Sun. Goldreich discussed the approximate equipartition between the energies of acoustic modes and resonant turbulent eddies with lifetimes comparable to the mode periods. He also drew attention to the role of this convective driving mechanism operating in pulsating white dwarfs (ZZ Ceti stars). In his discussion of the excitation and damping of p-modes, Nordlund reported the results of his numerical experiments to highlight the role of entropy fluctuations associated with convective downdrafts in driving the oscillations by overcoming the damping due to Reynolds stresses (i.e., turbulent pressure compared to which viscous and radiative damping are small). The study of line asymmetries and widths in the power spectrum of oscillations was stressed by Nigam for diagnosing the physics of excitation and damping of solar p-modes.

Helioseismology has made it possible to probe the variation with depth and latitude of angular velocity in the sun's interior. The inversions applied to the MDI splitting data yield a rotation profile which is in broad agreement with the general picture of solar internal rotation earlier inferred from the GONG, BBSO/BiSON splitting data, as can be seen from Fig. 3. The detailed inference about the internal rotation and large-scale flows from the MDI data was summarized by Schou: The observed latitudinal differential rotation persists through the convection zone with the angular velocity practically constant on surfaces of constant latitude, while the radiative interior, rotates almost uniformly at a value close to 0.93 of the equatorial angular velocity. There is a radial shear layer (tachocline) at the transition between

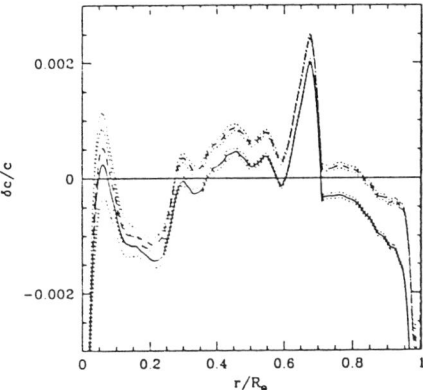

Figure 4. Relative difference in sound speed between the Sun and the standard solar model of Christensen-Dalsgaard et al. (1996) as inferred using different values for solar radius. The continuous line represents the results using $R_\odot = 695.78$ Mm, while the dashed line represents those using $R_\odot = 695.99$ Mm. The dotted lines show the 1σ error limits.

the convection zone and the radiative interior the structure of which is yet to be resolved. A quite striking feature of the rotation profile is the presence of bumps and wiggles suggestive of zonal bands of fast and slow rotation reminiscent of torsional oscillations migrating from high to low latitude during a solar cycle. A noteworthy observation reported by Schou was a hint of a jet-like feature at 75-degrees latitude located at a radial distance of $0.95 R_\odot$ and also a very slowly rotating structure near the polar regions, a polar vortex extending to a significant depth inside.

5. Global properties of the Sun

The fundamental mode (f-mode) provides a valuable diagnostic probe of the flows, rotation rate and magnetic fields present in the surface regions of the Sun. Interestingly, the accurately measured f-mode frequencies from the GONG/MDI datasets can also provide a handle to infer a reasonably accurate measure of solar radius (Antia 1997, Schou et al. 1997). The f-mode frequencies are largely independent of the stratification in the solar surface layers and asymptotically satisfy the dispersion relation (Gough 1977): $\omega^2 = gk(1+\epsilon)$, where ($g = GM/R^2$ and $k = \sqrt{\ell(\ell+1)}/R$) to give $\delta R/R \sim -2/3\, \delta\omega/\omega$. It turns out that the observed f-mode frequencies are approximately 0.05% higher than those predicted by the standard solar model with $R_\odot = 695.99$ Mm. This implies a decrease in (seismic) solar radius in the range of 200 − 300 km. Similar conclusions are reached by Brown and Christensen-Dalsgaard (1997) from measurements of the solar diameter. The possible errors resulting from uncertainty in the value of the solar radius will naturally be reflected in helioseismic inversions. Fig. 4 shows the relative difference in sound speed between the Sun and the current standard solar model of Christensen-Dalsgaard et al. (1996) with two values of the solar radius, 695.780 Mm and 695.99 Mm. It is clear that the difference in the sound speed due to a decrease in the solar radius by 210 km is more than the estimated inversion errors through the bulk of the solar interior.

With the availability of better data through the present solar activity we hope f-mode frequencies and their temporal variation can furnish us with independent estimates of not only the seismic radius of the Sun, but also its oblateness and their solar cycle related variations. An important outcome of the inferred differential rotation profile of the Sun is that the quadrupole moment, J_2 of the solar gravitational potential caused by the centrifugal force is about $1.7 \times 10^{-7} \, GM/R^3$, consistent with the prediction of General Relativity and also the radar ranging measurements.

Fröhlich presented a composite account of the total solar irradiance measurements, during the past 17 years from NIMBUS, ACRIM and VIRGO showing variations over all time scales ranging from days to years. Over the solar activity cycle, the total irradiance variability is seen to be about 0.1%, with the enhancement at the level of $\sim 0.2\%$, contributed by faculae, bright networks etc., while the decrease of 0.1 % is on account of the overall darkening effects of sunspots. The total irradiance changes appear to be well correlated with the p-mode frequency variation with the solar cycle. Theoretical explanations of the observed solar irradiance were discussed by Spruit. He advanced plausible arguments about the role of the convective envelope acting as a reservoir for thermal energy and the irradiance being modulated largely by the surface effects arising from sunspots and small-scale magnetic fields.

6. Small-scale structures in the Sun: local helioseismology

It can be readily appreciated that global helioseismology has furnished a considerable amount of information about the solar interior. Clearly, it is important to enquire if we can attempt a scaled-down version of the inversion techniques for probing the small-scaled localized structures.

The time-distance helioseismology is a newly developed technique to construct three-dimensional tomographic inversion of the solar internal layers. The remarkable results obtained with this powerful method were described by Duvall and Kosovichev in (i) tracking subsurface meridional flows, (ii) revealing flow-fields on supergranular scales, iii) mapping of converging and diverging flows round sunspots, (iv) providing an effective measure of the antisymmetric component of solar differential rotation (asymmetry in rotation profiles in the two hemispheres not revealed by traditional technique based on inversion of frequency splittings).

This is a very powerful method for probing the localized structures in the sun and has clearly a great potential, although the validity of the ray approximation and aspects relating to changes in path lengths and wave speeds need to be investigated further.

Goode described the application of local helioseismology to solar flares and supersonic seismic events occurring in dark intergranular lanes that pump energy into p-modes. The measurements of seismic response of such localized events will clearly provide valuable information about the underlying flare mechanism and also about the subphotospheric structure of active regions.

Deubner presented a great body of information on measurements of amplitudes and phases of intensity fluctuation and Doppler shift due to wave motions as a function of frequency and wave number. This information is evidently important for studying properties of solar atmospheric oscillations and for examining the chromospheric structure. An important diagnostic of the solar chromospheric thermal stratification was provided by Ayres who argued, from observations of the strong CO lines, for the inhomogeneous structure of the chromosphere with its lower layers made up of cool gas and pervaded by network of small-scale magnetic fields . The recently reported SOHO observations of the "Sun's magnetic carpet" in the form of a sprinkling of more than tens of thousands of magnetic field concentrations spread over the sun's surface, will almost certainly have a significant influence on the chromospheric structure. The calculations reported by Banerjee on the influence of magnetic fields on the nature of oscillations in a stratified atmosphere assume importance for studying the effect of magneto-atmospheric waves on spectral line profiles.

7. Asteroseismology

The objective of asteroseismology is to construct seismic stellar models to test theories of internal structure, evolution and dynamics of stars. Asteroseismology is still in its infancy with only a limited number of modes, highly reminiscent of the situation for the Sun in the 1980s when only a very few modes were available from whole-disk measurements, in contrast to the luxury of accessing millions of modes which the current helioseismic database is endowed with. But stellar astronomy has revealed the g-mode oscillations which have been elusive in the Sun, and has supplied p-modes with a range of quantum numbers (n, l, m), which characterize the modes, for a wide class of stars. In order to understand the physical processes inside stars measurements of oscillation frequencies of a variety of stars with different masses and ages are desirable. The new developments in asteroseismology summarized by Kurtz and the theoretical aspects outlined by Christensen-Dalsgaard provided a very admirable overview of this rapidly growing field. Table 1 summarizes some of the principal features of stellar pulsators discussed at the meeting.

There is a clear need for sets of oscillation frequencies for a wide range of stars in the H-R diagram with different masses, spectral types and at different stages of stellar evolution. The information about the frequencies, amplitudes and line shapes, and their associated temporal information evidently hold the key to the internal structure and dynamics of stars. A real asset of asteroseismology is the g-modes which have large amplitudes in deep interiors and are evidently better suited than p-modes to probe the stellar cores. Equally, the frequencies of g-modes are sensitive to the mean molecular weight gradient of the layers and, as stressed by Roxburgh, these modes have the potential of diagnosing the chemically inhomogeneous central regions of stars. Furthermore, the measurements of oscillation frequencies will enhance our knowledge about how the convective efficiency (the mixing length parameter) varies with the spectral type. The width of the main sequence, for example, is extended by overshooting from the convective cores, as also the ages are extended by overshooting. The resonant modes of stellar oscillations will have a broad range of characteristic spatial scales; only the relatively low-degree modes which penetrate into the stellar cores can be observed. Presently, it may not be possible to infer the internal constitution of stars, in anywhere near as much precision as has been possible for the Sun. Nonetheless, the measurements of oscillation frequencies will provide reasonable constraints on the sound speed profiles, especially in the deep interiors of stars and in some rapid rotators even the internal rotation profiles from the splitting data. The locations of convective boundaries and the helium ionization zone can indeed be established seismically from the characteristic oscillatory signals in the frequencies resulting from discontinuities in the sound speed and its derivatives (cf. Monteiro, Christensen-Dalsgaard and Thompson, these proceedings).

The asteroseismic measurements should provide information about the surface and internal differential rotation of stars. For example, the large-amplitude δ Scuti pulsators would enable us to determine the radial as well as the latitudinal variation of the angular velocity. The asteroseismological studies should be complemented by

TABLE 1. Stellar Pulsators

STAR (type)	PERIOD RANGE	MODE(s)	
Sun(G)	2–6 min	p	Stochastically driven in the outer convection zone (Goldreich)
Solar-type stars (Procyon, η Boo, Arcturus, α Cen)(G)	$\gtrsim 10$ min	$p\,(n \gg 1)$	No confirmed detection of sun-like oscillations. Spectrum of regularly spaced modes in outer layers (Bedding)
White dwarfs (ZZ Ceti) (DO, DA, DB)	2–15 min	$g\,(n \gg 1)$	Overstable g-modes driven by stochastic excitation in the shallow convection zone. Seismic determination of white dwarf masses and confirmation of gravitational diffusion of heavy elements. Radial velocity variation in WD(G29–38) observed to have time-scale nearly same as intensity pulsation period (Clemens, Kawaler)
Rapidly oscillating Ap stars (F, A)	5–15 min	$p\,(n \gg 1)$	Driving due to kappa-mechanism or overstable magnetic convection! Atmospheric $T(\tau)$ relation influenced by magnetic fields. Complementary information for seismic studies from Hipparcos (Mathews)
Delta Scuti stars (F5–A2)	0.5–7 hrs	p, g	Bridge between Sun and higher mass evolved stars. Multiperiodic with dozens of acoustic & gravity modes driven by opacity mechanism operating in H, He ionization zones. Seismic potential for probing core overshoot and internal differential rotation and age determination (Breger, Guzik)
Beta Cephei stars (B1 - B2)	2–6 hrs	p, g	Metal-opacity driven multi-mode pulsators. Potentially useful for inferring metallicity and ages of young star clusters, and also for probing internal rotation and convective penetration. (Eyers)
Slowly pulsating B stars (B3 - B9)	1–4 days	$g(n \gg 1)$	Metal-opacity driven pulsators. Over 100 new periodic B stars reported by Hipparcos. (Dziembowski, Eyers, Aerts)
γ Doradus-type stars (Early F)	1–2 days	$g(n \gg 1)$	Over 2 dozen early F-types with non-radial g-mode pulsations of low degree, probably stochastically driven. Relevant to the radial variation of solar-type stars for for unambiguous signature in planetary searches (Krisciunas)
Hot B subdwarfs (B)	120–500 sec	p, f	Newly discovered multiperiodic pulsators driven by metal opacity mechanism. Peculiar chemical abundance, diffusion (Stobie, Fontaine)

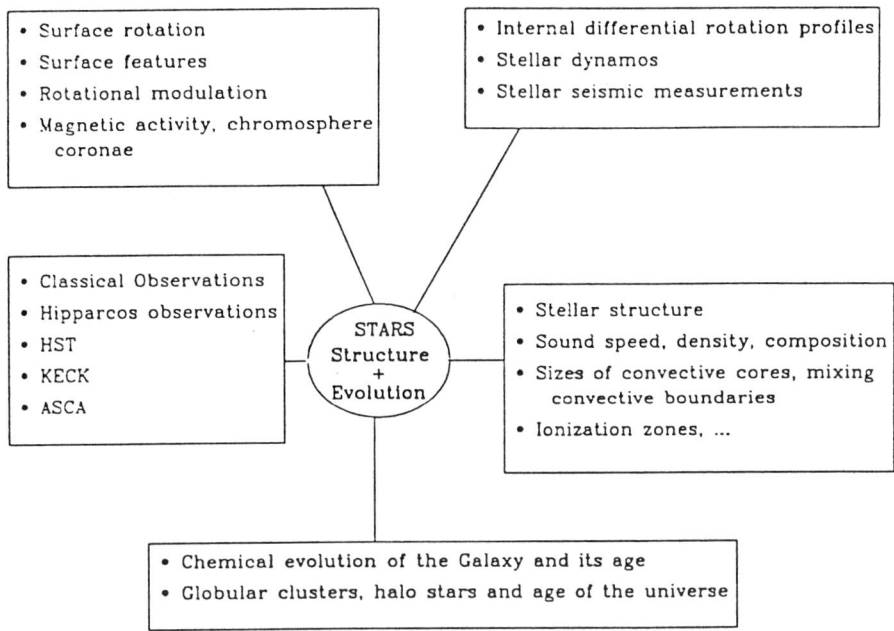

Figure 5.

monitoring activity in the sun-like and late-type stars. The rotational modulation observed in the far UV is a fairly sensitive measure of stellar rotation periods, as UV indicators are useful for tracking magnetic activity on stars. The rotation rate, and hence the activity, is found to decrease with age because of the loss of angular momentum via the magnetic wind, giving an indirect handle on the stellar age. Clearly, measurements of gross radial rotational shear are crucial for locating the seat of the magnetic dynamo operating in sun-like stars.

The information gained from the oscillation frequencies is, of course, complementary to the classical data about stars such as magnitude and colour-index. The H-R diagram for stellar oscillation frequencies introduced by Christensen-Dalsgaard (1988) shows the seismic parameters which are sensitive to different aspects of stellar structure. Thus, the large separation, Δ, is sensitive to the outer layers of stars, while the small separation, δ carries information about the core-structure. The data on Δ and δ can be profitably employed for improving our knowledge of convective core overshoot, especially the variation of the extent of penetration with stellar mass.

From values of the seismic parameters it should be possible to determine the mass and age of a star provided there is information available about the chemical composition and the convective efficiency. Recall, the low-degree data for massive stars contain information from the small separation, δ, about the discontinuity in the mean molecular weight and the density at the core boundary, a signal which is reflected in a discontinuity in the sound speed. Roxburgh emphasized the penetrative power of the seismic parameter, δ and noted how its oscillatory behaviour carries a signature of the convective core and its variation with frequency tracks the density profile in the central regions. Thus, stellar seismology would serve as a powerful diagnostic of the chemically inhomogeneous interior of stars which is a crucial input for stellar evolution.

Stars in a cluster or in a binary system are particularly valuable targets for seismology. The oscillation data along with the classical observations can be gainfully utilized for calibrating cluster parameters for improving estimates of mass and age of stars in clusters and for a better understanding of cluster distances. The hope is to have a better handle on the distance scale in the Galaxy and eventually the cosmological distances through designing more precise period-luminosity relations. The chemistry of the Galaxy and its chemical evolution would become amenable from the study of stars in various populations such as the disk, bulge and halo, through the seismic constraints imposed on helium abundance, heavy element diffusion, core sizes. Note that our current understanding of the chronology of the Universe is directed by stellar models with an uncertainty of a factor of 2 in the production rate of heavy elements. Consequently, at present asteroseismology, which seems to be the only tool available to look inside the stars, can assist in providing a better constraint on the age of the Galaxy, and hence on the value of the Hubble constant. Fig. 5 shows the central role of stellar structure and evolution in the context of astrophysical setting. The effectiveness of combining asteroseismic data with the classical observations with HST, KECK coupled with Hipparcos measurements for advancing the frontiers of astronomy is self-evident. What we need to develop are reliable methods for mode identification, and this can be accomplished with coordinated multicolour photometry and spectroscopy from networks distributed round the globe (cf WET, δ Scuti campaigns). For this purpose, the community must appeal for more multisite campaigns of long duration and for high precision photometric observations with satellite-borne instruments such as SOHO, COROT through this solar cycle and beyond. So be it!

I am grateful to H. M. Antia for insightful discussions and assistance. Thanks are also due to Ian Roxburgh, Franz Deubner, Don Kurtz and Douglas Gough for helpful advice and comments.

References

Antia, H. M. 1997, A&A (in press) astro-ph/9707226
Antia, H. M. and Chitre, S. M.: 1995, ApJ 442, 434
Antia, H. M. and Chitre, S. M.: 1997, MNRAS 289, L1
Bahcall, J. N.: 1989, Neutrino Astrophysics, Cambridge University Press, Cambridge
Bahcall, J. N. and Bethe, H. A.: 1990, Phys. Rev. Lett. 65, 2233
Bahcall, J. N. and Pinsonneault, M. H.: 1995, Rev. Mod. Phys. 67, 781
Brown, T. and Christensen-Dalsgaard, J.: 1997,
Deubner, F.-L.: 1979. A&A 44, 371.
Christensen-Dalsgaard, J.: 1988, in Advances in Helio- and Asteroseismology, eds. J. Christensen-Dalsgaard and S. Frandsen, Reidel, pp 295-298
Christensen-Dalsgaard, J., Däppen, W. et al. 1996, Science 272, 1286.
Gough, D. O., Kosovichev, A. G. et al: 1996, Science 272, 1296
Gough, D. O.: 1977, in The Energy Balance and Hydrodynamics of the Solar Chromosphere and Corona, Eds. R.-M. Bonnete and P. Delache, G. de Bussac, pp 3-36
Iglesias, C. A. and Rogers, F. J. 1996, ApJ 464, 943
Kosovichev, A. G.: 1996, Bull. Astron. Soc. India 24, 355
Leibacher, J. W. and Stein, R. F. 1971, Ap Lett 7, 191
Leighton, R. B., Noyes, R. W. and Simon, G. W.: 1962, ApJ 135, 474
Roxburgh, I. W.: 1996, Bull. Astron. Soc. India 24, 89
Schou J., Kosovichev A. G., Goode P. R., Dziembowski W. A., 1997, ApJ 489, L197
Shibahashi, H. and Takata, M.: 1996, PASJ 48, 377
Ulrich, R. K.: 1970, ApJ 162, 993

Index

α Cen A 285ff.
α Cir 242, 299f., 309
β Cephei stars 348
β Cephei
    asteroseismology 395
    seismic models 357
δ Scuti stars 250
    observations 323
    theory 331
δ Scuti 291
ϵ Per stars 349
γ Dor stars 293, 387
γ Dor 323
κ-mechanism 279
θ Tuc 337
10 TeV cosmic rays 465
12(DD)Lacertae
    rotation and pulsation 381
    rotational splitting 379f.
4 CVn 323, 335
51 Peg 233ff., 329
53 Per stars 349
9 Aurigae 325

A-type stars
    normal and peculiar 385
acoustic source 418
acoustic waves
    emission 213
    scattered 216
acoustic wavetrains 412
active region 173
adiabatic exponent 317
amplitude limitation 250
amplitude variations 323
angular momentum 40
aspect ratio 124
asteroseismology 245, 479
    "active" 261ff.
    definition 231f.
    space mission 301ff.

asymmetry
    spectral peaks 420
atmosphere
    isothermal 423ff.
atmospheric dynamics 427ff.
averaging kernel 129
avoided crossings 247

B stars
    amplitudes 355ff.
    instability strip 355
    nonlinear theory 359
    slowly pulsating 295ff.
Böhm-Vitense decrement 327
Balmer lines
    suppression of oscillation signal 375
Bayes's theorem 127
Be stars 351
beryllium 27ff., 135
big bang nucleosythesis 26ff.
bisector velocity 309
BiSON – LOI comparison 45
BiSON update 221
BiSON 169, 172

Ca K grains 427ff., 437
    formation 444
CaII resonance lines 437
canopy 406
carbon monoxide 403
CD-24° 7599
    rotational splitting 336
cell flashes 406
chemical composition 41
chromosphere 405, 421
CO Aur 389
CO off-limb emission 409
COmosphere 406
convection theory 115
convection zone 85

convection   73, 122, 217, 457
   large scale   59ff.
   observations   59ff.
convective boundaries in stars   315
convective cores   73ff.
convective envelope
   solar   81
convective flow   107
convective overshooting   75, 76, 333
corona   405
coronal heating
   simulation   467f., 469f.
COROT project   301ff.

data
   analysis   43
   for helioseismology   1ff.
   reduction (helioseismic)   13ff.
degradation of radiometers   90
differential response technique   80
differential rotation   142, 149
diffusion   27ff., 136, 281, 337
   impact on evolution   264
diffusive time scale   104
double-mode Cepheids
   Fourier decomposition   389
driving of oscillations   186
dynamic atmosphere diagnostics   441

EC 14026 stars   235ff.
   driving mechanism   369
   mode identification   364
   predictions   371
   theory   367
element segregation   135
equation of state   318
equivalent width
   cycle variability   459
ERBS satellite   90
error correlation   131
excitation
   overstability   249
   resonance   186
   stochastic   248
extrasolar planets   328

f-mode   165ff., 187
FG Vir   323
   metallicity   332f.
flow fields   175
flows
   large-scale   149
flux tubes   215, 406
frequencies
   combinational   214
   comparison   45
frequency separation
   large   246
   small   246
frequency splitting   142
frequency
   buoyancy   245
   dynamical   245
   Lamb   245

g-mode pulsations   323
g-modes   232
   δ Scuti   334
   solar   51, 55
GONG – LOI comparison   167
GONG   13, 49, 54
granular flow   451
granulation   106
   dynamics   451
granules   185, 217
GW Lib   233, 321f.

^{3}He/^{4}He ratio   135
H$_\alpha$ line   309
H$_\beta$ equivalent width
   fractional variation   378
   integrated light   376
Halo stars   31ff.
HB red variables   401
HD 128898   242
HD 134214   242
HD 164615   327
HD 224638   327
HD 24712   299f.
HeII ionization region   318
helium abundance   317

high-frequency interference peaks   416
high-frequency peaks   447
high-speed photometry   362
HIPPARCOS   232
    B stars   295
    survey   291
horizontal branch   365
horizontal velocities   458
hot subdwarfs
    pulsation   361
HR 1217   240
HR 2740   387
HR 3831   241
Hyades   28ff.

image geometry   15
infrared solar observations   403
instability strip   293
integral constraint   73, 76
internal gravity waves   456
internal structure
    solar   135
inversion   22, 25ff., 125
    oscillations   117
    resolution   129
    rotation   251
    trade-off curve   130
irradiance variations   103, 119
irradiance
    proxy model   98
    reconstruction   98
    total solar   89

jet
    high latitude   144

k-ϵ model   121

lambda-meter method   442
light elements   135
line asymmetries   195
line asymmetry
    $\ell = 2$ 170
line profile variations   387, 393
linearization in inversions   126

lithium abundance   25ff., 41, 135
LOI   43
long period variables   399
Lorentz force   169

magnetic activity   103, 224
magnetic field   423ff.
magneto-optical filter
    calibration   53
mass loss   337, 351
maximum likelihood estimation   43
McMath-Pierce telescope   408
MDI observations   157ff.
meridional flow   149
mesogranulation   457f.
metallicism   237f.
MHD oscillations
    observations   453
MHD waves   425
mixing length   74, 78
mixing processes   135
mixing-length theory   218
moat   108
mode identification   326, 332, 393
mode parameter estimation   17
modes
    low degree   173, 179
    mixed   247
molecular weight gradient   139
multicolour photometry   269

near-surface perturbation   174
neutrino
    solar   21, 42, 137
NGC 2516   325
NIMBUS-7 satellite   90
non-radial pulsations
    CV primary   321f.
    line profile   383
nonradial oscillations
    axisymmetric   278
NSO-FTS   408
numerical simulations   78, 107

OB stars   347
obliquely rotating core   37
optimal masks   179
optimally localized averages   129
oscillation
    excitation   457
oscillations in active regions
    observations   449
oscillations
    local excitation   461
    solar-like   285ff.
    solar   21
outbursts   351
overshoot layer   121
overstable convection   278

p-mode amplitudes   113
p-mode charateristics   153
p-mode excitation   183
p-mode frequency   49
p-mode spectrum   85
p-modes
    "Raman spectroscopy"   213
    amplitude modulation   113
    correlation   223
    damping   199ff.
    excitation   199ff.
    high n 448
    line profiles   229
    phase spectra   455f.
    power   227
    rotation modulation   114
    temporal behaviour   227
peak fitting   17
peak-bagging   132f.
penetration distance   124
penetration   73
penetrative convection   123
phase relations   427ff.
phase spectra from MDI   455
photometric campaigns   326
photometric facular index   99
photometric sunspot index   98
photosphere   187, 404
plage regions   413

Pleiades   28ff.
PMS 1.8 $M_\odot$ stars
    theory   397
Procyon campaign   319f.
pseudo-modes   448
pulsation axes   38

radial velocities   458
radiance
    latitudinal variation   111
radiation hydrodynamics   115
radiation-hydrodynamic modelling   438
radiometric accuracy   90
Raman scattering   216
rapid spectroscopy   269
regularized least squares   142
regularized least-squares   127
resonance condition
    Cherenkov   214
roAp stars   238ff., 277, 299f., 309
    amplitude gradients   277ff.
    kinematics   312
    magnetic properties   313
    phase-shift   270
    radial node   271
    singularity   311
Rossby wave
    instability   177f.
rotation (stars)   352
rotation axis   37
rotation near surface   145
rotation   25ff., 38, 335
rotational splitting   153, 248, 387

scattering centers   215
second overtone pulsators   389
seismic events   184
    2-D spatio-temporal evolution   461f.
    flares   191
seismic model
    solar   22, 81f.
SMM   90
SOHO   43, 91, 447
    integrated magnetic field   225
solar activity   174, 219

solar atmospheric structure
  diagnostics   427ff.
solar chromosphere
  basic physics   436
  observations   436
  simulations   437
solar core   420
solar cycle   146, 171
  magnetic field distribution   463, 465
solar magnetic field
  spatial distribution   463
solar observations
  resolved vs unresolved   473
solar oscillations   219
  excitation   197, 221
  three-dimensional   175
solar radius   165
solar rotation   141
  LOWL and GONG   181
  splittings   167
solar shadow   465
solar structure
  spherical and aspherical   157ff.
solar-like oscillations   316
solar-like stars
  equivalent width technique   375
solar-stellar seismology
  synthesis   471
solar-type stars
  oscillations   319
SPB stars   232
stellar oscillations   299f.
stellar p-mode spectra
  periodicity   315
stellar p-modes
  asymptotic description   391
stochastic excitation   223
sunspot blocking   107
sunspots   219, 413
super-resolution   130
superadiabatic convection   198
supergiants   350

tachocline   143f.
temperature inversion   282
temperature minimum   404
temperature profile   41
thermal shadows   413
thermal time scale   104
time series
  composite TSI   96
  modelling   57
time-distance seismology   149
transition zone   405
turbulence   196
  dynamics   451
turbulent pressure   452

UARS satellite   91
umbral oscillations   425

velocity–intensity spectra   429
vertical velocity (rms)   123
viscous dissipation   78

wave reflection   430
waveforms   428
white dwarf pulsations   253ff., 261ff.
white dwarfs   232f.
Whole Earth Telescope   262f.
window function   47
windows
  neutrinos vs seismology   472

ZZ Ceti stars   253ff.